COMBINATORIAL AND HIGH-THROUGHPUT DISCOVERY AND OPTIMIZATION OF CATALYSTS AND MATERIALS

CRITICAL REVIEWS IN COMBINATORIAL CHEMISTRY

Series Editors

BING YAN
School of Pharmaceutical Sciences
Shandong University, China

ANTHONY W. CZARNIK
Department of Chemistry
University of Nevada–Reno, U.S.A.

A series of monographs in molecular diversity and combinatorial chemistry, high-throughput discovery, and associated technologies.

Combinatorial and High-Throughput Discovery and Optimization of Catalysts and Materials
Edited by Radislav A. Potyrailo and Wilhelm F. Maier

COMBINATORIAL AND HIGH-THROUGHPUT DISCOVERY AND OPTIMIZATION OF CATALYSTS AND MATERIALS

Edited by

Radislav A. Potyrailo
Wilhelm F. Maier

CRC Press
Taylor & Francis Group
Boca Raton London New York

CRC Press is an imprint of the
Taylor & Francis Group, an **informa** business

A TAYLOR & FRANCIS BOOK

CRC Press
Taylor & Francis Group
6000 Broken Sound Parkway NW, Suite 300
Boca Raton, FL 33487-2742

First issued in paperback 2019

© 2007 by Taylor & Francis Group, LLC
CRC Press is an imprint of Taylor & Francis Group, an Informa business

No claim to original U.S. Government works

ISBN-13: 978-0-8493-3669-0 (hbk)
ISBN-13: 978-0-367-39059-4 (pbk)
Library of Congress Card Number 2006000149

Library of Congress Cataloging-in-Publication Data

Combinatorial and high-throughput discovery and optimization of catalysts and materials / edited by
   Radislav A. Potyrailo and Wilhelm F. Maier.
      p. cm. -- (Critical reviews in combinatorial chemistry)
   Includes bibliographical references and index.
   ISBN-13: 978-0-8493-3669-0 (alk. paper)
   ISBN-10: 0-8493-3669-4 (alk. paper)
   1. Materials. 2. Combinatorial chemistry. I. Potyrailo, Radislav A. II. Maier, Wilhelm F. III. Series.

TA403.6.C59 2006
620.1'1--dc22   2006000149

**Visit the Taylor & Francis Web site at
http://www.taylorandfrancis.com**

**and the CRC Press Web site at
http://www.crcpress.com**

Preface

- What problems have been solved and what are the remaining bottlenecks after a decade of enormous achievements in combinatorial materials science?
- How much acceleration in materials development is adequate in combinatorial experiments?
- Will the robotic manipulation of materials libraries replace the need for human-guided materials research?
- What is an adequate level of automation?
- Why does the combinatorial science field have only rare success stories of materials discoveries?
- Have complex materials become the subject of successful high-throughput studies?

This book attempts to provide answers to these and many other questions and a timely snapshot on the status of the combinatorial materials and catalysts development. It is clear that over the 10 years since the first publications in 1995 in modern combinatorial experimentation, the field has matured from the initial demonstrations of impressive dense libraries to testing and validation of combinatorially developed materials on traditional scales, to the interpretations of the collected data to extract new knowledge, and to develop new screening techniques to evaluate new materials properties on the small scale as discrete and gradient materials arrays.

Although many new technological advances improve research productivity through combinatorial and high-throughput experimentation tools, the human judgment and input remains the most important, the key aspect in the interpretation and practical application of newly discovered knowledge, which also helps in making well-supported decisions. As illustrated in this book, modern combinatorial science has matured to the point where it has produced an impressive set of tools that enable the scientist to accelerate the productivity of research in order to create new knowledge for more rational materials design. Combinatorial materials science has become an established, although by no means simple, research tool, applicable to an always broader variety of materials and problems. The discovery of knowledge is becoming a reality, although we still have to learn how to utilize this knowledge and turn it into a deeper understanding for the benefit of materials science. This will probably become a major subject of combinatorial materials science in the years to come.

We would like to take this opportunity to thank Bing Yan for inviting us to produce this book and Lindsey Hofmeister and Jill Jurgensen at Taylor & Francis for their help in its preparation. We would like to express our gratitude to the authors for their hard work in preparing their chapters on time and to the referees for reviewing them. Special thanks go to Susanne Kern-Schumacher at Universitaet des Saarlandes, who provided invaluable help in organizing and preparing the manuscripts. R.A.P. thanks the leadership team at GE Global Research for supporting the entire GE combinatorial chemistry effort. W.F.M. thanks his research team for support and for accepting a significant reduction in attention for the group. Last, but not least, we would like to thank our families, who have allowed this book to become part of them for quite a while.

Preface

- What problems have been solved and what are the remaining bottlenecks after a decade of enormous achievements in combinatorial materials science?
- How much acceleration in materials development is adequate in combinatorial experiments?
- Will the robotic manipulation of materials libraries replace the need for human-guided materials research?
- What is an adequate level of automation?
- Why does the combinatorial science field have only rare success stories of materials discoveries?
- Have complex materials become the subject of successful high-throughput studies?

This book attempts to provide answers to these and many other questions and a timely snapshot on the status of the combinatorial materials and catalysts development. It is clear that over the 10 years since the first publications in 1995 in modern combinatorial experimentation, the field has matured from the initial demonstrations of impressive dense libraries to testing and validation of combinatorially developed materials on traditional scales, to the interpretations of the collected data to extract new knowledge, and to develop new screening techniques to evaluate new materials properties on the small scale as discrete and gradient materials arrays.

Although many new technological advances improve research productivity through combinatorial and high-throughput experimentation tools, the human judgment and input remains the most important, the key aspect in the interpretation and practical application of newly discovered knowledge, which also helps in making well-supported decisions. As illustrated in this book, modern combinatorial science has matured to the point where it has produced an impressive set of tools that enable the scientist to accelerate the productivity of research in order to create new knowledge for more rational materials design. Combinatorial materials science has become an established, although by no means simple, research tool, applicable to an always broader variety of materials and problems. The discovery of knowledge is becoming a reality, although we still have to learn how to utilize this knowledge and turn it into a deeper understanding for the benefit of materials science. This will probably become a major subject of combinatorial materials science in the years to come.

We would like to take this opportunity to thank Bing Yan for inviting us to produce this book and Lindsey Hofmeister and Jill Jurgensen at Taylor & Francis for their help in its preparation. We would like to express our gratitude to the authors for their hard work in preparing their chapters on time and to the referees for reviewing them. Special thanks go to Susanne Kern-Schumacher at Universitaet des Saarlandes, who provided invaluable help in organizing and preparing the manuscripts. R.A.P. thanks the leadership team at GE Global Research for supporting the entire GE combinatorial chemistry effort. W.F.M. thanks his research team for support and for accepting a significant reduction in attention for the group. Last, but not least, we would like to thank our families, who have allowed this book to become part of them for quite a while.

Editors

Radislav A. Potyrailo is a project leader at the General Electric Global Research Center in Niskayuna, New York. He earned an engineering degree in optoelectronic instrumentation from Kiev Polytechnic Institute, Ukraine, in 1985 and a Ph.D. in analytical chemistry from Indiana University, Bloomington, in 1998. His research is focused on the design and implementation of high-throughput spectroscopic and sensor-based analytical instrumentation for screening of combinatorial libraries of materials, application of combinatorial methodologies for the development of functional materials, adaptation of high-throughput screening analytical instruments for other analytical applications, and development of GE lab-on-a-chip technologies. Over his scientific career, Dr. Potyrailo has co-authored two monographs on sensor technologies (1991, 1992), co-edited a book on high-throughput analysis (2003), published over 50 peer-reviewed technical papers, and given over 30 invited lectures at national and international technical meetings. He has been issued over 35 U.S. patents for sensors, analytical instrumentation, data analysis algorithms, functional materials, and combinatorial screening. He is the initiator and a co-organizer of the first Gordon Research Conference on Combinatorial and High Throughput Materials Science (2002); co-organizer of symposia on Combinatorial Methods in Materials Science at Pittcon (2000, 2001), MRS (2003, 2005) and ACS meetings (2004, 2005); and co-organizer of the Third Japan–U.S. Workshop on Combinatorial Materials Science, 2004, in Okinawa, Japan.

Wilhelm F. Maier holds a chair in technical chemistry at Saarland University, Saarbruecken, Germany. He received a degree in chemical engineering from Ohm Polytechnikum, Nuernberg, Germany, in 1971 and a diploma in chemistry from Philipps University, Marburg, Germany, in 1975, followed by a Ph.D. in organic chemistry in 1978. From 1981 to 1988 he was an assistant professor of physical organic chemistry at the University of California, Berkeley. In 1988 he moved to the University of Essen, Germany, as a professor of technical chemistry. In 1992 he became head of the Heterogeneous Catalysis Group at the Max Planck Institut für Kohlenforschung in Muelheim, Germany. In 2000 he moved to his present position at Saarland University. His research is concerned with the preparation, characterization, and application of heterogeneous catalysts and the use of high-throughput and combinatorial technologies for the discovery and optimization of new catalysts and materials. He has co-authored over 150 publications and holds 14 patents. He was awarded the Carl Duisberg Gedächtnispresis from the GDCh in Germany in 1993 and the Arnold Lectureship at Southern Illinois University, Carbondale, in 1996. In 2005 he received the award for the best publication of 2004 in QCS (Wiley-VCH) from the European Society of Combinatorial Sciences. He is a member of the GDCh, the MRS, and DECHEMA, as well as the editorial boards of the *International Journal of Nanotechnology* and *QCS*. He organized a summer school of catalysis in 2004 and was chairman of the Gordon Conference on Combinatorial and High-Throughput Materials Science of 2005.

Contributors

Jennifer Abrahamian
UOP LLC
Des Plaines, Illinois

Diego F. Acevedo
Universidad Nacional de Río Cuarto
Río Cuarto, Argentina

Arik G. Alexanian
National Academy of Sciences of Armenia
Ashtarak, Armenia

Nikolay S. Aramyan
National Academy of Sciences of Armenia
Ashtarak, Armenia

Karapet E. Avjyan
National Academy of Sciences of Armenia
Ashtarak, Armenia

James A. Bahr
North Dakota State University
Fargo, North Dakota

Jun Bao
University of Science and Technology of
    China
Hefei, China

César A. Barbero
Universidad Nacional de Río Cuarto
Río Cuarto, Argentina

Lorenz Bauer
UOP LLC
Des Plaines, Illinois

Maureen Bricker
UOP LLC
Des Plaines, Illinois

Kwang Ohk Cheon
Evident Technologies
Troy, New York

Bret J. Chisholm
North Dakota State University
Fargo, North Dakota

David A. Christianson
North Dakota State University
Fargo, North Dakota

Peter Claus
Technische Universität Darmstadt
Darmstadt, Germany

Mats Eriksson
Linköping University
Linköping, Sweden

Gerald Frenzer
Universität des Saarlandes
Saarbrücken, Germany

Christine M. Gallagher-Lein
North Dakota State University
Fargo, North Dakota

Chen Gao
University of Science and Technology of
    China
Hefei, China

Michael Gatter
UOP LLC
Des Plaines, Illinois

Ralph Gillespie
UOP LLC
Des Plaines, Illinois

Romen P. Grigoryan
Yerevan Physics Institute
Yerevan, Armenia

Nathan J. Gubbins
North Dakota State University
Fargo, North Dakota

Satoru Inoue
National Institute for Materials Science
Ibaraki, Japan

Virginie Jéhanno
Siemens AG
Munich, Germany

Min Ku Jeon
Korea Advanced Institute of Science and
    Technology
Taejon, South Korea

Chang Hwa Jung
Korea Advanced Institute of Science and
    Technology
Taejon, South Korea

Helmut Karl
Universität Augsburg
Augsburg, Germany

Ashot M. Khachatryan
National Academy of Sciences of Armenia
Ashtarak, Armenia

Jens Klein
hte AG
Heidelberg, Germany

Roger Klingvall
Linköping University
Linköping, Sweden

Peter Kolb
hte AG
Heidelberg, Germany

Tomoya Konishi
National Institute for Materials Science
Ibaraki, Japan

Gang Li
University of California
Los Angeles, California

Jing Hua Liu
Korea Advanced Institute of Science and
    Technology
Taejon, South Korea

Martin Lucas
Technische Universität Darmstadt
Darmstadt, Germany

Ingemar Lundström
Linköping University
Linköping, Sweden

Asif Mahmood
Korea Advanced Institute of Science and
    Technology
Taejon, South Korea

Wilhelm F. Maier
Universität des Saarlandes
Saarbrücken, Germany

María C. Miras
Universidad Nacional de Río Cuarto
Río Cuarto, Argentina

John R. Owen
University of Southampton
Southampton, United Kingdom

Radislav A. Potyrailo
General Electric Global Research
Niskayuna, New York

Arun Rajagopalan
Rensselaer Polytechnic Institute
Troy, New York

Krishna Rajan
Iowa State University
Ames, Iowa

Wolfgang Rossner
Siemens AG
Munich, Germany

Sabine Schimpf
Technische Universität Darmstadt
Darmstadt, Germany

Timm Schmidt
Gesellschaft für umweltkompatible
    Prozesstechnik mbH
Saarbrücken, Germany

Stephan Andreas Schunck
hte AG
Heidelberg, Germany

Stephen Schuyten
University of Notre Dame
Notre Dame, Indiana

Namsoo Shin
Sunchon National University
Chonam, South Korea

Joseph Shinar
Iowa State University
Ames, Iowa

Ruth Shinar
Iowa State University
Ames, Iowa

Kee-Sun Sohn
Sunchon National University
Chonam, South Korea

Grigorii Soloveichik
General Electric Global Research
Niskayuna, New York

Alan D. Spong
University of Southampton
Southampton, United Kingdom

Shigeru Suehara
National Institute for Materials Science
Ibaraki, Japan

Dong Jin Suh
University of Notre Dame
Notre Dame, Indiana

Eun Jung Sun
Korea Advanced Institute of
Science and Technology
Taejon, South Korea

Andreas Sundermann
hte AG
Heidelberg, Germany

Erik B. Svedberg
Seagate Technology
Pittsburgh, Pennsylvania

Shin-ichi Todoroki
National Institute for Materials Science
Ibaraki, Japan

Fawn M. Uhl
North Dakota State University
Fargo, North Dakota

Girts Vitins
University of Southampton
Southampton, United Kingdom

Dean C. Webster
North Dakota State University
Fargo, North Dakota

Berit Wessler
Siemens AG
Munich, Germany

Donald W. Whisenhunt, Jr.
General Electric Global Research
Niskayuna, New York

Eduardo E. Wolf
University of Notre Dame
Notre Dame, Indiana

Seong Ihl Woo
Korea Advanced Institute of Science and
Technology
Taejon, South Korea

Ronald J. Wroczynski
General Electric Global Research
Niskayuna, New York

Arsham S. Yeremyan
National Academy of Sciences of Armenia
Ashtarak, Armenia

Torsten Zech
hte AG
Heidelberg, Germany

Zhaoqun Zhou
Iowa State University
Ames, Iowa

Acknowledgments

The editors and contributors are grateful to the copyright holders for permission to reprint the following material:

CHAPTER 4
Figures 4.3, 4.4, 4.7–4.9: Koinuma, H. and Kawasaki, M., *Combinatorial Technology* (in Japanese), Maruzen, Tokyo, Japan, 2004.

CHAPTER 5
Figures 5.4, 5.6–5.10: Klingvall, R., Lundström, I., Löfdahl, M., and Eriksson, M., A combinatorial approach for field-effect gas sensor research and development, *IEEE Sensors Journal*, 5, 5, October 2005.

CHAPTER 8
Tables 8.4, 8.5: Spivack, J.L., Cawse, J.N., Whisenhunt, D.W. Jr. et al., *App. Cat. A: General*, 254, 5–25, 2003.

CHAPTER 9
Figures 9.1, 9.2: Senkan, S., Combinatorial heterogeneous catalysis — A new path in an old field, *Angew. Chem. Int. Ed.*, 40, 312, 2001.
Figure 9.4: Bergh, S., Guan, S., Hagemeyer, A., Lugmair, C., Turner, H., Volpe, A.F. Jr., Weinberg, W.H., Mott, G., Gas phase oxidation of ethane to acetic acid using high-throughput screening in a massively parallel microfluidic reactor system, *Appl. Catal. A: General*, 254, 67, 2003.
Figure 9.5: Desrosiers, P., Guram, A., Hagemeyer, A., Jandeleit, B., Poojary, D.M., Turner, H., Weinberg, H., Selective oxidation of alcohols by combinatorial catalysis, *Catal. Today*, 67, 397, 2001.
Figure 9.6: Akporiaye, D.E., Dahl, J.M., Karlsson, A., Wendelbo, R., Combinatorial approach to the hydrothermal synthesis of zeolites, *Angew. Chem. Int. Ed.*, 37, 609, 1998.
Figure 9.7: Lucas, M. and Claus, P., High throughput screening in monolith reactors for total oxidation reactions, special issue: "Combinatorial Catalysis," Maier, W.F., Guest Editor, *Appl. Catal. A: General*, 254, 35, 2003.
Figures 9.8–9.10: Claus, P., Hönicke, D., and Zech, T., Miniaturization of screening devices for the combinatorial development of heterogeneous catalysts, *Catal. Today*, 67, 319, 2001.
Figure 9.11: Hofmann, C., Schmidt, H.-W., and Schüth, F., A multipurpose parallelized 49-channel reactor for the screening of catalysts: Methane oxidation as the example reaction, *J. Catal.*, 198, 348, 2001.
Figure 9.12: Pérez-Ramírez, J., Berger, R.J., Mul, G., Kapteijn, F., and Moulijn, J.A., The six-flow reactor technology: A review on fast catalyst screening and kinetic studies, *Catal. Today*, 60, 93, 2000.
Figure 9.13: Moulijin, J.A., Pérez-Ramírez, J., Berger, R.J., Hamminga, G., Mul, G., and Kapteijn, F., High-throughput experimentation in catalyst testing and in kinetic studies for heterogeneous catalysis, *Catal. Today*, 81, 457, 2003.
Figure 9.14: Thomson, S., Hoffmann, C., Ruthe, S., Schmidt, H.-W., and Schüth, F., The development of a high throughput reactor for the catalytic screening of three phase reactions, *Appl. Catal. A: General*, 220, 253, 2001.

Figure 9.15: Claus, P., Hönicke, D., and Zech, T., Miniaturization of screening devices for the combinatorial development of heterogeneous catalysts, *Catal. Today*, 67, 319, 2001.

CHAPTER 12
Figure 12.3: Webster, D.C., Bennett, J., Kuebler, S., Kossuth, M.B., and Jonasdottir, S., High throughput workflow for the development of coatings, *JCT Coatings Tech.*, 1(16), 34–39, 2004.

CHAPTER 14
Figure 14.1: Gurau, B. et al., Structural and electrochemcial characterization of binary, ternary, and quaternary platinum alloy catalysts for methanol electro-oxidation, *J. Phys. Chem. B*, 102, 9997, 1998.

Figures 14.2, 14.3: Choi, W.C., Kim, J.D., and Woo, S.I., Quaternary Pt-based electrocatalyst for methanol oxidation by combinatorial electrochemistry, *Catal. Today*, 74, 235, 2002.

Figures 14.4, 14.5: Choi et al., Development of enchanced materials for direct-methanl fuel cell by combinatorial method and nanoscience, *Catal. Today*, 2004

Figures 14.6, 14.7: Chu, Y.H. et al., Evaluation of the Nafion effect on the activity of Pt-Ru electrocatalysts for the electro-oxidation of methanol, *J. Power Sources*, 118, 334, 2003.

Figures 14.8–14.10: Chen, G. et al., Combinatorial discovery of bifunctional oxygen reduction — water oxidation electrocatalysts for regenerative fuel cells, *Catal. Today*, 67,341, 2001

Figures 14.11–14.12: Guerin, S. et al., Combinatorial electrochemical screening of fuel cell electrocatalysts, *J. Comb. Chem.*, 6, 149, 2004.

Figures 14.13–14.15: Liu, R. and Smotkin, E.S., Array membrane electrode assemblies for high throughput screening of direct methanol fuel cell anode catalysts, *J. Electroanal.* Chem., 535, 49, 2002.

Figures 14.16, 14.17: Strasser, P. et al., High throughput experimental and theoretical predictive screening of materials — a comparative study of search strategies for new fuel cell anode catalysts, *J. Phys. Chem. B*, 107, 11013, 2003.

Figure 14.18: Jiang, R. and Chu, D., A combinatorial approach toward electrochemical analysis, *J. Electroanal. Chem.*, 527, 137, 2002.

Figure 14.19: Stevens, D.A., Domaratzki, R.E., and Dahn, J.R., 64-channel fuel cell for testing sputtered combinatorial arrays of oxygen reduction catalysts, 206th meeting of The Electrochemical Society, 2004, 1903.

Figures 14.20–14.22: Jayaraman, S. and Hillier, A.C., Screening the reactivity of PtxRuy and PtxRuyMoz catalysts toward the hydrogen oxidation reaction with the scanning electrochemical microscope, *J. Phys. Chem. B*, 107,5221, 2003.

Figures 14.23, 14.24: Fernandez, J.L. and Bard, A.J., Scanning electrochemical microscopy. 47. Imaging electrocatalytic activity for oxygen reduction in an acidic medium by the tip generation-substrate collection mode, *Anal. Chem.*, 75, 2967, 2003

Figure 14.25: Yamada, Y. et al., High-throughput screening of PEMFC anode catalysts by IR thermography, *Appl. Surf. Sci.*, 223, 220, 2004.

CHAPTER 17
Figures 17.1–17.3: Woo, S.I. et al., Current status of combinatorial and high-throughput methods for discovering new materials and catalysis, *QSAR Comb. Sci.*, 24, 138, 2005.

Figure 17.4: Xu, Y., *Ferroelectric Materials and Their Applications*. Elsevier Science Publications, Los Angeles, 1991, chap. 1.

Figure 17.5, 17.7: Lai, S., Current status of the phase change memory and its future, in IEDm'03 Technical Digest in IEEE International, Intel Corporation, RN2-05, 2003.

Figure 17.6: Nakayama, K. et al., Nonvolatile memory based on phase change in Se-Sb-Te glass, *Jpn. J. Appl. Phys.*, 42, 404, 2003.

Figures 17.8–17.10: Kyrsta, S. et al., Characterization of Ge-Sb-Te thin films deposited using a composition-spread approach, *Thin Solid Films*, 398-399, 379, 2001.

Figure 17.11: Bez, R. and Pirovano, A., Nonvolatile memory technologies: Emerging concepts and new materials, *Mater. Sci. Semicond. Process.*, 7, 349, 2004.

Figure 17.12: Takahashi, R. et al., Development of a new combinatorial mask for addressable ternary phase diagramming: Application to rare earth doped phosphors, *Appl. Surf. Sci.*, 223, 249, 2004.

CHAPTER 20

Figure 20.1: Zou, L., Savvate'ev, V., Booher, J., Kim, C.-H., and Shinar, J., Combinatorial fabrication and studies of intense efficient ultraviolet-violet organic light emitting device arrays, *Appl. Phys. Lett.*, 79, 2282, 2001.

CHAPTER 21

Figure 21.1: Xiang, X. D. et al., A combinatorial approach to materials discovery, *Science*, 268, 1738, 1995.

Figure 21.2: Wang, J. et al., Identification of a blue photoluminescent composite material from a combinatorial library, *Science*, 279, 1712, 1998.

Figures 21.3–21.6: Chen, L. et al., Combinatorial synthesis of insoluble oxide library from ultrafine/nano particle suspension using a drop-on-demand inkjet delivery system, *J. Comb. Chem.*, 6, 699, 2004.

Figure 21.7: Danielson, E. et al., A combinatorial approach to the discovery and optimization of luminescentmaterials, *Nature*, 389, 944, 1997.

Figure 21.8: Danielson, E. et al., A rare-earth phosphor containing one-dimensional chains identified through combinatorial methods, *Science*, 279, 837, 1998.

Figures 21.9, 21.10: Liu, X. N. et al., Combinatorial screening for new borophosphate VUV phosphors, *Appl. Surf. Sci.*, 223, 144, 2004.

Figures 21.11, 21.12: Sohn, K. S. et al., Combinatorial search for new red phosphors of high efficiency at VUV excitation based on the YRO4 (R=As, Nb, P, V) system, *Chem. Mater.*, 14, 2140, 2002.

Figures 21.13, 21.14: Hayashi, H. et al., Bright blue phosphors in ZnO–WO3 binary system discovered through combinatorial methodology, *Appl. Phys. Lett.*, 82(9), 1365, 2003.

Figures 21.15, 21:16: Mordkovich, V. Z. et al., Discovery and optimization of new ZnO-based phosphors using a combinatorial method, *Adv. Funct. Matter.*, 13(7), 519, 2003.

CHAPTER 22

Table 22.1, Figures 22.7–22.11, 22.14, 22.15: Kim, C.H. et al., Combinatorial synthesis of Tb-activated phosphors in the CaO-Gd203-Al2O3 system, *J. Electrochem. Soc.*, 149, H21, 2002.

Figures 22.1–22.6: Sohn, K.-S. et al., Combinatorial search for new red phosphors of high efficiency at VUV excitation based on the YRO4 (R=As, Nb, P, V) system, *Chem Mater.*, 14, 2140, 2002.

Figures 22. 12, 22.13: Sohn, K.-S., Lee, J.M., and Shin, N., A search for new red phosphors using a computational evolutionary optimization process, *Adv. Mater.*, 15, 2081, 2003.

CHAPTER 23

Figures 23.7–23.9: Alexanian, A.G. et al., *Meas. Sci. And Technol.*, 16, 167, 2005.

Table of Contents

SECTION 6
Optic Materials

Section 1

General Aspects of Combinatorial Materials Science

1 Combinatorial Materials and Catalysts Development: Where Are We and How Far Can We Go?

Radislav A. Potyrailo and Wilhelm F. Maier

CONTENTS

1.1 INTRODUCTION

The concept of combinatorial methods is well known in science. For example, *combinatorics* is a branch of mathematics that studies finite collections of objects that satisfy some predefined criteria, and is concerned with counting the objects in those collections and with deciding whether certain optimal objects exist.[1,2] In life sciences, combinatorial methods have found their applications in intelligent and systematic searching of large-parameter spaces for new candidate drug molecules in combination with high-throughput techniques, applied to evaluate activity.[3] It has been recognized that to screen all possible combinations of molecular building blocks for their target-binding properties is a time- and resource-consuming effort, even with the availability of ultra-high-throughput screening instrumentation.[4,5]

In materials science, materials properties depend not only on composition, but also on morphology, microstructure and other parameters that are dependent on material-preparation conditions. A typical example of such dependence is shown in Figure 1.1, which illustrates the large number of parameters important for selectivity, stability, and reproducibility of operation of formulated polymeric[6] and metal oxide semiconductor[7] sensors.

As a result of this complexity, a true combinatorial experimentation is rarely performed in materials science with a complete set of materials and process variables explored. Instead, carefully selected subsets of the parameters are often explored in an automated parallel or rapid sequential fashion. Combinatorial experimentation (the systematic strategy to explore parameter spaces by combination of building blocks or parameters) can clearly be differentiated from the more general high-throughput experimentation (HTE). However, the terms "combinatorial chemistry" and "combinatorial

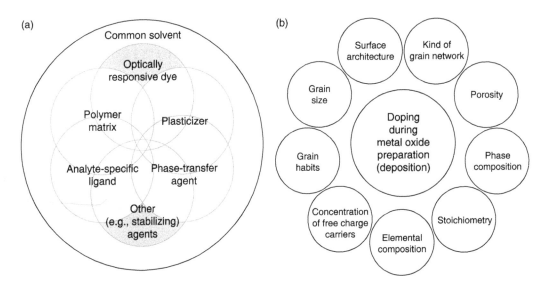

FIGURE 1.1 Parameters important for selectivity, stability, and reproducibility of operation of (a) formulated polymeric[6] and (b) metal oxide semiconductor[7] sensors.

materials science" are often somewhat misused for all types of automated parallel and rapid sequential materials evaluation processes, which are not combinations, but rather just high-throughput experiments such as composition spreads or parameter variations. An adequate definition of combinatorial materials science is a process that couples the capability for parallel production of large arrays of diverse materials together with different high-throughput measurement techniques for various physical properties and subsequent navigation in the collected data for identifying "lead" materials.[8–12] Schultz and co-workers have initiated the avalanche of applications of combinatorial methodologies in materials science with the publication of their first manuscript in 1995.[13] Since then, combinatorial materials science has enjoyed much success and rapid progress. An incomplete list of materials reported in conjunction with combinatorial and high-throughput screening techniques is presented in Table 1.1.[14–37]

The acceptance of combinatorial methodologies in materials science can be illustrated by the rate of growth of publications in this area. Our SciFinder reference search of scientific publications has revealed a very interesting trend shown in Figure 1.2. While the number of publications related to rational and combinatorial methods of materials development is increasing annually, over the last 5 years the growth rate of combinatorial-related publications exceeds that of the publications related to rational development of materials.[6] We believe that these data illustrate not only the acceptance of combinatorial methodologies in materials science, but also the demand to apply combinatorial methodologies for the generation of new knowledge for more rational materials design.

There are significant differences in motivation and application of combinatorial and high-throughput technologies in academia and industry. In industrial laboratories that practice but do not manufacture combinatorial technologies (e.g., BASF, Bayer, BP, Dow, DuPont, General Electric, General Motors, UOP, and a growing number of other industrial research labs), HTE is being applied to save time and costs. The rapid screening of parameter spaces allows fast assessment of the effect of additives, modifiers, and preparation conditions. Thus a lead material can be optimized rapidly and time to market is reduced. Due to harsh time limits and success pressure, often there are not enough opportunities to explore a wide range of unknown parameter spaces.

Examples of companies specializing in HTE include Symyx Technologies in the U.S., hte-AG in Germany, Avantium in the Netherlands, Accelergy in China, and some others. The integrated

TABLE 1.1

Examples of Materials Explored Using Combinatorial and High-Throughput Experimentation Techniques

Material explored	Reference
Heterogeneous catalysts	14
Homogeneous catalysts	15
Fuel cell anode catalysts	16
Polymerization catalysts	17
Enantioselective catalysts	18
Electrocatalysts	19
Polymers	20
Zeolites	21
Luminescent compounds	22
Magnetoresistive compounds	23
Metal alloy materials	24
Coating materials	25
Fuel cell materials	26
Solar cell materials	27
Agricultural materials	28
Ferroelectric/dielectric materials	29
Structural materials	30
Solid oxide fuel cells	31
Hydrogen storage materials	32
Organic light-emitting materials	33
Polymeric sensor materials	34
Metal oxide sensor materials	35
Formulated sensor materials	36
Organic dyes	37

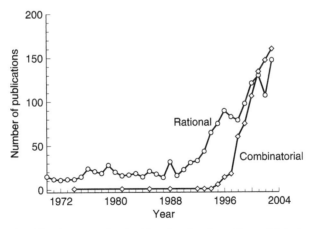

FIGURE 1.2 Number of publications associated with rational and combinatorial design of materials as determined from a SciFinder search.

technologies developed in these companies are of very high standards, but costly. In such dedicated operations, the combinatorial workflow is highly efficient. These HTE companies mainly operate on contract research, discovering and optimizing new solutions to old and new problems, and on the sale of specialized high-throughput technologies or software to industrial laboratories for in-house R&D. The discovery and commercialization of new materials may also become a profitable business for these HTE companies.

In academia, the lack of adequate capital and equipment, as well as permanent staff, often prevents the development of sophisticated and highly productive HTE workflows. Another weakness of academic laboratories is the relatively rapid turnaround of creative young scientists who leave the group after earning their degree or finishing their post-doctoral appointment. Although the use of HTE is increasing steadily in academia, most applications are targeted to accelerate the traditional research operations of the group, e.g., the optimization of reaction conditions to increase yields for each step of a natural-product synthesis. There are still only a few academic laboratories specializing in combinatorial or high-throughput technologies or in the discovery of new materials.

1.2   COMBINATORIAL MATERIALS SCIENCE (WHERE ARE WE?)

The power of the integrated materials-development workflow was realized first by Hanak.[38] The key aspects of this workflow included a complete compositional mapping of a multicomponent system in one experiment, a simple, rapid, nondestructive, all-inclusive chemical analysis, testing of properties by a scanning device, and computer data processing as shown in Figure 1.3a. Although attractive, at that time not all of the components of the workflow were readily available to fully implement this idea. Individual aspects for the accelerated materials development were reported in the literature, including combinatorial and factorial experimental designs,[39] parallel synthesis of libraries of materials on a single substrate,[40,41] and screening of materials for performance properties.[42] Automated loading and measurements of multiple samples has been practiced with auto samplers for various types of analytical instruments only since the early 1970s.[43,44] Computer data processing became available in the 1950s.[45,46] The quantitative approach to biochemical structure–activity relationships became the focus of research in the 1960s.[47,48]

At present, the components of the combinatorial materials-development workflow are almost unchanged with the exception of several new, important aspects added into a modern "combinatorial materials cycle" as illustrated in Figure 1.3b. These new aspects include planning of experiments, data mining, and scale-up. The aspects of combinatorial materials science that are driving the most research attention at present can be explored from reviewing publications in the area of combinatorial materials science. Figure 1.4 presents results of an analysis of publications in this area obtained using the SciFinder reference search.[49] The percent distribution of publications is broken down into categories similar to those shown in Figure 1.3. These categories were generalized from a more detailed search of keywords. The "Other" category included the remaining publications. Publications related to materials synthesis, measurement techniques/instrumentation, and experimental planning lead with 37, 34, and 11% of the total number of publications.

All elements of the workflow are intended to operate as an integrated system. Multiple combinatorial concepts originally defined as highly automated have been refined to have more human input, with only an appropriate level of automation on a subset of complete materials parameter spaces. The aspects of database and data mining are less developed. Certain aspects of experimental planning will depend on the throughput of the material fabrication and capabilities of performance testing. An adequate level of automation is an important issue. As has been recently shown, for a throughput of 50–100 coating formulations per day, it is acceptable to perform certain aspects of the process manually,[50,51] unlike the thousands of screened samples in the pharmaceutical industry

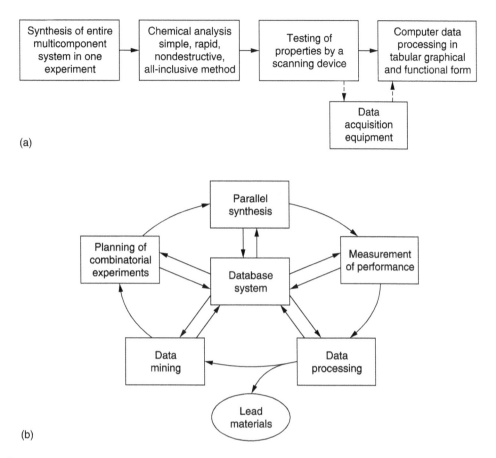

(a)

(b)

FIGURE 1.3 Concepts of the combinatorial materials-development workflow. (a) Initial concept proposed by Hanak in 1970.[38] (b) Modern "combinatorial materials cycle."[49]

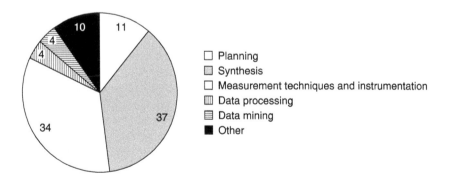

FIGURE 1.4 Comparison of the relative numbers of publications in various aspects of combinatorial chemistry and combinatorial materials science (shown as percent of total number of publications). Indicated categories are generalized from more detailed search keywords.[49]

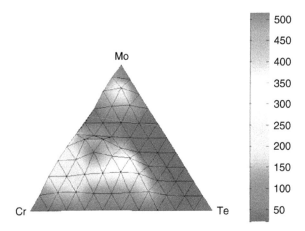

FIGURE 1.5 (See color insert following page 172) Catalytic activity of ternary mixed oxides for the oxidation of propene to acroleine as a function of chemical compositon.[55]

where automation is more desirable yet should be approached with caution.[5] Also, in high-throughput catalyst development, where commonly 208 materials per library are prepared by automated procedures with pipetting robots, the catalysts are routinely transferred by hand from the preparation vials into the wells in the library.[52] Here, manual operation is necessary to ensure comparable mass transport conditions during catalytic testing, which is not given, when the catalyst is synthesized directly in the well of the library. A balance of automated and manual operations in the combinatorial process map is attractive not only from the cost-effectiveness of the process, but also from the maintenance and customer acceptance points of view.

An important prerequisite for starting a combinatorial search for new materials is the materials synthesis procedure applied. It is well known that the preparation of a material of the same elemental composition by a variety of synthesis procedures (co-precipitation, impregnation, thin film deposition and annealing, sol–gel synthesis or simple variation of the recipe through solvent, temperature, sequence of additions, pre- or post-treatments) will result in a variety of different materials of identical composition, but different function. For example, it has been shown in the search for descriptors of catalytic properties that chemical composition of catalysts obtained by different synthesis procedures does not correlate with activity.[53] When optimization of chemical composition becomes the issue of HTE, great care has to be taken to ensure that the synthesis of materials of varying composition is carried out under identical synthetic conditions. This implies synthesis recipes that tolerate a broad variation in chemical composition. With such a set of materials, catalytic activity, measured under identical reaction conditions, correlates with chemical composition and can be called QCAR (quantitative composition activity relationship).[54] When such correlations are found, modeling of catalytic activity and prediction of optimal compositions becomes possible, as illustrated in Figure 1.5.[55] One essential aspect of such an approach is that the dependence of the materials function on chemical composition is continuous and not erratic.

To address numerous materials-specific properties, a variety of high-throughput characterization tools are required. Characterization tools are used for rapid and automated assessment of single or multiple properties of the large number of samples fabricated together as a combinatorial array or "library."[10,49,56] Typical library layouts can be discrete[13,39,41] and gradient,[38,40,57–61] as schematically depicted in Figure 1.6. A specific type of library layout will depend on the required density of space to be explored, available library-fabrication capabilities, and capabilities of HT characterization tools. These factors will also affect the dimensionality of the library, which can be one-, two-, or three-dimensional.

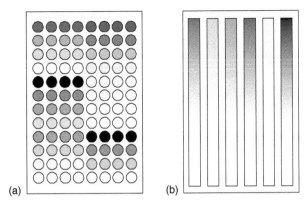

FIGURE 1.6 Typical layouts of (a) discrete and (b) gradient combinatorial libraries.

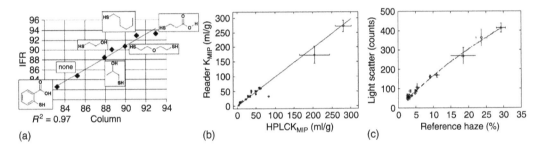

FIGURE 1.7 Correlation between evaluation of materials using traditional and combinatorial approaches. (a) Selectivity measurements of p,p'-bisphenol A in the microscale incremental flow reactor (IFR) and macroscale column reactor.[70] (b) Fast parallel assessment of rebinding capacity of an acrylate-based polymer library imprinted with the local anesthetic bupivacaine for use in pure aqueous systems using an UV microplate reader and a serial analysis by conventional HPLC.[77] (c) Scattered light intensity vs haze of different Taber-tested coating samples. A slight non-linear fit is accounted for in a developed model.[50]

While it is attractive to produce multiple small-scale samples of materials at once using combinatorial tools, it is important to validate the performance of the combinatorial system by reproducing materials with good performance in laboratory-scale synthesis and performance testing under conventional test conditions. In a reliable combinatorial workflow, relative materials performances correlate well with those reproduced by traditional-scale fabrication and testing. Thus, a correlation between the traditional-scale and combinatorial-scale materials is established using known materials. Examples of such correlations are demonstrated in Figure 1.7.

Literature data may contain an appreciable amount of error associated with different methods of fabrication and characterization of materials properties. For example, results of Taber-testing of coatings for their abrasion resistance in different laboratories collected during round-robin studies can differ by as much as 30% (interlaboratory coefficient of variation).[62] In another example, Henry's law constants can differ by as much as one or two orders of magnitude when reported by different laboratories.[63] These examples demonstrate the importance of a careful assessment of historic data assembled from different reference sources. Here high-throughput experimentation is clearly superior, since data are collected in an automated fashion under identical conditions, which results in a dramatic improvement of overall data quality.

TABLE 1.2
Functions of a Data Management System

Function	Current capabilities	Needs
Experimental planning	Composition parameters Process parameters Library design	Iterative intelligent experimental planning based on results from virtual or experimental libraries
Database	Entry and search of composition and process variables Operation with heterogeneous data Unification of numerical data between different instruments, databases, individual computers	Storage and manipulation (search) of large amounts (terabytes and more) of data
Instrument control	Operation of diverse instruments	Interinstrument calibration Full instrument diagnostics Plug-n-play multiple instrument configurations
Data processing	Visualization of compositions and process conditions of library elements Visualization of measured parameters Matrix algebra Cluster analysis Quantification Outlier detection Multivariate processing of steady-state and time-resolved data Third-party statistical packages	Advanced data compression Processing of large amounts (terabytes and more) of data
Data mining	Prediction of properties of new materials Virtual libraries Cluster analysis Molecular modeling QSAR	Identification of appropriate descriptors on different levels (atomic, molecular, process, etc.)

The most significant difference between high-throughput experimentation and conventional research is not the experimental technology, but its information technology components. HTE requires various computer programs to maintain an efficient workflow. Design and synthesis protocols for materials libraries are computer-assisted, materials synthesis and library preparation should be carried out with robotic help. Property screening and materials characterization are also software controlled. Materials synthesis data, as well as property and characterization data, have to be collected into a proper materials database. Data in such a database are not just stored; the database is connected to or contains proper statistical analysis and visualization tools as well as modeling and data-mining tools. For a "wealthy" industrial user, there are already several complete commercial software packages available from Accelrys, Avantium, hte AG, Symyx Technologies, and other companies.[64] For databases, commercial as well as open source solutions are available.[65]

In an ideal combinatorial workflow, one should "analyze in a day what is made in a day."[66] Such capability depends quite a bit on the adequate data management capabilities of the combinatorial workflow. Table 1.2 illustrates important functions of the data management system and demonstrates aspects that have been addressed in the literature and are still under development.

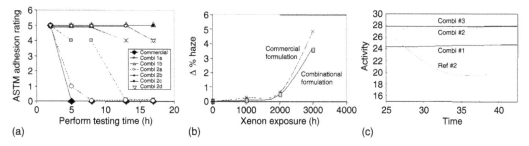

FIGURE 1.8 Examples of scale-up of combinatorial materials and catalyst leads. (a) Results of scaling up of combinatorially developed coatings with high adhesion performance. Performance test, boiling water. Adhesion varies from 5 (best) to 0 (worst). Combi 1a–b and 2a–d are laboratory-scale formulations based on two different formulations discovered using the combinatorial process.[51] (b) Comparison of Δ % haze weathering performance of combinatorial lead and commercial control formulations.[78] (c) Scaled-up combinatorial catalyst development in a pilot plant. Comparison of candidate catalyst leads from a combinatorial catalyst development study with a state-of-the-art reference catalyst. All three catalyst leads have better performance than the reference sample. Combi #3 will be commercialized.[69]

1.3 SCALE-UP OF COMBINATORIAL LEADS

Production of combinatorial leads on the laboratory scale reveals the extent of reliability and realism of the data obtained on the combinatorial scale. These include preparation and processing methods, characterization, and some others. It has become common practice, as can be seen in most publications, that, depending on the applications, the multilevel testing of combinatorial materials is also applied.[67] Materials developed on the combinatorial scale and validated on scale-up versions or in practical applications include catalysts,[68–72] phosphors,[73] formulated organic coatings,[51] sensing polymers,[74] and some others. Selected examples of such developments are demonstrated in Figure 1.8.

Unfortunately, successful applications of high-throughput methods are not always made public. Such materials are often described and characterized without any details of how they have been discovered. This not only applies to materials, but also to drugs or organic synthesis procedures. Therefore, it is rather difficult to reliably judge the impact of HTE on science and industry.

1.4 REMAINING CHALLENGES (HOW FAR CAN WE GO?)

In addition to the attractive aspects of combinatorial materials science such as speed and the high level of detail of screening of composition and process parameter spaces, it is important to keep in mind the remaining challenges and possible limitations of combinatorial techniques. For example, as recently outlined,[11,75] the performance of new synthetic, testing, and characterization tools may not necessarily be compatible with that of existing counterparts; thus care should be taken to establish the desired correlations and understand the sources of possible discrepancies. Also, a high degree of control should be maintained over synthesis and analysis of individual samples in the library in order not to introduce uncontrolled yet critical parameters that can change across the library.

Discovery clearly remains a big challenge. Integrated approaches, such as the one described recently,[76] document how complex such an approach is and how important information technology becomes for HTE. Clearly, such integrated approaches to discovery will not become standard in the average research laboratory and a reduction of complexity will be essential for the broader acceptance of HTE. Multiparameter modeling, multiparameter visualization and multiparameter data mining are also among the large challenges, where development has just started.

A major problem in materials science is the lack of knowledge connecting materials structure with the desired property. This is also one of the most important opportunities for HTE. With the development of intelligent software for data mining, knowledge discovery will become a typical and valuable product of HTE. Knowledge discovery addresses the discovery of correlations between a desired materials function and its structure, comprising chemical composition, microstructure, phase composition, porosity and pore size, dimensionalities, etc. Knowledge discovery is only possible with large data sets of comparable quality, preferentially collected under identical conditions (or carefully scaled to identical conditions), if the data are accessible in a suitable database and if appropriate software tools are available for data mining.

Discovery does not come automatically from large data sets; it has its own requirements. For example, to identify the optimal range of catalytic activity in the composition spread of Mo, Cr, and Te mixed oxides in Figure 1.5, all samples covering the search space have to be prepared by the identical synthesis procedure and activity data have to be collected under identical reaction conditions. Knowledge discovered by HTE, such as the catalytic activity shown in Figure 1.5, does not yet provide any understanding of the effects of these three elements on catalytic activity. Here theoretical calculations as well as conventional research are required to elucidate the interaction between the three oxides.

1.5   OUTLOOK

Since the publication of Schultz and co-workers[13] combinatorial materials science has demonstrated its applicability in numerous applications in academia and industrial and governmental R&D. Several companies producing HTE equipment have also been established. At present, only a few specialized companies have the capabilities and resources (software, high-throughput technologies, skilled personnel) to truly access the potential of combinatorial ideas for materials discovery and optimization. A new generation of scientists who were educated in combinatorially oriented academic laboratories already has started to impact combinatorial industrial R&D. High-throughput and combinatorial techniques are rapidly conquering the formulations laboratories in industry. Multiparameter formulation (detergents, cosmetics, polymer blends, paints, coatings, and others), the most complex and most poorly understood part of chemical production, apparently strongly benefits from HTE.

Combinatorial and high-throughput screening technologies will increasingly impact the development of materials in several distinct ways. The availability of the parallel synthesis and high-throughput characterization tools has already provided previously unavailable capabilities to perform multiple experiments on a much shorter timescale. In these experiments, the quality of data can be even better than in conventional one-at-a-time experiments because of the elimination of possible experimental artifacts due to an uncontrolled variation of experimental conditions. Time savings and data quality will continue to inspire scientists in academia and industry to explore more risky ideas which were previously too time-consuming to experiment with. The application of only small amounts of reagents needed for combinatorial experiments will continue to make this approach very cost-effective on a per-experiment basis. New combinatorial schemes will be developed where various reaction variables can be altered simultaneously and where important cooperative effects can be carefully probed to provide a better understanding of variation sources. This information should lead to the development of new desired materials performance models that will relate intrinsic and performance material properties. Advances in computational chemistry and informatics will provide more capabilities for virtual screening and more guided selection of space for experimental screening. Overall, gathered and analyzed data will provide new opportunities for structure–activity relationships analysis and reduction of gaps in materials-development knowledge, and for more cost- and time-effective materials design.

The true multidisciplinary aspects of combinatorial techniques will continue to impact the researchers as well. It is likely that in the future an effective combinatorial chemist and materials scientist will acquire skills as diverse as experimental planning, automated synthesis, basics of high-throughput materials characterization, chemometrics, and data mining. Perhaps everyone who will deal with massive amounts of data will be able to perform standard multivariate statistical and cluster analysis using available tools for rapid assessment of materials trends that are not visually noticeable due to several components involved in experimental design. These skills will provide a powerful vision in exploring multicomponent interactions and synergistic effects in sensor materials, impossible to imagine in the 20th century.

REFERENCES

1. Knobloch, E. and Berlin, W., The mathematical studies of G. W. Leibniz on combinatorics, *Historia Mathematica* 1, 409, 1974.
2. Erickson, M. J., *Introduction to Combinatorics,* Wiley, New York, 1996.
3. Czarnik, A. W. and DeWitt, S. H., *A Practical Guide to Combinatorial Chemistry*, American Chemical Society, Washington, DC, 1997.
4. Borman, S., Rescuing combinatorial chemistry, in *C&EN 2004*, 82 (40), 45–56.
5. Mullin, R., Drug discovery, in *C&EN 2004*, 82 (30), 23–32.
6. Potyrailo, R. A., Polymeric sensor materials: Toward an alliance of combinatorial and rational design tools, *Angew. Chem. Intl. Ed.*, 2006, in press.
7. Korotcenkov, G., Gas response control through structural and chemical modification of metal oxide films: State of the art and approaches, *Sens. Actuators B* 107, 209, 2005.
8. Takeuchi, I., Newsam, J. M., Wille, L. T., Koinuma, H. and Amis, E. J., Combinatorial and artificial intelligence methods in materials science, in *MRS symposium proceedings,* Materials Research Society, Warrendale, PA, 2002.
9. Xiang, X.-D. and Takeuchi, I., *Combinatorial Materials Synthesis*, Marcel Dekker, New York, 2003.
10. Potyrailo, R. A. and Amis, E. J., *High Throughput Analysis: A Tool for Combinatorial Materials Science*, Kluwer Academic/Plenum Publishers, New York, 2003.
11. Koinuma, H. and Takeuchi, I., Combinatorial solid state chemistry of inorganic materials, *Nature Materials* 3, 429, 2004.
12. Potyrailo, R. A., Karim, A., Wang, Q. and Chikyow, T., Combinatorial and artificial intelligence methods in materials science II, in *MRS symposium proceedings,* Materials Research Society, Warrendale, PA, 2004.
13. Xiang, X.-D., Sun, X., Briceño, G., Lou, Y., Wang, K. A., Chang, H., Wallace-Freedman, W. G., Chen, S. W. and Schultz, P. G., A combinatorial approach to materials discovery, *Science* 268, 1738, 1995.
14. Holzwarth, A., Schmidt, H.-W. and Maier, W., Detection of catalytic activity in combinatorial libraries of heterogeneous catalysts by IR thermography, *Angew. Chem. Int. Ed.* 37, 2644, 1998.
15. Cooper, A. C., McAlexander, L. H., Lee, D.-H., Torres, M. T. and Crabtree, R. H., Reactive dyes as a method for rapid screening of homogeneous catalysts, *J. Am. Chem. Soc.* 120, 9971, 1998.
16. Strasser, P., Fan, Q., Devenney, M., Weinberg, W. H., Liu, P. and Norskov, J. K., High throughput experimental and theoretical predictive screening of materials - a comparative study of search strategies for new fuel cell anode catalysts, *J. Phys. Chem. B* 107, 11013, 2003.
17. Lemmon, J. P., Wroczynski, R. J., Whisenhunt Jr., D. W. and Flanagan, W. P., High throughput strategies for monomer and polymer synthesis and characterization, *Polymer Preprints* 42 (2), 630, 2001.
18. Reetz, M. T., Becker, M. H., Kuhling, K. M. and Holzwarth, A., Time-resolved IR-thermographic detection and screening of enantioselectivity in catalytic reactions, *Angew. Chem. Int. Ed.* 37, 2647, 1998.
19. Reddington, E., Sapienza, A., Gurau, B., Viswanathan, R., Sarangapani, S., Smotkin, E. S. and Mallouk, T. E., Combinatorial electrochemistry: A highly parallel, optical screening method for discovery of better electrocatalysts, *Science* 280, 1735, 1998.
20. Brocchini, S., James, K., Tangpasuthadol, V. and Kohn, J., A combinatorial approach for polymer design, *J. Am. Chem. Soc.* 119, 4553, 1997.
21. Lai, R., Kang, B. S. and Gavalas, G. R., Parallel synthesis of zsm-5 zeolite films from clear organic-free solutions, *Angew. Chem. Int. Ed.* 40, 408, 2001.

22. Danielson, E., Devenney, M., Giaquinta, D. M., Golden, J. H., Haushalter, R. C., McFarland, E. W., Poojary, D. M., Reaves, C. M., Weinberg, W. H. and Wu, X. D., A rare-earth phosphor containing one-dimensional chains identified through combinatorial methods, *Science* 279, 837, 1998.

23. Briceño, G., Chang, H., Sun, X., Schultz, P. G. and Xiang, X.-D., A class of cobalt oxide magnetoresistance materials discovered with combinatorial synthesis, *Science* 270, 273, 1995.

24. Ramirez, A. G. and Saha, R., Combinatorial studies for determining properties of thin-film gold–cobalt alloys, *Appl. Phys. Lett.* 85, 5215, 2004.

25. Chisholm, B. J., Potyrailo, R. A., Cawse, J. N., Shaffer, R. E., Brennan, M. J., Moison, C., Whisenhunt, D. W., Flanagan, W. P., Olson, D. R., Akhave, J. R., Saunders, D. L., Mehrabi, A. and Licon, M., The development of combinatorial chemistry methods for coating development I. Overview of the experimental factory, *Prog. Org. Coat.* 45, 313, 2002.

26. Jiang, R., Rong, C. and Chu, D., Combinatorial approach toward high-throughput analysis of direct methanol fuel cells, *J. Comb. Chem.* ASAP article, 2005.

27. Hänsel, H., Zettl, H., Krausch, G., Schmitz, C., Kisselev, R., Thelakkat, M. and Schmidt, H., Combinatorial study of the long-term stability of organic thin-film solar cells, *Appl. Phys. Lett.* 81, 2106, 2002.

28. Wong, D. W. and Robertson, G. H., Combinatorial chemistry and its applications in agriculture and food, *Adv. Exp. Med. Biol.* 464, 91, 1999.

29. Chang, H., Gao, C., Takeuchi, I., Yoo, Y., Wang, J., Schultz, P. G., Xiang, X.-D., Sharma, R. P., Downes, M. and Venkatesan, T., Combinatorial synthesis and high throughput evaluation of ferroelectric/dielectric thin-film libraries for microwave applications, *Appl. Phys. Lett.* 72 (17), 2185, 1998.

30. Zhao, J.-C., A combinatorial approach for structural materials, *Adv. Eng. Mat.* 3, 143, 2001.

31. Lemmon, J. P., Manivannan, V., Jordan, T., Hassib, L., Siclovan, O., Othon, M. and Pilliod, M., High throughput screening of materials for solid oxide fuel cells, in *Combinatorial and artificial intelligence methods in materials science II. MRS symposium proceedings*, Potyrailo, R. A., Karim, A., Wang, Q. and Chikyow, T., Materials Research Society, Warrendale, PA, 2004, p. 27.

32. Olk, C. H., Combinatorial approach to material synthesis and screening of hydrogen storage alloys, *Meas. Sci. Technol.* 16, 14, 2005.

33. Zou, L., Savvate'ev, V., Booher, J., Kim, C.-H. and Shinar, J., Combinatorial fabrication and studies of intense efficient ultraviolet-violet organic light-emitting device arrays, *Appl. Phys. Lett.* 79, 2282, 2001.

34. Potyrailo, R. A., Morris, W. G. and Wroczynski, R. J., Multifunctional sensor system for high-throughput primary, secondary, and tertiary screening of combinatorially developed materials, *Rev. Sci. Instrum.* 75, 2177, 2004.

35. Frantzen, A., Scheidtmann, J., Frenzer, G., Maier, W. F., Jockel, J., Brinz, T., Sanders, D. and Simon, U., High-throughput method for the impedance spectroscopic characterization of resistive gas sensors, *Angew. Chem. Int. Ed.* 43, 752, 2004.

36. Dickinson, T. A., Walt, D. R., White, J. and Kauer, J. S., Generating sensor diversity through combinatorial polymer synthesis, *Anal. Chem.* 69, 3413, 1997.

37. Szurdoki, F., Ren, D. and Walt, D. R., A combinatorial approach to discover new chelators for optical metal ion sensing, *Anal. Chem.* 72, 5250, 2000.

38. Hanak, J. J., The "multiple-sample concept" in materials research: Synthesis, compositional analysis and testing of entire multicomponent systems, *J. Materials Science* 5, 964, 1970.

39. Birina, G. A. and Boitsov, K. A., Experimental use of combinational and factorial plans for optimizing the compositions of electronic materials, *Zavodskaya Laboratoriya (in Russian)* 40 (7), 855, 1974.

40. Kennedy, K., Stefansky, T., Davy, G., Zackay, V. F. and Parker, E. R., Rapid method for determining ternary-alloy phase diagrams, *J. Appl. Phys.* 36 (12), 3808, 1965.

41. Hoffmann, R., Not a library, *Angew. Chem. Int. Ed.* 40, 3337, 2001.

42. Hoogenboom, R., Meier, M. A. R. and Schubert, U. S., Combinatorial methods, automated synthesis and high-throughput screening in polymer research: Past and present, *Macromol. Rapid Commun.* 24, 15, 2003.

43. Burtis, C. A., Johnson, W. F. and Overton, J. B., Automated loading of discrete, microliter volumes of liquids into a miniature fast analyzer, *Anal. Chem.* 46, 786, 1974.

44. Gregory IV, R. P., Lowry, J. D. and Malmstadt, H. V., Multichannel pipet for parallel aliquoting of samples and reagents into centrifugal analyzer minidiscs, *Anal. Chem.* 49, 1608, 1977.

45. Anderson, F. W. and Moser, J. H., Automatic computer program for reduction of routine emission spectrographic data, *Anal. Chem.* 30, 879, 1958.

46. Eash, M. A. and Gohlke, R. S., Mass spectrometric analysis. A small computer program for the analysis of mass spectra, *Anal. Chem.* 34, 713, 1962.

47. Hansch, C., Quantitative approach to biochemical structure-activity relationships, *Acc. Chem. Res.* 2, 232, 1969.

48. Craig, P. N., Comparison of batch and time-sharing computer runs for correlating structures and bioactivity by the hansch method, *J. Chem. Doc.* 11, 160, 1971.

49. Potyrailo, R. A. and Takeuchi, I., Role of high-throughput characterization tools in combinatorial materials science, *Meas. Sci. Technol.* 16, 1, 2005.

50. Potyrailo, R. A., Chisholm, B. J., Olson, D. R., Brennan, M. J. and Molaison, C. A., Development of combinatorial chemistry methods for coatings: High-throughput screening of abrasion resistance of coatings libraries, *Anal. Chem.* 74, 5105, 2002.

51. Potyrailo, R. A., Chisholm, B. J., Morris, W. G., Cawse, J. N., Flanagan, W. P., Hassib, L., Molaison, C. A., Ezbiansky, K., Medford, G. and Reitz, H., Development of combinatorial chemistry methods for coatings: High-throughput adhesion evaluation and scale-up of combinatorial leads, *J. Comb. Chem.* 5, 472, 2003.

52. Saalfrank, J. W. and Maier, W. F., Directed evolution of noble-metal-free catalysts for the oxidation of CO at room temperature, *Angew. Chem. Int. Ed.* 43, 2028, 2004.

53. Klanner, C., Farrusseng, D., Baumes, L., Lenliz, M., Mirodatos, C. and Schüth, F., The development of descriptors for solids: Teaching catalytic intuition to a computer, *Angew. Chem. Int. Ed.* 43, 5347, 2004.

54. Scheidtmann, J., Klär, D. L., Saalfrank, J. W., Schmidt, T. and Maier, W. F., Qcar – influence of a sample on its neighbours in composition spreads of co-ni-mn-mixed-oxides and m1-m2-mixed-oxides, *QSAR Comb. Sci.* 24, 203, 2005.

55. Sieg, S., Stutz, B., Schmidt, T., Hamprecht, F. A., Maier, W. F., *J. Mol. Model.*, in press.

56. MacLean, D., Baldwin, J. J., Ivanov, V. T., Kato, Y., Shaw, A., Schneider, P. and Gordon, E. M., Glossary of terms used in combinatorial chemistry, *J. Comb. Chem.* 2, 562, 2000.

57. Bever, M. B. and Duwez, P. E., Gradients in composite materials, *Mater. Sci. Eng.* 10, 1, 1972.

58. Shen, M. and Bever, M. B., Gradients in polymeric materials, *J. Mater. Sci.* 7, 741, 1972.

59. Ilschner, B., Structural and compositional gradients: Basic idea, preparation, applications, *J. De Physique IV, Colloq. C7* 3, 763, 1993.

60. Pompe, W., Worch, H., Epple, M., Friess, W., Gelinsky, M., Greil, P., Hempel, U., Scharnweber, D. and Schulte, K., Functionally graded materials for biomedical applications, *Mater. Sci. Eng. A* 362, 40, 2003.

61. Kieback, B., Neubrand, A. and Riedel, H., Processing techniques for functionally graded materials, *Mater. Sci. Eng. A* 362, 81, 2003.

62. Morse, M. P., Abrasion resistance, in *Paint and Coating Testing Manual*, Koleske, J. V., American Society for Testing and Materials, Philadelphia, PA, 1995, p. 525.

63. *Chemical kinetics and photochemical data for use in atmospheric studies*, http://jpldataeval.jpl.nasa.gov/download.html Jet Propulsion Laboratory, JPL Publication No.02-25, 2003.

64. Adams, N. and Schubert, U. S., From science to innovation and from data to knowledge: Escience in the Dutch Polymer Institute's high-throughput experimentation cluster, *QSAR Comb. Sci.* 24, 58, 2005.

65. Frantzen, A., Sanders, D., Scheidtmann, J., Simon, U. and Maier, W. F., A flexible database for combinatorial and high throughput materials science, *QSAR Comb. Sci.* 24, 22, 2005.

66. Cohan, P. E., Combinatorial materials science applied – mini case studies, lessons and strategies, in *Combi 2002 – the 4th annual international symposium on combinatorial approaches for new materials discovery*, Knowledge Foundation, Jan. 23–25, 2002, San Diego, CA.

67. Potyrailo, R. A. and Pickett, J. E., High-throughput multilevel performance screening of advanced materials, *Angew. Chem. Int. Ed.* 41, 4230, 2002.

68. Li, G. Y., Highly active, air-stable combinatorial catalysts for the cross-coupling of aryl chlorides: the first homogeneous combinatorial catalysts for industrial applications, in *Combi 2002 – the 4th annual international symposium on combinatorial approaches for new materials discovery*, Knowledge Foundation, Jan. 23–25, 2002, San Diego, CA.

69. Bricker, M. L., Gillespie, R. D., Holmgren, J. S., Sachtler, J. W. A. and Willis, R. R., Scaling-up of catalysts discovered from small scale experiments, in *High Throughput Analysis: A Tool for Combinatorial Materials Science*, Potyrailo, R. A. and Amis, E. J., editors, Kluwer Academic/Plenum Publishers, New York, 2003, ch. 26.

70. Spivack, J. L., Webb, J., Flanagan, W. P., Sabourin, C., May, R. and Hassib, L., Design, use and results from combinatorial steady-state reactors for the optimization of highly selective heterogeneous catalysts in liquid systems, in *Combinatorial and artificial intelligence methods in materials science II. MRS symposium proceedings*, Potyrailo, R. A., Karim, A., Wang, Q., and Chikyow, T., editors, Materials Research Society, Warrendale, PA, 2004, p. 199.
71. Spivack, J. L., Cawse, J. N., Whisenhunt Jr., D. W., Johnson, B.F., Shalyaev, K. V., Male, J., Pressman, E. J., Ofori, J., Soloveichik, G., Patel, B. P., Chick, T. L., Smith, D. J., Jordan, T. M., Brennan, M. R., Kilmer, R. J. and Williams, E. D., Combinatorial discovery of metal co-catalysts for the carbonylation of phenol, *Appl. Catal., A* 254, 5, 2003.
72. Boussie, T. R., Diamond, G. M., Goh, C., Hall, K. A., LaPointe, A. M., Leclerc, M., Lund, C., Murphy, V., Shoemaker, J. A. W., Tracht, U., Turner, H., Zhang, J., Uno, T., Rosen, R. K. and Stevens, J. C., A fully integrated high-throughput screening methodology for the discovery of new polyolefin catalysts: Discovery of a new class of high temperature single-site group (iv) copolymerization catalysts, *J. Am. Chem. Soc.* 125, 4306, 2003.
73. Archibald, B., Brümmer, O., Devenney, M., Giaquinta, D. M., Jandeleit, B., Weinberg, W. H. and Weskamp, T., Combinatorial aspects of materials science, in *Handbook of Combinatorial Chemistry. Drugs, Catalysts, Materials*, Nicolaou, K. C., Hanko, R., and Hartwig, editors, W. Wiley-VCH, Weinheim, Germany, 2002, ch. 34, p. 1017.
74. Potyrailo, R. A., Combinatorial and high-throughput development of polymer sensor materials for optical and resonant sensors, *Polymer Preprints* 45 (2), 174, 2004.
75. Potyrailo, R. A., Analytical spectroscopic tools for high-throughput screening of combinatorial materials libraries, *Trends Anal. Chem.* 22, 374, 2003.
76. Farrusseng, D., Klanner, C., Baumes, L, Lenliz, M., Mirodatos, C. and Schüth F., Design of discovery libraries for solids based on qsar models, *QSAR Comb. Sci.* 24, 78, 2005.
77. Dirion, B., Cobb, Z., Schillinger, E., Andersson, L. I. and Sellergren, B., Water-compatible molecularly imprinted polymers obtained via high-throughput synthesis and experimental design, *J. Am. Chem. Soc.* 125, 15101, 2003.
78. Potyrailo, R. A., Ezbiansky, K., Chisholm, B. J., Morris, W. G., Cawse, J. N., Hassib, L., Medford, G. and Reitz, H., Development of combinatorial chemistry methods for coatings: High-throughput weathering evaluation and scale-up of combinatorial leads, *J. Comb. Chem.* 7, 190, 2005.

2 Expanding the Scope of Combinatorial Synthesis of Inorganic Solids: Application of the Split&Pool Principle for the Screening of Functional Materials

Stephan Andreas Schunk, Peter Kolb, Andreas Sundermann, Torsten Zech, and Jens Klein

CONTENTS

2.1   INTRODUCTION

The fields of combinatorial and high-throughput experimentation (HTE) have matured to accepted scientific and technological branches and are applied in research chains throughout the academic world and in industry. Especially challenging was the technological transition of the "combichem approaches" deriving from organic chemistry towards the world of inorganic materials synthesis and screening. For many scientists of the materials science community this transition was seen as one that could never be achieved successfully. Nevertheless academic and industrial groups achieved scientific solutions and technological standards that allow high-throughput screening of inorganic materials for diverse properties of interest. The number of true combinatorial approaches though, applied to inorganic materials sciences, is small, if not negligible.

The goals of this chapter are (1) to illustrate the background of important approaches in combinatorial sciences and high-throughput experimentation for organic and inorganic sciences and (2) to illustrate that true combinatorial approaches like the Split&Pool principle can be applied in the world of inorganic functional materials and may be of great value in accelerating the screening process in the search of compounds with respective target properties.

2.2   COMBINATORIAL SYNTHESIS OF ORGANIC COMPOUNDS

With increasing demands on HTE[1-6] in the field of inorganics and materials sciences for the provision of higher test capacities, faster analysis times, and materials characterization, as well as shorter product-to-market intervals, much effort is being made to develop new methodologies and technologies in which HTE can be used in the synthesis of materials. Qualitatively, these materials should fulfill the requirements of conventional lab-scale materials research (i.e., homogeneity of support materials, support interaction, porosity, scalability, etc.), quantitatively one postulates highly diverse materials libraries with thousands of members in very low amounts (i.e., milligram or microgram scale).

Between 1980 and 1995, organic, biochemical, and pharmaceutical chemists started innovative basic approaches to the syntheses of molecular entities and peptide-based molecules.[7-15] This work was based on Merrifield's[16] research on solid-phase synthesis (SPS) in 1963.

It is worthwhile putting the ground-breaking "one bead–one compound" concept into perspective by comparing it with other alternative approaches used in combinatorial chemistry and high-throughput experimentation.

R. Frank was one of the first scientists using and applying the potential of the combination of several different solid-phase bound substrates for a reaction with a single reagent.[17] He used cellulose paper disks ("filter-disk method") as solid supports for the synthesis of oligonucleotides (see Figure 2.1). By the covalent attachment to the solid carrier, a manual distribution of differently loaded paper disks into reactor vessels is possible. The experimental realization of this new method was limited and determined by the chosen chemistry and the application of oligonucleotides.

R.A. Houghten[7] published in 1985 the so-called "tea-bag method," which was originally developed for the multiple peptide synthesis in the range of circa 150 different peptides in 50-mg amounts (with a length of the peptide chains of approximately ten amino acids). For separation and isolation means of the polymeric support ("beads"), they are shrink-packed into net-like, polypropylene bags. Each bag contains between 50 and 100 mg of polymeric resin beads as supports, wherein the mean bead diameter has to exceed the mean pore diameter of the polymeric membrane of the bag (see Figure 2.2). A simple handwritten and solvent-resistant label on the outside of the bag served for coding. Washing steps, coupling, and cleaving of protecting groups are carried out in a common container (i.e., beaker or flask), wherein the coupling of different amino acids and the final cleavage of the resulting peptide are accomplished in a unique vessel for each or a selection of polymeric bags containing the support resin beads. One characteristic of the tea-bag

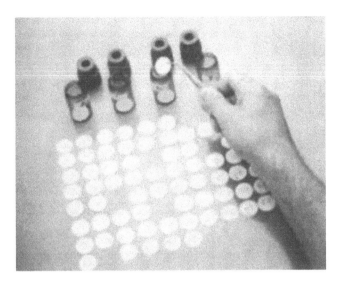

FIGURE 2.1 Filter-disk synthesis.[17]

method is that the information about the spatial localization of each reaction step is ensured by labeling the bags. The flexibility, the relatively high amounts of resulting peptides, and the low equipment investment are advantageous, but the intensity of manual lab work (washing, sorting, labeling, etc.) and the limitations regarding the number of manageable peptides shorten the scope of application of this special method.

Another principle, the "pin synthesis" reported in 1984 by H.M. Geysen et al.,[8] operates at much lower amounts of peptides (Figure 2.3). Geysen and co-workers performed a parallel synthesis of molecular libraries of peptides, in which the peptides were created one amino acid at a time, beginning with an attachment to an array of amino-functionalized polyethylene rods (i.e., 8 × 12 matrix, each pin: 4 mm diameter, 40 mm length). By knowing the specific reaction history of each pin, the resulting peptide on each pin is known. All reaction steps are performed by handling the array of pins either in separate wells for each pin or huge containers for the whole array. Because of the very small amount of produced peptide (ca. 300 nmol) and the lack of such sensitive test assays at this time, there was no need to separate the peptide from the top of the pin. By using this method the positional information of each material was known at each reaction step. This methodology has a large impact on immunochemical assays for peptide–antibody interactions.

In 1991, S.P.A. Fodor et al.[9] presented a new light-directed, spatially addressable parallel chemical synthesis principle (see Figure 2.4). They synthesized peptide libraries in the form of two-dimensional arrays, e.g., on a glass plate, each member of the combinatorial library being identified by its position on the array. By merging the techniques of solid phase chemistry and photolithography, an array of peptides can be displayed, which is spatially addressable. The base plate is first functionalized with amino groups. These are protected with a photolabile leaving group. By masking regions of the plate and removing the protecting group from exposed regions with light, amino groups become available for coupling at selected sites on the plate. By a smart masking technique, excellent spatial resolution can be achieved on sites of only 50 μm^2, thus making it possible to provide 40,000 discrete synthesis sites on a 1-cm^2 plate. The position of each peptide depends on the masking strategy. This method is especially interesting for currently developed DNA sequencing techniques and for diagnosing genetic mutations ("DNA-on-a-chip"). Material amounts are in the μmol or nmol range.

In 1992, R. Frank[18] again reported the logical further development of his above-described filter disk method: the "spot synthesis." Through simple merging of his support-based approach and an

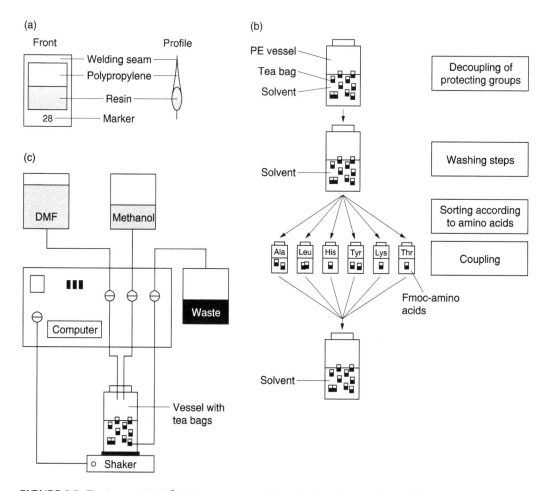

FIGURE 2.2 Tea-bag synthesis.[7] (a) Bag structure; (b) synthetic pathway; (c) setup for automated synthesis.

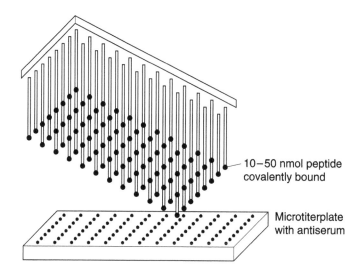

FIGURE 2.3 Pin synthesis.[8]

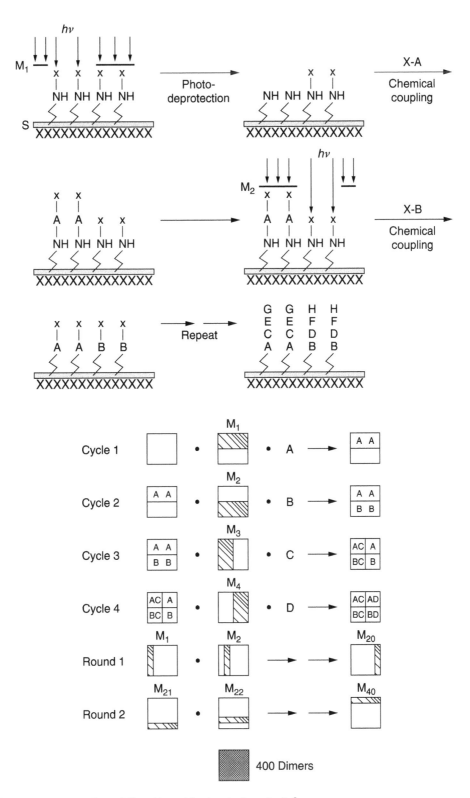

FIGURE 2.4 Light-directed, spatially addressable chemical synthesis.[9]

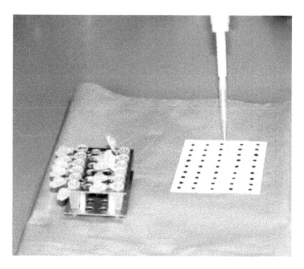

FIGURE 2.5 Spot synthesis.[18]

approach for positionally addressable parallel synthesis (i.e., Geysen's pin synthesis), Frank used a single cellulose sheet and an automated parallel synthesizer based on a pipetting robot to load different regions of the cellulose paper with different monomers. Upon dispensing a small drop of liquid, the droplet is adsorbed and forms a circular spot. This spot forms an open reactor for chemical conversions involving reactive linkers anchored to the membrane support. In this way a great number of separate synthesis regions can be arranged as an array on a large cellulose sheet (see Figure 2.5). Each spot is individually addressable. Common reaction steps to all synthesis regions can be easily applied to the whole membrane support (i.e., washing steps).

Among the first scientists using the term "combinatorial chemistry," A. Furka and co-workers[10] in 1988 published an absolutely novel and pioneering concept, merging some characteristics of the above-described methods invented by Houghton and Geysen, the "Split&Pool synthesis" (synonyms: "Mix&Split," "Split&Combine," "One bead–One compound," "Selectide process," etc.). A batch of resin beads (i.e., several million porous spherical beads, homogeneous in size and loading capacity) are divided ("split") into a number of aliquots of equal size (see Figure 2.1). To each portion a different monomer A, B, C, etc. (i.e., amino acids) is coupled. When coupling is completed on each aliquot of resin beads, excess reagent is removed by washing with an appropriate solvent and the aliquots are recombined ("pooled") and thoroughly mixed. This process of dividing the resins into aliquots, quantitatively coupling the monomers separately on to each aliquot, and washing the resins before recombining and intensively mixing, may be repeated several times depending on the size of the combinatorial library required and the capacity of the resins to accommodate all the members of the library.

As can be seen in Figure 2.6, if three monomers A, B, and C are used, two rounds of coupling result in a library of $3^2 = 9$ members, three rounds in a library of $3^3 = 27$ members, four rounds in a library of $3^4 = 81$ and so on. So, the total number of library elements N is $N = M^S$, where M corresponds to the number of monomers, and S corresponds to the number of split steps. It is important to ensure in this synthetic strategy that each bead in the batch of resins carries only a single member of the library, i.e., each bead behaves like a microreaction vessel with — in the optimal case — quantitative conversion. The practical limit of the library size depends strongly on the amount of beads, which can be handled and screened with an according assay or an ulterior test. This "Split&Pool synthesis" in the above-mentioned scientific fields allows the fast generation of a huge diversity of discrete molecules by consecutively building up covalent bonds at predefined positions in the molecule via functional groups.

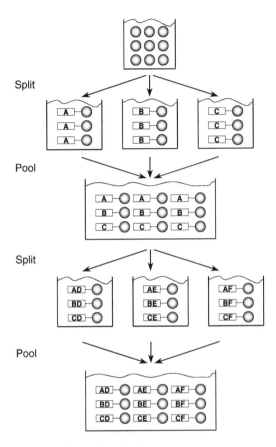

FIGURE 2.6 Schematic draft of a Split&Pool synthesis.

Principally, one can differentiate between a random Split&Pool and directed Split&Pool: in the first case[11,19] the chemical history of the beads or particles is lost. After each synthesis step, the resin beads from all vessels are pooled and randomly split for the next combinatorial step. The compound distribution in a collection of beads (library) is driven by statistical means. Depending on the factor of over-determination of a library, each compound is synthesized numerous times. In the case of "directed Split&Pool," a short excursion to tagging and encoding methods is necessary for the understanding, because these technologies are additional tools in the combinatorial synthesis to follow and retain the chemical history of each bead or compound within a Split&Pool synthesis.

Dealing with large libraries of molecules or entities is complicated by the loss of the "who's who" information: sample identity or synthesis history cannot be as easily assigned as in a single lab-scale synthesis in one flask. Encoding or tagging methodologies[20] can be divided into chemical and non-chemical encoding methods. In the latter case, positional encoding, as described by the light-directed synthesis of Fodor,[9] the multipin method of Geysen[8] or the common use of microtiter plates (i.e., 96-well plates) are obvious and simple examples for positional encoding technologies, the non-positional encoding is represented by the tea-bag approach of Houghton,[7] or the most advanced techniques of radio-frequency tags, published by IRORI.[21] In this class of non-chemical encoding means, sample identity can be recovered via their fixed position during the synthesis procedure or via "external" markers, escorting the sample within the whole process. Another example of a non-chemical encoding technology was described by Xiao et al. in 1997.[22] They presented a new laser optical synthesis chip (LOSC), which combines two functions: encoding the produced material by a laser-fabricated binary code, which can be read automatically by a pattern-recognition system, and serving as solid-phase support material. The most simple method of non-chemical encoding was published by J.W. Guiles.[23] Guiles and co-workers used colored glass beads and vessels with colored lids. By using eight different colors, a binary code for 64 materials is easily performed.

Chemical encoding means the "internal" or intrinsic anchoring of a sample's check mark within a combinatorial library. This chemical marker has to be invariably connected to the compound itself during the synthesis; it has to be chemically inert under the synthesis conditions as well, as it should not interfere during the property screening. Examples for chemical tags are oligonucleotide tags, haloaromatic tags, and many more.[20] A directed Split&Pool synthesis has to fulfill several issues: the chemical history has to be recorded, as described by the encoding technologies. The solid-phase support (i.e., resin bead) must be formulated in such a way that one unit provides the required or wished amount of the compound. The whole process of distributing solid-phase particles needs to be integrated in an overall workflow. The historical manual work in Split&Pool steps is tedious and error-prone.

For the realization of huge, highly diverse or focused materials libraries for solid state materials and inorganic chemistry, mainly two concepts have been investigated in academia and industry: (1) traditional lab bench chemistry, where parts or all steps can be automated with robotic systems for liquid dosing and handling of solids, or a combination of both as is applied at hte AG, or (2) the parallel synthesis on two-dimensional substrates by sputtering or other deposition methods, based on the above-described methodology and masking technology of Fodor et al.[9] or the methods described by J.L. Winkler and colleagues.[24]

2.3 PITFALLS AND CHALLENGES IN ORGANIC COMBINATORIAL SYNTHESIS

2.3.1 METHODS FOR MONOMER LINKAGE

The concept of combinatorial chemistry was generated from the need to synthetically access large libraries of compounds in order to screen diversity spaces and get information about possible structure–activity relationships within a given chemical space. Being tools that proved valuable, the synthesis of potential pharmaceuticals and other biologically active compounds combinatorial methods have been used for quite some time.

The successful application of the concept of combinatorial synthesis requires that high-throughput methods are present for all steps of the preparative process. The challenge from that prerequisite is 2-fold. Suitable instrumentation has to be used in order to ensure an automated, accurate delivery of reactants (see also Chapter 4). The key to further improve the overall efficiency of the synthetic process is the use of a suitable synthetic methodology that allows a minimization of the essential steps for product synthesis, separation, and characterization. At the same time, it is necessary in some cases to individually monitor the progress of a multitude of reactions taking place simultaneously in a limited spatial array and to avoid any uncontrolled reagent crossover between reaction centers.

The first approaches to these challenges were the now classical solid-phase syntheses pioneered by Merrifield.[17] This methodology was developed to allow the synthesis of peptides much faster than was possible using the conventional "one reaction–one pot" strategy. The fact that this synthetic method was finally accepted as a tool for the synthesis of not only peptides but also for libraries having a range of different functionalities was the result of enormous developmental work facing several general challenges.

One of the major demands is that any successful synthesis performed by a multitude of consecutive steps either requires that each intermediate can be isolated and purified or that each reaction proceeds with practically quantitative yield. Clearly, the first approach is not practical when the reaction proceeds with a substrate attached to a solid support. Thus, for a successful synthesis, the synthetic methodology has to be highly efficient. The most simple way to achieve complete turnover is to use the non-supported reagents in excess and remove the unreacted components by filtration or washing, leaving the functionalized immobilized substrate behind and ready for the next transformation.

Even if the synthetic methodology ensures a reliable completeness of each reaction step, further complications can arise from interactions between reaction centers. It has been argued[25] that, since the polystyrene resins normally used as supports are strongly hydrophobic and peptide chains are polar entities, interpeptide interactions are likely to occur unless very low substrate loadings are used. This, in turn, limits the amount of products that are available from a reaction employing a given amount of resin. Consequently, resins were developed for solid-phase synthesis, which emulated the environment of more polar substrates, polyamides[26] and polyether-functionalized polystyrenes (TentaGel)[27] being the most common.

2.3.2 Different Types of Linkers

As is the case with every multistep synthesis, failure in performing the very last step with sufficient selectivity can make the overall sequence unsuccessful. In the context of solid-phase synthesis, the very last step means detachment of the final product from the solid support. The linkage between the substrate and the support has to be resistant towards the reagents and conditions employed in substrate construction to avoid yield-reducing premature link scission and at the same time has to be labile under conditions which leave all the functionalities of the final, detached product intact.

The original linkage used by Merrifield was a benzylether functionality which was established by attaching the N-protected first amino acid in the targeted sequence via its carboxylate to chloromethyl functionalized, crosslinked polystyrene. After completion of the peptide synthesis, detachment from the resin can be accomplished by treatment with strong acids (trifluoroacetic acid or anhydrous HF), which also removes protecting groups present in the different amino acid fragments.

To enable the peptide synthesis to proceed under milder conditions, several strategies for the buildup of libraries and for the linkers were developed.

2.3.3 Monitoring the Progress of a Reaction and Identifying the Products

As outlined above, successful implementation of combinatorial methods in organic chemistry relies on efficient synthetic methodology. As the process of improving known organic reactions and finding new ones has been pursued ever since, the progress based in this area is merely an organic one. It was and is, however, strongly accelerated by the high requirements of methods applicable in combichem, mainly a very high efficiency at mild reaction conditions accompanied by a broad substrate range.

A more general challenge of combinatorial organic chemistry has at the same time been a point of much skepticism at the very beginning. The way of carrying out library syntheses on a solid support and the large number and the small amounts of library members makes the isolation and purification of intermediates and final products, and the monitoring of the reaction by standards impossible or at least difficult. Moreover, for very large libraries, even the recovery and identification of the products can be a difficult task when the solid-phase method developed by Merrifield is used, i.e., the synthesis of different library compounds which are present on different resin corns but in the same reaction vessel. One solution to this problem is to recover the product samples by using a much finer grid of spatially addressable reaction vessels. One invention in this area is small (cm range) reaction bags made of polypropylene mesh ("teabag") which, after filling with resin corns and closing, could be placed in reaction vessels to effect the desired functionalization.[28] As the beads are too big to penetrate the pores and leave the bag, bags with different peptide chains can be placed in one reaction vessel without the possibility of admixing the different products. A second milestone is the development of so-called multipins, which can be regarded as solid supports with a handle. These arrays can be generated by irradiating PE rods with high-intensity γ-irradiation and quenching the resulting radical species with acrylic acid, which thus becomes fixed on the PE support and can be used as an anchor group for attaching substrates.[8]

Another possibility for product identification consists in the use of tagging strategies. On each resin corn a molecular tag is generated, i.e., a structure with a unique, unambiguous composition designed for more rapid and unambiguous structure determination by standard chromatographic and/or spectroscopic methods.

This technique requires the mutual compatibility of reaction conditions for substrate synthesis on the one hand and tag synthesis on the other in order to avoid interference between tag buildup and substrate buildup. Furthermore, the methods and conditions for identification and characterization of the tag and thus of the substrate it encodes must not corrupt the substrates integrity. Finally, as any interference between tag and substrate can significantly influence the results of property screening of the substrate after synthesis, detachment of the tag must be attainable under conditions that leave the substrate fully intact.

One of the first realizations of this principle was the use of oligonucleotide tags in peptide synthesis.[29] Synthetic strategies for oligonucleotide and oligopeptide construction are orthogonal to each other. Additionally, by means of the polymerase chain reaction (PCR), an established, selective method for oligonucleotide sequencing, and thus decoding of the library, was available.

As methods for peptide sequencing are well established, the use of peptides as tagging entities is also attractive. This technique is especially useful when the substrate library is a non-peptide one.[9] Among the tags that offer greater flexibility in designing the encoded library, aromatic halogen compounds[30] and secondary amine units in the form of amide sidechains[31] are to be mentioned.

Even more flexibility is offered by using non-chemical tagging strategies, i.e., the tag is not generated by a sequence of chemical reactions, but by, for example, optical encoding via laser labeling[22] or radio-frequency marking of encodable and decodable microchips.[21] For these methods, practically no interference of the encoding–decoding process with the chemical functionality or integrity of the substrate is expected.

In summary, the challenges of combinatorial organic chemistry had a significant influence on the development of organic chemistry in general. From the beginning it is becoming increasingly clear that multidisciplinary approaches will have to be developed and applied in order to keep up with future requirements of combinatorial chemistry in terms of speed and reliability.

2.4 HIGH-THROUGHPUT SYNTHESIS OF INORGANIC COMPOUNDS

The synthesis of inorganic compounds, especially inorganic functional materials, is substantially different from the requirements for the synthesis applied in the field of organic chemistry. Whereas combinatorial organic synthesis in most cases targets the synthesis of single molecular entities, well-defined and characterizable via an arsenal of spectroscopic, chromatographic, and spectrometric methods, the complexity of inorganic solids makes different approaches necessary for synthesis and quality control of the library (see Table 2.1).

2.4.1 Approaches Used in the Generation of Organic Compounds

Generally, for organic entities two different approaches for the synthetic generation can be distinguished. The first approach is the synthesis on solid phase and the second the parallel synthesis not using polymer-bound reaction pathways. For both approaches a wealth of automation technology is available today and at first glance the need for specific development tools for high-throughput synthetic materials synthesis is not obvious. Yet one of the most clear requirements for materials synthesis is synthetic platforms that tolerate conditions usually applied for the synthesis of functional materials. For solution-based inorganic chemistry there is a good match in techniques applied in the automated solution-based synthesis of organic compounds; typical technology platforms supplied

TABLE 2.1
Differences in High-Throughput Experimentation for Molecular Entities in Pharmaceutical Research and Inorganic Solids Applied for Catalysis

Organic molecular entities	Inorganic functional materials
Discrete molecular entities	Three-dimensional structures, often multielement composites, potentially metastable materials
Finite number of "active centers," often well characterized, can in many cases be modeled, good understanding of chemistry on a molecular level	Only for few materials basic understanding of type and function of active centers
Purities of over 85% achievable via combinatorial synthesis	Pure substances often without catalytic effect, especially multicomponent systems with defect structures show greatest use in catalysis
Highly developed screening procedures for biological activity available	Characterization of numerous, often independent, parameters challenging
Descriptors for library diversity well developed	Basic development work for descriptors still required
Good availability of synthetic building blocks and methods	Adaptation of complete "unit-operations" for automated synthesis required

TABLE 2.2
Comparison of Synthetic Approaches Used in Organic and Inorganic High-Throughput Experimentation

Screening approach	Organic synthetic approach	Inorganic synthetic approach
Mass screening/Stage I	One bead–one compound One substrate–multitude of compounds	One substrate–multitude of compounds
Refined screening/Stage II	Multitude of beads–one compound Parallel solution-based synthesis	Parallel synthetic approaches based on synthetic "unit operations"

by a number of specialized companies can be used for the purposes of liquid handling and reacting liquids in appropriate vessels.[32–35] Still, as soon as solid formation takes place, most of the technology suppliers lack sufficient solutions that allow the automatic handling of the resulting precipitates. Organic crystalline solids can in many cases be obtained in the form of crystals in the micrometer range, which can be filtered off and easily washed with solvents. Common technical solutions for the implementation of crystallization reactions are offered[33] and usually consist of filtration devices operated over glass frits, Teflon frits or membranes. It is, however, understandable that precipitation steps with subsequent filtration can usually be avoided and typically solvent separation is performed under evaporative conditions, either thermally or via application of vacuum conditions, sometimes even vacuum centrifugation.[33,36] The precipitation of inorganic solids from solution is a very complex process aiming at more than the pure separation of a compound from solution.[37,38] Chemical conditions play a larger role in terms of performance properties of the inorganic solids obtained than for organic materials (see Table 2.2). For inorganic precipitates the workup of the resulting precipitate, like aging conditions or washing steps, are crucial and have to be well monitored. Common practice for a large range of functional inorganic materials is the thermal post-treatment, often under specific gas mixtures. An additional preparative aspect involved with inorganic solids is shaping procedures, even simple operations like sieving and tableting are far beyond the scope of the automatic synthesis platforms used for organic synthesis. Still all of these preparation procedures tune the properties of the inorganic material and detailed control over every single synthetic operation is essential.

2.4.2 THE TRANSITION TO INORGANIC MATERIALS

A step in order to avoid the above-mentioned challenges and yet be able to obtain large libraries of materials synthetically goes back to the work of Hanak.[39,40] Apart from avoiding solution-based methodologies and using sputtering techniques Hanak introduced the concept of employing sub-strate-bound libraries. The essential background of the conceptual approach is that, via deposition leading to efficient "bonding" (of either chemical or physical nature) of different elements on a support material, an integrated library format is created which accompanies a large-membered library attached to one single device. Usually, it is desired that the support materials show inertness in terms of interference with chemistries employed or properties being screened against in the screening process. It is obvious that this approach is different to the synthetic approaches used for organic entities where the compounds are either attached on solid phase (single or multiple beads) or obtained as solids or solutions in adequate formats (typically vials or microtiter plate formats).

In the original approach Hanak made use of another property that is unique to inorganic solids: while organic entities are well defined via their molecular structure and substitutional change is quantitized, inorganic materials represent extended atomic or molecular entities where compositional changes can be continuous or quasi-continuous. The author exploited this property by building continuous gradients of different elements over the substrate material and obtained a two-dimensional quasi-continuous library of materials that he could then screen for certain materials properties. If one compares this approach with organic synthesis on a solid phase it is interesting to see that there is a transition from the organic approach of where "one bead–one compound" or "plurality of beads–one compound" now translates in the inorganic world as "one substrate–plurality of compounds."

For a large number of applications this deposition technique on a substrate via sputtering, evaporation techniques, ablation techniques, and chemical vapor deposition has proven to be a viable way of synthesizing substrate-bound libraries.[41–44] Typical deposition techniques use several source materials and spatial resolution is achieved via masking techniques. A wealth of parameters like masking strategies, sequence of deposition, rate of deposition, and much more are the factors controlling the exact composition of the library members bound to the support. Usually, an annealing step is performed at the end of the deposition technique in order to create crystalline or amorphous materials from the deposited precursors. Starting from a gradient approach this deposition methodology has been largely refined and large libraries of materials can be obtained, the number of materials on one support can range up to library sizes of 25,000 library members.[45] Still, even if this technique is very effective, it is questionable whether or not, in accordance with organic synthesis strategies, it can be called a truly combinatorial synthetic technique as it follows more the logic of a massive parallel approach to library synthesis. It is interesting to see that epitactical growth of library candidates can even be sized down to epitaxy of single molecular layers.[46] This broadens the scope of deposition techniques, essentially because it allows not only precursor deposition on the substrate for further reaction to a single compound by annealing but also offers the potential of creating amorphous or crystalline solids with atomically/molecularly controlled spatial doping of the solid directly in the deposition step.

Deposition techniques like sputtering, chemical vapor deposition or epitactical methods are especially useful for the synthesis of "dense" and defect-free solid matter. For a large range of applications, like catalysis, usually hydrothermal, sol–gel or solution-based methodologies are employed as a synthetic concept for the generation of substrate-bound libraries. In contrast to the deposition techniques discussed before, the target solid materials can, in many cases represent metastable modifications, which still are highly interesting for a certain range of applications, in particular catalysis. Still, technologically many solutions have been generated for the efficient fabrication of such libraries. In order to achieve spatial resolution and allow accurate deposition volume control in the nanoliter range Xiang[47] and Mallouk[48] have employed ink jets as dispensers. Very high library densities can be achieved: up to approximately 100 library candidates per square inch of the support can be obtained with this technology. Groups of authors prepared their library members by using aqueous solutions of their precursors, thus obtaining either mixed oxides, oxide mixtures or via reduction alloys of metal compounds.

An alternative chemical pathway that in some cases is more advanced as it also may open the gate to different classes of materials is the sol–gel process[49]; especially the absence of counter-ions, easily evaporative solvents and metal sources which are well available offer elegant synthetic pathways in combination with a wide variety of synthetic protocols of sol–gel processing. As the authors describe their process, metal alkoxide sols are dispensed via liquid-dispensing robots, then gel formation is induced and the solid network obtained.[50,51] Depending on the dispensing device, high library member densities can be realized on the substrate, similar to the densities cited earlier.

Also synthetically prepared via sol–gel methods, but with much lower library densities on the support, was the library of multicomponent oxides of Mo-V-Nb-O$_x$ on a quartz wafer described by Liu et al.[52] The actual size of the quartz wafer was 7.5 cm in diameter with 132 library members on the support, the dispensing volumes of the precursor solutions were in the microliter range, so that the mixed oxides were deposited as thick films and used for catalyst screening.

Even the complex technical solutions for the issue of zeolite synthesis with the final products obtained on a single support was achieved.[53] A common design of the reactor with full implementation of the later support was integrated via a sandwich-based setup. While zeolite synthesis was achieved in small Teflon reactors a silicon wafer would serve as support for the bound samples after hydrothermal synthesis and calcination. The use of the support is here more sophisticated than in the examples given above. While above the support is the synthetic medium of deposition and does in general not require refined layout, other than dimples or wells in some examples,[52] here the support is a part of an integrated reaction setup and is only after the decisive synthetic steps being dismantled serving as pure support medium.

But the level of sophistication can not only rise due to demands on the synthetic side, elaborated multifunctional "substrates" are also used to meet demands connected to the screening process. One of the most challenging screening procedures is the screening for catalytic properties on a micro-level as control of fluids and heat requires a high level of accuracy. A nice example is the work by Claus and Zech,[54,55] where the layout of the support was equipped with an integrated fluidic manifold of binary layout, deposition regions for the library members in microchannels and product outlets for analysis of the gaseous products.

2.4.3 Alternative Approaches: Substrate-Free Synthetic Concepts

The number of publications dealing with substrate-free synthesis and combinatorial materials chemistry is far smaller than the number of publications of library synthesis and characterization of substrate-bound libraries. The reasons are multifold: on the one hand the "simple" parallel or fast sequential synthesis of solid materials can in many cases hardly be attributed to the term combinatorial and mere automation aspects are in many cases not considered to be of high enough scientific information content to find their way into the relevant journals. On the other hand the early work committed to high-throughput experimentation was usually connected to substrate-bound libraries and therefore in many cases substrate-bound libraries and materials combichem seem to be mentally connected. However, it is evident that, especially if the testing requirements go beyond a qualitative screen and the focus is on the application demands testing under conditions relevant on an industrial scale, synthetic procedures have to be adapted in order to ensure adaptability of the methods employed. The analogies with parallel synthesis of organic compounds are very obvious: as soon as the size of the synthetic effort is increased the demands with respect to the accuracy of the synthetic method employed go up as well. For organic chemistry the purity of the compounds becomes an essential issue and often separation steps are included in the synthetic effort combined with checks for the compound purity. In materials synthesis accurate control of parameters like pore volume, pore-size distribution, element distribution, formation of certain phases, etc., become an essential feature of the synthetic effort and therefore from a certain degree of complexity demand analytical control.

As stated above only a few authors mention the parallel or fast sequential synthesis of inorganic materials, the application focus is mainly a catalytic one. Senkan's group reports the successful synthesis of impregnated catalysts by using alumina extrudates as support materials.[56,57] The single supports of a size of 4 mm by 1 mm were impregnated with Pt/Pd/In as active metals using aqueous precursor solutions which were dispersed by an automated liquid-handling system. After the impregnation step, drying and calcination were performed. Impregnation and co-precipitation techniques for larger scales are also described by Baerns' group.[58–60] The relevant scales and testing procedures are in a range that can surely be called "close to conventional" and allows conclusions to the catalyst behavior on a larger scale. Hoffmann et al. report the automated synthesis of Au-based catalysts via pH-controlled co-precipitation and equilibrium adsorption techniques.[61] From a preparative point of view these techniques require a high level of sophistication and the complexity of automating these techniques has to be adapted to these requirements. The point that makes this preparative work extremely important is the fact that a lot of the techniques required to prepare these types of catalyst have large technical relevance and mimic unit operations used for the synthesis of certain catalyst types on an industrial scale.[37]

If one compares the approach of non-substrate-bound parallel or fast sequential synthesis with the substrate-bound approaches it becomes pretty apparent that the focus of the non-substrate-bound approaches is different, e.g., as stated above more directed towards the technical goals of the final application and the degree of parallelization is usually also moderate (6- to 16-fold).[59,61] Library handling is completely different and the idea of integrated formats does not play a role any longer to a large extent. Even if automated solutions for sequential or parallel sample handling can be envisaged, the single handling of samples can also become a bottleneck and has to be analyzed with respect to its time limitation within the workflow envisaged.

2.5   APPLYING COMBICHEM TO INORGANIC MATERIALS: THE SPLIT&POOL PRINCIPLE

As stated before, from a chemical point of view it is questionable whether a direct transition from the organic world and the combinatorial synthesis concepts apply in the domain of inorganic materials. For organic entities the synthetic concepts related to combinatorial chemistry can on the one hand be checked on whether or not identity was achieved during the synthetic operation and on the other hand the synthetic principle is easily translated into scalar or vectorial displays that represent the different chemical entities. The beauty of organic synthesis also derives from the fact that any synthetic step will lead to a chemically defined alteration of the entity that can again be represented by the named displays. There are different ways of representing these displays; as a useful example the reader is recommended to read the following publication on "smiles" or "broad smiles."[62] The translation of these scalar or vectorial displays into the inorganic world stops at the point of library member representation. Inorganic materials represent three-dimensional bodies, so chemical alterations may be spatially segregated, which adds a further dimension to the representation. On top of this, more factors, like influences of the matrix material as crystallinity, defects, phase composition, the pore system, connectivity of the pores, total pore volume, and surface area, and the nature of the surface may also be crucial in a number of applications.[63] Even if the multidimensionality of the inorganic materials could principally be represented by a vectorial or scalar display, in many cases the behavior of inorganic systems will not be a linear one. This is a great difference with respect to organic entities where synthetic operations in most cases scale linear with regard to alterations of the substrate. The synthetic operations can therefore be taken as a direct function of substrate modification.

For inorganic materials, changes of properties may in many cases not scale linear with synthetic procedures applied, and/or induce more than one effect at a time. A good example may be dopants added to inorganic materials that do not only act solely as additional elements being present in a sample altering the chemical composition of the bulk and the surface, but may also induce bulk or surface nucleation of different phases, or, depending on the nature of the inorganic solid, may segregate and just be present in the surface layer (see, for example, Somoriaj[64]). The complexity of multicomponent systems or mixtures of multicomponent systems in some cases even make it difficult to predict the implementations that a synthetic operation will have on a given system with regard to the properties of the material.

If we compare the chemistries involved for the creation of organic libraries and libraries of inorganic materials, one can conclude that for the synthesis of organic libraries series of well-defined synthetic steps lead to libraries of defined chemical entities, whereas for inorganic materials library creation series of well-defined synthetic steps lead to libraries of complex materials which are at first characterized by the synthetic pathway employed to synthesize the respective library member, but the single synthetic operation and overall pathway may not have the same effect over the whole library of complex materials. The question arises as to whether a true combinatorial approach revising the complexities of inorganic materials synthesis is possible after all? From a direct transition step of the organic to the inorganic synthesis principles, except for a very few examples, the direct answer has to be no. Still if one considers including the intelligent combination of synthetic steps in a combinatorial manner for the generation of complex inorganic materials libraries into the group of combinatorial synthetic chemistries, then synthetic approaches to inorganic materials can also be considered combinatorial. If the topic is regarded from this point of view the synthetic possibilities offer a wide range of perspectives for variation. Purity of compounds, potential aging of precursor solutions, sequence, and decent history of steps during addition of compounds offer a range of synthetic parameters that can be varied and used to create diverse libraries. Successful synthetic efforts for Split&Pool synthesis have been reported by two groups independently.[65–68]

As discussed above, among the different methodologies for the efficient synthesis of libraries of a high degree of diversity, the Split&Pool principle offsets other synthetic methodologies by far. With a relatively small number of operations, compared to standard parallel operations, a large number of variations of library members can be achieved. The essential steps of the synthetic effort are subsumed in Figure 2.6 and Figure 2.7, from a principal point of view the synthetic steps of splitting and pooling are identical, the steps wherein a synthetic operation is performed, of course, differ due to the different demands arising from the different chemistries required for inorganic solid-state chemistry. Here, of course, different considerations have to be taken than for the organic synthetic strategies. Before addressing the synthetic requirements for synthesizing Split&Pool libraries of inorganic materials we wish to draw the reader's attention to some mathematical considerations which are also of major importance for successful synthetic performance.

2.6 MATHEMATICAL BACKGROUND FOR SPLIT&POOL SYNTHESIS

2.6.1 BASIC CONSIDERATIONS

The most important feature of Split&Pool synthesis is the immense reduction of operations (e.g., dosing, drying, etc.) necessary to achieve certain library sizes. Formally, the number of operations increases proportionally to the number of Split&Pool steps (Equation 2.1).

$$N_{op}^{sp} = \sum_{i=1}^{s} n_i \cdot k \propto k \cdot s \qquad (2.1)$$

where N_{op}^{sp} = number of necessary operations, k = number of components, s = number of split and pool steps (for an explanation see Figure 2.7), and n_i = number of process steps for split step i.

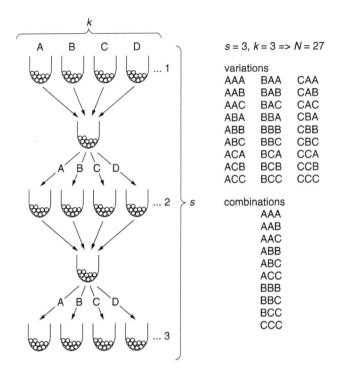

FIGURE 2.7 Left: Schematic representation of the Split&Pool procedure and explanation of symbols used in the text. The parameter k represents the number of features aiming to produce different materials. This could be different substances (e.g., A = Pd, B = Rh, etc.) but also concentrations (e.g., A = 0.01 mol/l, B = 0.03 mol/l, etc.) or post-processing steps like heat treatment, steaming, etc. The parameter s represents the number of successive Split&Pool steps. Right: example for a library if $s = k = 3$. Top: The sequence of operations is important; bottom: the sequence is not important.

In contrast, the library size (i.e., the number of produced materials) is given by the number of possible variations and therefore grows exponentially (Equation 2.2):

$$N_{op}^{sp} = k^s \tag{2.2}$$

Equation 2.2 holds true at least for organic (especially peptide) synthesis, where the sequence of steps/reagents definitely influences the number of distinguishable products.

In inorganic synthesis the situation is more complicated. k does not necessarily correspond to different precursors or building blocks as in organic synthesis. Instead, it represents any reaction parameter that can change the properties of bulk materials. For example, operations A, B, and C could correspond to calcination steps at different temperatures and their sequence could determine the materials properties. As a consequence, there may as well be "commutative" reactions/process steps. Certain reaction steps can be applied in an arbitrary order without yielding distinguishable materials. This introduces a higher symmetry into the parameter space and the size of the resulting library will be of course substantially smaller than predicted by Equation 2.2. In the extreme case of all operations being commutative, the resulting library size has to be described as the number of combinations (Equation 2.3).

$$N_{lib}^{sp} = \binom{k+s-1}{s} = \frac{(k+s-1)!}{s!(k-1)!} \tag{2.3}$$

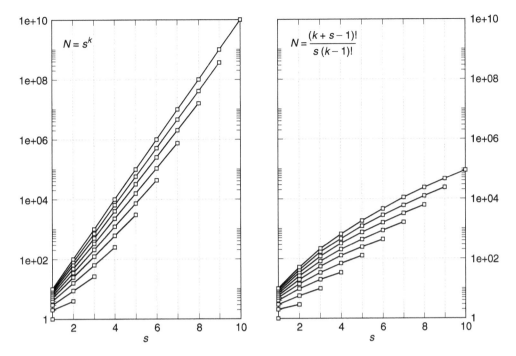

FIGURE 2.8 Plots of library size vs number of Split&Pool steps for two extreme situations. Left: the sequence of operations is important. Right: The sequence of operations is not important. Each curve corresponds to one value of k (i.e., the number of features). k ranges from 2 (lowest curve) to 10.

Note that the latter equation is also valid if the Split&Pool synthesis is carried out according to protocols with a fixed sequence of steps (i.e., reagents/process steps are applied only once like A, B, C in the first step, D, E, F in the second and so on).

An example for the resulting libraries in both cases is given in Figure 2.7 for a typical Split&Pool synthesis. A comparison of the corresponding maximum library sizes to be achieved in both extreme cases is shown in Figure 2.8.

In real-world applications allowing for a permutation of sequences, the number of distinct materials is expected to be somewhere in between. As an example one can imagine that certain metals form alloys and thus make the sequence of impregnation steps unimportant, while others do not, and form, for example, layers covering each other. This behavior can change depending on the conditions applied for annealing. Since the chemical structure of multicomponent inorganic materials is often unpredictable an a priori estimation of the diversity (i.e., number of points in activity space) or even the size (i.e., number of points in structure space) of an inorganic Split&Pool library is difficult.

2.6.2 CHALLENGES OF INORGANIC SPLIT&POOL SYNTHESIS

In addition to the challenge of predicting the "theoretical" maximum diversity of an inorganic Split&Pool library, some points have to be considered to achieve it. One important question is how redundant does the library have to be planned in order to realize completeness. Since the Split&Pool method is subject to statistical processes, the exact composition of the target library is undetermined and only expectation values can be given. Therefore, a certain excess of starting materials/support beads must be introduced into the synthesis to cover situations where statistical deviations from these expectation values occur (i.e., every possible material is synthesized at least once).

In more detail, the split step is described by a hypergeometric distribution:

$$P(x) = \frac{\binom{X}{x}\binom{N-X}{n-x}}{\binom{n}{N}}$$
(2.4)

N: total number of samples in the pool, n: number of samples to be picked from the pool, X: number of type "X" samples in the pool, x: number of type "X" samples among the n selected.

The hypergeometric distribution describes a random selection without repetition among objects of two distinct types (e.g., "samples of type X" and "not of type X"). Because the number X depends on the result of the previous split step, Equation 2.4 can be applied to calculate conditional probabilities for each split step (i.e., the distribution of samples of a certain type among the split sets) for all values of X. This can be used for numerical simulations of the Split&Pool method by iterative application of Equation 2.4. Using a computer it is possible to trace how deviations from the expectation value in one split step propagate through the following steps and affect the resulting library.

An illustration of this issue is given in Figure 2.9 for a simple example of a $3 \times 3 \times 3$ library (as depicted in Figure 2.7).

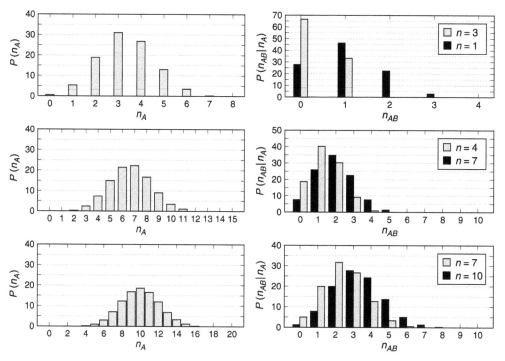

FIGURE 2.9 Analysis of distributions during a three-level Split&Pool synthesis using three substances/features. The left plot in each row shows the probability $P(n_A)$ of finding n_A beads of type "A" among 1/3 of the total number of beads after the first split step. The right plot shows the conditional probability $P(n_{AB} \mid n_A)$ of finding n_{AB} beads of type "AB" after the second split step subject to the result of the first split step. We give here the distributions for two possible results of the first split step – an ideal case (n_A = average of the distribution) and a "worst" case (n_A = average – standard deviation). All probabilities are calculated for a hypergeometric distribution. Each row represents a different level of redundancy. Top: the synthesis is performed using 30 beads (no redundancy, since 27 different materials are expected), middle: 60 beads, bottom: 90 beads (three times as many beads as needed).

The example demonstrates how the result of the last split step depends on the total number of samples and fluctuations occurring in early split steps.

As the numerical simulation shows, three times as many beads as the expected library size should be used in order to have 95% certainty of synthesizing a complete library, which is in line with findings published in the recent literature for peptide synthesis.[65] For a solid state synthesis this error can be eliminated almost completely by using the well-known "direct divide" strategy, which directly divides the samples from each reaction vessel into the next set of vessels without intermediate pooling step or directed synthesis using tagging techniques. Here, the only remaining error source is the uniform distribution of samples among the reaction vessels, which has only technical but in principle no statistical limitations.

The statement made above is based on the assumption of "perfect" synthesis steps not introducing additional errors. From organic chemistry applications it is known that the Split&Pool method may suffer from the insufficient purity of synthesized compounds. Impurities are introduced by incomplete conversion or side reactions and usually accumulate during the Split&Pool synthesis. In materials science applications the reliability of the individual process steps enters in a similar way, although there is no simple Boolean probe for the success of a reaction/process step. The simple expression "the molecule was formed/not formed" has to be replaced by a more fuzzy, macroscopic measure like "5 ± 0.5 mmol of A have been deposited." These fluctuations will also accumulate during the Split&Pool synthesis and one has to remember that if the concentration levels are too narrow a certain amount of redundant samples will result while other features are lost. As a consequence, the library size (i.e., the number of materials prepared) has to be increased. This ensures that enough samples with a feature parameter close to the corresponding expectation value are prepared and therefore all target regions of the parameter space are covered (see Figure 2.10).

On the other hand, increasing the number of samples to achieve the theoretical diversity of Split&Pool libraries is no panacea. Although it is possible to synthesize a large, highly redundant library to ensure every possible variation has been made, one has to take into account that more samples will have to be tested. The probability of selecting at least one sample of type "X" converges faster to 100% for ten samples than for 1000 samples, even if the relative amount of type "X" samples is the same (e.g., 50%) in both cases. This slightly paradoxical situation can be rationalized by the fact that the hypergeometric distribution describing the selection merges into the

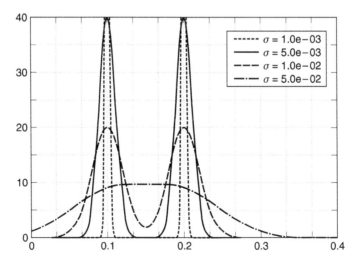

FIGURE 2.10 Effect of errors introduced by the synthesis procedure. For a numerical simulation, two concentration levels of 0.1 and 0.2 are assumed. These are superimposed by a Gaussian of standard deviation β. If the standard deviation becomes as large as 1/10 of the level separation, a significant overlap of the two Gaussians occurs.

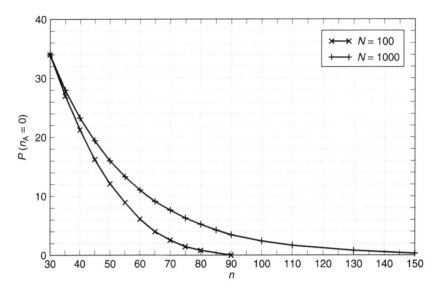

FIGURE 2.11 Probability of missing one combination of a library if only subsets are tested. As an example, a 3 × 3 × 3 library containing 27 distinct samples is assumed. In one case, $N = 100$ samples have been prepared (i.e., 4 of type A) and $n = 30 - 100$ samples are selected for testing, in the other $N = 1000$ samples (i.e., 40 of type A). $P(n_A = 0)$ gives the probability of missing sample A. The comparison shows that for a highly redundant library more samples have to be tested to achieve the same level of certainty.

binomial distribution (Equation 2.5) for small subsets selected from large pools because the individual statistical events of picking one sample become mutually independent.

$$P(x) = \binom{n}{x} \left(\frac{X}{N} \right)^x \left(\frac{N - X}{N} \right)^{n-k} \tag{2.5}$$

N: total number of samples in the pool, n: number of samples to be picked from the pool, X: number of type "X" samples in the pool, x: number of type "X" samples among the n selected.

For example, the event of picking one sample of type X from a small pool decreases the chance of picking a second one significantly, while it does not matter for a large pool of samples.

As is demonstrated in Figure 2.11 by a numeric evaluation of the corresponding hypergeometric distributions, one has to test about 80 samples of a 100-sample pool to have 99% certainty that all 27 possible variations are in the test set (assuming uniform distribution). For a 1000-sample pool (containing the same 27 variations in uniform distribution) more than 150 tests would have to be performed to reach the same level of certainty. Thus, the use of redundant libraries can only become a costly alternative to the development of a robust synthesis protocol.

2.7   BEYOND THEORETICAL CONSIDERATIONS: APPLICATION EXAMPLES OF THE COMBINATORIAL SYNTHESIS OF FUNCTIONAL INORGANIC MATERIALS VIA THE SPLIT&POOL METHODOLOGY

The synthetic approach of inorganic materials via the Split&Pool methodology is closely connected to the question of synthesizing materials on or within physical entities that will contain the library member throughout the synthetic process. If physical entities are employed as "carriers" that show sufficient chemical inertness towards the screening process, a separation of the final library compound may not be necessary at the end of the synthetic steps. We will later see that in some cases it may also be desirable to employ "carriers" that may even become a part of the desired material (and subsequently the "carrier" entity a library member as a whole), so that the separation of the created compound is not an issue of interest any longer.

Generally, solution-based approaches for the generation of inorganic Split&Pool libraries have substantial advantages over approaches where solid phases are introduced as chemical sources during the different synthetic steps. Solution chemistry potentially offers a wide range of synthetic opportunities that can be exploited not only for the purpose of parallel synthesis, but also for synthetic steps for Split&Pool library creation (see, for example, Matijevic,[38] Baerns et al.,[60] and Schüth et al.[61]].

For "carrier"-free synthetic approaches, techniques employing different vessels like cans, or alternatively "containers" that are able to absorb and contain considerable amounts of solutions with regard to their own volume are a central component of the synthetic effort. Figure 2.12 illustrates a possible approach to a substrate-free synthesis using vessels or cans as the alternative "carrier" system. Starting from solutions (V_1) these can be "split" into vessels or cans, becoming a library of different identity by either adding components (E1 or E2) or by altering the solution with any form of physical treatment (heat, evaporation of solvent, or the like). The pooling step in Figure 2.12, of course, in many cases becomes absurd if all the solutions of the first Split step are united again. Still, if only some "carriers" or parts of the solutions from the "carriers" are united during the pooling step, this step may very well make sense from a synthetic point of view and can be usefully employed, although in many cases the number of pooling steps may be smaller than the number of splitting steps. Via addition of further components (E4 to E6) the complexity of the mixture is increased and the final performance of a precipitation step P leads to a solid library member L1 to L4. The elegance of the procedure lies in the fact that encoding is eased if "carriers" in the form of vessels or cans are employed which can be equipped with tags. For many applications a shaping step of the resulting library members will be essential. Obtaining a powder as a product leaves a range of methods for shaping open which can be adapted according to the demands of screening procedure employed.

Another possible option for the application of the Split&Pool principle for obtaining "carrier"-free libraries in the form of powders is the use of combustible "carriers" that can contain solutions. A variety of organic polymers is available for this purpose; the ones which proved to be of highest value were acrylates or hydrophilized polyolefins, suitable for the needs of an approach based on aqueous chemistries or other highly polar solvents. In this case the similarity with regard to organic synthetic approaches seems very high, from a point of view of the polymer properties discrete functional groups which bind to the entities added during the synthetic steps are obsolete, the only compatibility that is demanded is that the carrier is capable of absorbing the solvent.

The alternative approaches connected to synthetic work based on the use of "carrier" systems has a similar background as the one described above. The synthetic "backbone" can be purely of inorganic, organic, or of composite nature. As many steps in the generation of functional inorganic materials may involve a thermal treatment of the given material, a clear preference is on such materials that tolerate thermal treatment steps without deterioration of the shaped entity. This indicates why fully inorganic materials are preferred in a large number of cases as "carriers."

For a large range of applications impregnation or equilibrium adsorption methods, including subsequent drying or calcination steps for solvent removal, can be employed as a synthetic approach to obtain large libraries of materials. Carriers that can be employed are available or can be customized. At hte Aktiengesellschaft standard "carrier" materials of pure and mixed oxides in the form of beads are kept in stock for synthetic operations. A wide range of chemistries can be performed from a synthetic point of view based on these "carrier" materials. Standard materials include Al_2O_3, SiO_2, TiO_2, and ZrO_2 and a variety of other mixed oxides. Usually, a spherical shape is preferred as it eases transfer operations and minimizes potential damage during transfer.

As an alternative to impregnation procedures, coating procedures are another possibility for the application of different chemical compositions to a "carrier" body. A sequential procedure can though lead to an onion-like structure of the deposited components which may not in all cases be desirable,

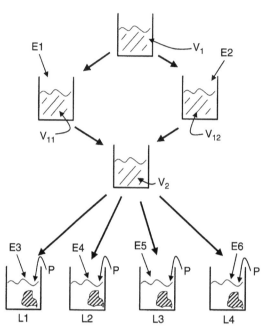

FIGURE 2.12 Potential pathway for obtaining "carrier"-free inorganic solids via a solution/precipitation-based Split&Pool methodology approach.

but for some applications one may also exploit this concept. In combination with impregnation procedures, coating procedures may prove to be most powerful and efficient.

2.7.1 SYNTHETIC EXAMPLE

In order to illustrate the principle of Split&Pool synthesis a synthetic example of a library based on alumina "carriers" impregnated with different metal solutions was chosen as a case study. The potential application focus on such a library is its use in a catalytic application. The inorganic library of approximately 3000 Mo-Bi-Co-Fe-Ni/γ-Al$_2$O$_3$ library candidates was synthesized applying the following metal salts in aqueous solution as metal precursors in the Split&Pool synthesis: Ammoniumheptamolybdate [(NH$_4$)$_6$Mo$_7$O$_{24}$ $\times$ 4 H$_2$O] with a molarity of 0.025, Bi(NO$_3$)$_3$ $\times$ 5 H$_2$O with a molarity of 0.075, Co(NO$_3$)$_2$ $\times$ 6 H$_2$O with a molarity of 0.25, Fe(NO$_3$)$_2$ $\times$ 9 H$_2$O with a molarity of 0.25, and Ni(NO$_3$)$_2$ $\times$ 6 H$_2$O with a molarity of 0.1. As ceramic "carrier" the sieve fraction with 1-mm diameter of γ-Al$_2$O$_3$ beads (Condea) were employed. The synthesis started with 2 g of the alumina "carrier" beads, corresponding to approximately 3000 single, uniform beads. The dividing steps of the starting library (Split step) in four equal portions into porcelain dishes were achieved by weighing on a lab bench balance LA 1200 from Sartorius. The impregnation steps (adsorption impregnation with 80% solution of the water uptake, water uptake of the Al$_2$O$_3$ support: 588 μl) were performed on a MultiProbe IIEX (Packard) robotic system, which was additionally equipped with a Titramax 100 shaker (Heidolph), on which the porcelain dishes with the appropriate part of the alumina beads were shaken continuously while adding the precursor solutions. After evaporating the solution and waiting for 20 min, after each impregnation step each library part has been dried for 16 h at 80°C followed by a final calcination step at 400°C subsequent to all the impregnation steps. A sketch of the Split&Pool steps and a synthesis scheme is shown in Figure 2.6 and Figure 2.7.

For optical analysis via imaging of the final Split&Pool library an AX 70 microscope was used (Olympus). The apparatus was equipped with an automated *xyz* table and autofocus control. The

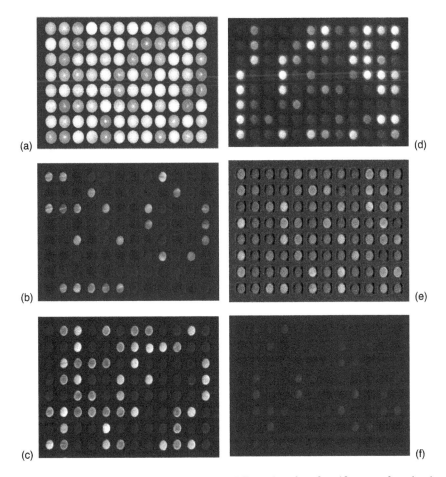

FIGURE 2.13 Optical microscope picture (a) and μ-XRF results of an 8 × 12 array of randomly selected materials (1 mm), coincidentally selected from the library, described in Figure 2.2 (b) Ni distribution; (c) Fe distribution; (d) Mo distribution; (e) Bi distribution; (f) Co distribution.

software package analysis 3.0 (SIS) is used for hardware control and analysis means. Pictures were taken in the multi-imaging alignment mode at a magnification of 5.

For verification purposes of the success of the different deposition steps in combination with the Split&Pool methodology X-ray fluorescence was chosen as an analysis tool. The elemental analysis was performed by X-ray fluorescence analysis on an Eagle II μProbe from Roentgenanalytik with Rh-Kα radiation. An essential feature is the small diameter of the measurement spot: The X-ray beam is focused by a multicapillary system to a spot size of 50 μm on the sample surface. The XRF analysis of the 8 × 12 catalyst library selection (Figure 2.13) is routinely accomplished automatically by an elemental mapping at a pattern of 512 × 400 points, equally distributed over the rectangular library field, each point (50-μm diameter) being measured for 300 ms.

In the above-described Split&Pool synthesis example (Figure 2.13) five different chemical precursor solutions, varying the metal atom (Mo^{VI}, Bi^{III}, Fe^{II}, Co^{II}, Ni^{II}), at four different concentration levels (0, 0.002, 0.1, 1 wt%) were used. From the underlying mathematics it is obvious that the resulting Split&Pool library includes $4^5 = 1024$ different members, covering all possible element combinations of the above-mentioned metal oxides and concentrations. As the pool of 3000 library members exceeds the number of chemically different members (only 1024 compositionally different materials), the library has a certain redundancy. Each material is approximately represented with three library members of the S&P library. For ease of discussion we decided to specify this factor, which represents the redundancy of the library as a function of chemically identical library members as the factor of over-determination for a certain library. According to the total number of approximately 3000 Al_2O_3 beads, the factor of over-determination of the materials library in this case study is 2.9. Realistically, in many cases of routine Split&Pool synthetic operation where redundancy of the library can be assumed, a factor of over-determination between 1.2 and 1.5 from a statistical point of view versus screening efficiency is sufficient, a reproducible synthesis procedure is assumed. Still, an over-determination of a factor of 3 is in good accord with the theoretical considerations described above.

For materials characterization and a consistent proof of the synthetic principle we have chosen μ-XRF with a spot size of 50 μm as the analytical tool allowing acceptable speed, resolution, and data quality for the synthesis approach, designed for primary screening (stage-1 screening). The elemental mapping over an 8 × 12 matrix with a 512 × 400 resolution (0.3 s/point) produces a high-resolution XRF mapping in approximately 17 h. This data quality is not required for routine analysis of Split&Pool libraries. An elemental mapping over a 96-fold or 384-fold single-bead reactor (SBR) used for storage and analysis of the library members with sufficient data quality can be performed over night. Figure 2.13 represents the XRF mapping results (Figure 2.13b–f) for 96 randomly selected materials as well as a picture taken by optical microscopy (Figure 2.13a), composed of several frames from the above-mentioned selection of beads. The photograph of the 96 beads already illustrates the homogeneity of the multiple impregnation procedure showing a homogeneous coloring of all the library members. The reddish-colored beads correspond to materials with a high iron content, the bluish-colored beads are materials with a high cobalt content. For other library members no obvious correlation between color and composition could be observed.

The visualization of the XRF elemental mapping results was color coded (Figure 2.13b–f). For the ease of the reader the intensity of the colors is proportional to the element concentrations: the more intense the colors are, the higher is the amount of the corresponding metal in the material. The colors were chosen arbitrarily. As one can see in Figure 2.13, each material is different in chemical composition. The four concentration levels (0, 0.002, 0.1, 1 wt%), as described in the experimental part, could be estimated from the mapping results (Figure 2.13b–f), considering that the differentiation between 0 wt% and 0.002 wt% could only be achieved with our equipment with a high degree of calibration. So, only three different levels in concentration can be recognized in Figure 2.13b–f. Recognizing the level of precursor concentration is obviously sufficient to determine the synthesis history of interesting library members for the given synthetic example.

The routine workflow (Figure 2.14) for the identification of outstanding candidate compositions of library members of a Split&Pool library is fairly straightforward: a detailed post-identification of single leads or hits takes place after the screening in our single-bead reactor system (SBR).[70] Figure 2.14 displays the routine workflow, integrating the newly developed Split&Pool synthesis, screening the materials in the single-bead reactor system, identifying materials with the best catalytic performances, optionally storing or archiving the whole library in the reactor as "materials chip" for further treatment (activation, pretreatment, or testing in another target reaction), and evaluating the whole data set, i.e., with a combination of evolutionary algorithms and neural networks for modeling means and for predicting a new library generation, which has to be synthesized in the next step. Next to the presented new synthesis tool, the central part of this workflow is our single-bead reactor system (SBR). This multifunctional architecture fulfills several needs, which are describe elsewhere[70]: besides the basic feature as, of course, screening reactor, it acts as "zero-background" carrier for the characterization part (i.e., μ-XRF, μ-XRD, etc.); because of its chemical inertness it can be used for synthetic pretreatment steps (activation, hydrothermal treatment, reductions, etc.), and due to its advantageous format it serves as a storage and archiving device with the positional information of each material.

In Figure 2.15, the routinely employed 384-fold and the prototype 105-fold single-bead reactor can be seen. As a size comparison, a conventional match is shown; for the ease of the reader it

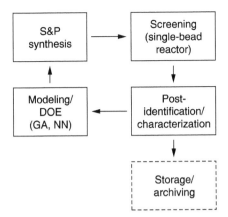

FIGURE 2.14 Typical workflow in primary screening, including the S&P synthesis principle for inorganic solids, i.e., heterogeneous catalysts.

should be remarked that the 384-fold single-bead reactor is approximately half the size of a credit card. The flexible concept of the single-bead reactor allows the adaptation of a number of analysis techniques. Analysis can take place either by sequential techniques that retrieve samples from single library members, typically via a transducing capillary or via integral techniques monitoring all samples at a time. Suitable fast sequential analysis techniques are conventional MS, GC/MS, GC, or dispersive or non-dispersive IR. Though mostly sequential techniques are used in the current screening setups, it is also feasible to adapt integral analysis techniques that allow true parallel analysis by applying e.g., IR-thermographic or PAS (photoacoustic spectroscopy) techniques. It should be remarked that the IR-transparent architecture of the silicon-based reactor contributes to the versatility in conjunction with IR thermography.

2.8   OUTLOOK AND FUTURE

The transition from approaches used for the synthesis and testing of organic compounds in comparison to approaches for synthesis and screening of inorganic materials make it obvious that there is an essential gap between efficient synthetic strategies for organic and inorganic entities. With our work we tried to bridge this gap and establish tools that have an efficiency comparable to the organic synthetic tools that are in place for producing highly diverse libraries of inorganic materials. We could demonstrate for the application of the fully combinatorial synthesis of solid state materials via the Split&Pool method leads to diverse libraries and can successfully be employed for materials screening. High-throughput analytical techniques evidence the versatility, reproducibility, and high performance of the new technique, developed by hte Aktiengesellschaft. Within a few synthetic steps combined with splitting and pooling operations, highly diverse materials libraries for one of our stage-1 reactor systems (SBR) for catalytic tests can be generated.

The extension of the concept to other applications synthesizing different classes of inorganic solid-state materials that can be produced via impregnation, coating, or other techniques listed above are pretty straightforward. As mentioned above, a challenging topic under investigation is synthesis techniques that produce support-free inorganic bulk materials in the form of powders or preferably in spherical form with a homogeneous particle diameter. The results presented for catalytic materials have an exemplary character: as a matter of course the development of new sensor materials, adsorbents, materials with special electronic, magnetic or optical properties, or other functional solid-state materials are possible. This new approach of coupling pooled and parallel strategies has far-reaching implications for future synthesis and screening strategies.

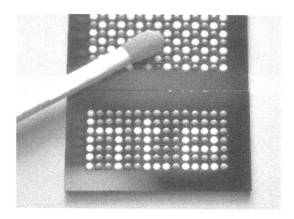

FIGURE 2.15 (See color insert following page 172) Single-bead reactor architecture (384-fold and 96-fold).

ACKNOWLEDGMENTS

We would especially like to thank our co-workers Oliver Laus, Andreas Strasser, and Uwe Vietze for stimulating input and helpful discussions.

REFERENCES

1. Taylor, S. J. and Morken, J. P., Thermographic selection of effective catalysts from an encoded polymer-bound library, *Science* 280, 267, 1998.
2. Moates, F. C., Somani, M., Annamalai, J., Richardson, J. T., Luss, D. and Willson, R.C., Infrared thermographic screening of combinatorial libraries of heterogeneous catalysts, *Ind. Eng. Chem. Res.* 35, 4801, 1996.
3. Holzwarth, A., Schmitdt, H.-W. and Maier, W. F., IR-thermographische erkennung katalytischer aktivität in kombinatorischen bibliotheken heterogener katalysatoren, *Angew. Chem.* 110, 2788, 1998.
4. Orschel, M., Klein, J., Schmidt, H.-W. and Maier, W. F., Erkennung der selektivität von oxidationsreaktionen auf katalysatorbibliotheken durch ortsaufgelöste Massenspektrometrie, *Angew. Chem.* 111, 2961, 1999.
5. Cong, P., Doolen, R. D., Fan, Q., Giaquinta D. M., Guan, S., McFarland, E. W., Poojary, D.M., Self, K., Turner, H. W. and Weinberg, W. H., Kombinatorische parallelsynthese und hochgeschwindigkeitsrasterung von heterogenkatalysator-bibliotheken, *Angew. Chem.* 111, 508, 1999.
6. Hoffmann, C., Wolf, A. and Schüth, F., Parallele synthese und prüfung von katalysatoren nah an konventionellen testbedingungen, *Angew. Chem.* 111, 2971, 1999.

7. Houghten, R. A., General method for the rapid solid-phase synthesis of large numbers of peptides: specificity of antigen–antibody interaction at the level of individual amino acids, *Proc. Natl. Acad. Sci. USA* 82, 5131, 1985.
8. Geysen, H. M., Meloen, R. H. and Barteling, S. J., Use of peptide synthesis to probe viral antigens for epitopes to a resolution of a single amino acid, *Proc. Natl. Acad. Sci. USA* 81, 3998, 1984.
9. Fodor, S. P. A., Leighton Read, J., Pirrung, M. C., Stryer, L., Tsai Lu, A. and Solas, D., Light-directed specially addressable parallel chemical synthesis, *Science* 251, 767, 1991.
10. Furka, Á., Sebestyen, F., Asgedom, M. and Dibo, G., Cornucopia of peptides by synthesis, Abstr. 14th. *Int. Congr. Biochem.*, Prague, 5, 47, 1988.
11. Lowe, G., Combinatorial chemistry, *Chem. Soc. Rev.* 24, 309, 1995.
12. Pirrung, M., Spatially addressable combinatorial libraries, *Chem. Rev.* 97, 473, 1997.
13. Balkenhohl, F., von dem Busche-Huennefeld, C., Lansky, A. and Zechel, C., Combinatorial synthesis of small organic molecules, *Angew. Chem.* 35, 2288, 1996.
14. Jung, G. and Beck-Sickinger, A. G., Multiple peptide synthesis methods and their applications, *Angew. Chem. Int. Ed.* 104, 375, 1992.
15. Terret, N. K., *Combinatorial Chemistry*, Oxford University Press, Oxford, UK, 1998.
16. Merrifield, R. B., Solid phase peptide synthesis. I. The synthesis of a tetrapeptide, *J. Am. Chem. Soc.* 85, 2149, 1963.
17. Frank, R., Heikens, W., Heisterberg-Moutsis, G. and Blöcker, H., A new general approach for the simultaneous chemical synthesis of large numbers of oligonucleotides: segmental solid supports, *Nucl. Acid. Res.* 11, 4365, 1983.
18. Frank, R., SPOT synthesis: an easy technique for the positionally addressable, parallel chemical synthesis on a membrane support, *Tetrahedron*, 48, 9217, 1992.
19. Lam, K. S., Salmon, S. E., Hersh, E. M., Hruby, V. J., Kazmierski, W. M. and Knapp, R. J., A new type of synthetic peptide library for identifying ligand-binding activity, *Nature* 354, 82, 1991.
20. Krämer, T., Atonenko, V. V., Mortezaei, R. and Kulikov, N. V., *Handbook of Combinatorial Chemistry*, Nicolaou, K. C., Hanko, R. and Hartwig, W., Eds. Wiley-VCH, Weinheim, 2002, p. 170.
21. Nicolaou, K. C., Xiao, X. Y., Parandoosh, Z., Senyei, A. and Nova, M., Radiofrequency encoded combinatorial chemistry, *Angew. Chem. Int. Ed.* 3, 2289, 1995.
22. Xiao, X.-Y., Zhao, C., Potash, H. and Nova, M. P., Combinatorial chemistry with laser optical encoding, *Angew. Chem. Int. Ed.* 36, 780, 1997.
23. Guiles, J. W., Lanter, C. L. and Rivero, R. A., Eine neue markierungstechnik für die kombinatorische chemie nach der "Misch-und-Sortier"-Methode, *Angew. Chem.* 110, 967, 1999.
24. Winkler, J. L., Fodor, S. P. A., Buchko, C. J., Ross, D., Aldwin, L. and Modlin, D. N., Combinatorial strategies for polymer synthesis, WO 93/09668, 20.11.1992.
25. Scott, L. T., Rebek, J., Ovsvanko, PL. and Sims, C. L., Organic chemistry on the solid phase. Site-site interactions on functionalized polystyrene, *J. Am. Chem. Soc.* 99, 625, 1977.
26. Atherton, E., Clive, P. L. J. and Sheppard, R. C., Polyamide supports for polypeptide synthesis, *J. Am. Chem. Soc.* 97, 6584, 1975.
27. Bayer, E., Dengler, M. and Hemmasi, B., *Int. J. Pept. Protein Res.* 25, 178, 1981.
28. Houghten, R. A., General method for the rapid solid-phase synthesis of large numbers of peptides: specificity of antigen–antibody interaction at the level of individual amino acids, *Proc. Natl. Acad. Sci. USA* 82, 5131, 1985.
29. Nielsen, J., Brenner, S. and Janda, K. D., Synthetic methods for the implementation of encoded combinatorial chemistry, *J. Am. Chem. Soc.* 115, 9812, 1993.
30. Ohlmeyer, M. H. J., Swanson, R. N., Dillard, L. W., Reader, J. C., Asouline, G., Kobayashi, R., Complex synthetic chemical libraries indexed with molecular tags, *Proc. Natl. Acad. Sci. USA* 90, 10922, 1993.
31. Ni, Z.-J., Maclean, D., Holmes C. P., Murphy, M. M., Ruhland. B. and Jacobs, J. W., Versatile approach to encoding combinatorial organic syntheses using chemically robust secondary amine tags, *Med. Chem.* 39, 1601, 1996.
32. http://www.gilson.com
33. http://www.chemspeed.com
34. http://www.zinsser-analytic.com

35. http://www.accelab.de

36. Brümmer, H., Markert, R. and Schwemler, C., *GIT Laborfachzeitschrift* 43, 598, 1999.

37. Stiles, A. B., *Catalyst manufacture-laboratory and commercial preparations*, Marcel Dekker, New York, 1983.

38. Matijevic, E., Monodispersed metal (hydrous) oxides—A fascinating field of colloid science, *Acc. Chem. Res.* 14, 22, 1981.

39. Hanak, J. J., The 'Multiple-Sample Concept' in materials, *J. Mater. Sci.* 5, 964, 1970.

40. Hanak, J. J., *J.Vac. Sci. Technol.* 8, 172, 1971.

41. Schultz, P. G. and Xiang, X. D., *Curr. Opin. Sol. State Mat. Sci.* 3, 153, 1998.

42. Liu, D. R. and Schultz, P. G., Generating new molecular function: a lesson from nature, *Angew. Chem Int. Ed.* 38, 36, 1999.

43. McFarland, E. W. and Weinberg, W. H., Combinatorial approaches to materials discovery, *Trends Biotechnol.* 17, 107, 1999.

44. McFarland, E. W. and Weinberg, W. H., *Mater. Technol.* 3, 153, 1998.

45. Service, R. F., The fast way to a better fuel cell, *Science* 280, 1690, 1998.

46. Koinuma, H., Aiyer, H. N. and Matsumoto, Y., Kombinatorische materielle festkörperwissenschaft und technologie, *Sci. Techn. Adv. Mater.* 1, 1, 2000.

47. Sun, X. D., Wang, K. A., Yoo, Y., Wallace-Freedman, W. G., Gao, C. and Xiang, X. D., Solution-phase synthesis of luminescent materials libraries, *Adv. Mater.* 9, 1046, 1997.

48. Reddington, E., Sapienza, A., Gurau, B., Viswanathan, R., Sarangapani, S., Smotkin, E. S. and Mallouk, T. E., Combinatorial electrochemistry: a highly parallel, optical screening method for discovery of better electrocatalysts, *Science,* 280, 173, 1998.

49. Wur, J. L., Devenney, M., Danielson, E., Poojary, D. and Weinberg, W. H., *Mater. Res. Soc. Symp. Proc.* 560, 65, 1999.

50. Wur, J. L., Devenney, M., Danielson, E., Poojary, D. and Weinberg, W. H., *Mater. Res. Soc. Symp. Proc.* 560, 65, 1999.

51. Rantala, J. T., Kololuma, T. and Kivimaki, L., *SPIE-Int. Soc. Opt. Eng.* 3941, 11, 2000.

52. Liu, Y., Cong, P., Doolen, R. D., Turner, H.W. and Weinberg, W.H., High-throughput synthesis and screening of V-Al-Nb and Cr-Al-Nb oxide libraries for ethane oxidative dehydrogenation to ethylene, *Catal. Today* 61, 87, 2000.

53. Klein, J., Lehmann, C. W., Schmidt, H. W. and Maier, W. F., Combinatorial approach to the hydrothermal synthesis of zeolites, *Angew. Chem. Int. Ed.* 37, 609, 1998.

54. Claus, P., Hönike, D. and Zech, T., Miniaturization of screening devices for the combinatorial development of heterogeneous catalysts, *Catal. Today* 67, 319, 2001.

55. Zech, T., *Miniaturisierte screening systeme für die kombinatorische heterogene katalyse*, Fortschritt-Berichte VDI, VDI Verlag, Düsseldorf 2002.

56. Senkan, S. M. and Öztürk. S., Discovery and optimization of heterogeneous catalysts by using combinatorial chemistry, *Angew. Chem. Int. Ed.* 38, 791, 1999.

57. Senkan, S. M., Krantz, K. and Öztürk, S., High-throughput testing of heterogeneous catalyst libraries using array microreactors and mass spectrometry, *Angew. Chem. Int. Ed.* 38, 2794, 1999.

58. Rodemerck, U., Ignaszewski, P., Lucas, M., Claus, P. and Baerns, M., Parallelisierte synthese und schnelle katalytische testung von katalysatorbibliotheken für die oxidationsreaktionen, *Chem. Ing. Tech* 71, 873, 1999.

59. Buyevskaya, O.,Wolf, D. and Baerns, M., Ethylene and propene by oxidative dehydrogenation of ethane and propane – 'Performance of rare-earth oxide-based catalysts and development of redox-type catalytic materials by combinatorial methods', *Catal. Today* 62, 91, 2000.

60. Buyevskaya, O. V., Brückner, A., Kondratenko, E. V., Wolf, D. and Baerns, M., Fundamental and combinatorial approaches in the search for and optimisation of catalytic materials for the oxidative dehydrogenation of propane to propene, *Catal. Today* 67, 369, 2001.

61. Hoffmann, C., Wolf, A. and Schüth, F., Parallel synthesis and testing of catalysts under nearly conventional testing conditions, *Angew. Chem. Int. Ed.* 38, 2800, 1999.

62. Hinze, J. and Welz, U., "Broad Smiles," In: Gasteiger, J., Ed, *Software development in chemistry*. Gesellschaft Deutscher Chemiker (GDCh), Frankfurt/Main, 1996, Ch 10, p. 59.

63. Schüth, F., Hoffmann, C., Wolf, A., Schunk, S., Stichert, W. and Brenner, A., High throughput experi-mentation in catalysis. In: *Combinatorial methods in organic Chemistry,* Jung W., Ed. Combinatorial Chemistry, VCH-Wiley, Weinheim, 1999, p. 464.
64. Somoriaj, G. A., *Introduction to surface chemistry and catalysis,* Wiley Interscience, New York, 1994.
65. Burgess, K., Liaw, A. I. and Wang, N., Combinatorial technologies involving reiterative division/coupling/recombination: statistical considerations, *J. Med. Chem.* 37, 2985, 1994.

3 Informatics-Based Optimization of Crystallographic Descriptors for Framework Structures

Arun Rajagopalan and Krishna Rajan

CONTENTS

3.1 INTRODUCTION

Over the years there have been numerous attempts at classifying zeolites and other network structures. Zeolites and other network structures have complicated structures that have to be described using multiple types of descriptors. Researchers have tried different approaches for the development of such descriptors, based on experiments and theory. Each of these tries to take into account (separately) the topology and the symmetry of the structure being investigated. The symmetry aspects are usually dealt with in terms of space groups while the topological aspects are dealt with by using combinations of planar nets, chains, secondary building units, coordination sequences, and ring statistics.[1–4] Our approach is informatics based, one that combines insights from both science and statistics to develop descriptors that can be used to study structure relationships. Here we use Wigner–Seitz cells in an attempt to encode simultaneously both the topological and symmetry properties of the network.[5] We use these cells as a basis for development of "secondary" descriptors, which are then used to study structure relationships.

3.2 WIGNER–SEITZ CELLS AND ZEOLITE CLASSIFICATION SCHEMES

3.2.1 WIGNER–SEITZ CELLS

In this section we describe the usage of Voronoi polyhedra and the state-of-the-art classification schemes of zeolites, to place our work in context.

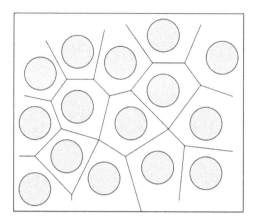

FIGURE 3.1 Voronoi polyhedra in two dimensions.

Voronoi polyhedra are convex polyhedra used to describe the structure of liquids and glasses (Figure 3.1). The equivalent construction for crystals is the Wigner–Seitz (WS) cell. A WS cell can be constructed for each identical point in a lattice. We define it as the space containing all points nearer to that lattice point compared to its neighboring lattice points. The construction consists of drawing a vector from the selected lattice point to one of its neighbors. The perpendicular bisector plane of that vector is a face of the polyhedron. If we construct the planes for all the neighboring lattice points we get the WS cell. It is now a primitive cell with all the rotational and translational symmetry of the lattice.

The Voronoi diagram (Figure 3.1) and its dual, the Delaunay triangulation, have been used in a large number of fields ranging from astronomy and biology to robotics and computer science. It was first used by Dirilecht to map the regions of gravitational influence of the stars.[6] Competition among plants can be modeled based on plant polygons or areas potentially available to a tree. They are used to minimize small angles in finite element meshes. In pattern recognition, one-dimensional descriptors for two-dimensional shapes such as medial axes can be calculated. The avoidance of obstacles in a robot's path is another problem that can be tackled.[7]

Voronoi polyhedra have been used as a tool in identifying channels and crypts in atomic assemblies. As given by David,[8] Voronoi polyhedra[9,10] help in categorizing the chemical identity of the atoms which bind these channels, and help in quantifying geometric characteristics of these channels.

Voronoi polyhedra have also been used to study aspects of aqueous solvation,[11] and to study cluster surfaces.[12–14] Richards[15] used Voronoi polyhedra for studying molecular volumes of proteins. Recently, Baranyai and Ruff[16] have also modified the Voronoi polyhedron definition to study adjacency (neighboring identities) in molten alkali halides.

Wigner–Seitz cells are a particular form of Voronoi cells used in physics. In 1933, Wigner–Seitz cells were used for describing the packing of spheres.[17] They are also used to describe the band structure of materials through Brillouin zones. Other applications of Wigner–Seitz cells include simulations of grain growth and molecular packing in biochemistry.[18,19]

3.2.2 Current Zeolite Classification Schemes

The Atlas of Zeolite Structures[20] describes all known frameworks of zeolites in terms of secondary building units (SBUs) and also gives building schemes based on nets and chains.

Another source, the Consortium of Theoretical Frameworks (CTF), has extensive data on two-dimensional nets in their catalog. Basing their work on this data, Han and Smith[2–4] systematically generated four-connected three-dimensional nets by conversion of:

(a) Edges of a vertical stack of congruent three-connected 2D nets into chains
(b) Some horizontal edges of a vertical stack of three-connected 2D nets into vertical zigzag chains
(c) Some edges of a congruent stack of horizontal three-connected 2D nets into a three-repeat zigzag–straight saw chain

Their studies list hundreds of 3D nets together with detailed topological information.

Friedrichs et al.[21] have systematically enumerated all possible tilings of uninodal, binodal, and trinodal four-connected networks. For this they start with the Delaney symbol of the tilings and mutate the "gene" under multiple constraints. These include feasibility under Euclidian space (it has to be one of the 230 space groups) as well as suitable bond lengths and bond angles. They then apply this technique to zeolites and zeolite-like networks.

An example of another approach to the problem of classification of nets is by Lord et al.[22] who have used the space groups of the nets as a starting point. They describe the abstract group isomorphic to the space group in terms of their generators. In this compact notation, each space group can be described in terms of the minimal set of generators required. Every element of the group can be obtained as a product of the group elements in the generators set. The relations are equations expressing the unit elements as products of generators. This builds on the work of Coxeter and Moser[23] who presented sets of generators and relations for each of the 17 Euclidian plane groups. Lord et al. use this information as a basis for the topological classification of periodic tilings of two-dimensional Euclidian space. They assign a symbol string encoding all the topological properties. The method is extended to triply periodic nets in three-dimensional Euclidian space, using a set of standard generators and relations for each of the 230 space groups. Thus, in Lord's scheme, the symmetry groups are taken first and used to encode and classify three-dimensional nets. This is in contrast to the approach by Friedrichs et al. in which the starting point is the encoding of the topological properties of the nets.

For example, the sodalite network has cubic symmetry Im3m. The generators and relations for Im3m are

$$\textbf{Im 3 m } \{M_1, M_2, R\} \ B = (RM_1)^2$$

$$(RM_1)^6 = (RM_2)^4 = (M_1M_2)^2 = (RM_1RM_2)^4 = E$$

where $M_1 = $ m x, x, z $M_2 = $ m x, y, 0 R $= 2$ x, ¼, x $+$ ½ and the sodalite network can be represented as SOD (R M_2)M_1.

3.3 CRYSTALLOGRAPHIC ENUMERATION SCHEMES

3.3.1 BACKGROUND

The present work can now be placed in context with the previous research on both network structures and Voronoi polyhedra. Voronoi polyhedra and Wigner–Seitz cells have been used for describing the local regions of influence in computer simulations and to describe the packing of molecules. Here we use it to describe the topology as well as symmetry of the network on a unit cell level. We also go beyond classification by generating descriptors based on Wigner–Seitz cells and then using them to predict properties on a larger scale.

The work of Friedrichs et al. is concerned with all mathematically possible four-connected uninodal, binodal, and trinodal three-dimensional networks, whereas ours is a different method of classification — it does not take into account binodality or trinodality, but with the overall WS cell based on the space group and lattice information. While our system covers all known zeolites, Friedrichs' scheme covers only a small part so far. We also provide a way to navigate among the different categories by choosing descriptors (old and new) such as *c/a* and its variants.

It is interesting to note that the work of Friedrichs as well as of Lord can be extended by our methods. They use two different methods of encoding information of topology and symmetry in sequences of symbols, manipulatable by "mutation." We can treat these as structure descriptors, which using the right data-mining method can be used to predict properties as well as to study larger-scale structural information.

Thus our work describes the local topology and symmetry of the network, which circumscribes or determines the possible overall topologies available to the network. We chose them as a basis for descriptors, which give better information and sense of local atomic environment. This topology–symmetry information is extracted from the spacegroup and lattice parameters by means of primary and "secondary" descriptors and used in statistical techniques — principal component analysis (PCA) and partial least squares (PLS) to study structure relationships.

Two case studies are presented. The pore size of zeolites (a critical parameter in catalysis) and M–O bond lengths in zeolite analogs and their relationship with descriptors of topology–symmetry are studied. The aim here is to identify a selection of formulations which represent statistically guided combinations of structural descriptors (i.e., "secondary descriptors") which can help in developing a more robust understanding of the key crystallographic parameters governing the development of framework structures such as zeolites.[24]

3.3.2   Wigner–Seitz Cell-Based Secondary Descriptors

Conventional approaches to characterizing zeolites and zeolite analogs provide an array of geometrical parameters (e.g., space group, tortuosity, lattice parameter, etc.). All these parameters, are "primary" descriptors, as they provide a direct description of geometrical features of the zeolite. These descriptors were the starting motivation for our study. In addition, we can develop "derived" or "secondary" descriptors, which can be simple mathematical transformations of the original descriptor.

For the two case studies shown here, we focus on the role of crystallographic metrics. In the first case, the crystallographic geometric factors that are important for analyzing the average ring size are identified and the relationships between the geometric factors analyzed using principal component analysis (PCA). In the second case, the zeolite analog dataset is analyzed using "secondary" descriptors and partial least squares is used to predict the M–O bond lengths.

As seen in Figure 3.2, variations in the cell parameter ratios such as the c/a ratio reflect strongly in the shape of the Wigner–Seitz cell representations of crystal structure. The various types of Wigner–Seitz cells possible are shown in Table 3.1 and Table 3.2. It is interesting to note that this parameter is in fact widely used in predicting structural stability in metallic and covalently bonded inorganic structures. The Wigner–Sietz cell-based descriptions (c/a and other ratios) are very valuable in representing the symmetry of the band structure and hence providing a better link between bonding and pore size. The asymmetry of the crystal structure as identified by these ratio-type formulations reinforce the fact that certain scalar parameters which describe distortions in the lattice symmetry are more sensitive to structural changes. While the lattice constants themselves do not lend much physical value, the c/a ratio and related Wigner–Seitz cell descriptors are very sensitive structural parameters to hybridization effects on bonding and electronic screening effects. The potential now exists to extend this work further to identify the role of allotropic transformations in such structures. This type of information can be developed in an empirical fashion or can be statistically guided and can serve as a useful tool for interpretation of structure relationships.

3.4   INFORMATICS STRATEGY

Two datasets have been used in this analysis — one for analysis of pore sizes, and the other for M–O bond length calculation. The algorithm used was SIMPLS. Cross-validation by "leave one out"

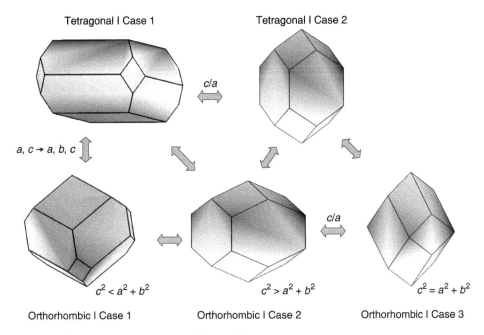

Tetragonal I Case 1

Tetragonal I Case 2

c/a

$a, c \rightarrow a, b, c$

c/a

$c^2 < a^2 + b^2$

$c^2 > a^2 + b^2$

$c^2 = a^2 + b^2$

Orthorhombic I Case 1

Orthorhombic I Case 2

Orthorhombic I Case 3

FIGURE 3.2 Effect of parameter ratios on Wigner–Seitz cell geometries.

TABLE 3.1
The Five Topologies of Wigner–Seitz Cells[37]

W–S cell topology	Number of faces	Number of vertices
I	14	24
II	12	18
III	12	14
IV	8	12
V	6	8

TABLE 3.2
The 24 Types of Wigner–Seitz Cells

Crystal system	Topology				
	I	II	III	IV	V
Cubic	1	0	1	0	1
Tetragonal	1	1	0	0	1
Trigonal	1	0	1	0	0
Orthorhombic	2	1	1	1	1
Monoclinic	2	2	1	1	0
Triclinic	1	1	1	0	0
Hexagonal	0	0	0	1	0

method was used in both cases to validate the models. Prediction error sum of squares (PRESS) was used to calculate the number of latent variables in each case.

For the first case study, the data for the frameworks were collected from the database of the Structure Commission of the International Zeolite Association.[20] In this study, the pore size is the structure metric we are exploring. The secondary descriptors of the lattice parameters are our structural parameters. In this section, we show the value of secondary or derived descriptors of topology–symmetry in the above design process. We use a standard multivariate technique (PCA) as our tool to demonstrate the use of these descriptors. The descriptors used were the crystal lattice parameters (a, b, c, alpha, beta, gamma), topological classifications, space group descriptors, and derived descriptors. The derived descriptors are transformations of the lattice parameters (exp), combinations (a/b, alpha/beta), and transformations of the combinations (Table 3.3). The statistical reasons behind these transformations are the increased correlations and predictive capability between the geometric descriptors and ring size descriptors. Even more importantly, the physical rationale in the different transformations and combinations can yield different valid measures of size, anisotropy, and asymmetry of the Wigner–Seitz cells. The combination of parameterizations explored in this study may be viewed as our "combinatorial" experimentation strategy for searching for better descriptors for the framework structure. The ratio of lattice parameters provides the simplest scalar metric for changes in Bravais lattice symmetry. The choice of the other formulation combinations

TABLE 3.3
The Enumeration Scheme for Descriptors

ID	Topological descriptors	ID	Space group descriptors	ID	Lattice and secondary descriptors
1	SBU1 nodes	6	Space group number	19	a
2	SBU1 connections	7	Symmorphic or not	20	b
3	SBU1 one-connected nodes	8	Presence of screw axis	21	c/a
4	SBU1 two-connected nodes	9	Presence of glide plane	22	a/c
5	SBU1 three-connected nodes	10	Crystal system	23	$\exp(a/c)$
		11	Bravais lattice	24	b/a
		12	P	25	a/b
		13	B	26	$\exp(a/b)$
		14	C	27	b/c
		15	I	28	γ
		16	F	29	γ/β
		17	R		
		18	Point group	30	$\dfrac{\gamma\beta}{\alpha^2}$
				31	$\exp\left(\dfrac{\gamma\beta}{\alpha^2}\right)$
				32	$\dfrac{c\gamma}{a\alpha}$
				33	$\dfrac{a\alpha}{c\gamma}$
				34	$\dfrac{b\beta}{c\gamma}$
				35	$\dfrac{b\gamma}{c\beta}$

of parameters was guided by the use of scalar metrics that are used in the study of describing homogeneous distortions in crystalline structures. The latter approaches are well established in fields such as the study of shear-type transformations. The logic of using these types of derived descriptors was that we wish to be guided by some physical basis to start our descriptor development program. The next step is to use the appropriate methods to explore this multivariate data set to lay some statistical foundations for the use of these descriptors.[25-33]

We have used PCA primarily as a tool to seek patterns. Figure 3.3 shows the distribution of PCA scores for framework structures — a large number have high space group numbers (and hence higher symmetry). It also shows the directions of increasing *c/a* and *b/c* ratios. In Figure 3.4 and Figure 3.5, the relationships between various descriptors in relation to the principal components (PC) are illustrated. The first principal component is a strong function of the *b*c/*(beta*gamma) ratio [#34], *b/c* ratio [#27] and gamma [#28] in terms of primary and some secondary parameters. These are the "asymmetry" parameters. The first principal component is also a function of the symmetry parameters such as space group number [#6], point group [#18], Bravais lattice [#11], and crystal system [#10]. The second principal component is a strong linear function of c/a [#21], *c*a/*(gamma*alpha) [#32], and a [#19] – all topology–symmetry descriptors. The third component is strongly dependent on *a/b* [#25] and *b*c/*(gamma*beta) [#35]. Thus, the first component is a measure of the asymmetry due to *c/a* ratio as well as asymmetry due to the hexagonal unit cell. The second component measures the anisotropy in the *a–b* plane. The third principal component again is a function strongly dependent on topology–symmetry descriptors. The fourth principal component is a function of purely topological descriptors — namely the secondary building unit

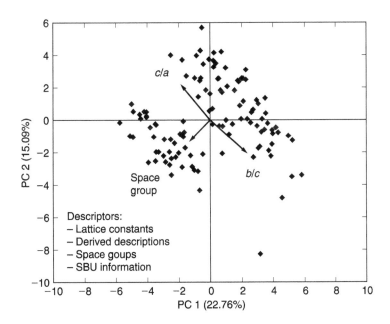

FIGURE 3.3 The PCA score plot for all the framework structures studied. The labeling is not included for clarity of the diagram. The vectors point toward the directions of increasing normalized values of c/a, b/c, and space group number. The space group number for each space group, as given in the International Tables, is a good scalar index for the symmetry of the framework. The numbers vary from 1 (Triclinic) to 230 (cubic system).

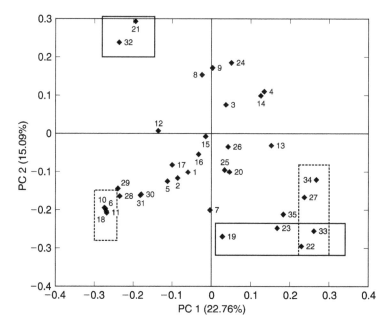

FIGURE 3.4 The loading plots for the first pair of principal components. The relationships between each of the principal components and the loading variables can be seen in the plot. The enumeration scheme is described in Table 3.3. The first principal component is a strong function of the b*c/(beta*gamma) ratio [#34], b/c ratio [#27] and gamma [#28] in terms of primary and some secondary parameters. These are the "asymmetry" parameters. The first principal component is also a function of the symmetry parameters such as space group number [#6], point group [#18], Bravais lattice [#11], and crystal system [#10]. The second principal component is a strong linear function of c/a [#21], c*a/(gamma*alpha) [#32] and a [#19] – all topology-symmetry descriptors. Thus, the first component is a measure of the asymmetry due to c/a ratio as well as asymmetry due to the hexagonal unit cell. The second component measures the anisotropy in the a–b plane.

descriptors. Thus the first four components show the relative contributions of symmetry, topology–symmetry, and topology descriptors to the distribution.

Having suggested a new set of descriptor formulations in the first case study, the next step is to test whether in fact these secondary descriptors do indeed help in identifying appropriate chemistries of new zeolites. While using these proposed secondary crystallographic descriptors in selecting zeolite chemistries for potential combinatorial experiments is clearly the subject of a future study, for the purpose of this study we will take advantage of prior studies reported in the literature to explore the potential value of our findings. Zeolite analogs and microporous structures based on aluminophosphates, and substituted aluminophosphates are being studied for industrial processes such as catalysis and gas separation,[34] and are used in our second case study.

Feng et al.[34] have noted that the average bond length (M–O) for each metal-atom site is linearly dependent on the occupancy factor of Co. Using the compound data from their study we explored our ability to predict the M–O bond length experimentally detected from X-ray diffraction studies. The compounds are synthesized using amines and are zeolite analogs based on cobalt phosphate with varied compositions and structure types. These structures, in the T^{2+}/T^{5+} system (where T refers to tetrahedrally coordinated atoms), are related to low-Si/Al molecular sieves. We have extended our treatment using secondary descriptors from zeolite frameworks to zeolite analogs with an array of compositions. In Figure 3.6 and Figure 3.7, we show the results of increasing the dimensions of our descriptor space by including our secondary descriptors on its effect on their predictive capability.

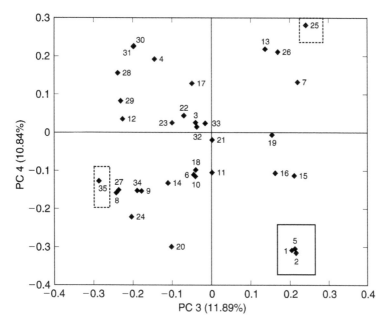

FIGURE 3.5 The loading plots for the second pair of principal components. The relationships between each of the principal components and the loading variables can be seen in the plot. The enumeration scheme is described in Table 3.3. The third principal component is strongly dependent on a/b [#25] and b*c/ (gamma*beta) [#35]. The third principal component again is a function strongly dependent on topology–symmetry descriptors. The fourth principal component is a function of purely topological descriptors — namely the secondary building unit descriptors. Thus the first four components show the relative contributions of symmetry, topology–symmetry, and topology descriptors to the distribution.

Because of their importance in calculating Co/Al ratios, we use the average M–O distances of all unique metal-atom sites in each structure in our analysis. The data analysis used four lattice constant descriptors (a, b, c, alpha), 12 space group descriptors and four secondary descriptors. The two graphs show the prediction with only four lattice constants (Figure 3.6, Q^2 value of 0.173), and with lattice constants, space group and secondary descriptors (Figure 3.7, Q^2 value of 0.796). The number of latent variables chosen was four for the first case and five for the second case. The secondary descriptors included interactions only among the lattice constants, as the space group descriptors were not well suited for interaction parameters, particularly if they are to be physically interpretable. This shows the wide range of applicability of the methodology proposed. The case study illustrates the value of secondary descriptors in improving the performance of standard statistical techniques.

Note that in Figure 3.6 and Figure 3.7, the "predicted" M–O bond lengths are based on our own statistical calculations. We used the lattice parameters, and Wigner–Seitz scalar metrics for the 35 compounds to calculate the corresponding M–O bond lengths. These "predicted" values are compared to the "measured" values in the partial least squares plots.

The "measured" M–O bond lengths were the data gathered from experiments of Feng et al. They used X-ray diffraction on single crystals and Reitveld refinement techniques in order to calculate the atomic positions, occupancy factors, and M–O bond lengths. They found that for Co/Al mixed sites, for a well-refined structure, bond length = 1.74 + 0.19 occupancy of cobalt with an error of 0.01–0.02 Angstrom. Feng et al. point out that occupancy factors are more likely to be correlated with thermal parameters and are unreliable when guest amines are seriously disordered and not

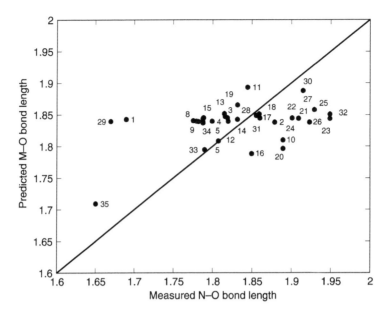

FIGURE 3.6 The PLS prediction of M–O bond length with only four lattice constants. When four latent variables are used, a Q^2 value of 0.173 is obtained. M–O bond lengths are given in Angstroms.

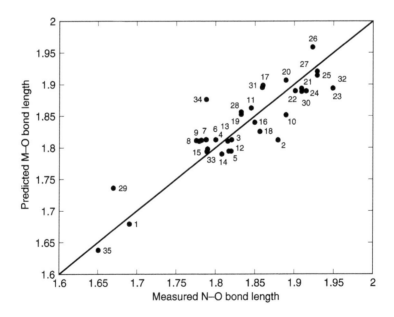

FIGURE 3.7 The PLS prediction with 12 space group descriptors, four lattice constants, and four secondary descriptors. When five latent variables are used, as shown above, a Q^2 value of 0.799 is obtained. M–O bond lengths are given in Angstroms.

included in the refinement. Hence they use metal–oxygen bond distances as a more reliable way of calculating the Co/Al ratio for each individual Co/Al site.

The three extremal data points in the M–O bond PLS plots belong to analcime, feldspar, and faujasite analogs. In the analcime (#1) and faujasite (#35) analogs, Co, Al, and P occupy the same site as there is only one unique T-atom site. Feng et al. also acknowledge that in ACP-FL2 (#29), a feldspar analog, the distribution of Co, Al, and P is also random even though two unique T-atom sites are available. They note the low probability of a completely random distribution of Al, Co, and P sites, and attribute the apparent disorder observed in these three structures to the potential presence of different crystal domains in which the metal-atom sites and P sites are exchanged. Feng et al. point out that, in general, for Al-containing cobalt phosphates, tetrahedral Co and Al atoms occupy the same sites that alternate with ordered P sites even though there is strong crystallographic evidence of pure Co^{2+} sites in thomsonite analogs and pure Co^{2+} and Al^{3+} in some other structures. In the analcime analog, as the Al–O, Co–O, and P–O bond lengths are significantly different, the disorder of tetrahedral atoms affected the positional ordering of bridging oxygen atoms. The refined T–O distance was 1.69 (3) Angstrom, in agreement with the weighted average (1.69 Angstrom) of P–O (3×1.52 Angstrom), Co–O (2×1.93 Angstrom) and Al–O (1×1.74 Angstrom) bonds.[35] The R-factors for these three structures (10.2, 4.74, and 15.5 for analcime, feldspar, and faujasite analogs, respectively) indicate the reasonable accuracy of estimates of the bond lengths in these compounds.

The increase in quality of the fit is notable when we add the secondary descriptors and space group descriptors. The three data points (#1, #29, #35) are no longer outliers in the final multivariate model that includes the secondary descriptors. This implies that the effect of same-site substitution has been taken into account inherently by the model, essentially due to the addition of the space group and secondary descriptors.

3.5   CONCLUSIONS AND FUTURE WORK

An analytical foundation for statistically identifying key crystallographic descriptors for zeolite frameworks and zeolite analogs was presented in this chapter. A new set of descriptor formulations containing various aspects of information about framework structures was introduced. Selected crystallographic structure data for all known zeolite structure types were analyzed using principal components analysis in order to understand the statistical distribution and hence help design new zeolites and zeolite analogs. Critical crystallographic ratios (based on Wigner–Seitz cells) and transformed variables were calculated. The principal components of the framework descriptors were used to identify the important linear combinations of critical descriptors. These new secondary descriptors were tested against experimental findings of zeolite-analog structures, demonstrating the value of these secondary descriptors as a guide for future combinatorial experiments.

It is interesting to note that only when the proper secondary descriptors, based on c/a metrics of Wigner–Seitz cells, were included, that a higher predictive capability was obtained. The use of non-linear models in the future may increase the accuracy of prediction. In general, primary descriptors are property, structural, and process variables that are measurable, or computable with a physical meaning to attribute to them. Thus, primary descriptors like the lattice parameters are not optimized statistically but are used because they can be measured comparatively easily and have a well-established scientific significance. The ability to use data-mining methods effectively is undermined if we choose to use only these unoptimized variables. The secondary descriptors fill this gap well. These are essentially descriptors derived from the primary variables in order to enhance the data-mining performance. By using secondary descriptors from zeolite frameworks with no compositional information to zeolite analogs with different compositions we have demonstrated that our selection of secondary descriptors can capture the complexities of chemistry and other geometrical parameters and hence provide a basis of future combinatorial experiments. This improvement provides an optimistic framework to develop this approach of calculating secondary descriptors, which warrants further study for developing better informatics tools for catalysis design.

The results point toward extending this investigation in three areas: use of other linear and non-linear techniques, development of better descriptors for symmetry and topology of networks and space groups, and better-designed crystallographic databases.

The information about symmetry included in the 230 space groups can be encoded in multiple ways. These conventions like the Hermann–Mauguin international symbols (full or short forms) or Schoenflies notations are convenient for compact identification of the most important, critical, and distinguishing characteristics of the space groups.[36] On the other hand, when we try to use the same notation for data mining, we run into trouble, as they are not designed for this purpose. In our work, we have tried to avoid the problem by using a series of descriptors to mark the absence or presence of symmetry elements, such as glide planes, and also by using space group numbers as nominal indicators of their symmetry. We are investigating the use of better space group descriptors for data mining based on the following two concepts:

1. Multiple notations are preferable to single notations as they contain different types and amounts of information in complementary formats.
2. Numerically or sequentially representable notations (rule based) are preferable.

While the present work has focused on PCA and PLS as linear multivariate statistical techniques for finding relationships, and identifying patterns in the data, this is just the beginning. There is a whole host of methods (linear and non-linear) that can be used for analyzing data sets with unique characteristics. Non-linear KPLS, support vector machines, association rule mining, and self organizing maps are a few such techniques available for analysis of materials datasets.

To complete the cycle, the results of statistical computations and scientific simulations on different material systems will be validated by combinatorial experiments. These experiments will also generate new data for analysis and discovery. After each cycle, the experiments will be more focused, insights better, and calculations more accurate than the cycle before.

ACKNOWLEDGMENTS

The authors gratefully acknowledge support from the National Science Foundation International Materials Institute Program for the Combinatorial Sciences and Material Informatics Collaboratory: CoSMIC-IMI; grant # DMR 0231291.

REFERENCES

1. Wells, A. F., *Three Dimensional Nets and Polyhedra*, Wiley, New York, 1977.
2. Han, S. and Smith, J. V., Enumeration of four-connected three-dimensional nets. I. Conversion of all edges of simple three-connected two-dimensional nets into crankshaft chains, *Acta Cryst. A*, 55, 332, 1999.
3. Han, S. and Smith, J. V., Enumeration of four-connected three-dimensional nets. II. Conversion of edges of three-connected 2D nets into zigzag chains, *Acta Cryst. A*, 55, 342, 1999.
4. Han, S. and Smith, J. V., Enumeration of four-connected three-dimensional nets. III. Conversion of edges of three-connected two-dimensional nets into saw chains, *Acta Cryst. A*, 55, 360, 1999.
5. Rajagopalan, A., *Topology and Symmetry Guided Design of Framework Structures*, Ph. D. Thesis, Rensselaer Polytechnic Institute, Troy, New York, 2004.
6. Okabe, A., Boots, B. and Sugihara, K., *Spatial Tesselations: Concepts and Applications of Voronoi Diagrams*, Wiley, Chichester, UK, 1992.

7.   Drysdale, S., Geometry in action, http://www.ics.uci.edu/~eppstein/geom.html, 2001.
8.   David, C. W., Voronoi polyhedra as structure probes in large molecular systems – VII. Channel identification, *Comput. Chem.*, 12, 3, 207, 1988.
9.   Voronoi, G., Nouvelles applications des parametres continus a la theorie des formes quadratiques, *J. Reine U. Angew Math.*, 134, 198, 1908.
10. O'Keefe, M., A proposed rigorous definition of cooordination number, *Acta Cryst. A*, 35, 772, 1979.
11. David, E. E. and David, C. W. Voronoi polyhedra as a tool for studying solvation structure, *J. Chem. Phys.*, 76, 4611, 1982.
12. David, E. E. and David, C. W., Voronoi polyhedra and cluster recognition, *J. Chem. Phys.*, 77, 3288, 1982.
13. David, E. E. and David, C. W., Voronoi polyhedra and solvent structure for aqueous solutions (III), *J. Chem. Phys.*, 77, 6251, 1982.
14. David, E. E. and David, C. W., Voronoi polyhedra for studying solvation structure (IV), *J. Chem. Phys.*, 78, 1459, 1983.
15. Richards, F. M., The interpretation of protein structure: total volume, group volume distributions and packing density, *J. Mol. Biol.*, 82, 1, 1974.
16. Baranyai, A. and Ruff, I., Statistical geometry of molten alkali halides, *J. Chem. Phys.*, 85, 365, 1986.
17. Wigner, E. and Seitz, F., On the constitution of metallic sodium, *Phys. Rev.*, 43, 804, 1933.
18. Aboav, D. A., The arrangement of grains in a polycrystal, *Metallography*, 3, 383, 1970.
19. Vander Voort, G. F., *Practical Applications of Quantitative Metallography. ASTM Special Technical Publication,* Philadelphia, PA, 839, 85, 1982.
20. Baerlocher, C. H. and McCusker, L. B., Database of zeolite structures, http://www.iza-structure.org/databases/
21. Friedrichs, O. D., Dress, A. W. N., Hudson, D. H., Klinowski, J. and McKay, A. L., Systematic enumeration of crystalline networks, *Nature*, 400, 644, 1999.
22. Lord, E. A., Mackay, A. L. and Ranganathan, S., New geometries for new materials, http://metalrg.iisc.ernet.in/~lord , 2004.
23. Coxeter, H. S. M. and Moser, W. O. J., *Generators and Relations for Discrete Groups*, Springer-Verlag, Berlin, 1980.
24. Rajagopalan, A., Shuh, C., Li, X. and Rajan, K., "Secondary" descriptor development for zeolite framework design: an informatics approach, *Appl. Catalysis A*, 254, 147–160, 2003.
25. Jolliffe, I. T., *Principal Component Analysis*, Springer-Verlag, Berlin, 1986.
26. Preisendorfer, R. W., *Principal Component Analysis in Meterology and Oceanography*, Elsevier, Amsterdam, 1988.
27. Shum, H., Ikeuchi, K. and Reddy, P., Principal component analysis with missing data and its application to polyhedral object modeling, *IEEE Trans. on Pattern Analysis and Machine Intelligence*, 17(9), 854, 1995.
28. Bajorath, J., Selected concepts and investigations in compound classification, molecular descriptor analysis, and virtual screening, *J. Chem. Inf. Comput. Sci.*, 41 (2), 233, 2001.
29. Suh, C., Rajagopalan, A., Li, X. and Rajan, K., Application of principal component analysis materials science, *Data Sci. J.*, 1 (1), 19–26, 2002.
30. Wichern, D. W. and Johnson, R. A., *Applied Multivariate Statistical Analysis*, 5th Edn., Prentice Hall, Upper Saddle River, New Jersey, 2002.
31. Wold, S., *Food Research and Data Analysis*, Martens, H. and Russwurm, H., Eds., Applied Science Publishers, London, 1983.
32. Wold, S., *Chemometrics: Mathematics and Statistics in Chemistry*, Kowalski, B., Ed., Reidel, Dordrecht, Netherlands,1984.
33. Wold, H., *Multivariate Analysis,* Krishnaiah, P. R., Ed., Academic Press, New York 1966, p. 391.
34. Feng, P., Bu, X. and Stucky, G. D., Hydrothermal syntheses and structural characterization of zeolite analogue compounds based on cobalt phosphate, *Nature*, 388, 735, 1997.
35. Shannon, R. D., Revised effective ionic radii and systematic studies of interatomic distances in halides and chalcogenides, *Acta Cryst. A*, 32, 751, 1976.
36. Fischer, W. and Koch, E., *International Tables For Crystallography A*, Hahn, T., Ed., Kluwer, Dordrecht, Netherlands, 1983, Ch. 14.
37. Delaunay, B. N., Neue Darstellung der geometrischen Kristallographie, *Z. Kristallograph.*, 84, 109, 1932.

4 Combinatorial Study of New Glasses

Tomoya Konishi, Shigeru Suehara, Shin-ichi Todoroki, and Satoru Inoue

CONTENTS

4.1 INTRODUCTION

The term "new glass" has a more profound meaning than the literal sense of the word. New functions and properties beyond our imagination have resulted from advanced technologies such as controlled production and new processes. Today, various types of new glasses are playing important roles in modern industries; for instance, an optical fiber amplifier, a zero expansion glass, an integrated circuit (IC) photo mask, laser rods, and so on.[1] Functions and properties of a glass are often determined by its composition. Fortunately, glass is different from other ceramic materials in that the composition is continuously changeable regardless of stoichiometry, owing to the random network structure and the inorganic nature. This is a great advantage to materials design because the property of the glass is tunable by changing the composition. Moreover, glass provides a suitable matrix for various functional ingredients.

The first step to developing new glasses is to determine a glass-forming region in a given multicomponent system. In order to do this, conventionally, researchers melt a batch of material in a container using an electric furnace and pour the melt onto a quenching plate to test the glass-forming tendency for each composition. The process is very simple but time consuming for a lot of tests, limiting the total efficiency of research progress. The problem will be solved by adopting combinatorial methodology, which is exactly intended to mine out valuable compositions from myriad combinations of components.

Combinatorial methodology often involves some techniques to synthesize samples rapidly, because it requires many more samples than conventional one-by-one methodology in order to screen out compositions for glass-forming and certain interesting properties. In our laboratory, we adopted automation and parallelization to the preparation of glass samples as well as the rules of glass formation. The detail of these techniques will be described in Section 4.2. Next, some examples of new glasses discovered through the combinatorial methodology will be disclosed.

4.2 EXPERIMENTAL

4.2.1 Prediction of Glass-Forming Compositions

Since the number of possible combinations of components is nearly infinite, it is a good idea to predict glass-forming compositions by using the rules deduced theoretically or proposed empirically by the pioneering researchers as summarized in Figure 4.1. Zachariasen's rule,[2] Sun's rule,[3] and Stanworth's rule[4] are commonly accepted by researchers and give clear ideas about glass formation in oxide systems. Inoue and MacFarlane showed that the experimental glass-forming tendency in the chlorofluoro mixed halide ZBLAN (Zr-Ba-La-Al-Na) system correlated well with the degree of localization of the chlorine ions from random distribution among the glass structures composed by MD simulation.[5] Recent advancement in the performance of computers should promote studies on screening glass-forming systems by MD simulations.[6]

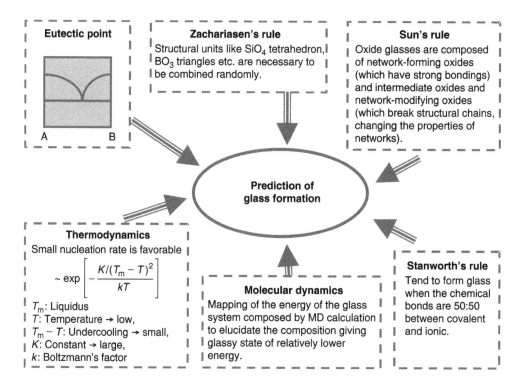

FIGURE 4.1 Rules for prediction of glass formation (reproduced from Inoue et al.[6]).

4.2.2 INSTRUMENTAL DETAILS

Because combinatorial methodology requires a lot of samples, it is essential to develop instruments or apparatuses for efficient synthesis of samples. In our laboratory, the authors originally designed and developed the following instruments and apparatuses to accelerate preparation of glass samples. Figure 4.2 shows the overview of combinatorial research on glass in our laboratory.[7]

4.2.2.1 Automatic Batch Preparation Apparatus

As a starting material for glass, a glass batch is prepared by weighing and blending the reagents that compose the formula of the glass. The weight of each reagent is determined suitably for the glass composition. The authors developed an automatic batch preparation apparatus to automate a sequence of weighing reagents.[6–8] This machine dispenses four kinds of reagents into 24 crucibles, one by one, according to a recipe. This machine is useful to prepare a large quantity of glass batches with various compositions to make a multicomponent glass sample library.

Figure 4.3 shows a photograph of the apparatus. The apparatus was designed as an automatic powder dispenser with a robot. The apparatus is composed of four reservoir tanks of reagents attached to a revolving arm, an electric balance, and a personal computer. Crucibles are placed in a 3×4 or 4×6 matrix on the electric balance. Rotational movement of the revolving arm selects the reagent to dispense, and translational movement of the arm selects the crucible into which reagents are to be placed. A preset amount of reagent powder is directly poured into a crucible from the nozzle attached to the bottom of each tank. The computer is connected to the valve of the nozzle and the electric balance. While dispensing the reagents, the computer monitors the weight of the reagent in the crucible and suitably regulates the powder flow rate not to feed too much over the target weight. The computer controls the valve, taking account of the fluidity of the powder. Owing to this feeding mechanism, weighing error is about several tens mg. Because the capacity of crucibles is several tens g, the error corresponds to no more than 1%. To prevent the powder from absorbing moisture and congesting in the nozzle, the apparatus is placed in an air-conditioned room. The apparatus is also equipped with an ionizer to remove static electricity at the nozzle.

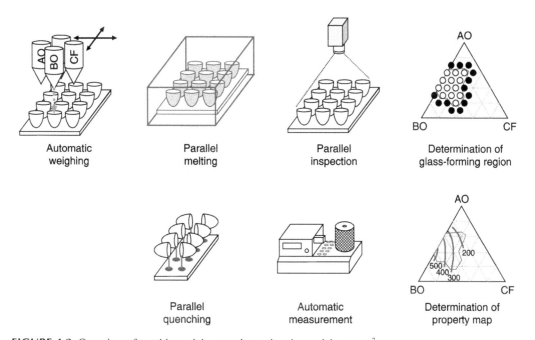

FIGURE 4.2 Overview of combinatorial research on glass in our laboratory.[7]

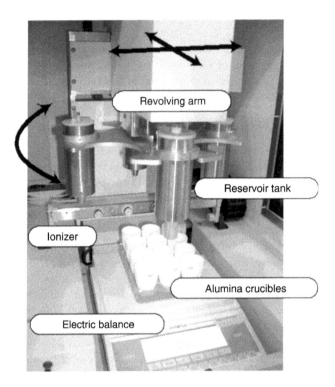

FIGURE 4.3 Automatic batch preparation apparatus.[9]

In order to obtain uniform glass, the prepared batch must be well mixed in advance of melting. The authors assembled a tool to shake multiple batches instead of a pestle and a mortar to increase efficiency.[7] Figure 4.4 shows a picture of the tool. Batches in 12 or 24 crucibles are agitated by rotating them slowly by hand.

Here, this apparatus does not prepare batches much quicker than a well-trained researcher; however, the apparatus contributes much to the efficiency of the whole research process because it automatically works on many batches without failure. Meanwhile, the researcher can concentrate on other tasks.

4.2.2.2   Multisample Glass-Melting Furnace

This apparatus is "a robotic furnace" in the sense that it simultaneously carries as many as 18 crucibles into a furnace to melt at once, and takes them out to cast the melts on a quenching substrate.

Figure 4.5 shows a photograph of the apparatus. This apparatus is composed of a furnace with molybdenum silicide heaters capable of 1300°C, a pair of crucible holders which can bear nine crucibles of 30 ml capacity, a pair of crucible rotators, and stainless-steel quenching plates which can be preheated to 200°C. The maximum temperature is limited to 1200°C for ordinary use because the crucible holders are made of silica. Figure 4.6 shows a schematic diagram of the apparatus movements. Once the crucibles with batches inside are mounted onto holders, they are automatically carried into the furnace. After a certain period of melting and homogenizing, the bottom of the furnace comes down with the pair of crucible holders. The crucible rotators receive the crucible holders, respectively, from the base plate of the furnace, and tilt the crucibles to pour the melt onto the quenching plate. The crucible rotators can pour the melt while moving aside to obtain a strip of samples for dilatometry. Because the movements of the pair of crucible rotators are synchronized by a computer operation, all the melts are quenched at exactly the same rate. Because there are

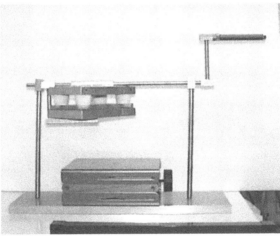

FIGURE 4.4 Tool to homogenize multiple batches in 12–24 crucibles by rotating handle.[9]

compositions where samples sometimes form glass and sometimes do not form glass according to cooling rates, it is very important to guarantee a common cooling rate for all the compositions, especially when a researcher wants to determine a glass-forming region in a composition diagram. Moreover, the same cooling rate for repeating tests contributes to the excellent reproducibility.

As described above, since a small change in the cooling rate may affect glass formation, it has been said that making glass is craftsmanship. The benefit of this apparatus is not only that a researcher can boost sample preparation by 18 times but also that everyone can prepare glass with the skill of an experienced glassmaker.

4.2.2.3 Multisample Glass-Forming Tester

This apparatus performs batch-melting, heat-treatment, and quenching on 24 samples at once to test their glass forming. It is capable of handling more than 200 samples a day to aid high-throughput determination of a glass-forming region and rapid preparation of a sample library to investigate compositional dependency of a certain property.[10]

Figure 4.7 and Figure 4.8 show a photographic and a schematic depiction of the apparatus, respectively. This apparatus includes a furnace with silicon carbide heaters capable of 1500°C, a transportation

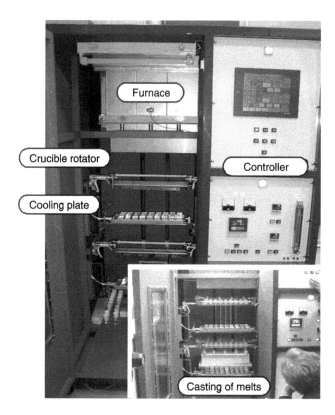

FIGURE 4.5 Multisample batch-melting furnace.

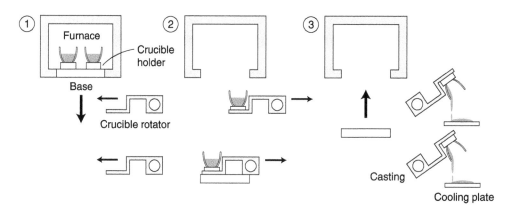

FIGURE 4.6 Schematic illustration of mechanical movements of the multisample glass-melting furnace.

system for 24 crucibles, and a cooling chamber equipped with a pair of gas blowers and a water-cooled chill. The core tube of the furnace and the crucible transporter are made of graphite for the flat temperature profile in the furnace. In order to prevent the graphite from being burned away at high temperatures, the inner atmosphere is substituted with nitrogen or argon gas. As shown in Figure 4.9, the crucible holder can bear 24 crucibles of 20 mm in diameter. It can transport them between the furnace and the cooling position in 2 s.

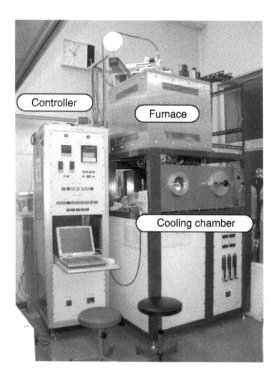

FIGURE 4.7 Multisample glass-forming tester.[9]

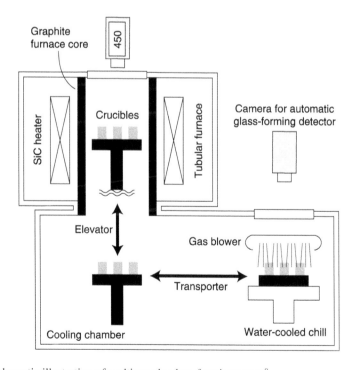

FIGURE 4.8 Schematic illustration of multisample glass-forming tester.[9]

FIGURE 4.9 Holder and 24-carbon crucible, 20 mm in diameter.[9]

First, graphite crucibles with samples inside are mounted on a graphite holder in the cooling position. The transporter carries them into the furnace to start heat treatment under certain temperature and time conditions. During the heat treatment, the temperature of each sample can be measured one by one with a radiation thermometer through silica windows. As soon as the samples are finished with the heat treatment, they are transported back to the cooling position. The melts are not cast on a quenching plate but cooled in crucibles by gas-blowing from the top and water-cooled from the bottom. The cooling rate is ~10°C/s which is comparable with that of casting a melt on a quenching plate. The appearance of the cooled samples is observed through a view port attached to the top of the cooling chamber. Photographs of cooled samples are captured into a computer with a video camera, and automatic glass-forming detection is performed by using image-processing technology as described in the next section.

4.2.2.4   Automatic Glass-Forming Detector

A melt is quenched quickly to form glass or to devitrify according to the composition and the cooling rate. This apparatus determines glass forming or devitrification of samples using image-processing techniques. A photograph of a sample on a black plate is captured into a computer with a charge-coupled device (CCD) camera as shown in Figure 4.10. If the sample forms glass, the image will be dark because the black plate can be seen through the transparent sample; otherwise it will be rather bright because of light scattering from devitrification. The computer collects statistics on the brightness of the image to decide whether the sample is glass-formed or not. Figure 4.11 shows a photograph of a glass sample partially devitrified and line profiles of pixel brightness. It should be taken into account for calculation that the brightness of highlighted spots is greater even in the glassy parts, as indicated by dashed lines. The accuracy of glass-forming detection was almost the same as that of visual inspection. The detection results are directly stored into a glass-samples database along with the photographs of the samples.

4.3   NEW GLASSES DISCOVERED THROUGH A COMBINATORIAL METHOD

4.3.1   Reddish-Colored Glass: The P_2O_5-TeO_2-ZnO System

Tellurite glasses have attracted much attention because they have various interesting properties such as low glass-transition temperature (~400°C),[11] pale green to dark red–brown coloration,[12] high refractive indices (~2.0), thermochromism,[13] and so on. In spite of these attractive properties,

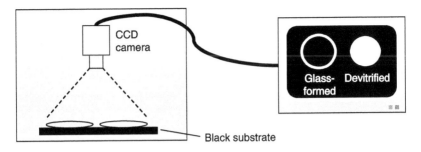

FIGURE 4.10 Schematic setup for automatic glass-forming detector.

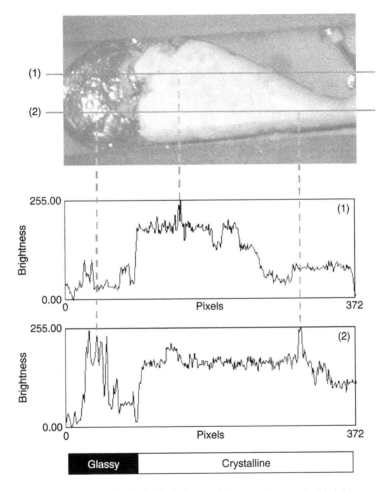

FIGURE 4.11 Photograph of partially devitrified glass and line profiles of pixel brightness.

few studies have been made on the compositional dependence of basic properties such as glass-forming and thermal properties, which are essential for the design of glass materials. By using combinatorial methodology, the authors investigated the glass-forming region, the compositional dependence of the glass-transition temperature and the coloring property for the P_2O_5-TeO_2-ZnO (PTZ) system.[14,15] The coloring origin was also considered from the viewpoint of microstructure and valence states of component cations.

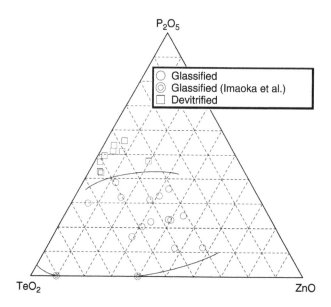

FIGURE 4.12 Glass-forming region of P_2O_5-TeO_2-ZnO system on the basis of analyzed composition. Glass-forming region of binary TeO_2-ZnO glass by Imaoka and Yamazaki[15] is also shown. Boundary lines are drawn as a guide to the eye.

Batches for PTZ compositions were prepared from reagent-grade H_3PO_4, TeO_2, and ZnO with the automatic batch-preparation apparatus. The batches were melted and quenched automatically with the multisample glass-melting furnace. The batches in alumina crucibles were heated at 1000 or 1100°C and kept at the temperature for 20 min to attain uniform melts. The melts were then quenched by casting them onto a cooling plate. The glass transition temperature and the softening temperature during heating at 10°C/min were measured on the glass-forming samples with an automatic differential thermal analyzer (DTA) equipped with an automatic sample feeder. Scanning electron microscopy (SEM) analysis was conducted on a fractured surface to observe the microstructure. The valence states of the component cations were estimated by X-ray photoelectron spectroscopy (XPS) analysis.

Figure 4.12 shows the glass-forming region for the PTZ system melted at 1000°C plotted on the basis of analyzed composition. The glass-forming region is located where the P_2O_5 content is less than 40 mol%.

Thermal properties were measured for the glass-forming samples, and the glass transition temperatures are given in Figure 4.13. The glass transition temperature increased continously with the increasing ZnO content. Because ZnO may bridge between PO_4 and TeO_4 chains as an intermediate oxide,[17] it is assumed that ZnO strengthens the glass network structure to increase the glass transition temperature.

Figure 4.14 shows photographs of as-quenched PTZ samples placed on a ternary compositional chart. The samples generally exhibited a yellowish to reddish color with the hue depending on the composition and the heating temperature. The color became deeper with increasing TeO_2 content and increasing heating temperature. Figure 4.15 shows SEM micrographs of yellowish-colored $30P_2O_5$-$50TeO_2$-$20ZnO$ glass and reddish-colored $30P_2O_5$-$10TeO_2$-$60ZnO$ glass. Both glasses had a similar microstructure of nanometer-sized fine grains scattering uniformly in the matrix. Figure 4.16 shows the XPS spectra of the two glasses. While peaks of Zn and P are singlets indicating a single chemical bonding state, Te peaks are doublets with a shoulder indicating two bonding states: one is Te^{4+} and the other is Te or Te^{2-}.[18] It is unlikely that Te^{4+} and Te^{2-} co-exist stably, so the particles observed with the SEM are assumed to be colloidal metallic Te. From the

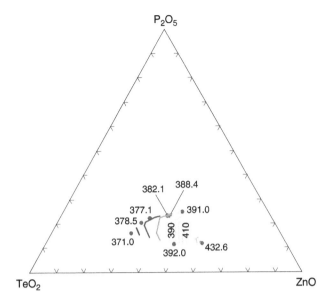

FIGURE 4.13 Glass transition temperature of P_2O_5-TeO_2-ZnO glasses melted under 1000°C plotted on ternary diagram. Contour lines are drawn as a guide to the eye.

results of SEM and XPS analyses, the coloring was considered to be due to colloidal coloration by precipitated metallic Te particles. This is consistent with the result that the reddish color became deeper with the increasing TeO_2 content.

It was found that PTZ glass exhibited a reddish color without introducing any coloring reagent. Moreover, the color depth was controllable merely by the composition. This phenomenon is advantageous for industrial production of warm-colored optical glass for optical filters since it has been difficult to give such color to glass, conventionally, by using a suspension of colloidal gold with the size and density precisely controlled.

4.3.2   LOW-MELTING AND LEAD-FREE GLASS: THE B_2O_3-TeO_2-BaF_2/BaO SYSTEM

Low-melting glasses have glass transition temperatures below 600°C, which is much lower than that of fused silica: 1200°C. They are utilized conventionally for packaging and welding, and recently also for patterning cell walls on plasma display panels and soldering amplification repeaters of a submarine fiber optic cable system. The PbO-B_2O_3-SiO_2 system is typical of conventional low-melting glasses. The reason why lead has been considerably used in spite of the current issue of environmental pollution is that it lowers melting temperature without increasing thermal expansion coefficient or decreasing water resistibility. As a candidate for alternative lead-free and low-melting material, the authors determined a new glass-forming region in B_2O_3-TeO_2-$[xBaF_2/(100 - x)BaO]$ pseudo-ternary systems and measured their thermal properties such as softening temperature and thermal expansion coefficient.[19]

The batch for each composition was prepared from reagent-grade H_3BO_3, TeO_2, $BaCO_3$, and BaF_2 with the automatic batch preparation apparatus. The batches were melted and quenched automatically in the 18-sample glass-melting furnace. The 5-g batches in alumina crucibles were kept in air at 1000°C for 20 min to attain uniform melts. The melts were quenched by automatic operation by pouring them onto a cooling plate. For the samples without devitrification, softening points were determined by differential thermal analysis at a heating rate of 10°C/min. Thermal expansion coefficients were measured with a thermo-mechanical analyzer. For these measurements, each

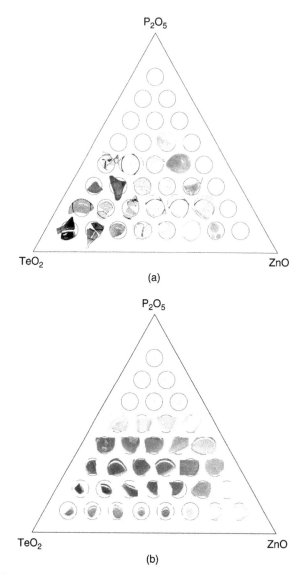

FIGURE 4.14 (See color insert following page 172) Appearance pictures of as-quenched P_2O_5-TeO_2-ZnO samples placed on ternary diagram. Batch-melting temperatures were (a) 1000°C and (b) 1100°C.

sample was cut into a block ($5 \times 5 \times 15$ mm³) and mechanically polished. The heating rate was 5°C/min and the load was 5 g. Chemical compositions were determined by an inductively coupled plasma (ICP) spectrometer in order to clarify the compositional deviation from the batch composition during melting.

Figure 4.17 shows the glass-forming regions by batch compositions for B_2O_3-TeO_2-$[xBaF_2/(100 - x)BaO]$ systems, where $x = 0$, 25, 50, 75, 100. The glass-forming region is widest when the substitution amount of BaF_2 for BaO is 25%. Generally, glass-forming ability is modified by suppressing nucleation of crystallization or phase separation. For example, lowering the liquidus temperature decreases the driving force for nucleation; namely, the degree of supercooling from the liquidus temperature to the crystalline or phase-separation temperature. As will be described later, substituting BaF_2 for BaO lowers the liquidus temperature by increasing ionicity in the glass

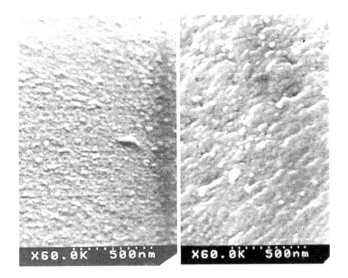

FIGURE 4.15 FE-SEM micrographs of yellowish-colored $30P_2O_5$-$50TeO_2$-$20ZnO$ glass (left) and reddish-colored $30P_2O_5$-$10TeO_2$-$60ZnO$ glass (right).

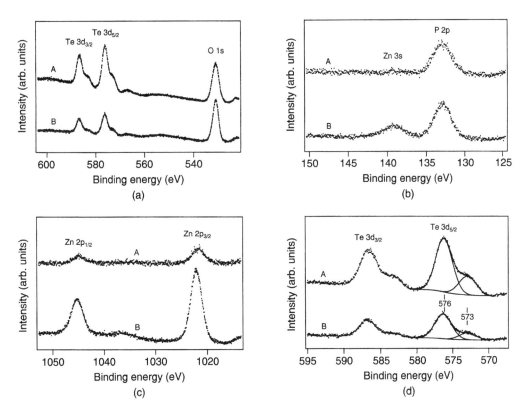

FIGURE 4.16 Core-level photoelectron spectra of (a) Te 3d and O 1s; (b) Zn 3s and P 2p; and (c) Zn 2p for yellowish $30P_2O_5$-$50TeO_2$-$20ZnO$ glass (denoted by A) and reddish $30P_2O_5$-$10TeO_2$-$60ZnO$ glass (denoted by B). (d) Deconvolution of Te $3d_{5/2}$ peak into Te^{4+} (at 576 eV) and Te, Te^{2-} (at 573 eV) components are also plotted.

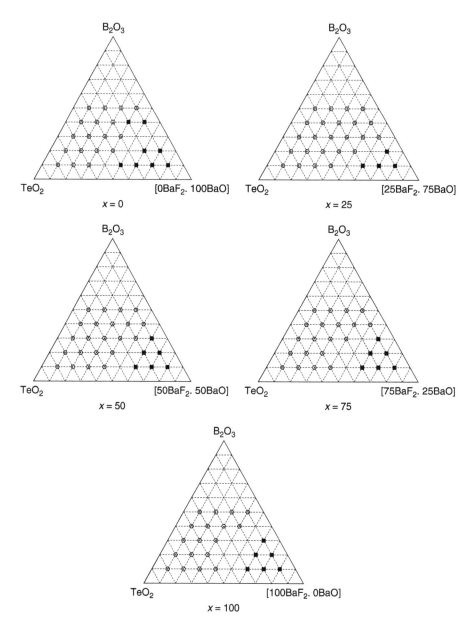

FIGURE 4.17 Glass-forming regions in B_2O_3-TeO_2-$[xBaF_2/(100 - x)BaO]$ systems by batch composition. Open circle markers indicate glass-formed compositions. Open triangle markers indicate devitrified compositions. Filled square markers indicate unmelted compositions.

network structures. When the substitution amount is more than 25%, on the other hand, forming the glass network structure of B_2O_3 and TeO_2 becomes difficult.

Figure 4.18 shows the result of ICP compositional analysis on $30B_2O_3$-$40TeO_2$-$30[xBaF_2/(100 - x)BaO]$ glasses as a function of substitution amount of BaF_2 for BaO, x. As can be clearly seen from the figure, during melting fluorides promoted the evaporation of boron and the erosion of alumina crucibles into the melts. Because tellurium content and barium content are almost constant, it can be understood that 5 mol% of B_2O_3 was substituted with Al_2O_3 when 30 mol% ($x = 100$) of BaO was substituted with BaF_2.

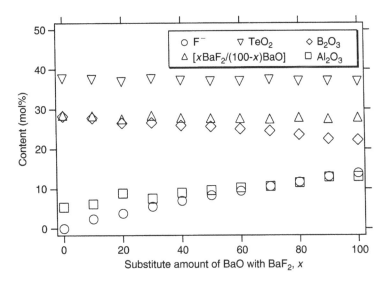

FIGURE 4.18 Composition of 30 B_2O_3-40TeO_2-30[xBaF$_2$/(100 − x)BaO] glasses by ICP analysis.

Figure 4.19 shows softening temperatures of 30B_2O_3-40TeO_2-30[xBaF$_2$/(100 − x)BaO] glasses as a function of substitution amount of BaF$_2$ for BaO, x. In this study, softening temperature was determined as the minimum point of a glass-transition endothermic curve in a DTA diagram. The glasses have softening temperatures at 460 ± 10°C. The softening temperatures decreased with increasing substitution amount of BaF$_2$ for BaO. Generally, softening temperature or glass-transition temperature is closely related to the number of non-bridging oxygen atoms, single-bond energy, electron negativity difference, ionic radius, and coordination number. Because electron negativity of F (4.0) is larger than that of O (3.5), the ionicity in M–F (M represents B or Te) bonds is larger than that in M–O bonds. The decrease in softening temperature was brought by the increase in ionic bonds with its single-bond energy less than that of covalent bonds.

Figure 4.20 shows thermal expansion coefficients of 30B_2O_3-40TeO_2-30[xBaF$_2$/(100 − x)BaO] glasses as a function of substitution amount of BaO with BaF$_2$, x. A thermal expansion coefficient generally tends to increase with the increasing ionic bonds. In this study, however, a significant change in thermal expansion coefficient was not observed regardless of the substituting amount of BaO with BaF$_2$. Although the reason is not yet known, one explanation is that the effect of the decrease in the thermal expansion coefficient by the covalent alumina and the effect of the increase in the thermal expansion coefficient by the ionic fluorides are compensating each other. As a result, the softening temperature of the 30B_2O_3-40TeO_2-30[xBaF$_2$/(100 − x)BaO] glass could be continuously tailored from 450 to 470°C without increasing thermal expansion coefficient within error of ± 25 × 10⁻⁷/°C.

Figure 4.21 shows thermal expansion coefficients versus softening temperatures of 30B_2O_3-40TeO_2-30[xBaF$_2$/(100 − x)BaO] glasses along with data for various low-melting glasses reproduced from the literature.[20] The softening temperatures and the thermal expansion coefficients of the present glass system are comparable to those of typical lead-containing glasses, showing the potential of alternative low-melting glasses compared to conventional lead-containing glasses.

4.3.3 HOST GLASS FOR DIVALENT RARE EARTH: THE NA$_2$O-B$_2$O$_3$-SIO$_2$:SM^{2+} SYSTEM

Fluorescence emission from rare earth ions is widely used for optical devices such as fiber optic amplifiers, laser rods, and light sources for optical telecommunications. It is known that their

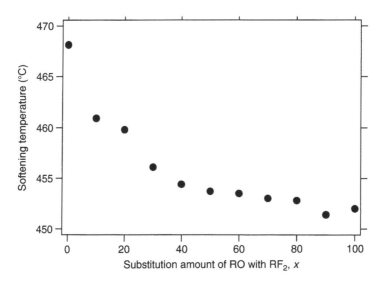

FIGURE 4.19 Softening temperatures of $30B_2O_3$-$40TeO_2$-$30[xBaF_2/(100 - x)BaO]$ glasses as a function of substitution amount of BaO with BaF_2, x.

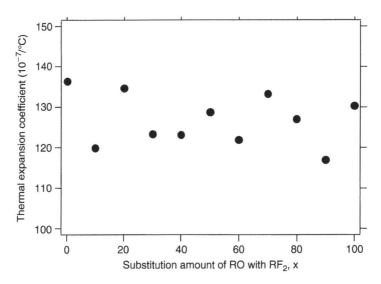

FIGURE 4.20 Thermal expansion coefficients of $30B_2O_3$-$40TeO_2$-$30[xBaF_2/(100 - x)BaO]$ glasses as a function of substitution amount of BaO with BaF_2, x.

valence states produce different emission patterns: for instance, divalent rare earth ions (Ln^{2+}) exhibit broad emission lines from the 5d–4f electron transition, where 5d orbitals are broadened by the surrounding ligand field, while trivalent rare earth ions (Ln^{3+}) show sharper emission lines from the 4f–4f electron transition, where 4f orbitals are shielded by the outer 5s orbitals not to be affected by the ligand fields. It is known that thermal reduction of rare earth in a glass matrix is closely related to composition and local structure.[21] Here, we present a study on thermal reduction properties of samarium ions in various compositions of host soda borosilicate glasses.[22]

Compositions of 43 samples prepared are plotted in Figure 4.22. Each sample was doped with 1 mol% of Sm_2O_3. A 10-g batch for each composition was prepared from reagent-grade

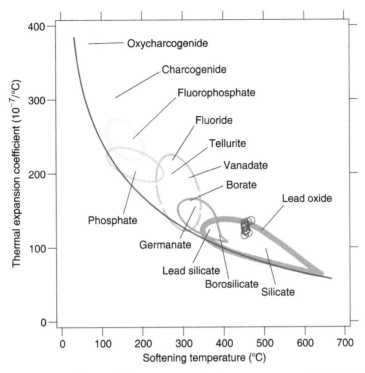

FIGURE 4.21 Thermal expansion coefficients versus softening temperatures of $30B_2O_3$-$40TeO_2$-$30[xBaF_2/(100-x)BaO]$ glasses. Data for other low-melting glasses are reproduced from Konishi et al.[19]

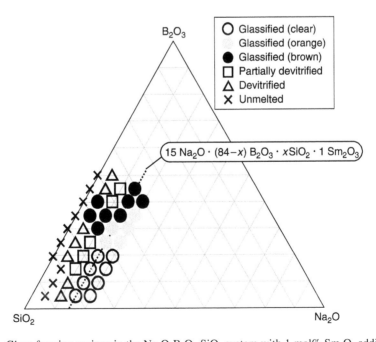

FIGURE 4.22 Glass-forming regions in the Na_2O-B_2O_3-SiO_2 system with 1 mol% Sm_2O_3 addition.

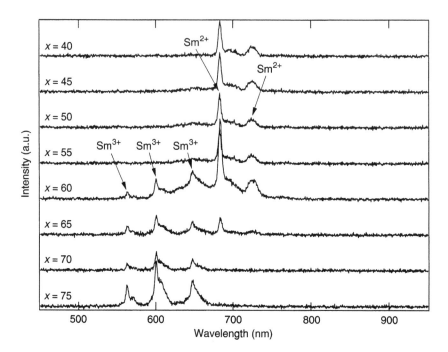

FIGURE 4.23 Fluorescent emission spectra of $15Na_2O \cdot (84 - x)B_2O_3 \cdot xSiO_2 \cdot Sm_2O_3$ (mol% by batch) glasses excited by a mercury lamp.

Na_2CO_3, H_3BO_3, SiO_2, and Sm_2O_3 powders using the automatic batch preparation apparatus. The multisample glass formation tester, which can treat 24 samples at once, was used for melting the batches in a reductive N_2 atmosphere of 0.1 MPa. After being melted at 1350°C for 80 min in graphite crucibles, the melts were cooled to room temperature at the average cooling rate of 10°C/s. The fluorescence emission spectra of the samples were measured under the irradiation of a mercury lamp.

The glass-forming region of 1 mol% of the Sm_2O_3-doped Na_2O-B_2O_3-SiO_2 system is shown in Figure 4.22. The samples exhibited clear to brown colors depending on the composition. Figure 4.23 shows emission spectra of $15Na_2O \cdot (84 - x)B_2O_3 \cdot xSiO_2 \cdot 1Sm_2O_3$ (mol% by batch) glasses. In the region of $40 < x < 50$, the emission peaks attributed to $^5D_0 \rightarrow {}^7F_J$ ($J = 0$, 1, 2) transitions of Sm^{2+} were observed at 685, 700, and 725 nm. At $x = 70$ and 75, the observed emission peaks of 565, 600, and 650 nm were assigned $^4G_{5/2} \rightarrow {}^6H_J$ ($J = 5/2$, 7/2, 9/2) transitions of Sm^{3+}. For the in-between region of $55 < x < 65$, emission peaks by both Sm^{2+} and Sm^{3+} were observed. This result indicates that the thermal-reduction behavior of Sm^{3+} ions significantly depends on the host glass composition. In other words, there are three types of host composition groups where Sm^{3+} ions are totally, partially, or hardly reduced to Sm^{2+}. Figure 4.24 shows the valence-state distribution map of samarium ion with respect to host glass composition. This distribution trend is similar to that of coloration of the samples shown in Figure 4.22. At a glance, Sm^{3+} ions seem to be easily reduced in a Na_2O-poor region and hardly reduced in a Na_2O-rich region, regardless of the SiO_2 content. Cheng et al. report that a rare earth in $BaO \cdot 4B_2O_3$ glass is easily reduced, because the reduced rare earth is prevented from oxygen attack in the cage of the three-dimensional and continuous network structure of the surrounding host glass.[21] According to this model, the rare earth is surrounded by three-dimensional SiO_4 and BO_4 structure units when the B_2O_3 content is rich. On the other hand, when the Na_2O content is rich instead of B_2O_3, the rare earth is coordinated by non-bridging

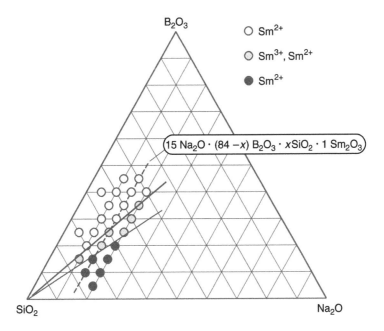

FIGURE 4.24 Ionic valence distribution of samarium ions in the Na_2O-B_2O_3-SiO_2 system. Dashed line traces compositions in Figure 4.17.

oxygen instead of a glass network structure. This offers a simple method of controlling the valence state of samarium ions merely by changing the host glass composition.

4.3.4 FLUORESCENT GLASS: THE WO_3-P_2O_5-ZnO SYSTEM

At first, we started to research alternative X-ray protection glass without containing lead oxide for the purpose of environmental protection. In the course of preparation of new glasses containing tungsten or molybdenum as an alternative X-ray absorber to replace lead, we accidentally discovered that some glass in the WO_3-P_2O_5-ZnO (WPZ) system exhibits fluorescence emission under irradiation of an ultraviolet (UV) lamp. Because it is interesting that this glass does not need special thermal treatment or the addition of an emitting center such as a rare earth and dye, the basic properties were investigated.

Batches for WPZ samples were prepared from reagent-grade WO_3, $NH_4H_2PO_4$, and ZnO by using the automatic batch preparation apparatus. A 5-g batch for each composition was put in an alumina crucible, melted with the multisample glass-melting furnace at 1200°C for 20 min in air, and then quenched on a stainless-steel plate. At the composition where fluorescence was observed, samples were also prepared by melting at 1250 and 1300°C. The fracture surface of the fluorescence-emitting glass was observed with a field emission scanning electron microscope (FE-SEM). Fluorescence spectra were measured on the powder sample with a fluorescence spectrometer (F-4500, Hitachi High-Technologies Co., Japan). Powder X-ray diffraction analysis (LabX XRD-6100, Shimadzu Co., Ltd., Japan) was performed on the sample after heat treatment at crystallization temperature.

The glass-forming region in the WPZ system melted at 1200°C is plotted on the basis of batch composition in Figure 4.25. The glass-forming region is located where the P_2O_5 content is below 60 mol%. Photos of as-obtained samples placed on the WPZ ternary composition diagrams taken

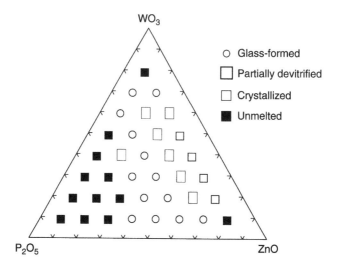

FIGURE 4.25 Glass-forming region of the WO_3-P_2O_5-ZnO system melted at 1200°C on the basis of batch composition.

with or without irradiation by a UV lamp are shown in Figure 4.26. As can be seen from the figure, only $10WO_3$-$20P_2O_5$-70ZnO glass emitted fluorescence on the 10 mol% increment diagram. In order to investigate the emitting compositions more precisely, we prepared samples at the compositions denoted by grid points within a 2.5 mol% interval around the $10WO_3$-$20P_2O_5$-70ZnO composition. It was found that emitting glasses were transparent with a yellowish haze and their compositions were in the region of ±5 mol% around the $10WO_3$-$20P_2O_5$-70ZnO (mol%). Figure 4.27 shows broad emission lines around 650 nm of the $10WO_3$-$20P_2O_5$-70ZnO glasses melted at 1200, 1250, and 1300°C. As the melting temperature increases, the emission intensity decreases and the yellowish haze becomes fainter. Glass melted at 1300°C is almost clear to the eye. The FE-SEM micrograph of the $10WO_3$-$20P_2O_5$-70ZnO glass in Figure 4.28 shows that nanometer-sized fine particles incorporated into the matrix are the cause of the haze. Growth in the particle size was attempted by heat treatment at 560°C, and the crystallization temperature was measured by differential thermal analysis for 25 min and by XRD analysis. The XRD pattern of the glass in Figure 4.29 indicates the presence of $Zn_3(PO_4)_2$ crystals which are used as phosphors. Because glass powder that does not contain a rare earth or dye is low-cost and stable against photobleaching, it has a potential application in fluorescent paints for outdoor use.

4.4 CONCLUSION

In this chapter, the instrumental details of the combinatorial study of new glasses and some newly discovered glasses were described. Particularly, a novel fluorescence-emitting glass without containing a rare earth or dye was accidentally discovered in the WPZ system. Without comprehensive synthesis covering a wide range of compositions by using the combinatorial method, the authors could not easily detect the narrow compositional region of fluorescence emission. This is a good example to show that combinatorial methodology is a powerful tool to increase not only efficiency in the research process but also the chances of serendipity. The authors expect that combinatorial studies on new glasses will greatly increase the entries in glass databases such as Interglad[23] and SciGlass.[24]

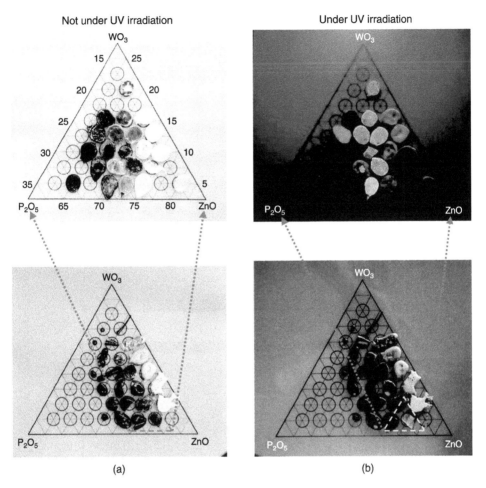

FIGURE 4.26 (See color insert following page 172) Photos of samples placed on WO$_3$-P$_2$O$_5$-ZnO ternary composition diagrams taken (a) without irradiation of UV lamp and (b) under irradiation of UV lamp.

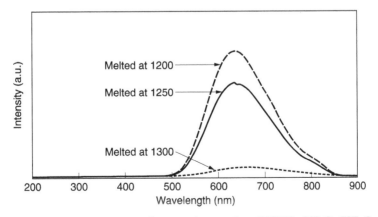

FIGURE 4.27 Fluorescence spectra measured on powder samples of 10WO$_3$-20P$_2$O$_5$-70ZnO glass melted at 1200, 1250, 1300°C by excitation at 330 nm.

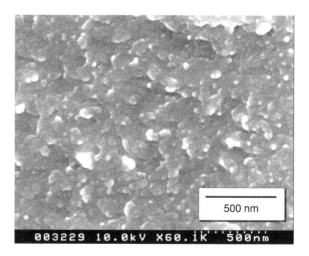

FIGURE 4.28 FE-SEM micrograph of fracture surface of $10WO_3$-$20P_2O_5$-$70ZnO$ glass.

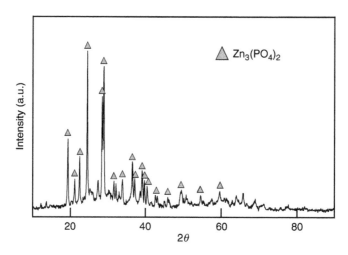

FIGURE 4.29 Powder X-ray diffraction pattern of $10WO_3$-$20P_2O_5$-$70ZnO$ glass after heat treatment at 560°C for 25 min.

REFERENCES

1. New Glass Forum (http://www.ngf.or.jp/).
2. Zachariasen, W. H., The Atomic Arrangement in Glass, *J. Am. Ceram. Soc.*, 54, 3841, 1932.
3. Sun, K. H., Fundamental Condition of Glass Formation, *J. Am. Ceram. Soc.*, 30, 277, 1947.
4. Stanworth, J. E., On the Structure of Glass, *J. Soc. Glass Technol.*, 30, 54T, 1946; ibid., 32, 154T, 1948; ibid., 32, 366T, 1948.
5. Inoue, S., MacFarlane, D. R., Molecular dynamics study of glass formation in the system ZrF_4-BaF_2-$BaCl_2$-MCl_2 (M = Mg, Ca, Sr), *J. Non-Cryst. Solids*, 95, 585, 1987.
6. Inoue, S., Todoroki, S., Matsumoto, T., Hondo, T., Araki, T., Watanabe Y., Combinatorial methodologies for determination of glass-forming region, *Appl. Surf. Sci.*, 189, 327, 2002.
7. Inoue, S., Todoroki, S., Konishi, T., Araki, T., Tsuchiya, T., Combinatorial glass research system, *Appl. Surf. Sci.*, 223, 233, 2004.

8. Matsumoto, T., Inoue, S., Todoroki, S., Hondo, T., Araki, T., Watanabe, Y., The apparatus for automatic batch melting in the crucibles, *Appl. Surf. Sci.*, 189, 234, 2002.
9. Koinuma, H., Kawasaki, M., Combinatorial technology (in Japanese), Maruzen, Tokyo, Japan, 2004.
10. Inoue, S., Todoroki, S., Matsumoto, T., Hondo, T., Araki, T., Tsuchiya, T., *Mat. Res. Symp. Proc.*, 700, 201, 2002.
11. Mizuno, Y., *New Glass* 6(3), 258, 1991.
12. Hasegawa, Y., Kawakubo, S., *Glastech. Ber.*, 30, 332, 1957.
13. Inoue, S., Shimizugawa, Y., Nukui, A., Maeseto, T., Thermochromic property of tellurite glasses containing transition metal oxides, *J. Non-Cryst. Solids*, 189, 36, 1995.
14. Konishi, T., Hondo, T., Araki, T., Nishio, K., Tsuchiya, T., Matsumoto, T., Suehara, S., Todoroki, S., Inoue, S., Investigation of glass formation and color properties in the P_2O_5–TeO_2–ZnO system, *J. Non-Cryst. Solids*, 324, 58, 2003.
15. Konishi, T., Hondo, T., Araki, T., Nishio, K., Tsuchiya, T., Matsumoto, T., Suehara, S., Todoroki, S., Inoue, S., *Mat. Res. Symp. Proc.*, 754, 465, 2003.
16. Imaoka, M., Yamazaki, T., *J. Ceram. Soc. Jpn.* (Yogyo-Kyokai-Shi), 76(5), 32, 1968.
17. Muruganandam, K., Seshasayee, M., Structural study of $LiPO_3$–TeO_2 glasses, *J. Non-Cryst. Solids*, 222, 131, 1997.
18. Ikeo, N., Iijima, Y., Nimura, N., Sigematsu, M., Tazawa, T., Matsumoto, S., Kojima, K., Nagasawa, Y., *Handbook of X-ray Photoelectron Spectroscopy*, JEOL, Tokyo, 1991.
19. Konishi, T., Matsumoto, T., Araki, T., Tsuchiya, T., Todoroki, S., Inoue, S., Optimization of host glass composition to make soda borosilicate glasses doped with reduced rare earth ions, *Appl. Surf. Sci.*, 223(1–3), 238, 2004.
20. Yamane, M., *New Glass*, Japanese Standards Association, Tokyo, Japan, 1989, p. 151.
21. Zheng, Q., Pei, Z., Wang, S., Su, Q., Lu, S., *Mat. Res. Bull.*, 34, 1837, 1999.
22. Konishi, T., Suehara, S., Todoroki, S., Inoue, S., *Proc. XX International Congress on Glass*, sr007H00313DIS, 2004.
23. International Glass Database (Interglad), *New Glass Forum*, Tokyo, Japan (http://www.ngf.or.jp/).
24. SciGlass, ESM Software, Inc., Ohio (http://www.esm-software.com/sciglass/).

5 A Combinatorial Method for Optimization of Materials for Gas-Sensitive Field-Effect Devices

Mats Eriksson, Roger Klingvall, and Ingemar Lundström

CONTENTS

5.1 INTRODUCTION

Gas sensors are needed in situations where knowledge of the chemistry in a certain atmosphere is required. Analytical instruments, such as mass spectrometers and gas chromatographs, exist which can be used to obtain very detailed information about chemical composition, but in many applications a small, cheap, and mobile solution is needed. This is where gas sensors come in. Perhaps the first commercially available gas sensor was the miner's safety lamp from the beginning of the 20th century, which was initially intended for safe illumination of mines. The metal net that surrounded the flame prevented it from igniting gas in the mine. An additional feature was that the flame of the safety lamp went out when the concentration of oxygen was low. By suitably reducing the flame of an ordinary safety lamp it could be used for gas detection.[1] A further development of the lamp could be used for more accurate gas detection. The concentration of combustible gases could be indicated by observing the height of the cap above the flame.[2] The chemical state of the air in the mine was thus converted to an optical signal that was interpreted by the human eye and brain.

Today gas sensors convert chemical information into an electrical signal. This is useful since the electrical signal can then be recorded and it can also be used for real-time control purposes in an application. Several different detection principles have been developed for chemical sensing.[3] The first commercially successful modern gas sensor was a Japanese invention, a semiconducting metal oxide, which changed its conductivity in the presence of gas leakage from gas stoves.[3] Another example is that ion-conductive materials can be used to measure the lambda value (i.e., the air-to-fuel ratio divided by the air-to-fuel ratio for stoichiometric conditions) in car exhausts.[3] The sensor signal is used to optimize the air-to-fuel mixture in the car engine in order to give good working conditions for the catalytic converter. Other types of gas sensors are capacitive, optical, and mass-sensitive gas sensors.[3]

One type of gas sensor which has a good sensitivity and selectivity to hydrogen is the gas-sensitive field-effect transistor, sometimes called the chemFET or the gasFET.[4–6] Even though the commercially available gas-sensitive field-effect devices are produced as transistors, a basic device that is often used for laboratory studies is a simple metal–insulator–semiconductor (MIS) structure, i.e., a MIS capacitor. This device is schematically shown in Figure 5.1. The hydrogen detection principle for this device is based on the following three steps:

1. Dissociation of the hydrogen molecule on the surface of a catalytic metal (such as Pd, Pt, Ir, etc.)
2. Transport of the so-formed hydrogen atoms through the metal film
3. Trapping at the metal insulator interface where a polarization occurs which can be detected through the electrical properties of the device

Also shown in Figure 5.1 is the "back reaction" where adsorbed hydrogen atoms react with adsorbed oxygen atoms in the catalytic water-forming reaction. This reaction determines the hydrogen-sensitive concentration region in air, which for a Pd-SiO_2-Si device typically is 1–1000 ppm. For concentrations higher than 1000 ppm a saturation of the response occurs.

By changing the properties of the catalytic film (such as choice of metal, porosity of the metal film, and operating temperature) the selectivity to different gases changes. For example, by choosing a porous Pt film a sensitive and selective ammonia sensor is achieved.[7,8]

It has been realized that the properties of the gas-sensitive layer are best optimized if the metal at the metal–insulator interface and the metal at the surface are optimized separately. Thus, by forming

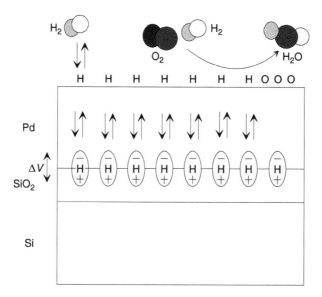

FIGURE 5.1 The hydrogen response mechanism of a MIS device. Hydrogen molecules dissociate on the metal surface. Hydrogen atoms diffuse through the metal film and are trapped at the metal–insulator interface where a polarization occurs, which can be detected through the electrical properties of the device. Also shown is the catalytic water-forming reaction, between adsorbed hydrogen and oxygen atoms, which has a large influence on the hydrogen response.

a metal double layer, one metal with preferred properties towards the gas phase and one metal with good interfacial properties towards the insulator, we have the freedom to optimize the two interfaces independently. One problem which turns out to be critical here is to choose the right thickness of these two layers. This is where the combinatorial method comes in. In order to optimize the thickness of the two layers we need:

1. A device where the thicknesses of the two layers are varied independently
2. A method to read out the gas response for all different thickness combinations

The next section describes how this can be achieved.

5.2 COMBINATORIAL METHOD

5.2.1 Gas-Sensitive Devices with Continuously Varying Properties in Two Dimensions

A new approach in the optimization of a double metal layer is to vary the thickness continuously along one direction for one of the metals and in the orthogonal direction for the second metal layer, which is grown on top of the first.[9] Before going into the two-dimensional case we start with an explanation of how a one-dimensional layer with a thickness gradient can be produced (see Figure 5.2). The metal layer is grown by evaporation in a vacuum chamber. A metal source is heated by bombarding it with an electron beam. Due to the high temperature, metal atoms desorb from the metal source and stick to the first surface they strike. The substrate onto which the metal layer is grown (an oxidized silicon wafer in this case) is placed above the metal source and a fixed shadow mask (not shown in Figure 5.2) defines the area where metallization occurs. To produce a thickness gradient we put a second, movable shutter in front of the substrate. By moving this shutter with constant speed in one direction a metal film with a linear thickness gradient is obtained.

Now that a linear thickness gradient of one metal has been grown it is straightforward to continue by growing a second layer on top of the first layer with orthogonal direction. We simply rotate the substrate 90° and grow the second layer on top of the first layer in exactly the same way as the first layer, but now with a different metal. In this way we achieve a device where each point along the surface has a unique combination of thicknesses. In one corner both metals will be thin, in another corner the bottom layer is thick and the top layer is thin, vice versa in a third corner and

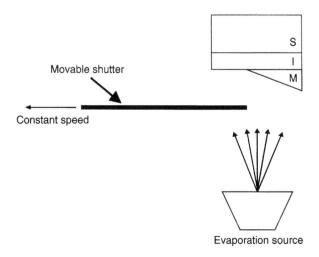

FIGURE 5.2 Growth of a linear thickness gradient by means of evaporation in vacuum and a movable shutter.

finally in the fourth corner both metal layers are thick. In between, all possible thickness combinations are present.

5.2.2   READOUT OF THE GAS RESPONSE WITH LATERAL RESOLUTION

In order to study the gas response with a lateral resolution along the one-dimensional and two-dimensional thickness gradients described in the previous section, a technique that measures the gas response locally at a certain point of the device is needed. The technique we use is called the scanning light pulse technique (SLPT) and was initially used to study insulator–semiconductor interfaces with lateral resolution.[10] It was realized that the technique was also well suited for field-effect gas sensor studies at the beginning of the 1990s.[11]

There are different versions of the technique, but the simplest version is to use a pulsed light beam, e.g., produced by a laser diode where the light intensity can be modulated (either electrically or with a mechanical shutter).[10–12] The light is focused and directed onto the sample. If the metal film is not too thick (in our case less than 100 nm), enough light intensity is transmitted through the metal layer and the insulator layer and ends up in the semiconductor. If the semiconductor is p-doped and the bias voltage on the metal layer is positive, a depletion layer is present in the semiconductor surface layer. The light generates electron–hole pairs that are separated in this depletion layer. If the light intensity is pulsed, an electric current pulse appears in the outer electric circuit (see Figure 5.3). The charge content of this current pulse increases with increasing depth of the depletion layer and therefore with increasing positive bias voltage as shown in Figure 5.4. To be more precise, the position of the current–voltage curve along the voltage axis is dependent not only on the applied voltage, but also on the work function difference between the metal layer and the semiconductor. When hydrogen atoms are trapped at the metal–insulator interface the polarized layer that is induced will effectively act as a change of the metal work function. This means that the current–voltage characteristics will shift along the voltage axis (in negative direction) when hydrogen is trapped at the interface. This shift is the same as that measured with gasFETs[5,12] only that now we have a MIS capacitor instead of a transistor.

The important difference between the gas-induced shift measured with the SLPT and that measured with a gasFET is that with the SLPT the *local gas response* is measured at the point where the light beam illuminates the device. The reason for this is that with SLPT the detected signal originates only from the illuminated spot (and not from the whole gate area as for the gasFET). Therefore by scanning the light (or the sample), a map of the gas response variation along a device in one or two dimensions can be produced.

Figure 5.5 shows the gas response to ammonia and hydrogen for a one-dimensional thickness gradient of (a) Pd and (b) Pt. As can be seen the gas response is clearly dependent on both the metal film thickness and on the choice of metal. Without trying to explain all the observations that can be made in these diagrams we merely emphasize a few of them.

The large difference in hydrogen response between Pt and Pd for the thick parts of the films has been shown to be due to differences in the catalytic activities on the metal surfaces.[13] The barrier for OH formation is lower on Pt than on Pd, which contributes to a higher hydrogen concentration on the Pd surface than on the Pt surface and therefore a higher concentration also at the metal–insulator interface and therefore a higher response.

The thickness dependence for the ammonia response along the Pt thickness gradient correlates very well with the variation of the porosity of the metal film, which has been observed by scanning electron microscopy studies.[7] In the thin end a large fraction of pores are present and their concentration gradually decreases as the Pt film gets thicker. It is believed that this correlation is due to a response mechanism where ammonia molecules dissociate at the "triple points," where both catalytic metal, insulator surface, and gas phase are present.[7] The free hydrogen atoms that are formed can

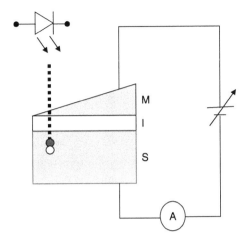

FIGURE 5.3 Light-induced current. Light from a laser diode is pulsed and focused on the metal surface of a MIS device. Part of the light reaches the semiconductor where electron–hole pairs are formed in the depletion region. The separation of these charges results in a current pulse in the outer circuit that can be measured.

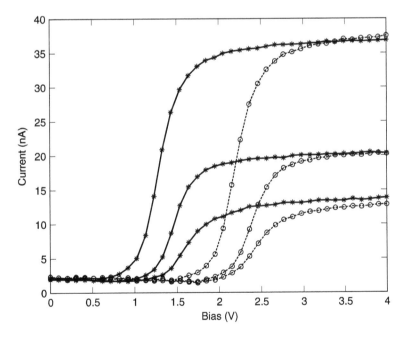

FIGURE 5.4 Current–voltage characteristics for a MIS device with a metal thickness gradient. The open circles are measurements performed in synthetic air. Measurements have been taken for three different thicknesses giving different saturation currents. The different values of the saturation current are due to decreasing light intensity reaching the semiconductor for increasing metal film thickness. The curves labeled with stars are measured with 1000 ppm hydrogen in synthetic air. The gas response is the voltage shift of the curves that occurs in the hydrogen-containing atmosphere and is of the order of 1 V for all three curves in this case.

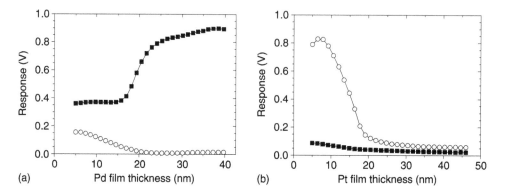

FIGURE 5.5 Response to 250 ppm hydrogen (filled squares) and to 250 ppm ammonia (open circles) for MIS devices with thickness gradients of (a) Pd and (b) Pt.

then diffuse along the metal–insulator interface and are detected in the same way as described for the hydrogen detection mechanism.

The situation becomes a little more complicated when the metal is in the form of a double layer. In the following section we will present results from two-dimensional thickness gradients, but will not attempt to fully explain the differences for different thickness combinations, simply because they are not yet fully understood. We will instead show the potential of the technique and can already state that there are several reasons why it is important to study these double layers. First of all, what we want to do is to optimize both thicknesses of the double metal layer. The two-dimensional layer combined with the SLPT gives us the opportunity to screen a very large number of possible combinations by using only one single device. This reduces the sample production effort quite considerably. Secondly the problems with batch-to-batch differences that are often observed for these kinds of devices are reduced, since all combinations were produced in a single run of only one component. Furthermore, several properties vary with metal film thickness, such as the catalytic properties, which vary along a metal thickness gradient due to differences in the microstructure. The porosity also varies as well as the stability of the metal film. Therefore the two-dimensional thickness gradient spans a parameter space that has a large potential impact on the gas response.

One important property of the light beam in an SLPT measurement is its width (or diameter), because this will in many cases limit the lateral resolution, i.e., the smallest pixel size, that can be obtained in a measurement. We have studied the width of the light beam by scanning it over a step in metal thickness and measuring the corresponding light pulse-induced current. The derivative of this signal gives a curve with a peak (see Figure 5.6), and the full width at half maximum (FWHM) of this peak is used as a measure of the beam width. A beam width of about 40 µm is obtained in this way. This is not the narrowest beam that can be obtained. In principle it should be possible to produce a beam width close to the wavelength if such a resolution is needed. It is important to note that, depending on the experimental situation, other properties than the beam diameter may limit the lateral resolution. The wavelength and the intensity of the light, the diffusion of charge carriers in the semiconductor and the diffusion of the chemical species to be detected may also limit the lateral resolution that can be obtained with this technique.[9]

5.3   THE COMBINATORIAL METHOD APPLIED TO DOUBLE LAYERS OF Rh + Pd AND Pt + Pd AS GAS-SENSITIVE LAYERS

Figure 5.7 shows gas response images for a MIS device with a double metal layer exposed to (a) hydrogen and (b) ammonia. The double metal layer is a thickness gradient of Rh in one direction with a thickness gradient of Pd, in the orthogonal direction, on top. The metal thickness is indicated

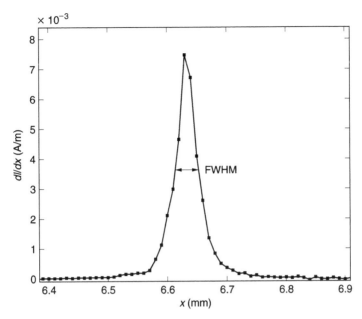

FIGURE 5.6 The current, measured as the light beam is scanned across a step in metal thickness, has been differentiated. The full width at half the maximum value (FWHM) is used as an estimation of the diameter of the light beam.

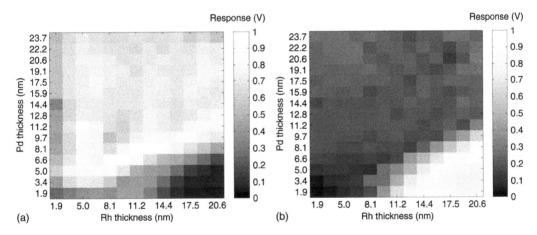

FIGURE 5.7 Response images for a double metal layer, Rh + Pd, with varying metal film thicknesses. The response is grayscale coded as indicated by the bars next to the images. (a) 500 ppm hydrogen and (b) 500 ppm ammonia in synthetic air. $T_{sensor} = 140°C$.

along the two axes. It is striking how complementary the two response images are. The lower right corner shows a high ammonia response and a low hydrogen response. The rest of the image shows the opposite: high hydrogen response and low ammonia response. This can be seen as a first test of the selectivity of the sensor (that should of course be complemented with response images for other gases). The selectivity is a measure of the sensor's ability to respond to a certain gas but not to other gases. Related to this is also the cross-sensitivity, which gives information about how stable the gas response is in the presence of other gases. If we are looking for a selective ammonia sensor we should choose a thickness combination from the lower right corner of Figure 5.7. The rest of the image

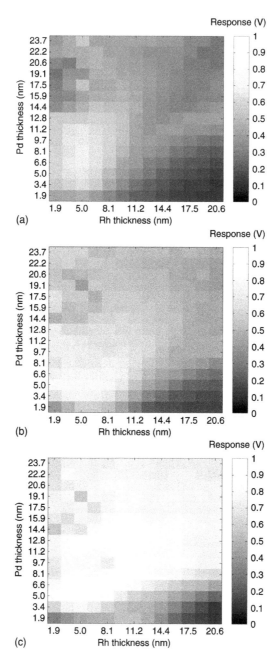

FIGURE 5.8 Response images for a double metal layer, Rh + Pd, for (a) 10 ppm, (b) 100 ppm, and (c) 1000 ppm hydrogen in synthetic air. $T_{sensor} = 140°C$.

shows a large hydrogen response, indicating greater freedom to choose the optimum thickness combination for a hydrogen sensor.

Another important gas sensor property is the (differential) sensitivity. This is defined as $\partial(\Delta V)/\partial C$ where ΔV is the response and C is the concentration of the test gas. A qualitative measure of the sensitivity can be obtained from Figure 5.8, where response images for hydrogen concentrations of 10, 100, and 1000 ppm hydrogen in synthetic air are shown. Figure 5.8a (10 ppm)

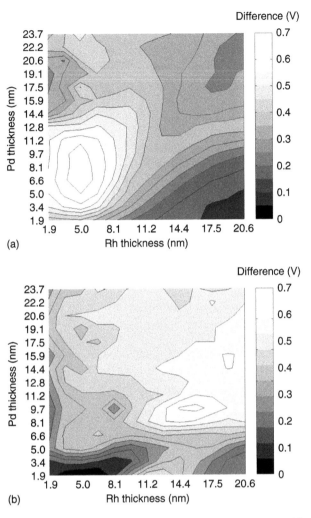

FIGURE 5.9 Difference images shown as contour plots. (a) Difference between 10 and 0 ppm hydrogen. (b) Difference between 1000 and 100 ppm hydrogen.

shows that a thickness combination from the lower left corner of the image is sensitive at low concentrations. Therefore a low detection limit can be expected in this region. The sensitivity at other concentrations is difficult to discern from these images. Therefore difference images have also been plotted in Figure 5.9. In this case the images are plotted as contour plots where Figure 5.9a is the contour plot of Figure 5.8a giving the difference between 10 and 0 ppm hydrogen. Again the maximum sensitivity is, of course, found in the lower left corner. Figure 5.9b shows the difference between 1000 and 100 ppm hydrogen. Here the maximum sensitivity is found in the mid right part of the image. This shows that the thickness combination with maximum sensitivity varies with concentration.

Another issue when dealing with gas sensors is the stability of the gas response. Figure 5.10a shows the same hydrogen response image as in Figure 5.8c (1000 ppm). Figure 5.10b shows the response image to the same hydrogen concentration after the MIS device has been treated to an accelerated aging process at elevated temperature. The response image looks quite different after the aging process, in particular the lower right corner has changed from a rather insensitive region to a much more sensitive region. (It should be noted that the response image of Figure 5.10a was measured with a device without any previous annealing.)

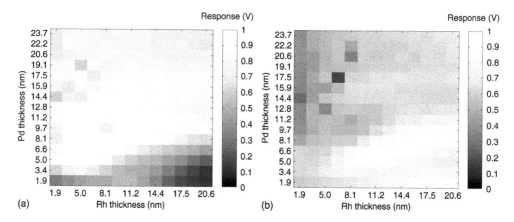

FIGURE 5.10 Stability test. Response to 1000 ppm hydrogen (a) before and (b) after an accelerated annealing.

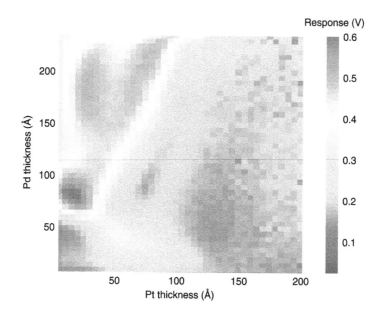

FIGURE 5.11 (See color insert following page 172) Response image for a double metal layer, Pt + Pd, with varying metal film thicknesses (Pd on top) and for 100 ppm hydrogen. $T_{sensor} = 140°C$.

Finally, the resolution of the response images presented so far is much lower than the light beam allows for. Figure 5.11 shows the response image to 100 ppm hydrogen for a Pt + Pd double layer. This image contains about 2000 pixels as compared with about 200 for the previous images. A remarkable improvement of details and sharpness is obtained. It can also be observed that some of the features from the one-dimensional measurements in Figure 5.5 can be recognized in the corresponding areas of Figure 5.11.

5.4 CONCLUDING REMARKS AND PERSPECTIVE

The samples used in all images presented here have been rather large, 16×16 mm^2, which means that it has not been necessary to use the highest possible resolution achievable for the system. We have been working with 100 pixels per cm^2 for the Rh + Pd images and 900 pixels per cm^2 for the Pt + Pd image. The beam width as measured in Figure 5.6 (about 40 μm) indicates that it is straightforward to reach a pixel density of 60 000 pixels per cm^2 if that is needed. As indicated above, 40 μm is by no means the ultimate beam width diameter. A beam diameter of 1 μm should be possible to achieve, resulting in 100 million pixels per cm^2. This will, however, put extreme demands on the scanning of the SLPT in terms of measurement speed per pixel.

A related possibility is that the physical size of the samples can be drastically reduced without losing any detail in the images. If change in film thickness of 2 pixels per nm, as in the Pt + Pd image (Figure 5.11), is considered acceptable, the size of the sample could be reduced from 16 × 16 mm^2 to about 2 × 2 mm^2 for a 40-μm light beam or less than 0.1 × 0.1 mm for a beam width around 1 μm. The price that has to be paid for this reduction in size is that it puts a very high demand on the metallization and on the mechanical stability of the SLPT setup.

The use of combinatorial methods for gas sensors in general and for field-effect gas sensors in particular is presently at a very early stage.[9,14,15] In the case of field-effect devices we can, for example, foresee the possibilities to study different degrees of alloying in an alloy gradient, different degrees of porosity for a well-defined "porosity gradient," different degrees of surface additives in a "surface modification gradient," and different operating temperature in a temperature gradient. The potential to use combinatorial methods for optimization and also for gas response mechanism studies is obvious and hopefully will be further explored in the near future.

REFERENCES

1. Clowes, F., A New portable miner's safety-lamp with hydrogen attachment for delicate gas testing; with exact measurements of flame-cap indications furnished by this and other testing lamps, *Proc. R. Soc. Lond.*, 1892–1893, 52, 484.
2. Clowes, F., An apparatus for testing the sensitiveness of safety-lamps, *Proc. R. Soc. Lond.*, 1891–1892, 50, 122.
3. Göpel, W., Hesse, J., Zemel., J. N., Eds., *Sensors – A comprehensive survey*, vol. 2, VCH Verlagsgesellschaft, Weinheim, Germany, 1991.
4. Lundström, K. I., Shivaraman, M. S. and Svensson, C. S., A hydrogen sensitive Pd-gate MOS transistor, *J. Appl. Phys.*, 46, 3876, 1975.
5. Lundström, K. I., *Sensors*, Vol. 2, VCH Verlagsgesellschaft, Weinheim, Germany, 1991, p. 467.
6. http://www.appliedsensor.com
7. Löfdahl, M., Utaiwasin, C., Carlsson, A., Lundström I. and Eriksson, M., Gas response dependence on gate metal morphology of field-effect devices, *Sens. Actuators B*, 80, 183, 2001.
8. Spetz, A., Helmersson, U., Enquist, F., Armgarth, M., Lundström, I., Structure and ammonia sensitivity of thin platinum or iridium gates in metal-oxide-silicon capacitors, *Thin Solid Films*, 177, 77, 1989.
9. Klingvall, R., Lundström, I., Löfdahl, M. and Eriksson, M., A combinatorial approach for field-effect gas sensor research and development, *IEEE Sensors Journal*, 5, 5, October 2005.
10. Engström, O. and Carlsson, A., Scanned light pulse technique for the investigation of insulator-semiconductor interfaces, *J. Appl. Phys.*, 54, 5245, 1983.
11. Lundström, I., Erlandsson, R., Frykman, U., Hedborg, E., Spetz, A., Sundgren, H., Welin, S., Winquist, F., Artificial 'olfactory' images from a chemical sensor using a light-pulse technique, *Nature*, 352, 47, 1991.
12. Löfdahl, M. and Lundström, I., Monitoring of hydrogen consumption along a palladium surface by using a scanning light pulse technique, *J. Appl. Phys.*, 86, 1106, 1999.
13. Löfdahl, M., Eriksson, M., Johansson, M. and Lundström, I., Difference in hydrogen sensitivity between Pt and Pd field-effect devices, *J. Appl. Phys.*, 91, 4275, 2002.
14. Taylor C. J. and Semancik, S., Use of microhotplate arrays as microdeposition substrates for materials exploration, *Chem. Mater.*, 14, 1671, 2002.
15. Aronova, M. A., Chang, K. S., Takeuchi, I., Jabs, H., Westerheim, D., Gonzales-Martin, A., Kim, J., Lewis, B., Combinatorial libraries of semiconductor gas sensors as inorganic electronic noses, *Appl. Phys. Lett.*, 83, 1255, 2003.

Section 2

Catalysis: Development and Discovery

6 Use of Combinatorial Heat Treatments to Accelerate the Commercialization of Materials for Use in Catalysis

Maureen Bricker, Lorenz Bauer, Michael Gatter, Jennifer Abrahamian, and Ralph Gillespie

CONTENTS

6.1 INTRODUCTION

The Council for Chemical Research has recognized the need for a methodology that can increase catalyst innovation while continuing to decrease cycle times (Catalysis Roadmap, Vision 2020).[1]

In the 1990s, UOP researchers believed that the adaptation of combinatorial science to the chemical industry, through the creation of tools and methods, would increase laboratory productivity on the same order of magnitude as was reported in the pharmaceutical industry in the 1980s. We are not alone in this belief, and we and others in both academic and industrial environments have worked to create a combinatorial toolbox for use in the invention of materials, catalysts, and adsorbents.[2–29]

99

SINTEF was the first company in the world[18] to perform combinatorial hydrothermal synthesis of molecular sieves, which are materials used extensively as a basis for catalysts and adsorbents in the chemical industry. The partnership between SINTEF and UOP began in 1997.[30–33] This alliance has developed a range of combinatorial test platforms and tools for synthesis of materials and catalysts. These tools apply to the following four areas:

1. Material synthesis, including parallel inorganic synthesis
2. Catalyst preparation, including metal loading and finishing materials in either oxidation, oxygen/chlorine, and/or hydrogen
3. Catalyst evaluation, including probe reaction assays for catalyst acidity, basicity, metal function (dehydrogenation), and bifunctional (acid/metal) activity and commercial assays for propane dehydrogenation
4. Informatics, including flexible data management, analysis and visualization, mining and predictive performance modeling

The approach that was co-developed with SINTEF was to design the preparation and testing systems on a modular basis. As much as possible, these synthesis and characterization components are run with a high degree of automation. This process has been described elsewhere,[33,37] and the program has also been assisted by the funding of NIST.[43] It is important to have a balanced capability to make and analyze the property of interest at the same rate, so that no bottlenecks occur.

Today, many UOP projects have used the combinatorial approach in some stage of research and development.[34–36,38] The tools of combinatorial chemistry can be applied in an early stage to help select materials, they can be used later in development to determine the sensitivity of a variable that will influence the large-scale manufacture, and they can be used to help diagnose commercial problems. In this work, three case studies have been chosen to show that these tools apply to practical problems, such as selection of a supplier's material in a process, the research goal of developing an application for a new material in catalysis, and the optimization of a material when a material structure-type is desired. In these case studies, the steaming capability of the heat treatment unit (HTU) is used.

6.2   CATALYST PREPARATION MODULE

The combinatorial preparation and evaluation of catalysts has some unique challenges that set it apart from other applications of combinatorial chemistry. Catalysts are extremely sensitive to the preparation scheme used for their production. Catalyst preparations can involve five to ten steps or more. Each step can have a number of key input variables. The net result is millions of different possibilities for even a simple catalyst problem. Even with the higher preparation and testing rates brought about by applying high-throughput methods, experiments must be carefully designed to efficiently evaluate a subset of the possible experimental space. The catalysis expertise of the experiment designer is an important factor in determining the success of the combinatorial catalysis effort. Many books, articles, and software are available to guide the experimenter in the statistical design of experiments.[29]

Catalyst preparations are often scale- and equipment-dependent. A catalyst that works well when prepared in small-scale equipment may perform poorly when prepared in commercial-scale equipment. It is important to have high-throughput preparation methods that will successfully translate to the methods used in large-scale preparations. Care must be taken during the development of high-throughput equipment to account for the scale effect. Likewise the scale-down of difficult preparations to small preparations must be done so as to make this practice of small-scale preparations acceptable. The translation between laboratory preparations and commercial production is important to UOP, where we invent, develop, and commercialize materials for use in adsorbents and catalysts.

6.3   HEAT TREATMENTS: AN OVERVIEW

Heat treatment is a critical unit operation that is investigated repeatedly throughout the steps of a material commercialization. This chapter will present an overview of the tools developed for parallel heat treatment of materials. Treatment of materials in various gas environments as a function of temperature is critical to obtaining an optimum catalyst formulation. These tools are commonly used at UOP. A few case studies that show how the HTU applies to the commercialization work process will also be presented.

The schedule of heating a catalyst in a gas environment is an extremely important area of material development, because this is where the primary chemistry takes place. Depending on the conditions chosen for heat treatment, many fundamental changes in the material can occur, such as evaporation of water and removal of volatile compounds used in the formulation, crystallographic phase transformations, loss of surface area, and binder hardening. The surface of a material is created during the heat treatments. The final chemistry of the surface is related to the number of defect sites, the level of hydroxyl group, and the metal dispersion. In the case of a zeolite, the heat treatment chosen can alter the Si/Al ratio in the zeolite structure, change the unit cell size, and affect the crystallinity of the zeolite. All of these changes can influence the targeted catalytic process. The catalytic transformations and adsorption preferences for polar and non-polar molecules are created on the surface of the material.

A dynamic environment exists on the surface of a material at high temperature. Hydroxyl groups form and react, mobile cations react with an oxide layer, the oxide layer anneals, micropores can be created or lost, and molecules from the firing atmosphere of oxygen, chlorine, hydrogen, and combustion by-products interact chemically with surface atoms. The strength, the brittleness, and the fracture rate can all be influenced by the methods chosen for the finishing of a material. Understanding and controlling the heat treatment of a material is important to the successful commercialization of all catalysts and adsorbents.

6.3.1 HEAT TREATMENT OF MOLECULAR SIEVES FOR ACID CATALYSTS

A zeolite is a crystalline, porous aluminosilicate. The metal atoms are surrounded by four oxygen anions to form an approximate tetrahedron consisting of a metal cation at the center and oxygen anions at the four apexes. The tetrahedrons can stack in regular arrays, and channels form. These materials are also known as molecular sieves because the channels that are formed can allow some molecules to flow into and out of the channels while larger molecules are excluded. Many structure forms are known, and a handful of zeolites is useful for catalysis.

In catalysts with a zeolite component, the zeolitic thermal chemistry can cause changes in alumina content of the zeolite structure and stability of the phase. The final composition and properties of the catalyst are determined by the treatments in a gas environment as a function of time and temperature. Acidic molecular sieves have found wide application in the oil-refining and petrochemical industries, particularly in conversion processes involving carbocation mechanisms.[39] The most common zeolites used for acid-catalyzed reactions are Y, MFI, and mordenite. The structures of these common zeolites are shown in Figure 6.1, Figure 6.2 and Figure 6.3. Many sources are available to view and evaluate the diversity of zeolite structures.[40]

Typically, an acidic zeolite is not used by itself as a catalyst. In a catalyst formulation, the zeolite is bound with another material (alumina, silica) in an intimate mixture, which is formed into a hardened material. A typical development goal would be to find the proper level of zeolite, the best binding materials, and the optimum conditions of heat treatment (time, temperature, gas environment) that produce a formulation that can be commercialized.

6.3.2 HEAT TREATMENT OF ALUMINA-BASED MATERIALS

Alumina is a common support material. However, the thermal decomposition of various alumina hydrates can result in a number of metastable crystalline variations of alumina, which will eventually form corundum or α-alumina upon further thermal exposure. The conditions chosen for the heat treatment of alumina will affect the phase, the pore structure, and the surface area of the

FIGURE 6.1 Faujasite structure.

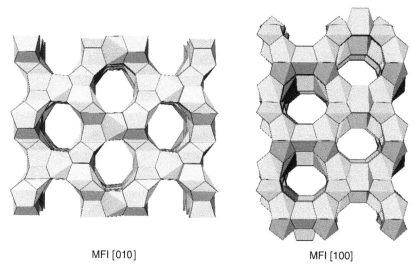

MFI [010]   MFI [100]

FIGURE 6.2 MFI structure.

support. A phase diagram of the changes of alumina as a function of starting material is shown in Figure 6.4. Other sources of information about the phase changes of alumina have been published by Alcoa.[41,42]

In a catalyst formulation, the interaction between the binder and the zeolite are important. Both components can change, and they can affect one another. Sometimes a binder must be changed if there is an adverse interaction with the zeolite. In most cases, even if the thermal stability of one component of a catalyst is known, there needs to be a new study if a new component is added. For instance, a binder that is known to work well with one zeolite will need to be fully re-evaluated for use with another.

6.3.2.1 Test Design

Once a catalyst project has been initiated and the preparation methods identified, the researcher must begin to design the experimental space that will be examined. The variables can include the major catalyst components, their concentration ranges, the reagents that will be used, the order in

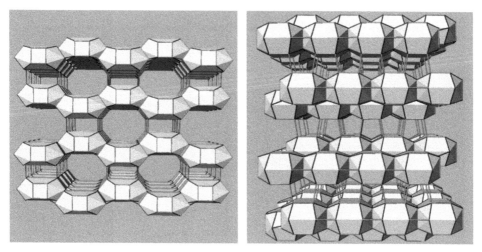

FIGURE 6.3 Mordenite structure.

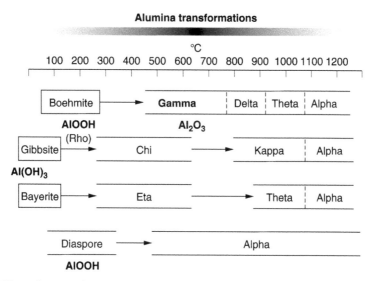

FIGURE 6.4 Phase diagram of the transition aluminas.

which experimental steps will be carried out, and which of the many possible heat treatments will be chosen for study.

6.3.2.2 Using Heat Treatments for Follow-Up Studies

Sometimes when the initial development of a new catalyst has been completed, and a prototype has been developed, additional studies can be done to predict the performance of the catalyst in a real process. Experimentation can reveal information about stability of phases, surface area of the support, and the dispersion of the metal on the support. That knowledge can reveal further information about the longevity of the catalysts and suggest the type of engineering process that will be best for a particular formulation. For instance, if a catalyst is developed that shows mild deactivation with certain types of aging, one might consider a fixed-bed catalytic process for that material. Depending

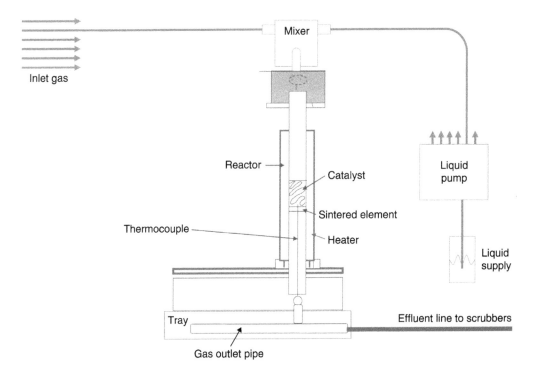

FIGURE 6.5 Overview of the heat treatment system.

on the cost of the catalyst, it may need two or three regeneration processes in its expected lifetime. On the other hand, if the catalyst shows that it deactivates rapidly but can be restored in activity, a different catalytic process will be designed.

6.4   THE HEAT TREATMENT UNIT

At UOP, we have developed a flexible HTU capable of performing experimental studies in which time, temperature, gas composition, and flow studies can be independently varied. A schematic of one reactor within one row of the reactor array is shown in Figure 6.5. The gases come into a mixing zone and can be combined with a liquid stream. The mixed gas enters the top of the reactor. If a fluidized bed is desired, the mixed gas can be introduced at the bottom of the reactor. The top and bottom of the reactor are held in a Teflon block, and the reactor is sealed at its top and bottom. The reactor has a fritted material at the bottom of the isothermal zone; the material to be treated sits on the frit. There is a thermocouple within the reactor to monitor and control temperature. An individually controlled heater surrounds the reactor. The effluent gas is collected in a common line and sent to scrubbers.

The modular arrangement of the heat treatment that is used is shown in Figure 6.6. The reactor is placed within a heater. The reactor/heaters are placed on as big a grid as is required for the experiment.

This system has been validated for drying, calcination, oxidation, oxygen/chlorine, hydrogen reduction, and steaming unit operations. The unique benefit of a heat treatment unit able to treat materials under a variety of conditions is the acceleration of a project – the ability to conduct many treatments at once. If the samples generated in a heat treatment study can then be evaluated in a high-throughput assay or performance test, the experimenter has a powerful tool for material development.

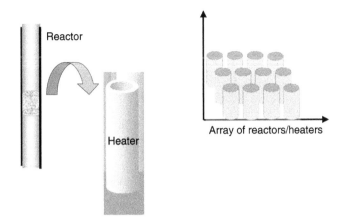

FIGURE 6.6 Modular arrangement of the heaters and reactors in the heat treatment unit.

6.5 EXAMPLES OF USING PARALLEL HEAT TREATMENTS TO SOLVE COMMERCIALLY RELEVANT PROBLEMS

The examples that follow show how parallel methods can be used to solve a wide variety of practical problems encountered in commercial research and development in addition to their use in the synthesis of new materials.

6.5.1 CASE STUDY I: FINDING THE OPTIMUM FINISHING TEMPERATURE FOR A PROTOTYPE CATALYST MATERIAL

Case study I is a study to evaluate the steam stability of a catalyst prepared with two different suppliers' zeolite. This material had a certain Si/Al ratio in the fresh state. The material was studied after it had been bound with silica at the targeted recipe. The final prototype formulation was tested after various steaming conditions. The factors in the experimental design were time, temperature, and steam. Five levels of each variable were examined. The time ranged from 12 to 200 h, the steam varied from 0 to 10 wt%, and the temperature varied from 325 to 610°C. The specific ranges are shown in Figure 6.7. In this study, two catalyst responses were observed: NMR framework alumina and alumina activity of the samples from a combinatorial high-pressure test. The overall plan of the experiment and the work strategy are shown in Figure 6.7.

The objective of the study was to determine which conditions of steaming caused irrevocable deactivation of the catalysts. All of the catalysts were prepared at 1-ml scale. The analysis: Test performance (model reaction) and the changes in the Si/Al ratio were measured with nuclear magnetic resonance (NMR). The steamed materials were analyzed in the solid NMR by performing magic angle spinning (MAS) to measure the ^{27}Al nuclei. During this analysis, the samples were packed in a rotor and spun, and the signal was acquired for 1 h. When aluminum was removed from the framework of the zeolite, the pattern changed. For example, Figure 6.8 shows the Al spectra from Supplier 1's material where the steam level was held constant at 6 mol% and the temperature increased. Both the intensity of the signal and the distribution of the peaks change as the temperature of steaming is increased.

The NMR analytical response data were received on the entire set of samples, and the analysis shows the amount of Al in the framework of the zeolite.

A plot of the change in framework alumina by NMR for the entire experimental design is shown in Figure 6.9, where the data are shown in a variable plot, made using JMP software. The first-level

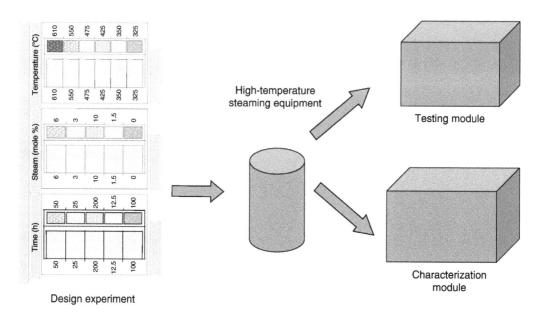

FIGURE 6.7 Case study I: Application of the HTU for a UOP project.

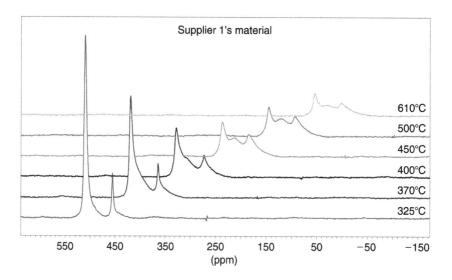

FIGURE 6.8 Case study I: Changes in ^{27}Al MAS NMR signal with increasing temperature (6% steam).

variable is the steam level, so there is a major division of the five steam levels in the x-direction. For each steam level, there is a division of the data at each temperature, and the temperature data are divided into two time periods. Time 1 is 4 h and Time 2 is 24 h. Showing the results in this way allows a great deal of data to be displayed, and the consistency of the trends at each steam level can be observed. The data show that as the temperature increases, the amount of Al in the framework decreases. The data also show that as the steam level is increased, the amount of Al in the framework is decreased.

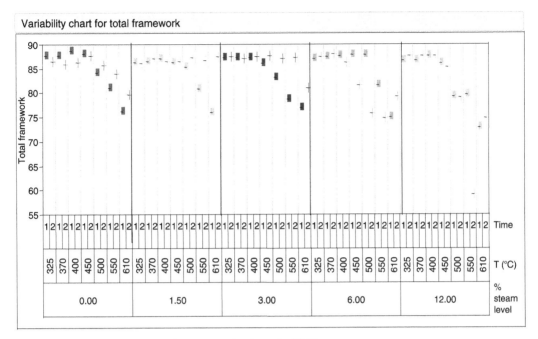

FIGURE 6.9 Case study I: Total framework aluminum by NMR.

A parallel test unit was used to evaluate the performance of the steamed samples. Figure 6.10 shows the response for the conversion of toluene at 400°C, at 25 psig. This plot shows the difference in performance of catalysts made with two different suppliers' zeolite. The data show that conversion drops with increasing bed temperature. There is also an effect of decreasing conversion with increasing steam levels. This is consistent with a drop in the aluminum in the framework of the zeolite. One supplier's product was certified for use in the catalyst and the other supplier was not, based on the difference in steam stability that was shown in the high-throughput screening test.

One goal of this work was to validate the steaming approach, and we worked with a group that had a few of the experimental points for the first supplier's materials. In the HTU work, the steam stability was evaluated over a wide range of conditions. The HTU data agreed with the previous studies – it duplicated their original data and generated a wider response surface for the evaluation of the material. When the second supplier provided their material for evaluation, the HTU was used to generate the second set of samples. Material 1 was compared to material 2 on the same basis.

6.5.2 CASE STUDY II: EVALUATION OF 37 NEW ACIDIC MATERIALS FOR APPLICATION

Case study II was conducted to determine which of 37 new acidic materials should be targeted for further work in scaleup. For this study, 627 steamed samples were generated and analyzed. The factors in the DOE were time, temperature, and steam. Four temperature levels, three steam levels, and two time levels of each variable were explored. The same design was applied to each material. The specific ranges are shown in Figure 6.11. The responses for this study were phase retention and crystalline intensity from high-throughput X-ray diffraction (XRD). One of the most important characteristics of crystalline solids is their regular and repeatable arrangement of atoms. All crystalline materials have distinct peaks in their XRD patterns, and peak intensity is directly related to the crystallinity of the material. Changes in peak intensity can be measured after each treatment. The retention or loss of peak intensity shows the critical conditions for a material and help determine the failure mode of a material.

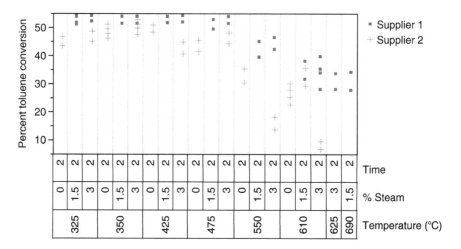

FIGURE 6.10 Case study I: Microreactor activity of catalysts prepared with different precursors.

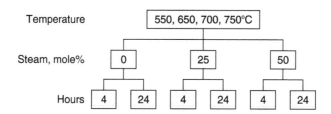

FIGURE 6.11 Case study II: Steaming design to analyze zeolite hydrothermal stability.

The preparation of the samples for XRD involved equilibrating the samples overnight in a 50% humidity chamber. Groups of samples were then run on a GADDS instrument through either 4–38° or 15–48° 2-theta, depending on the structure of the starting material. Each material was collected for 2 min counting time.

The first analysis was to take each material family and evaluate the normalized spectrum to determine peak intensity retention of the sample after an XRD scan. The pattern of the steamed material was compared to the pattern of the parent material. Figure 6.12 shows two types of materials: Figure 6.12a shows a material that lost crystallinity with steaming and Figure 6.12b shows a material that retained the crystalline phase upon steaming. The results of this primary screen showed which materials were robust.

The relative intensities of the stable materials were then compared across the steaming conditions. A map of the crystallinity retention, or intensity, as function of temperature and steam level is then generated. For example, intensities are plotted for a series of zeolites with the same framework structures but different Si/Al ratios (see Figure 6.13). The data show that the material with the lowest Si/Al ratio is the least thermally stable, as it has the greatest crystallinity loss over the calcination space.

Based on the crystallinity differences, only the more thermally stable higher Si/Al ratios from this evaluation were then tested in traditional pilot plants (Figure 6.14). Two materials from this study were evaluated against reference materials. One material was below the reference performance, and one material was better than the reference. The better of these two materials has been selected as a candidate for commercialization.

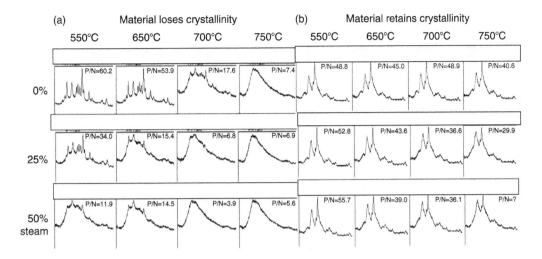

FIGURE 6.12 Case study II: Primary XRD screen to evaluate crystallinity retention.

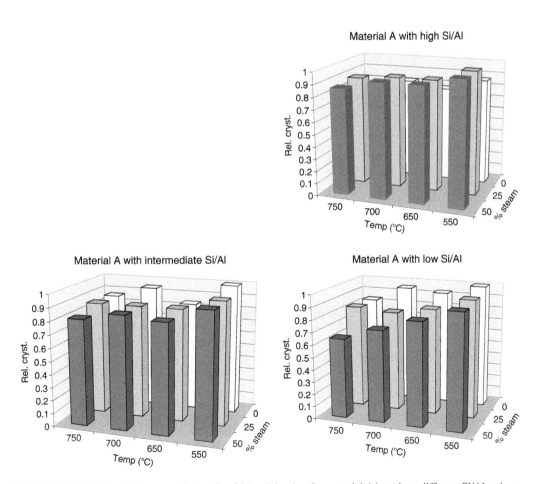

FIGURE 6.13 XRD relative crystallinity after 24-h calcination for material-14 made at different Si/Al ratios.

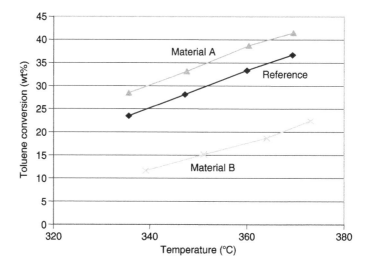

FIGURE 6.14 Test performance for the more hydrothermally stable material-14.

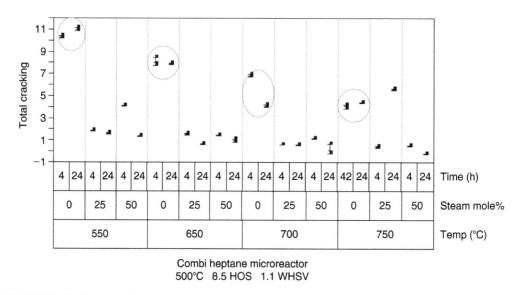

Combi heptane microreactor
500°C   8.5 HOS   1.1 WHSV

FIGURE 6.15 Heptane microreactor screening test to evaluate acid strength of steamed materials.

There was also a parallel screening test to evaluate the acidity of selected stable materials after treatment. The heptane cracking test was used to determine whether the acid sites had been changed with the steaming. One material that did not show great XRD changes with steaming was evaluated for cracking activity. These data are shown in Figure 6.15. For this data set, the plot shows that the material loses the activity if steam levels of 25 mol% or greater are applied, even though the XRD pattern shows that the material is thermally stable.

6.5.3   CASE STUDY III: EVALUATION OF SYNTHESIS METHODS TO FIND A MORE STABLE ZEOLITE

Case study III shows the use of the HTU to evaluate different synthesis methods to produce a more stable zeolite material. Usually, a particular zeolite structure is chosen for a targeted application. The

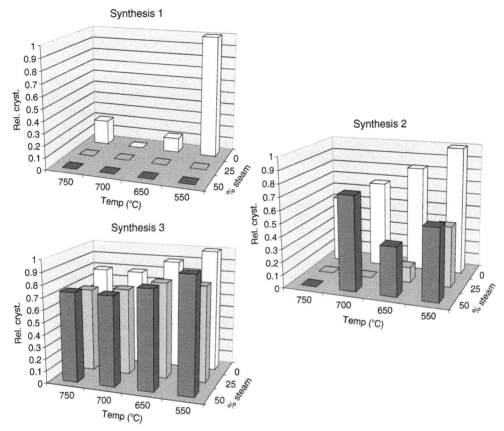

FIGURE 6.16 XRD relative crystallinity after 24-h calcination for material-4 made via different synthetic routes.

reasons can vary, but often a particular channel size is desired for an application. Sometimes, the initial synthesis is successful and the material has promise for a particular application, but the material does not show the desired thermal or hydrothermal stability. In that case, the experimenter will try alternate synthesis methods or a change in the Si/Al composition in synthesis. In this case study, both the synthesis organic templating agent and the Si/Al ratio were varied. Figure 6.16 shows the XRD intensity map as a function of calcination conditions for three materials prepared with the same structure. These data show that the hydrothermal stability of the structure can be controlled with a coupling of the synthesis method and the Si/Al ratio.

6.6 CONCLUSION

The HTU has been used in a wide variety of programs to accelerate studies of materials, and has allowed better mapping of both the thermal and the hydrothermal performance of materials. This unit has great utility at the early stages of research to find basic properties of new material discoveries. The power of the technique is enhanced if there is a corresponding fast analysis tool to prevent bottlenecks from developing. The HTU can also be used to solve very practical commercial problems. It has been used to determine the optimum finishing of a prototype material, to guide the formulation procedure for manufacturing, and to help establish tolerances on process targets in production. It has been used for the certification of raw materials to be used in manufacturing. In addition to oxidation and steaming studies, the unit has been demonstrated for reduction work and studies of impure gas streams. The HTU has been used for a diverse range of studies and new applications are expected.

When a system for parallel treatment of the material is coupled with a high-throughput analytical technique, the full impact of the experiment comes back to the experimenter in a timely way. The full experiment can be interpreted, and in many cases the data can be modeled. In the past, using the traditional work process, a steaming study might have involved ten samples. All ten would be analyzed, but the sequential nature of the study with competing priorities would make the study hard to complete. In the HTU the full DOE can be done in a few days, the analyses can be done on the entire set at the same time, and the experimenter can get reactivity data on the full set. Now the entire study can be presented and understood by a greater audience.

ACKNOWLEDGMENTS

We would like to acknowledge the support of NIST for the development of the combinatorial techniques at UOP (ATP Cooperative agreement number 70NANB9H3035). We would like to thank the UOP XRD group, Judy Triphahn and Andrzej Ringwelski and the UOP NMR group of Sesh Prabakhar, Linda Laipert, and Al LeComte. We also thank the dedicated work of Susan Koster for the preparation of materials and both Charles McGonegal and Larry Ranes in the area of combi reactor testing. We thank Steve Wilson for providing images of different zeolite structures for use in this publication and James Rekoske and Robert Larson for providing the materials and the technical discussion on case study I.

REFERENCES

1. American Chemical Society, American Institute of Chemical Engineers, Chemical Manufacturers Association, Council for Chemical Research, and Synthetic Organic Chemical Manufacturers Association. Technology Vision 2020 "The U.S. Chemical Industry" December 1996.
2. Mittach, A. and Bosch, C., United States Patent 993,144.
3. Hanak, J. J., The effect of grain size on the superconducting transition temperature of the transition metals, *Phys. Lett.*, 30A(3), 201, 1969.
4. Hanak, J. J. and Bolker, B. T. F., Calculation of composition of dilute cosputtered multicomponent films, *J. Appl. Phys.*, 44(11), 5142, 1973.
5. Gittleman, J. I., Cohen, R. W. and Hanak, J. J., Fluctuation rounding of the superconducting transition in the three dimensional regime, *Phys. Lett.*, 29A(2), 56, 1969.
6. Hanak, J. J., The "Multiple-Sample Concept" in materials research: synthesis, compositional analysis and testing of entire multicomponent systems, *J. Material Sci.*, 5, 964, 1970.
7. Hanak, J. J. and Yocum, P. N., DC-electroluminescent flat panel display, *Govt. Rep. Announce. (U.S.)*, 73(21), 128, 1973.
8. Hanak, J. J., Electroluminescence in ZnS: Mnx: Cuy rf-sputtered films, *J. Proc. 6th Intern. Vacuum Congress Japan, J. Appl.Phys, Suppl* 2(1), 809, 1974.
9. Xiang, X.-D., A combinatorial approach to material discovery, *Science*, 268, 1738, 1995.
10. Moates, F. C., Infrared thermographic screening of combinatorial libraries of heterogeneous catalysts, *Ind. Eng. Chem. Res.*, 35, 4801, 1996.
11. Burgess, K. and Porte, A. M., Accelerated synthesis and screening of steroselective transition metal complexes, *Advance. Catalytic Proc.*, 2, 69, 1997.
12. Danielson, E., A Rare-Earth phosphor containing one-dimensional chairs through combinatorial methods, *Science*, 279, 837, 1998.
13. Hideomi, K. and Nobuyoki, M., Combinatorial chemistry of inorganic materials, *J. Chem., (Japan)*, 4, 70, 1998.
14. Reddington, E., Combinatorial electrochemistry: a highly parallel, optical screening method for discovery of better electrocatalysts, *Science*, 280, 1735, 1998.
15. Hoffman, C., Wolf, A. and Schuth, F., Parallel synthesis and testing of catalysts under nearly conventional testing conditions, *Angew. Chem. Int. Ed.*, 38(18), 2800, 1999.

16. Klein, J., Combinatorial material libraries on the microgram scale with an example of hydrothermal synthesis, *Angew. Chem. Int. Ed.*, 38(24), 3369, 1998.
17. Orschel, M., Detection of reaction selectivity on catalyst libraries by spatially resolved mass spectrometry, *Angew. Chem. Int. Ed.*, 38(18), 2791, 1999.
18. Akporiaye, D. E., Dahl, I. M., Karlsson, A. and Wendelbo, R., Combinatorial approach to the hydrothermal synthesis of zeolites, *Angew. Chem. Int. Ed.*, 37(5), 609–611, 1998.
19. Senkan, S., High-throughput screening of solid-state catalyst libraries, *Nature*, 394, 350, 1998.
20. Senkan, S., High-throughput testing of heterogeneous catalyst libraries using array microreactors and mass spectrometry, *Angew. Chem. Int. Ed*, 38(18), 2794, 1999.
21. Ozturk, S. and Senkan, S., Discovery of new fuel-lean NO reduction catalyst leads using combinatorial methodologies, *Appl. Catal. B*, 38(3), 243–248, 2002.
22. Senkan, S. and Ozturk, S., Discovery and optimization of heterogeneous catalysts by using combinatorial chemistry, *Angew. Chem. Int. Ed.*, 38(6), 791, 1999.
23. Senkan, S., Combinatorial heterogeneous catalysis – A new path in an old field, *Angew. Chem. Int. Ed.*, 40, 312, 2001.
24. Jandeleit, B., Combinatorial materials science and catalysis, *Angew. Chem. Int. Ed.*, 38, 2494, 1999.
25. Rodemerck, U., Parallel synthesis and fast screening of heterogeneous catalysis, *Topics in Catalysis*, 13(3), 249, 2000.
26. Choi, K., Combinatorial methods for the synthesis of aluminophosphate molecular sieves, *Angew. Chem. Int. Ed.*, 38(19), 2891, 1999.
27. Bein, T., Efficient assays for combinatorial methods for the discovery of catalysts, *Angew. Chem. Int. Ed.*, 38(3), 323, 1999.
28. Basaldella, E. I., Kikot, A. and Tara, J. C., Effect of aluminum concentration on crystal size & morphology in the synthesis of a Na/Al zeolite, *Materials Letters*, 31, 83–86, 1997.
29. Montgomery, D. C., *Design and Analysis of Experiments*, 4th ed, John Wiley and Sons, New York, 1997, p. 604.
30. Lewis, G. J., Mixed alkali templating in the Si/Al=3 and 10 systems: a combinatorial study, *Stud. Surf. Sci. Catal.*, 135, 597, 2001.
31. Akporiaye, D., Combinatorial chemistry – the emperor's new clothes, *J. Microporous Mesoporous Materials*, 48(1–3), 367, 2001.
32. Holmgren, J. S., Application of combinatorial tools to the discovery and commercialization of microporous solids: facts and fiction, *Stud. Surf. Sci. Catal.*, 135, 461–470, 2001.
33. Plassen, M., Automating a combinatorial hydrothermal synthesis and characterization, *NIWeek*, 2001, Austin, Texas, 2001.
34. Nayar, A., Laser-activated membrane introduction mass spectrometry for high-throughput evaluation of bulk heterogeneous catalysts, *Anal. Chem.*, 74(9), 1933, 2002.
35. Chan, B., The split-pool method of synthesis of solid state material combinatorial libraries, *J. Combi. Chem.* 4(6), 569, 2002.
36. Kuechl, D. E., Optimization of material synthesis using combinatorial methods, *Abstracts, 222nd ACS National Meeting, Chicago*, 2001, American Chemical Society, Washington D. C.
37. Bricker, M. L., Scaling-up of catalysts discovered from small scale experiments, in: *High Throughput Analysis: A Tool for Combinatorial Material Science*, Potyrailo, R. A., Amis, E. J., Eds., Kluwer Academic/Plenum Publishers, New York, NY, November 2003, Ch 26.
38. Bricker, M. L., Strategies and applications of combinatorial methods and high throughput screening to the discovery on non-noble metal catalyst, *App. Surf. Sci.*, 223(1–3), 109–117, 2004.
39. Meyers, R. A., *Handbook of Petroleum Refining Processes*, McGraw-Hill Inc., Mexico, 1986.
40. Baerlocher, C. and McCusker, L. B., Database of zeolite structures: http://www.iza-structure.org/databases.
41. Wefers, K. and Bell, G. M., Oxides and hydroxides of aluminum, *Technical Paper No. 19*, Aluminum Company of America, Pittsburgh, PA, 1972.
42. Hart, L. R. D., *Alumina Chemicals: Science and Technology Handbook*, The American Ceramics Society, 1990.
43. ATP award: Cooperative Agreement # NANB9H303

7 Infrared Thermography and High-Throughput Activity Techniques for Catalyst Evaluation for Hydrogen Generation from Methanol

Eduardo E. Wolf, Stephen Schuyten, and Dong Jin Suh

CONTENTS

7.1 INTRODUCTION

Combinatorial methods and high-throughput experimentation (HTE) have resulted in successful discoveries of drugs[1-3] and new materials.[4-7] Many laboratories in industry and academia have started to apply combinatorial methods to catalytic research.[7-18] Interestingly, one of the first applications of high-throughput techniques was precisely in the field of catalysis. In a review article written in 1950,[19] A. Mittash relates that in the search for an ammonia synthesis catalyst "One of my associates developed a laboratory scale reactor which makes it possible to work easily at pressures of 100 odd atmospheres, a reactor of which we soon have two dozen and more in continuous operation." The year was 1909! Mittash wrote that due to the lack of knowledge in the field, "a purely empirical search for suitable catalysts has to be employed ..., and that he and his associates had to carry out about 20,000 small scale tests and to investigate 3000 preparations as potential catalysts" completed in a period of 2 years – nearly 100 years ago! It is interesting that in the recent rediscovery of the high-throughput methodology only in a few articles is it recognized that such an empirical approach was the norm in industrial research in the past. In fact, it has been the authors' experience that parallel reactors are ubiquitous in industrial laboratories. The new twist in this area is the miniaturization of the reactors and advances in analytical detection systems that have given the field new impetus. In 1986 there was an article describing a parallel microreactor[20] that preceded many of the parallel reactor systems reported in the last 5 years. It is also interesting that some researchers report the use of high-throughput techniques to test 1000 catalysts a day without much consideration of pre-existing knowledge, whereas those who invented HTE lamented that the lack of pre-existing knowledge forced them to proceed in such a brute force approach. While the success of Mittash's work was truly remarkable, we should not forget that it was all initiated by Haber's insight into thermodynamics that led him and Mittash to realize that high pressure was the key to increasing the yield. So, even though the knowledge of catalysis was almost non-existent at the time, thermodynamics was sufficient to guide the experimentalists in the proper path of discovery.

Successful application of combinatorial and HTE methods to heterogeneous catalysis requires careful consideration not only of material composition, but also of the surface structure. The surface structure of metals supported on porous substrates is difficult to determine a priori, and it depends critically on the method of preparation and often on the reaction environment. In addition, catalytic activity is seldom additive, and simple combinations of materials do not always lead to the expected improvements. Combinatorial studies based in scanning the activity of subgroups of elements or compounds need some guidance rules to design new families of catalysts based on either pre-existing results or results generated in a preliminary search. Some authors advocate the use of computer algorithms such as neural networks or evolutionary algorithms to formulate new catalysts.

In our group we use HTE to accelerate the acquisition of results, but instead of relying on an algorithmic approach for the selection of a new generation of catalysts, we, as well as many other groups, follow a knowledge-based approach. Starting with an exhaustive study of the existing literature, a reaction model is synthesized and used to select a family of materials to be tested. Figure 7.1 shows schematically this inverted pyramid approach that we feel is an efficient way to utilize HTE methodologies in an academic setting. The key that differentiates the knowledge-based approach from the purely empirical approach is the model-based starting catalyst library. This increases the certainty of a "hit" but decreases the extent of the search. Extensive high-throughput studies can lead to the generation of vast amounts of results that need to be analyzed by computer methods. In the knowledge-based-medium throughput approach, new formulations are based on some understanding of structural factors leading to higher activity and selectivity. In our case, kinetics results and Monte Carlo simulations are used to analyze the effect of operational and structural variables in the reaction rates.[21,22] This approach bridges both the power of HTE with the power of today's catalysis know-how, thus decreasing the search space significantly.

In the knowledge-based approach the first level of experimentation is the use of an in situ spatially resolved technique, such as infrared thermography (IRT), to test a descriptor of catalytic activity of an array of up to 50 different catalysts. The second level of experimentation is to use a parallel reactor that can simultaneously test the activity of ten samples selected from the best activities observed in the IRT studies. Finally, the best catalyst is studied in a single-flow fixed-bed recycle reactor to get the reaction rate parameters. This catalyst is characterized to determine its bulk and surface structure. The knowledge gained from all these results is used to formulate a revised model of the surface and the reaction. The revised model is used in a second experimental iteration to optimize the catalyst. Strictly speaking this is a medium-throughput approach rather than HTE.

The IRT technique used in our work was first introduced at Notre Dame by Pawlicki and Schmitz.[23] Our group previously used IRT to display the formation of temperature patterns during auto-oscillations occurring during CO oxidation on Rh/SiO_2-supported catalysts to show that they depended on the catalyst preparation.[24,25] Moates et al.[26] employed IRT to study an array made of 16 metal-supported catalysts that were prepared by a conventional method of impregnating pellets of γ-Al_2O_3 with aqueous solutions of metal salt precursors. Holzwarth et al.[27] used infrared thermography to evaluate the catalytic activity of an array with 37 transition metal-supported catalysts

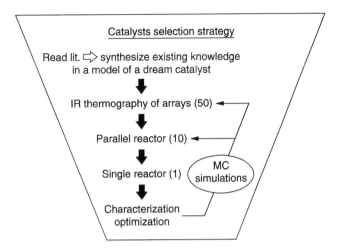

FIGURE 7.1 The proposed inverted pyramid experimental approach.

for the hydrogenation of hexane and the oxidation of isooctane and toluene. The parallel reactor used in the second stage of experimentation was designed and developed by ISRI Inc.[28] and tested by our group. The recycle reactor is a standard tubular quartz reactor with a diaphragm pump installed in a recycle loop. A flow meter and a precision valve are used to regulate the recycle flow rate and maintained at a ratio of 20 with the single-pass flow.

In this chapter we describe the first iteration in our studies of catalytic pathways to generate hydrogen from methanol. The use of hydrogen as a localized energy source for fuel cell-powered vehicles or small electronic devices is a topic of current interest. Fuel cells are efficient, with zero point of use emissions of NO_x, CO, volatile organic compounds (VOCs), and particulate matter. Hydrogen can be extracted from sources such as natural gas, water, biomass, or other more complex hydrocarbons. Despite the numerous advantages that fuel cells provide, there are still significant difficulties with the use of hydrogen, including transportation, storage, and handling. An attractive alternative solution to problems associated with storing molecular hydrogen involves on-board catalytic production of hydrogen from a high-energy liquid fuel such as methanol.[29]

Table 7.1 lists the main methanol-reforming reactions along with the corresponding heats of reactions. The first reaction (Equation 1) is the basic methanol decomposition reaction yielding hydrogen and carbon monoxide. This reaction is not suitable for fuel cell use because proton exchange membrane (PEM) cells require hydrogen feed containing less than 20 ppm CO.[29] Steam reforming (Equation 2) has the highest hydrogen to carbon ratio but is highly endothermic, not suitable for applications where a heat source is unavailable or bursts of energy may be needed. Partial oxidation (Equation 3) is exothermic, with a higher reaction rate than steam reforming and reduced tendency to form CO.[30] Finally, combined methanol reforming (Equation 4) is a combination of steam reforming and partial oxidation. This reaction offers a better balance between hydrogen to carbon ratio and heat of reaction, but is still endothermic. Given the choices, partial oxidation of methanol reaction seems to be a good candidate for on-board hydrogen generation. Development of an effective partial oxidation–reforming catalyst seems possible given what is known in the literature mostly involving Cu as the main component. A good catalyst will operate at low temperatures, resist deactivation, have high hydrogen and carbon dioxide selectivity, low water and CO by-product formation, and be cost effective.

As stated above, there are four overall reaction schemes involved in reforming methanol to hydrogen. While this research will focus on the partial oxidation reaction, products that are reactants in other reactions are also involved in the oxidative decomposition reaction network such as the water

TABLE 7.1
Methanol Reforming Reactions (All Heats of Reaction in kJ/mol)

Reaction	ΔH^{*}	ΔH^{**}	$\Delta H(l)/H_2$	H_2/C	Eqn.
$CH_3OH \rightarrow 2H_2 + CO$	91	128	64	2	1
$CH_3OH + H_2O \rightarrow 3H_2 + CO_2$	50	131	44	3	2
$CH_3OH + 1/2O_2 \rightarrow 2H_2 + CO_2$	−192	−155	−77	2	3
$4CH_3OH + 1/2O_2 + 3H_2O \rightarrow 11H_2 + 4CO_2$	−44	238	22	2.75	4

* All species are gas phase.
** Includes heat of vaporization for CH_3OH and H_2O.

gas shift reaction. For example, Peppley et al.,[31,32] when studying the steam-reforming reaction network showed that three reactions, including water gas shift and methanol decomposition, must be considered to represent the results accurately. Similarly, Cubeiro and Fierro,[33] in their study of methanol partial oxidation proposed that several reactions participate including combustion, decomposition, steam reforming, and water gas shift. Clearly, all four overall reforming reactions are closely linked, and must be considered when studying the partial oxidation reaction.

Methanol decomposition (Equation 1) to form hydrogen and CO is the most elementary methanol-reforming reaction. The most common catalysts used for this reaction are copper, nickel, and palladium. Catalyst compositions which have been previously studied include Pd on $CeZrO_2$, La_2O_3, SiO_2, ZrO_2, CeO_2, Pr_2O_3, and TiO_2,[33,34] various Cu catalysts with Zn, Ni, Cr, Mn, ZnO, Si, Ba, Mn, CuO and MnO on activated carbon,[35] Ni on various supports,[36,37] and Pt supported by CeO_2.[38] Alkali metals, especially sodium and potassium, increased the dispersion and surface area of copper metal in the catalyst. The promoted copper in the catalyst yielded an increase in conversion and selectivity for methanol decomposition.

The reforming reaction with the highest H_2/C ratio is steam reforming (Equation 2). Unfortunately, it is also the most endothermic, requiring energy to break the methanol and water apart, and vaporize the fuel mixture. Similar to methanol decomposition catalysts, the majority of steam-reforming catalysts include Cu along with Zn and Zr.[39,40] Since CO is not a primary product in the methanol steam-reforming reaction, minimizing CO formation could allow use of this reforming reaction in fuel cell systems. Iwasa et al.[41] reports of a tradeoff between high-selectivity Cu/ZnO catalysts, which deactivate at temperatures above 573 K, and group 8–10 metals, which exhibit good thermal stability, but poor selectivity. With a Zn modified Pd on CeO_2, activated carbon catalysts and use of co-precipitation catalyst preparation, he reports both high activity (>1000 cm^3/g/h), selectivity (99%), and resistance to deactivation.

The partial oxidation of methanol (Equation 3) has been studied over a variety of catalyst compositions including Pd on ZnO or ZrO_2,[33,42] Cu/ZnO,[43] Cu/ZnO/Al$_2$O$_3$,[44,45] and other transition metals.[46] Catalyst compositions with Cu/Zn/Zr/Pd show high activity for partial oxidation. In general, group VIII noble metals are active in methanol reforming, although they are more likely to produce CO via methanol decomposition (Equation 1) than traditional Cu/ZnO catalysts. Cubeiro and Fierro[33] reported hydrogen selectivity of 96% at a methanol conversion of 70% over a Pd catalyst supported by ZnO. However, CO selectivity was still high (>20%) under the reaction conditions used. In contrast, CO selectivity of less than 10% is reported over Cu-containing catalysts. These concentrations of CO are still several orders of magnitude higher than what is needed (<20 ppm) for PEM fuel cell operation.

Catalysts studied for the combined methanol-reforming reaction are very similar to those studied in steam and oxidative reforming, including Cu/Zn/Al/Zr-oxide catalysts. Velu et al.[47] reported very good results on a Cu/Zn/Zr catalyst showing 80% conversion with CO_2 selectivity at 99.8% with only traces of CO detected (<1000 ppm). None of the catalysts reported so far, however, including Cu/ZnO, the traditional catalyst for methanol synthesis, is robust enough for on-board hydrogen production in a variety of conditions. In this work, we have prepared a series of Cu/Zn/Pd and Cu/Zn/Pd/M catalysts with different compositions and metal combinations by the co-precipitation method and evaluated their reaction characteristics by the selective combinatorial catalysis experimentation.

Clearly this is a complex multivariable problem that, if studied in a sequential approach, will take many man-hours to unravel in a meaningful way. Consequently, we propose to use the medium-throughput approach to quickly evaluate the activity and selectivity of CuZnO catalysts. The reaction model consists of a dual catalyst system: a multicomponent oxide to promote the methanol conversion reactions and a noble metal to promote the partial oxidation reactions.

7.2 EXPERIMENTAL METHOD AND PROCEDURE

7.2.1 IRT REACTOR

The experimental system employed in IRT studies has been described elsewhere for research involving the preferential oxidation of CO (PROX).[48,49] It includes electronic flow controllers to provide gases at specific flow rates, temperature control, the IR-cell reactor and the IR camera (AGA Thermovision 782). A new IR reactor, shown in Figure 7.2, has an array containing up to 50 wells and is commercially available.[28] In addition to inlets and outlets for the continuous flow of gases, the reactor has heaters that are connected to a temperature controller to maintain the temperature of the reactor constant. The IR camera has an infrared detector and an optical scanner that scans over the whole object constructing an image in one second. The signal collected is displayed on a TV monitor and also via computer video capture. During reaction, the temperature of the reactor is increased step by step and the IR intensity is recorded after the image does not change with time. Maier et al.[27] describes the error that can occur in IRT measurements due to the different emissivities of the catalysts. To correct for emissivity effects, a blank experiment is conducted at the various temperatures in the absence of one of the reactants (CO), and the IR intensity emitted at a given base temperature in the absence of reaction is recorded. Based on the temperature calibration with the blank experiment, the actual temperature difference with and without reaction on each spot of the catalysts is obtained and used as a descriptor of catalytic activity.

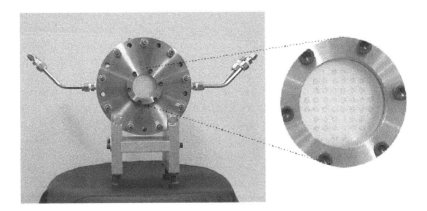

FIGURE 7.2 The IRT reactor with an array of 50 samples (left). The parallel flow COMBI reactor (right).

7.2.2   PARALLEL REACTOR

We also utilize a ten-port microreactor manufactured by ISRI (COMBI Reactor™)[28] which allows one activity and selectivity evaluation under continuous-flow conditions of the best catalysts obtained from the IRT studies. The COMBI Reactor is interfaced on line with a gas chromatograph (GC) that provides composition analysis of the effluent from each of the ten wells. Each well in the COMBI has a capacity of up to 1–2 g of catalyst, thus representing a more scalable result than the one obtained from the catalyst array that only uses a few mg of material. The GC provides sequential analysis of the effluent trapped from each reactor. The system also has programmable electronic flow controllers to meter various gases to the reactor and a temperature control to maintain the reactor temperature constant. The most active catalyst from the COMBI studies is then studied in detail in a single-flow microreactor to obtain kinetic results and compare their consistency with the proposed model. The model is then modified accordingly and new formulations/preparations are chosen to optimize the activity of the materials.

7.2.3   CATALYST PREPARATION AND REACTION

The catalysts were prepared in a parallel preparation system by carbonate co-precipitation from aqueous nitrate solutions of Cu, Zn, and Pd using sodium carbonate as a precipitating agent. The precipitate was aged with stirring, then filtered, and extensively washed. The resulting precipitate was dried overnight at 383 K and calcined in air at 723 K for 5 h. Prior to reaction, the catalyst was reduced in hydrogen at 573 K for 2 h. The gaseous mixture used for the reaction consisted of methanol vapor and oxygen diluted in nitrogen. The inlet concentration of methanol was managed by using a saturator at the temperature of 273 K. Reactant flow rate in this instance was 140 cc/min of nitrogen saturated with methanol at 0°C, resulting in a stream containing 4.0% methanol. Oxygen is also added to the reactant stream at a rate of 2.8 cc/min, yielding an O_2/CH_3OH ratio of 0.5. Reaction effluent was analyzed online by two gas chromatographs connected in series equipped with thermal conductivity detectors and molecular sieve 5A and Hayesep Q columns.

7.3   RESULTS AND DISCUSSION

7.3.1   MODEL OF THE IDEAL CATALYST

As shown in Figure 7.1, we first start with an exhaustive search of the existing literature of the reaction under study to formulate a hypothetical model of the catalyst that guides us to search for families of materials that might be active. The literature results suggested a model of a dual-catalyst system containing a methanol-reforming multicomponent catalysts and an oxidation catalyst. This model led us to the selection of two groups of materials that have the strongest potential to decompose methanol to hydrogen at low temperatures with the minimum formation of water. The first group contains Cu/ZnO, which selectively produces hydrogen and carbon dioxide. The second group contains Pd, which exhibits very high activity for the decomposition of methanol to hydrogen and carbon monoxide. A series of binary Cu/Zn and ternary Cu/Zn/Pd catalysts were first investigated to determine the optimal compositions and to check the existence of a synergistic effect between the two groups. Next we attempted to obtain further improvement in catalyst performance by incorporation of a fourth component to the ternary catalysts.

7.3.2   IRT REACTOR RESULTS

A total of 32 catalysts were examined by IRT in only four experiments. We first started with the large reactor with 50 spots, although not all the spots were used. The results of this first screening

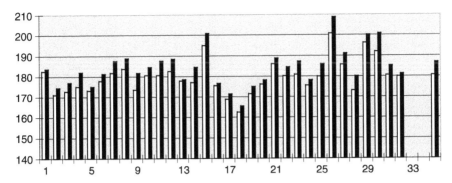

FIGURE 7.3 IRT results for oxidative methanol decomposition in the large IRT reactor for 32 Cu/Zn, Cu/Zn/Pd catalysts, and alumina blanks. Reaction conditions: methanol partial pressure 0.036 atm, O_2/methanol = 0.5, N_2 balance, Reactor in N_2 set at 175°C. The blank bar is before feeding the reactants, and the dark bars after feeding the reactants.

at 175°C are presented in Figure 7.3. The reactor temperature was increased in the presence of air until reaching the set point temperature, then the methanol-containing stream was added. The temperature of each spot was measured before adding the reactants into the reactor (blank bar) and after addition (dark bar).

It can be seen that there are temperature differences among the various spots as well as a ΔT on each spot before and after adding methanol. The differences in temperatures between the spots in the blank run are due to a temperature gradient in the holder. After addition of the reactants, the temperature increased in each spot, even on those containing an alumina blank. It should be noted that dehydration of methanol can occur on alumina increasing the temperature. The ΔT measured at each spot in the large reactor, while it shows activity trends, is not too accurate due to the temperature gradient in the holder. Consequently, we selected a smaller group of catalysts that showed the highest ΔT and re-ran them in the smaller nine spot IRT reactor. These results are presented in Figure 7.4 as a function of temperature for Cu/Zn and Cu/Zn/Pd catalysts. Figure 7.4 shows a bar diagram of the temperature difference (ΔT) obtained from the IRT images of the catalysts with and without reaction at 423–498 K as an indicator of catalytic activity reflecting reaction exothermicity.

Figure 7.4 shows that there is a significant difference in ΔT upon Pd addition whereas differences due to effect of the Cu/Zn ratio are relatively small. Catalyst 7 (Cu/Zn = 7/3, 1% Pd) and catalyst 8 (Cu/Zn = 8/2, 1% Pd) show the highest ΔT and thus are the most promising catalysts. The IRT results for Cu/Zn/Pd with 2% Pd (not shown) exhibited no further increase in the ΔT with increasing Pd contents over 1%. Hence we decided to keep the Cu/Zn ratio and Pd content at 7/3 or 8/2 and 1%, respectively, to study the quaternary catalysts. From additional IRT results for the quaternary catalysts Cu/Zn/Pd/M (not shown), some potential promoters were quickly identified. Although IRT is a quick way to screen the most promising catalysts the measured ΔT does not reflect selectivity, particularly for complex reaction systems. Some of the reactions are exothermic (oxidation and water gas shift reaction) and the others are endothermic (decomposition and steam reforming). Therefore it is important to note that the IRT results show the net exothermicity of all the reactions involved, some of which are not selective for hydrogen production. For this reason, some of the catalysts without Pd that were less active, were studied again in the parallel reactor to verify the trends observed in the IRT results.

7.3.3 PARALLEL-FLOW REACTOR

Further work proceeded using the parallel-flow COMBI reactor to verify the IRT results under continuous-flow conditions typical of catalytic exploratory results. The results for methanol conversion

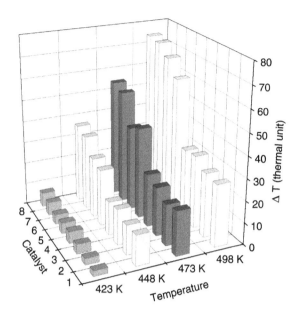

FIGURE 7.4 IRT results for oxidative methanol decomposition in the small nine spot reactor over various Cu/Zn and Cu/Zn/Pd catalysts. Reaction conditions: methanol partial pressure 0.036 atm, O_2/methanol = 0.5, N_2 balance. Catalysts compositions areas follows: #1: Cu/Zn = 5/5; #2: Cu/Zn = 6/4; #3: Cu/Zn = 7/3; #4: Cu/Zn = 8/2; #5: Cu/Zn = 5/5-1%Pd; #6: Cu/Zn = 6/4-1%Pd; #7: Cu/Zn = 7/3-1% Pd; #8: Cu/Zn = 8/2-1%Pd.

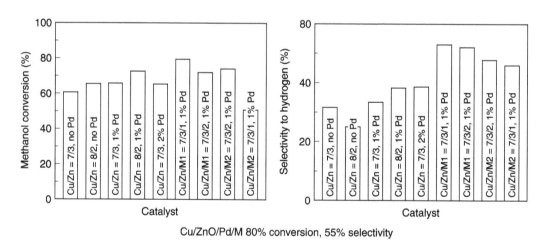

Cu/ZnO/Pd/M 80% conversion, 55% selectivity

FIGURE 7.5 Conversion and selectivity obtained in the parallel flow COMBI reactor. Reaction conditions: catalyst, 50 mg; total flow rate, 329.6 cc/min; methanol partial pressure, 0.036 atm; O_2/methanol = 0.5; N_2 balance.

and selectivity to hydrogen obtained in the parallel flow COMBI reactor for various catalysts selected from IRT results are presented in Figure 7.5.

The results were in fair accord with the corresponding IRT results. The addition of up to 1% Pd to Cu/Zn increased the conversion of methanol and selectivity to hydrogen. IRT results showed that further improvement in catalyst performance was obtained by adding Ce and Zr oxide promoters into the Cu/Zn/Pd catalysts. The selectivity and conversion exhibit the usual inverse correlation for most catalysts: i.e., the higher the conversion the lower the selectivity. However, the addition of Pd and metal oxides led to increased conversion and selectivity.

One issue of significant concern in this reaction is the catalyst stability over long periods of time. To this effect we performed time on stream (TOS) experiments with the various catalysts listed in Table 7.1. As shown in Figure 7.6a, the conversion actually increased with time on stream, which is not usual, because deactivation often occurs with TOS due to carbon deposition (coking). The downside of this activity increase is a decrease in selectivity towards hydrogen formation (Figure 7.6b). Clearly the catalyst is changing due to the reaction, probably due to changes in oxidation state. For example, XRD results show a change from Cu^{2+} to Cu^0 after reduction in hydrogen. TOS results show the used catalyst more active for the total oxidation of hydrogen and other carbon oxides to H_2O and CO_2, thus decreasing the hydrogen selectivity. Since the catalysts were first reduced in hydrogen, its oxidation state might change with TOS, leading to the formation of oxides sites that are less selective for the methanol partial oxidation reaction pathway. Additional studies using in situ EXAFS will help clarify these oxidation state changes.

7.3.4 SINGLE-FLOW REACTOR

Finally, several of the most active and selective catalysts were studied in a single-flow fixed-bed reactor to confirm the trends observed in the parallel reactor. The results showed similar conversion-selectivity results to those obtained in the COMBI reactor. There are some differences between the two types of reactors due to the fact that heat and mass transfer of the fixed-bed plug flow is different from that of the parallel reactor, but the trends are the same.

The results show that addition of Pd increases both conversion and selectivity (sample 3 vs 7) and that addition of promoters Zr (sample 13) and Ce (sample 15) separately to the Pd-containing catalysts increase further the methanol conversion and the selectivity. The presence of both promoters in the same sample (sample 20), however, does not improve the selectivity any further. It should be noted here that the oxygen conversion is 100%, indicating that the improvement in selectivity might be due to secondary reactions such as the water gas shift reaction.

Preliminary XPS results for the catalysts listed in Table 7.2 are shown in Table 7.3. Note the catalysts promoted with Zr (13 and 20) show no discernible palladium peak due to the Pd 3d doublet overlapping with the intense Zr 3p peak. It can be seen that despite more than twice as much copper used in preparation, XPS and XRD data on the calcined catalyst particles indicates that the surface Zn concentration is much higher than the bulk concentration relative to copper. This suggests that Zn may interact with Cu during preparation precipitating it early and encapsulating it in the bulk. This will block and decrease the XPS signal from Cu. Catalyst 15, containing Ce, shows enrichment of palladium on the surface relative to catalyst 7, which contains no Ce. Catalytic activity was higher for catalysts containing Ce, presumably due to enrichment of Pd on the surface. These preliminary results indicate that a key variable is not only the materials involved but also how they are prepared which defines the nature of their surfaces. There are many factors involved in the preparations, such as the solubility of the salts used as well as the precipitation sequence of the preparation.

BET results indicate that these oxides have low surface area of the order of 15–80 m^2/g suggesting that increasing the total area could improve the activity and selectivity. The preparation methods employed in this first iteration were taken from the literature and have not been optimized. These preliminary characterization results point out to the need for a detailed examination of the preparatory variables used before incorporating new components.

It should be emphasized that the results presented here were obtained in a fast-track experimental program comprising 4 weeks of experiments. A similar program using a sequential approach would have taken several months. As pointed out in Figure 7.1, it is now necessary to characterize some of the most active catalysts to elucidate the main causes for the superior catalytic performance and then to design a new model of the ideal catalyst. It is clearly demonstrated in Table 7.2, however, that a quaternary catalyst containing Zr is the most active and selective of all the catalysts studied.

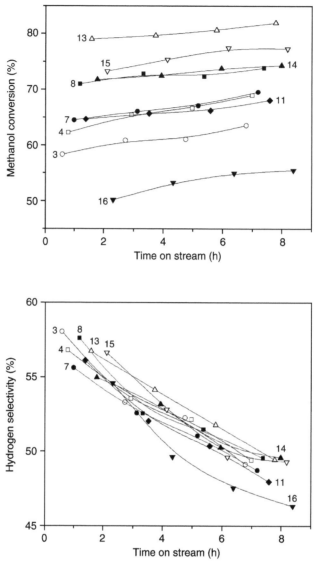

FIGURE 7.6 Conversion and selectivity vs TOS obtained in the parallel flow COMBI reactor. Reaction conditions: catalyst, 50 mg; total flow rate, 329.6 cc/min; methanol partial pressure, 0.036 atm; O_2/methanol = 0.5; N_2 balance, Reactor temperature set at $T = 250°C$. Catalyst compositions (Cu/Zn/Ce/Zr-Pd) are as follows: #3 = 7/3/0/0-0; #4 = 8/2/0/0-0; #7 = 7/3/0/0-1; #8 = 8/2/0/0-1; #11 = 7/3/0/0/2; #13 = 7/3/1/0-1; #14 = 7/3/2/0-1; #15 = 7/3/0/1/-1; #16 = 7/3/0/2-1.

Further work using multiple characterization techniques will help to optimize the catalysts needed for attaining a selective and stable catalyst for hydrogen production from methanol.

7.4 CONCLUSIONS

It has been shown that a knowledge-based combinatorial approach is an effective tool to accelerate the discovery of new catalytic phenomena and catalytic materials. An inverted pyramidal approach

TABLE 7.2
Oxidative Decomposition of Methanol Over Selected Cu/Zn, Cu/Zn/Pd, and Cu/Zn/Pd/M Catalysts Using a Single-Flow Fixed-Bed Reactor[a]

Catalyst Number	Composition Cu/Zn/Ce/Zr-Pd	Conversion %		Selectivity %		
		MetOH	O_2	H_2	CO_2	CO
3	7/3/0/0-0	70	100	24	52	48
7	7/3/0/0-1	87	100	38	55	45
13	7/3/0/1-1	95	100	57	67	33
15	7/3/1/0-1	95	100	67	70	30
20	7/3/1/1-1	95	100	52	60	40

[a] Reaction temperature is 210°C, Cu/Zn/Ce/Zr are in mole fraction. Pd is in wt%. Zr and Ce are promoter metal oxides.

TABLE 7.3
XPS Results of Selected Catalysts

Catalyst Number	Composition Cu/Zn/Ce/Zr-Pd	XPS fraction			Ratio	
		Cu	Zn	Pd	Cu/Zn	Cu/Pd
3	7/3/0/0-0	25.9	40.5	0	0.64	—
7	7/3/0/0-1	15	21.7	0.059	0.69	254
13	7/3/0/1-1	20	21	nd	0.95	—
15	7/3/1/0-1	14.8	17.5	1.67	0.85	8.9
20	7/3/1/1-1	10.5	17.2	nd	0.61	—

using simple and tested analytical techniques has been demonstrated for the oxidative decomposition of methanol. Addition of Pd to Cu/Zn up to 1% increased the conversion of methanol and selectivity to hydrogen. Addition of Zr and Ce promoters to Cu/Zn/Pd increased the conversion of methanol and selectivity to hydrogen. Stability with time on stream is still an unresolved problem and while conversion increases the selectivity decreases with TOS. Further work is under way to characterize and verify the working hypothesis of the selected groups of catalysts.

REFERENCES

1. Gordon, E. M. and Kerwin, J. F., *Combinatorial Chemistry and Molecular Diversity in Drug Discovery,* Wiley, New York, 1998.
2. Terret, N. K., *Combinatorial Chemistry,* Oxford University, Oxford, 1998.
3. Wilson, S. R. and Czarnik, A. W., *Combinatorial Chemistry,* Wiley, New York, 1997.
4. Xiang, X-D., Sun, X., Briceno, G., Lou, Y., Wang, K-A., Chang, H., Wallace-Freedman, W. G., Chen, S-W. and Schultz, P. G., Combinatorial approach to materials discovery, *Science,* 268, 1738, 1995.
5. Briceno, G., Chang, H., Sun, X., Schultz, P. G. and Xiang, X-D., Class of cobalt oxide magnetoresistance materials discovered with combinatorial synthesis, *Science,* 270, 273, 1995.

6. Danielson, E., Golden, J. H., McFarland, E. W., Reaves, C. M., Weinberg, W. H. and Wu, X. D., Combinatorial approach to the discovery and optimization of luminescent materials, *Nature,* 389, 944, 1997.

7. Danielson, E., Devenney, M., Giaquinta, D. M., Golden, J. H., Haushalter, R. C., McFarland, E. W., Poojary, D. M., Reaves, C. M., Weinberg, W. H. and Wu, X, D., A rare-earth phosphor containing one-dimensional chains identified through combinatorial methods, *Science,* 279, 837, 1998.

8. Jandeleit, B., Schaefer, D. J., Powers, T. S., Turner, H. W. and Weinberg, W. H., Combinatorial materials science and catalysis, *Angew. Chem. Int. Ed.,* 38, 2494, 1999.

9. Pescarmona, P. P., van der Waal, J. C., Maxwell, I. E. and Maschmeyer, T., Combinatorial chemistry, high-speed screening and catalysis, *Catal. Lett.,* 63, 1, 1999.

10. Wolf, D., Buyesvskaya, O. V. and Baerns, M., An evolutionary approach in the combinatorial selection and optimization of catalytic materials, *Appl. Catal. A, Gen.* 200, 63, 2000.

11. Senkan, S. M., High-throughput screening of solid-state catalyst libraries, *Nature,* 394, 350, 1998.

12. Hoffmann, C., Wolf, A. and Schüth, F., Parallel synthesis and testing of catalysts under nearly conventional testing conditions, *Angew. Chem. Int. Ed.,* 38, 2800, 1999.

13. Snively, C. M., Oskarsdottir, G. and Lauterbach, J., Parallel analysis of the reaction products from combinatorial catalyst libraries, *Angew. Chem. Int. Ed.,* 40, 3028, 2001.

14. Maier, W., Combinatorial chemistry – challenge and chance for the development of new catalysts and materials, *Angew. Chem. Int. Ed.,* 38, 1216, 1999.

15. Serra, J. M., Chica, A., and Corma, A., Development of a low temperature light paraffin isomerization catalysts with improved resistance to water and sulfur by combinatorial methods, *Appl. Catal. A, Gen.* 239, 35, 2003.

16. Schlögl, R., Combinatorial chemistry in heterogeneous catalysis: A new scientific approach or "the king's new clothes? *Angew. Chem. Int. Ed.,* 37, 2333, 1998.

17. Berger, R. J., Perez-Ramirez, J., Kapteijn, F. and Moulijn, J. A., Catalyst performance testing: Radial and axial dispersion related to dilution in fixed-bed laboratory reactors, *Appl. Catal. A,* 227, 321, 2002.

18. Holmgren, J., Bem, D., Bricker, M., Gillespie, R., Lewis, G., Akporiaye, D., Dahl, I., Karlsson, A., Plassen, M. and Wendelbo, R., Application of combinatorial tools to the discovery and commercialization of microporous solids: facts and fiction, *Stud. Surf. Sci. Catal.,* 135, 113, 2001.

19. Mittasch, A., *Advances in Catalysis,* Academic Press, New York, 1950, Vol II, p. 81.

20. Creer, J. G., Jackson, P., Pandy, G., Percival, G. G. and Seddon, D., *Appl. Catal.,* 22, 85, 1986.

21. Gracia, F. J. and Wolf, E. E., Monte Carlo simulations of the effect of crystallite size on the activity of supported catalysts, *Chem. Eng. J.,* 82, 291, 2001.

22. Gracia, F. J., Miller, J. and Kropf, J. A., In-situ FTIR, EXAFS and activity studies of the effect of crystallite size on Pt/Al$_2$O$_3$ catalysts during CO oxidation, *J. Catal.,* 220, 382, 2003.

23. Pawlicki, P. C. and Schmitz, R. A., Spatial effects on supported catalysts, *Chem. Eng. Prog.,* 83(2), 40, 1987.

24. Kellow, J. and Wolf, E. E., Infrared thermography and FTIR studies of catalyst preparation. Effects on surface reaction dynamics during CO and ethylene oxidation on Rh/SiO$_2$ catalysts, *Chem. Eng. Sci.,* 45, 2597, 1990.

25. Qin, F. and Wolf, E. E., Infrared thermography studies of unsteady-state processes during CO oxidation on supported catalysts, *Chem. Eng. Sci.,* 49, 4263, 1994.

26. Moates, F. C., Somani, M., Annamalai, J., Richardson, J. T., Luss, D. and Wilson, R. C., Infrared thermographic screening of combinatorial libraries of heterogeneous catalysts, *Ind. Eng. Chem.,* 35, 4801, 1996.

27. Holzwarth, A., Schmidt, H-W. and Meier, W. F., Detection of catalytic activity in combinatorial libraries of heterogeneous catalysts by IR thermography, *Angew. Chem. Int. Ed.,* 37, 2644, 1998.

28. http://www.in-situresearch.com

29. Kulprathipanja, A. and Falconer, J.L., Partial oxidation of methanol for hydrogen production using ITO/Al$_2$O$_3$ nanoparticle catalysts, *Appl. Catal. A,* 261, 77, 2004.

30. Brown, L. F., A comparative study of fuels for on-board hydrogen production for fuel-cell-powered automobiles, *Int. J. Hydrogen Energy,* 26, 381, 2001.

31. Peppley, B., Amphlett, J. C., Kearns, L. M. and Mann, R. F., Methanol-steam reforming on Cu/ZnO/Al$_2$O$_3$, Part 1: the reaction network, *Appl. Catal. A,* 179, 21, 1999.

32. Peppley, B., Amphlett, J. C., Kearns, L. M. and Mann, R. F., Methanol-steam reforming on Cu/ZnO/Al$_2$O$_3$, Part 2: A comprehensive kinetic model, *Appl. Catal. A,* 179, 31, 1999.

33. Cubeiro, M. L. and Fierro, J. L., Partial oxidation of methanol over supported palladium catalysts, *Appl. Catal. A*, 168, 307, 1998.
34. Kapoor, M. P., Ichihashi, Y., Kuraoka, K. and Matsumura, Y., Catalytic methanol decomposition over palladium deposited on thermally stable mesoporous titanium oxide, *J. Mol. Catal. A-Chem.*, 198, 303, 2003.
35. Tsoncheva, T., Vankova, S. and Mehandjiev, D., Methanol decomposition on copper and manganese oxides supported on activated carbon, *Reac. Kin. Catal. Lett.*, 80, 383, 2003.
36. Matsumura, Y., Tanaka, K., Tode, T., Yazawa, T. and Haruta, M., Catalytic methanol decomposition to carbon monoxide and hydrogen over nickel supported on silica, *J. Mol. Catal. A*, 152, 157, 2000.
37. Matsumura, Y., Tode, T., Yazawa, T. and Haruta, M., Catalytic methanol decomposition to carbon monoxide and hydrogen over Ni/SiO$_2$ of high nickel content, *J. Mol. Catal. A*, 99, 183, 1995.
38. Liu, Y. Y., Hayakawa, T., Suzuki, K. and Hamakawa, S., Highly active copper/ceria catalysts for steam reforming of methanol, *Appl. Catal. A*, 223, 137, 2002.
39. Chin, Y. H., Dagle, R., Hu, J. L., Dohnalkova, A. C. and Wang, Y., Steam reforming of methanol over highly active Pd/ZnO catalyst, *Catal. Today*, 77, 79, 2002.
40. Zhang, X. R. R., Shi, P. F., Zhao, J. X., Zhao, M. Y. and Liu, C. T. Production of hydrogen for fuel cells by steam reforming of methanol on Cu/ZrO2/Al2O3 catalysts, *Fuel Process. Techn*, 83, 183, 2003.
41. Iwasa, N., Mayanagi, T., Nomura, W., Arai, M. and Takezawa, N. Effect of Zn addition to supported Pd catalysts in the steam reforming of methanol, *App. Catal. A-l*, 248, 153, 2003.
42. Agrell, J., Germani, G., Järås, A. and Boutonnet, M., Production of hydrogen by partial oxidation of methanol over ZnO-Supported palladium catalysts prepared by microemulsion technique, *Appl. Catal. A*, 242, 233, 2003.
43. Agrell, J., Boutonnet, M., Melián-Cabrera, I. and Fierro, J. L., Production of hydrogen from methanol over binary Cu/ZnO catalysts Part I. Catalyst preparation and characterization, *Appl. Catal. A*, 253, 201, 2003.
44. Alejo, L., Lago, R., Peña, A. and Fierro, J. L., Partial oxidation of methanol to produce hydrogen over Cu-Zn-based catalysts, *Appl. Catal. A*, 162, 281, 1997.
45. Navarro, R. M., Peña, M. A. and Fierro J. L., Production of hydrogen by hartial oxidation of methanol over a Cu/ZnO/Al$_2$O$_3$ catalyst: Influence of the initial state of the catalyst on the start-up behavior of the reformer, *J. Catal.*, 212, 112, 2002.
46. Wang, Z. F., Xi, J. Y., Wang, W. P. and Lu, G. X., Selective production of hydrogen by partial oxidation of methanol over Cu/Cr catalysts, *J. Mol. Catal.*, 191, 123, 2003.
47. Velu, S., Suzuki, K., Kapoor, M. P., Ohashi, F. and Osaki, T., Selective production of hydrogen for fuel cells via oxidative steam reforming of methanol over CuZnAl(Zr)-oxide catalysts, *Appl. Catal. A*, 213, 47, 2001.
48. Gracia, F., Li, W. and Wolf, E. E., The preferential oxidation of CO: selective combinatorial activity and infrared studies, *Catal. Lett.*, 91, 235, 2003.
49. Li, W., Gracia, F. J. and Wolf, E. E., Selective combinatorial catalysis; challenges and opportunities: the preferential oxidation of carbon monoxide, *Catal. Today*, 81, 437, 2003.

8 New Catalysts for the Carbonylation of Phenol: Discovery Using High-Throughput Screening and Leads Scale-Up

Donald W. Whisenhunt, Jr. and Grigorii Soloveichik

CONTENTS

8.1   INTRODUCTION

8.1.1   Diphenylcarbonate (DPC) and the Search for a One-Step Process

Diphenylcarbonate (DPC) is a key component in the manufacture of Lexan® polycarbonate via the melt process. Polycarbonate has historically been manufactured by the interfacial process practiced by both GE and Bayer AG.[1] This process involves the reaction of phosgene with Bisphenol A (BPA) in a biphasic system with an amine catalyst and monohydroxyphenolic chain stopper (Scheme 8.1a).[2]

GE and Bayer are the world's largest producers of polycarbonate. Worldwide production is in excess of 4 billion pounds with GE producing over 1 billion pounds per year.[3] Polycarbonate is used in numerous applications including electronics, architectural, automotives, and safety.[2] This product represents a key research priority for GE.

Over the last 30 years the use of phosgene has come under stricter environmental regulations. Responding to this, a non-phosgene route to polycarbonate was developed and is in practice at the GE Advanced Materials plant in Cartagena, Spain. This non-phosgene route, known as the "melt" process, involves the reaction of DPC with BPA in a base-catalyzed solvent-less process (Scheme 8.1b).[2]

(8.1a)

Bisphenol A (BPA)   Phosgene

(8.1b)

LEXAN polycarbonate

Bisphenol A (BPA)   Diphenylcarbonate (DPC)

Currently, DPC is prepared from the transesterification of dimethylcarbonate (DMC) with phenol. DMC must first be prepared from the copper-catalyzed carbonylation of methanol. This two-step process is shown in Scheme 8.2.[4,5] Ideally, DPC can be prepared directly from phenol, carbon monoxide, and oxygen which is a simpler (one-step) energetically favorable reaction (Scheme 8.3).

(8.2)

$2 \ MeOH + CO + 1/2 O_2$ Dimethylcarbonate (DMC)   Diphenylcarbonate (DPC)

(8.3)

2 $+ CO + 1/2 O_2$ Catalysts / heat, pressure   Diphenylcarbonate (DPC) $+ H_2O$

In the mid-1970s Chalk discovered that para-substituted phenols can be oxidatively carbonylated using stoichiometric amounts of a Ru, Rh, Os, Ir, or Pd salt. The palladium salts performed the reaction best (Scheme 8.4). Yields of carbonylation products were high but the reaction yielded 1 mole of Pd^0 for every mole of product.[6,7]

(8.4)

$$2 \text{ phenol (OH, para-R)} + CO \xrightarrow[\text{CH}_2\text{Cl}_2]{\substack{\text{Pd(II) salt} \\ \text{base (R}_3\text{N)}}} \text{Carbonate} + \text{Salicylate}$$

Carbonate

Salicylate

In the late 1970s Hallgren and Matthews[8] determined that the ratio of carbonate to salicylate was controlled by the nature of the para-substituent, the Pd source, and the order of reagent addition. From this work they proposed the reaction mechanism shown in Scheme 8.5 where the carbonate is formed from the reductive elimination from the acyl Pd(II) carbonyl complex. This species was proposed to derive from a dimeric Pd(I) species.

(8.5)

$$1/2\, O_2 + 2H^+ \rightarrow M^{(n-2)+} ; \quad H_2O \leftarrow M^{n+} ; \quad Pd^{2+} ; \quad Pd^0 ; \quad 2\, \text{phenol} + CO ; \quad DPC + 2H^+$$

Unfortunately, the direct carbonylation of phenol has a number of hurdles that must be overcome in order to be commercially viable. First, unlike aliphatic alcohols, the carbonylation of aromatic alcohols is difficult due to the increased acidity of the aromatic alcohol and their oxidative instability.[8–10] Second, the reaction needs a catalyst to proceed at a sufficient rate and the best catalysts to date have been palladium salts. In order to make the reaction catalytic in palladium the reduced Pd^0 species must be reoxidized to Pd^{2+} in order to continue the catalytic cycle. Third, the oxidation of Pd^0 to Pd^{2+} using oxygen (the oxidant of choice)[11,12] is slow,[13] so additional co-catalysts, both organic (e.g., benzoquinone as in the 1,4 oxidation of conjugated dienes) and inorganic (transition metals) must be used. Researchers have struggled with this reaction for 40 years.

Up until 1995 the bulk of the patents issued on a one-step DPC process were to GE. These systems were based on palladium salts with cobalt or manganese as the inorganic co-catalyst (IOCC). Many systems used an organic co-catalyst (OCC) as well, usually benzoquinone. The final components of the catalyst "package" were a bromide source, usually in the form of a quaternary ammonium salt, and a base. The best systems had fewer than 500 catalyst turnovers (mol DPC/mol Pd catalyst) and a rate of <1.0 g/mol DPC/h.[14–17]

The state of the art in this area in 1995 still left open a number of issues. In particular, the cost of palladium was dictating that higher catalyst turnover numbers were needed, a high level of

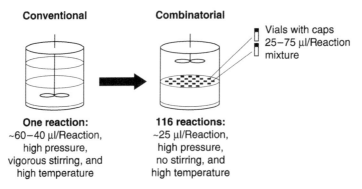

FIGURE 8.1 Conventional vs combinatorial reactor setup.

TABLE 8.1
Factors for One-Step DPC Synthesis (1*15*15*15*9*4*3*3*3 = 3,280,500 Potential Experiments)

Factors in typical screening experiment	Possibilities
Principal metal catalyst	1
Inorganic co-catalysts	15
Metal ligands	15
Organic co-cataylsts	15
Anion	9
Associated cation	4
Reaction time	32
Reaction temperature	3
Reaction pressure	3

---

palladium recovery was necessary and that the selectivity of phenol and CO going to DPC needed to be increased. All efforts up till then, and even to date, have failed to replace palladium as the main catalysts in this reaction. Efforts to reduce the conversion of CO to CO_2 and the conversion of phenol into non-DPC products continue.

Starting in 1995 GE developed a major research program to improve and commercialize a one-step DPC synthesis process. This effort has included catalyst development, engineering, and rigorous economic analysis. It was clear that the main problem to solve was to identify an active co-catalyst capable of quickly reoxidizing Pd(0), which is generated at the step of DPC formation. A working hypothesis was that the redox chain consisting of two (or more) redox components, like in natural redox chains, could provide better reaction rate and Pd turnover number. Typical one-step DPC reactions are run in an autoclave at high pressure (11.5 MPa, 1700 psig) and at elevated temperatures (100°C) for 1–2 h with vigorous stirring (Figure 8.1). The setup and tear-down time of the reactor limits the number of reactions that can be run per day to 1–2 per reactor. This limited throughput spurred the investigation into how to run more reactions.

In 1997 a group at GE was put together to tackle the problem of increased throughput for the one-step DPC reaction. This effort was necessary since the number of variables involved in this reaction is huge (Table 8.1). Recent discoveries of new co-catalysts for this reaction[18] have shown that an expanded search of the periodic table is warranted. The time required for traditional-scale reactions would make this search cost-prohibitive.

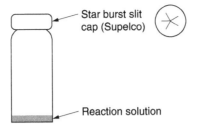

Star burst slit
cap (Supelco)

Reaction solution

FIGURE 8.2 Final reaction vials.

8.1.2 HIGH-THROUGHPUT CATALYSIS DEVELOPMENT

The concept of creating more samples to speed up research has been around for many years. In the 1970s Hanak developed a number of methods for creating compositional gradient arrays for the "multiple-sample concept."[19–40] The idea of creating an array of materials for rapid analysis has been very popular in undergraduate qualitative analysis courses for more than 15 years.[41] These undergraduate courses started using these techniques because they could be done on a small scale with reduced cost and risk. The techniques also greatly speed up the time required to carry out the experiments. The idea of running multiple chemical reactions at the same time is quite old as discussed in reference 42. The pharmaceutical industry was the first to adopt combinatorial chemistry and high-throughput screening for routine research. Numerous groups have more recently extended these techniques into the area of materials science,[43] for metal catalysis for organic reactions,[44,45] and solid-phase synthesis of organic molecules.[46] Currently most large chemical and material science companies are using high-throughput techniques to carry out research.[47]

In order to screen catalysts for the one-step synthesis of DPC a number of issues had to be addressed: (1) a miniaturized reactor system that was capable of working under high pressure and elevated temperature had to be designed; (2) the pressurized gas (a mixture of O_2 and CO) had to be delivered to all the samples without cross-contamination; (3) all samples needed to be exposed to the same temperature and pressure at the same time without gradients; and (4) the samples had to be processed and analyzed quickly and the data had to be stored in such a way that they could easily be retrieved.

It was determined that the reaction could be run on a small amount of reaction mixture at the bottom of a 2-mL gas chromatography vial. The reaction mixture volume was generally 25–75 μL. The relatively large reactor headspace coupled with the high surface to volume ratio of the solution prevented O_2 diffusion from being limiting.[48] Many of these small vials could be put in a holder and placed in a standard 1-gallon pressurized autoclave. The use of starburst septa allowed the reactant gas to fill the vials but reduced the cross-contamination between vials (see Figure 8.2).

A number of experiment factors ultimately determine if a catalyst "package" is successful including catalyst turnover number (TON), selectivity (both phenol and CO), recovery of palladium, etc. It was not possible to assess all of these factors using high-throughput screening. Due to the cost of palladium it was determined that the catalyst turnover number should be the figure of merit for the small-scale screening reactions. The goal of this effort was to rank order catalysts based on turnover number, optimize the best formulations and scale them up to a standard batch scale (60 g) and finally to a continuous 1-gallon bench-top unit (BTU).

8.2 EXPERIMENTAL

8.2.1 MATERIALS

The following salts were obtained and used without further purification, Bi(TMHD)$_3$ (Strem), Ce(acac)$_3$·xH$_2$O (Aldrich), Co(acac)$_2$ (Aldrich), Cr(acac)$_3$ (Aldrich), Cu(acac)$_2$ (Aldrich),

Eu(acac)$_3$·xH$_2$O (Aldrich), Fe(acac)$_3$ (Alfa), Ir(acac)$_3$ (Aldrich), Mn(acac)$_3$·4H$_2$O (Aldrich), Ni(acac)$_2$ (Aldrich), Pb(acac)$_2$ (Aldrich), Rh(acac)$_3$ (Alfa), Ru(acac)$_3$ (Alfa), SbBr$_3$ (Strem), Sn(acac)$_2$Br$_2$ (Aldrich), TiO(acac)$_2$ (Aldrich), V(acac)$_3$ (Strem), VO(acac)$_2$ (Strem), WCl$_6$ (Strem), Yb(acac)$_3$·xH$_2$O (Strem), Zn(acac)$_2$·xH$_2$O (Aldrich), Zr(acac)$_4$ (Aldrich), Cd(acac)$_2$ (Strem), In(acac)$_3$ Strem, Gd(acac)$_3$·xH$_2$O (Strem), Ca(acac)$_2$·xH$_2$O (Strem), Cs(acac) (Aldrich), La(acac)$_3$· xH$_2$O (Strem), ReO$_3$ (Strem), PbO (Aldrich), and Pd(acac)$_2$ (Aldrich) where TMHD is 2,2,6,6-tetramethyl-3, 5-heptanedionate and acac is acetylacetonate. Phenol was obtained from GE Advanced Materials (Mount Vernon, Indiana) and was used without further purification.

8.2.2 Analytical

Concentrations of reaction products were determined either by GC (gas chromatography, Agilent 6890 dual column with EZ-Flash technology from Thermedics Detection) or by LC (liquid chromatography, Agilent 1100 series). Internal standards were used for quantification with retention times calibrated against authentic samples. Custom EXCEL™ macros were written to pick peaks, generate standard curves, and determine the absolute concentration of 10–14 analytes per reaction mixture.

8.2.3 Equipment

The small-scale reactions were run in GC vials from Agilent. The pressurized autoclaves were from Autoclave Engineers and PARR. Automated liquid handlers were from Gilson and Hamilton. The automated weighing station was from Bohdan (now part of Mettler-Toledo).

8.2.4 Small-Scale Reactions

In general, reaction solutions were prepared by dissolving or slurrying the individual reaction components in molten (~60°C) phenol. These stock solutions were then used to prepare mixtures (~750 μL) or "mix vials." These mix vials were heated and stirred prior to being delivered to the final reaction vials. The reaction vials were at room temperature causing the reaction mixtures to freeze once delivered. This setup is shown in Figure 8.3 on a Hamilton liquid handler. Disposable tips were used to transfer the liquids to avoid contamination. Each of the final reaction vials was weighed before addition of the solution and after addition to determine exactly how much solution had been delivered.

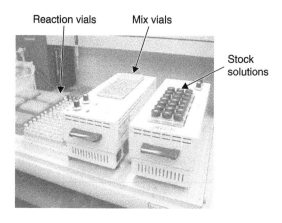

FIGURE 8.3 Sample preparation setup on Hamilton liquid handler.

Once all the solutions were formulated and the vials weighed they were transferred to a 58-well aluminum reaction block (Figure 8.4). This block acted as a vial holder for the reaction vials. Vials could be stacked two deep allowing 116 reactions to be run at the same time. The block was mounted in a standard 1-gallon autoclave and pressurized with a CO/O_2 (93:7) mixture at 5–10 MPa (700–1500 psig). The reactor was heated to 100°C and held at this temperature for 2–3 h. The reactor was then allowed to cool, generally overnight. The pressure was released and the block removed.

The vials were then reweighed to determine if any vials lost a significant mass ($> 15\%$). Data from reaction vials with significant weight loss were generally not used. The vials were placed on a Gilson 4-probe SPE robot. Internal standard in MTBE was added to dissolve the reaction mixtures. A small aliquot was passed through the SPE column to remove all metals and phenolic polymers. The eluent was transferred to another GC vial and BTSFA was added to derivatize the samples for gas chromatography.

The samples and standards were analyzed by gas chromatography. The peaks for DPC, phenol and numerous by-products were measured and quantified (for an example trace see Figure 8.5). The catalyst turnover number was calculated for each sample (mol DPC produced/mol Pd charged).

FIGURE 8.4 Reaction block with 58 wells.

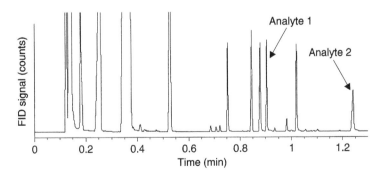

FIGURE 8.5 Typical GC trace for small-scale reactions. Analyte 1 and 2 are internal standards.

8.2.5   BATCH-SCALE REACTIONS

The reactants were weighed out as either liquids or solids and charged to a 450 ml Hastalloy C autoclave with a stirring paddle and dip tube. To remove water from the reaction activated molecular sieves (1/16-inch pellets 3 Å) were suspended in a Teflon™ basket in the reactor headspace. The reactor was sealed and charged with a CO/O_2 (93:7) mixture at 11.2 MPa (1600 psig) with stirring. The reactor is heated at 100°C for 2–4 h with samples being drawn at various time points.

8.2.6   BENCH-TOP UNIT REACTION

The bench-top unit (BTU) uses a 1-gallon reactor that is continuously fed with liquid phenol containing soluble catalyst and designed for producing about 1 kg/h of DPC. The gases (carbon monoxide and oxygen) are also continuously fed into the reactor at 800 psig. The liquid product is discharged from the reactor and separated from any entrained and dissolved gases in a gas–liquid separator. The majority of the liquid product is then fed to a dehydration column to remove water and recycled back to the reactor. Un-recycled liquid is sent to a second gas–liquid separator and then to a product storage drum, while the gases are vented to the atmosphere.

8.3   RESULTS AND DISCUSSION

8.3.1   VARIABILITY IN SMALL-VIAL REACTIONS

The preparation, screening, and analysis of the small-scale samples included many steps with many opportunities for errors. The GE Six Sigma[49] methodology was used to analyze and optimize each step of the process to reduce noise in the final figure of merit, Pd TON. This proved difficult using the classical Mitsubishi catalyst system (palladium, lead oxide, quaternary bromide in neat phenol).[18] The system always produces a heterogeneous system upon mixing the PbO with a bromide source. Through the use of Six Sigma and nested experimentation[50] it was determined that the variability in this system was due to the time between sample preparation and running the reaction. It was also determined that certain palladium salts were less sensitive to this time effect. The data in Table 8.2 illustrates that aging the reaction mixture at 70°C when using $Pd(acac)_2$ as catalyst can reduce the TON from ~2000 to ~900. This effect is significantly reduced when using $Pd(dppb)Cl_2$. However, the use of this phosphine ligand reduced the maximum TON from ~2500 to ~2000. For most small-scale experiments $Pd(acac)_2$ was used and the mixtures were cooled to room temperature as quickly as possible after mixing.

TABLE 8.2
Effect of Aging Reaction Mixtures for Two Palladium Catalysts

Age Pd stock (h)	Age Reaction Mixture (h)	$Pd(acac)_2$ avg. (TON)	RSD (%)	$Pd(dppb)Cl_2$ avg. (TON)	RSD (%)
4	4	801	9.1	1926	3.7
4	2	891	9.6	2036	6.5
4	0	2510	6.7	2132	2.0
2	2	841	4.6	2000	5.1
2	0	2330	7.4	1879	7.3
0	0	2368	2.0	2018	29.8

Reaction conditions (0.25 mM Pd, 12 equiv. PbO, 5.6 equiv. $Ce(acac)_3$ and 400 equiv. tetrabutylammonium bromide – TBAB). Dppb = bis(diphenylphosphino)butane. (Adapted from Soloveichik et al.[51] With permission.)

8.3.2 CORRELATION WITH BATCH-SCALE REACTIONS

To use the small-scale reactions to screen for effective catalyst systems the criteria for what a "hit" is had to be determined. To be a "hit" a catalyst system must perform better than a control system. For the work here better was defined to be 25% better than the control. The control systems changed depending on the experiment but one example is shown in Table 8.3. These data are a comparison between small-scale reactions and batch-scale reactions on the same systems. The Pd(acac)$_2$, PbO, TEAB (tetraethylammonium bromide) was used as the control and the addition of Ce(acac)$_3$ with increasing TEAB was examined. This experiment demonstrated that the small-scale reactions correlated very well ($R^2 = 0.98$) (Figure 8.6) with the large-scale reactions and that a 25% increase in TON could be used to determine if a system had been improved. In this case the addition of Ce(acac)$_3$ and the increase in TEAB ratio showed improvement.

8.3.3 IDENTIFICATION OF NEW CO-CATALYSTS

The successful addition of Pb, Co, and Mn to the Pd/Br system as a co-catalyst dramatically increased the turnover number. This indicated that a broad search for other co-catalysts was needed.

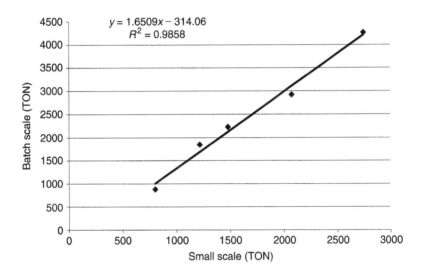

FIGURE 8.6 Plot of small-scale reactions vs batch-scale reaction results.

TABLE 8.3
Correlation of Small-Scale Experiments with Batch-Scale Experiments

Pd(acac)$_2$ (mM)	PbO (Pb:Pd)	Ce(acac)$_3$ (Ce:Pd)	TEAB (Br:Pd)	Small-Scale (TON)	Batch-Scale (TON)
0.29	54	7.62	308	2743	4267
0.31	52	7.37	148	2073	2921
0.26	61	8.71	94	1478	2230
0.28	57	7.96	83	1213	1848
0.31	51	—	145	800	878

Reaction conditions (100°C, 3 h, CO/O$_2$ 90:10 at 1500 psi). TEAB = tetraethylammonium bromide.

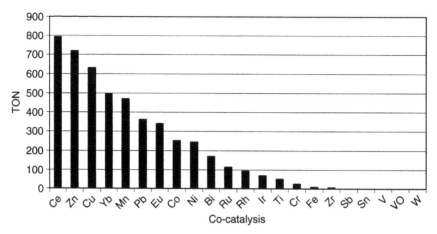

FIGURE 8.7 Results of single metal co-catalysis screening. TON is the moles of DPC produced/moles of Pd charged. Reaction conditions [Pd(acac)$_2$ = 0.25 mM, HegBr:Pd = 60, Co-catalyst:Pd = 14, phenol as solvent at 100°C under CO/O$_2$ (90:10) at 1500 psig for 3 h. HegBr = hexaethylguanidinium bromide].

Under identical conditions 19 metal co-catalysts were screened (Figure 8.7). These screening experiments led to a number of surprising results. It has been assumed and discussed elsewhere[48] that the main mechanism for DPC production with a co-catalyst is oxidation of Pd^0 species by the oxidized co-catalyst followed by the re-oxidation of the co-catalyst by O$_2$ (Scheme 8.5). The results here would indicate that this is not the only mechanism possible since non-redox active metals like Zn gave significant turnovers.

The next screening step for new co-catalysts was to examine whether combinations of co-catalysts could not only improve the turnover number but could the effect be synergistic. We define a synergistic combination as a pair of metal co-catalysts that give a turnover number greater than the sum of the turnover values obtained as single metal co-catalysts. All binary combinations were tested under the same conditions (Figure 8.8). A number of synergies were discovered through this screening experiment (Figure 8.9 and Table 8.4), but most were no better and some were significantly worse than the individual runs. There appear to be some poisons for this reaction. The V, V = O, W, Sb, and Sn salts used for these experiments consistently gave no turnovers. The most interesting synergies involved co-catalysts like Ti, Fe, and Zr, which alone had very low TON but significantly improved other catalysts like Pb, Cu, and Mn.

The final step in this screening series was to look at ternary systems (Table 8.5). The addition of a third co-catalyst to the system continued to improve the performance. From Table 8.5, 15 systems were found that had TONs higher than any of the binary combinations. By examining the frequency of co-catalysts that appear in the top ternary system it can be seen that the top three metals are Pb, Ti, and Mn. The results from these studies were very unpredictable. The addition of Fe (TON = 9 for single metal) to the Pb/Ti system increased the TON from 1068 to 1631. However, the addition of Fe to the Cu/Ti system reduced the TON by 100. With a solid mechanistic understanding still unclear it was necessary to carry out multiple experiments to determine the best catalyst systems.

8.3.4   OPTIMIZATION OF THE PB/TI SYSTEM

Despite the success of the ternary systems the use of three metal co-catalysts in addition to palladium made the economics of metal separation of recovery unfavorable. The high-throughput system was then used to optimize the top binary systems. As an example, the results of the

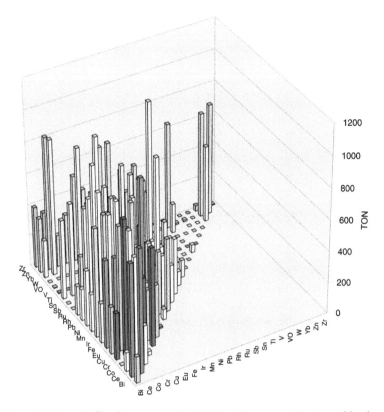

FIGURE 8.8 (See color insert following page 172) TON for binary co-catalysts combinations. Single metal co-catalysts are plotted along the diagonal.

Pb/Ti system are shown in Table 8.6. From these data it is clear that Pb is required for a significant number of TONs. Also, higher Br levels in general gave higher TON. There is a significant interaction between the Pb and Ti levels. The experimental space is shown pictorially in Figure 8.10.

8.3.5 ADDITION OF BASE AND VARIATION PROCESS PARAMETERS

Many of the earlier systems studied at GE involved the addition of base to the reaction mixture. Initial proof-of-principle studies indicated that the addition of base could be helpful to certain systems. A complete screening series similar to the experiments above was carried out in the presence of 150 equivalents of NaOH or NaOPh. At the same time an exhaustive study of the effect of temperature, pressure, and percent O_2 was undertaken. Lower temperature, pressure, and O_2 concentration all impact on the economics of this process favorably. However, the reduction of both pressure and percent O_2 caused all systems to perform very poorly (last column of Table 8.7). No attempt was made to optimize the parameters here but instead the goal was to determine which systems were less sensitive to lower pressure, temperature, and O_2 level (Table 8.7). The most resilient and best performing systems where Pb/Ce/base and Pb/Ti/base. The addition of base most affected the performance of the Pb/Ti and Ce/Fe systems. For most other systems the effect was minimal.

8.3.6 OPTIMIZATION OF THE PB/TI/BASE SYSTEM BATCH SCALE

The Pb/Ti/base system was optimized using design of experiments (DOE). Pd TON is described by a good 2FI model (Adj R-Squared 0.949, Pred R-Squared 0.919, C.V. 5.9, Adeq Precision 24.6).

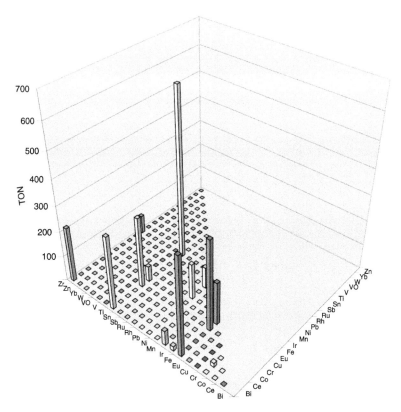

FIGURE 8.9 (See color insert following page 172) Synergies determined from binary co-catalyst combinations. Synergy is defined as the difference between the TON of the pair of co-catalysts and the sum of the TON for co-catalysts alone.

TABLE 8.4
Synergies Determined for Co-Catalyst Pairs

Co-Catalyst-1	Co-Catalyst-2	TON	Synergy
Ti	Pb	1068	656
Fe	Bi	563	385
Fe	Cu	995	355
Ti	Bi	509	290
Ti	Cu	960	279
Zr	Bi	393	219
Fe	Eu	520	170
Pb	Fe	514	143
Pb	Mn	916	85
Zr	Pb	439	72
Ti	Eu	453	62
Mn	Bi	696	57
Ir	Bi	254	17
Cr	Ce	832	16

Reaction conditions ([Pd(acac)$_2$ = 0.25 mM, HegBr:Pd = 60, Co-catalyst-1:Pd = Co-catalyst-2:Pd = 14, phenol as solvent at 100°C under CO/O$_2$ (90:10) at 1500 psig for 3 h. HegBr = hexaethylguanidinium bromide. Synergy is calculated as the difference between the TON of the pair minus the sum of TON of the individual co-catalysts. (Adapted from Spirak et al.[48] With permission.)

TABLE 8.5
TON Values for Ternary Systems

Co-Catalyst-1	Co-Catalyst-2	Co-Catalyst-3	TON
Ti	Pb	Fe	1631
Mn	Bi	Ti	1421
Pb	Ce	Ti	1401
Ti	Pb	Zn	1310
Cu	Ti	Ce	1278
Fe	Cu	Ce	1244
Pb	Mn	Ti	1237
Mn	Eu	Pb	1235
Ti	Pb	Yb	1206
Mn	Eu	Cu	1186
Mn	Bi	Eu	1154
Cu	Ti	Zn	1145
Fe	Cu	Zn	1141
Pb	Mn	Fe	1098
Pb	Mn	Cu	1094
Mn	Bi	Yb	1058
Pb	Mn	Yb	1048
Mn	Bi	Fe	1003
Pb	Ce	Fe	1002
Ti	Pb	Eu	990
Pb	Ce	Eu	975
Cu	Ti	Yb	965
Fe	Cu	Bi	929
Fe	Cu	Eu	923
Pb	Mn	Ce	921
Fe	Cu	Yb	899
Pb	Ce	Yb	896
Pb	Mn	Zn	888
Mn	Eu	Ce	882
Mn	Bi	Cu	879
Ti	Pb	Bi	873
Pb	Ce	Zn	860
Pb	Ce	Cu	858
Mn	Bi	Ce	817
Cu	Ti	Fe	812
Mn	Eu	Ti	791
Mn	Eu	Zn	782
Cu	Ti	Eu	778
Fe	Cu	Mn	690
Mn	Bi	Pb	683
Pb	Ce	Bi	635
Cu	Ti	Mn	620
Mn	Eu	Yb	464
Mn	Bi	Zn	452
Fe	Cu	Pb	449
Cu	Ti	Bi	426
Cu	Ti	Pb	393
Mn	Eu	Fe	238

Reaction conditions ([Pd(acac)$_2$ = 0.25 mM, HegBr:Pd = 60, Co-catalyst-1:Pd = Co-catalyst-2:Pd = Co-catalyst:Pd = 14, phenol as solvent at 100°C under CO/O$_2$ (90:10) at 1500 psig for 3 h. HegBr = hexaethylguanidinium bromide. (Adapted from Spirak et al.[48] With permission.)

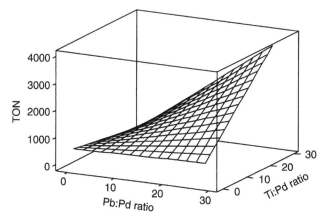

FIGURE 8.10 Contour plot of the interaction of the Pb and Ti levels for the binary Pb/Ti system.

TABLE 8.6
Optimization of the Pb/Ti System

Pb:Pd Ratio	Ti:Pd Ratio	HegBr:Pd Ratio	TON
28	5.6	120	2068
14	2.8	120	1325
14	14	60	1224
2.8	14	120	1191
5.6	28	120	983
28		120	892
14		120	871
5.6	28	30	862
14	2.8	30	553
28	5.6	30	479
14	14	30	455
28		30	307
		30	268
	14	120	212
	28	120	64
	14	30	32
	28	30	10

Reaction conditions [Pd as Pd(acac)$_2$ = 0.25 mM, Pb as PbO, Ti as TiO(acac)$_2$, phenol as solvent at 100°C under CO/O$_2$ (90:10) at 1500 psig for 3 h. HegBr = hexaethylguanidinium bromide].

Base was actually detrimental for Pd TON as well as Ti (Br is good) for this experiment (see Table 8.8). But the presence of Ti provides a high reaction rate and decreases the induction period in product formation.

8.3.7   BENCH-TOP UNIT PERFORMANCE OF TOP SYSTEMS

The results of batch-scale optimizations have been verified in a continuous reactor (more than 20 h) using the bench-top unit (BTU). The data for three optimized systems with Pb, Pb-Ti and

TABLE 8.7
Summary of Catalyst Systems With and Without Base at Various Temperatures, Pressures and O_2 Levels

Co-Catalyst-1	Co-Cat-1:Pd	Co-Catalyst-2	Co-Cat-2:Pd	Co-Catalyst-3	Co-Cat-3:Pd	NaOPh:Pd	T=100.6°C p=1580 psi O_2=7.83% TON	T=100.1°C p=1520 psi O_2=7.42% TON	T=85.2°C p=160 psi O_2=7.39% TON	T=85.3°C p=1520 psi O_2=3.46% TON	T=95.0°C p=200 psi O_2=3.8% TON
—	—	—	—	—	—	—	25	35	82	0	0
—	—	—	—	—	—	150	48	76	165	242	90
Pb	24	—	—	—	—	—	1408	1780	1611	433	137
Pb	24	—	—	—	—	150	1291	993	852	888	61
Mn	10	—	—	—	—	—	53	100	98	3	12
Mn	10	—	—	—	—	150	44	85	119	217	200
Cu	24	—	—	—	—	—	698	1482	1665	442	378
Cu	24	—	—	—	—	150	747	1779	1459	1275	291
Pb	24	Ce	5.6	—	—	—	1795	2055	2259	1081	408
Pb	24	Ce	5.6	—	—	150	2434	2317	1968	2279	395
Pb	24	Ti	5.6	—	—	—	1022	1162	1334	631	438
Pb	24	Ti	5.6	—	—	150	2822	2509	2606	2908	319
Pb	24	Mn	5.6	—	—	—	1228	565	1180	1240	785
Pb	24	Mn	5.6	—	—	150	814	863	684	1242	434
Cu	12	Fe	5.6	—	—	—	142	1108	1406	470	440
Cu	12	Fe	5.6	—	—	150	1325	1673	1556	1474	284
Cu	12	Mn	10	—	—	—	213	480	549	127	432
Cu	12	Mn	10	—	—	150	354	787	787	999	327
Cu	5.6	Ni	10	—	—	—	593	1456	1480	124	358
Cu	5.6	Ni	10	—	—	150	534	991	1262	997	297
Ce	5.6	Mn	10	—	—	—	304	339	458	29	106
Ce	5.6	Mn	10	—	—	150	284	291	417	747	702
Ce	5.6	Zn	50	—	—	—	492	566	597	180	287
Ce	5.6	Zn	50	—	—	150	224	255	267	584	175
Ce	5.6	Fe	10	—	—	—	452	474	505	17	33
Ce	5.6	Fe	10	—	—	150	1223	1851	1615	1626	409
Zn	50	Ti	5.6	—	—	—	1018	1070	1160	346	334
Zn	50	Ti	5.6	—	—	150	1764	1568	1295	1222	180
Bi	28	Ti	28	Mn	14	—	832	817	1213	693	1219
Bi	28	Ti	28	Mn	14	150	799	887	756	939	598

Reaction conditions ([Pd(dppb)Cl$_2$] = 0.25 mM, TEAB:Pd = 300, phenol as solvent at 3 h. HegBr = hexaethylguanidinium bromide.

TABLE 8.8
Batch Optimization of the Pb/Ti/Base System

No	Ti (eq)	MBr (eq)	NaOH (eq)	Pd TON	Selectivity (%)
1	2	198	260	3386	52.3
2	2	200	113	3940	59.8
3	2	323	104	6229	69
4	2	335	211	6012	79.4
5	2	445	116	7855	80.9
6	5	201	217	3860	54.5
7	5	312	238	5557	65.4
8	5	449	205	6691	59.2
9	8	198	125	4289	72
10	8	198	234	3380	47.5
11	8	199	268	4202	55.4
12	8	322	113	5395	63.9
13	8	447	114	6946	65.1
14	8	448	250	5817	55.5
15	8	457	253	6127	70.4

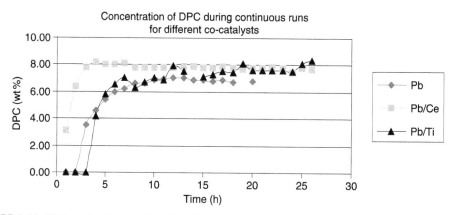

FIGURE 8.11 DPC production as a function of co-catalysts system during continuous BTU runs.

Pb-Ce as co-catalysts are presented in Figure 8.11. One can see that bimetallic co-catalysts perform better than the pure lead co-catalyst, which is consistent with results of the smaller-scale runs. Steady-state conditions are reached faster for Pb-Ce as co-catalyst, probably due to reduction of an induction period found for systems with lead co-catalysts. Steady-state DPC concentrations in the BTU runs are close to those obtained for batch reactions under similar conditions (no molecular sieves, lower pressure). The use of rapid and inexpensive screening experiments allowed for the optimum selection of systems to run at the BTU scale.

8.3.8 MECHANISM OF DPC FORMATION

DPC synthesis by oxidative carbonylation of phenol includes two catalytic reactions: palladium-catalyzed oxidative carbonylation and oxidation of Pd^0 with oxygen (Scheme 8.6). The first catalytic cycle is common for other oxidative carbonylation reactions. Palladium(II) complexes being dissolved in phenol react with CO to generate DPC and metal Pd. Formation of the intermediate acyl complex can proceed via two possible mechanisms: nucleophilic attack of coordinated CO by OPh^- or CO insertion into the Pd-OPh bond. Coordination of the phenoxy group to the acyl complex is followed by reductive elimination. At this stage possible ortho-metallation yields phenylsalicylate (PS). Apparently, Pd^0 species are the dormant state of the palladium catalyst (palladium black can be used as a catalyst) and Pd^0 oxidation is the rate-determining step. Catalytic carbonylation is first order in oxygen and zero order in CO. The presence of bromide ions eases the reoxidation due to reducing the Pd^0 oxidation potential by 300–400 mV. High bromide concentrations improve reaction selectivity, possibly preventing ortho-metallation of the coordinated phenolate group. However, if the concentration of Br^- ions is too large, the oxidation product is inactive coordinatively saturated $PdBr_4^{2-}$ ion. Because of these two opposite tendencies, the dependence of Pd TON vs bromide concentration has a maximum.

(8.6)

OH^- CO

$L = CO, PhOH, Br^-$

$(PhO)Pd^{II}BrL_2$

[O]

Br^-

$PdBr_4^{2-}$

$(PhOH)Pd^0(Br)L_2^-$

$(PhOC)Pd^{II}BrL_2$ O   PhOH

Br^- CO

$Pd^{II}(OH)_2$ HBr

$(PhOH)Pd^0(CO)L_2$ H_2O $Pd^{II}(OPh)_2L_2$ $(PhOC)Pd^{II}(OPh)L_2$

CO   PhOH, $-H_2$

$Pd^0(CO)_4$ Pd^0L_2 O   PS via orthometallation

PhOH, CO Pd_{black} DPC

The catalyst package deactivates with time. Possible reasons for catalyst deactivation might be aggregation of Pd^0 clusters (EXAFS of active systems shows the presence of small size clusters) or conversion of palladium into an inactive Pd(II) form, for instance, hydroxide, bromide depletion (nearly all Br^- is converted to BrPhOHs at the end of the reaction), attainment of a pH threshold (pH decreases during run and acid kills the catalyst), and finally co-catalyst conversion into an inactive form.

If the palladium catalytic cycle is understood in general terms, the oxygen catalytic cycle (depicted by [O] in Scheme 8.6) is unclear. It turns out that many transition metals and some non-transition metals serve as co-catalysts. Evidently, the mechanism of oxygen activation is different for the different co-catalysts. Some co-catalysts, for example, Mn(II), Co(II), and Cu(I) salts can be oxidized by oxygen to Mn(IV), Co(III), and Cu(II) and in an oxidized form can reoxidize Pd(0) to Pd(II). However, the activity of many metals cannot be explained simply in the terms of the redox potentials as many authors have suggested. For example, lead(II) cannot be oxidized by molecular oxygen and cannot, in turn, oxidize Pd(0). Therefore, an alternative mechanism of Pd(0) reoxidation should include reaction of oxygen with different types of species. The possible candidates for redox mediators are bromide ions, which have the right redox potential, or polynuclear lead complexes that are oxidized easier than mononuclear complexes. The formation of polynuclear lead species with short Pb–Pb interatomic distances in lead-based catalytic systems for DPC synthesis was found by EXAFS.[51] Such species are present only in reaction mixtures that are catalytically

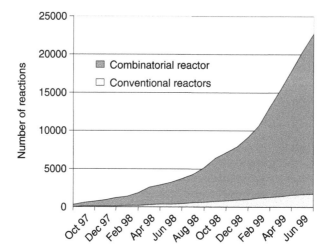

FIGURE 8.12 Number of reaction runs in the combinatorial system vs the number of conventional reaction runs over a 2-year period.

active. However, this is only one possibility and finding the mechanism of oxygen activation requires further work.

8.4 CONCLUSIONS

The use of high-throughput experimentation has allowed our group to test over 10,000 unique catalyst combinations. The leading catalyst formulations have been scaled up all the way to a 1-gallon continuous reactor. The use of HTS technology has proven invaluable for exploring large chemical spaces. The number of chemical reactions that could be carried out was ten times more than the use of conventional reactors alone (see Figure 8.12). A broad patent coverage was established for this chemical transformation.[52–68] The ability to explore this chemical reaction in detail has also led to a more complete understanding of the complex mechanism and kinetics of this commercially important reaction.

ACKNOWLEDGMENTS

We would like to thank the following colleagues for experimental help: James Spivack, James Cawse, Bruce Johnson, Kirill Shalyaev, Jonathan Male, Eric Presman, John Ofori, Dick Battista, Phil Moreno, Ben Patel, Timothy Chuck, David Smith, Tracey Jordan, Michael Brennan, Richard Kilmer, Eric Williams, and Yan Gao.

REFERENCES

1. Clagett, D. C. and Shafer, J. S., *Polym. Eng. Sci.*, 25, 457, 1985.
2. LeGrand, D. G and Bendler, J. T., *Handbook of Polycarbonate Science and Technology, Plastics Engineering*, Marcel Dekker, New York, 2000, pp. 10–12.
3. Ring, K. L. and Toki, G., *Chemical Economics Handbook*, SRI International, Menlo Park, CA, 2001, 580,1100H.
4. Illuminati, G., Romano, U. and Tesei, R., US Patent 4,182,726 1980 to Snam Progetti, S.p.A.
5. Romano, U. and Tesei, R., US Patent 4,045,464 1997 to Anic S.p.A.
6. Chalk, A. J., US Patent 4,096,169 1978 to General Electric Co.

7.   Chalk, A. J., US Patent 4,187,242 1980 to General Electric Co.
8.   Hallgren, J. E. and Matthews, R. O., *J. Organomet. Chem.*, 175, 135–142, 1979.
9.   Hallgren, J. E., Lucas, G. M. and Matthews, R. O., *J Organomet. Chem.*, 204, 135–138, 1981.
10. Hallgren, J. E. and Lucas, G. M., *J. Organomet. Chem.*, 212, 135–139, 1981.
11. Baeckvall, J. E., *Acc. Chem. Res.* 16, 335–342, 1983.
12. Baeckvall, J. E. and Gogoll, A., *J. Chem. Soc. Chem. Commun.*, 1238, 1987.
13. Tsuji, J., *Palladium Reagents and Catalysts – Innovations in Organic Synthesis*, Wiley, New York, 1995, pp. 19–20.
14. Hallgren, J. E., US Patent 4,096,168 1987 to General Electric Co.
15. Joyce, R. P., King, J. A. Jr. and Pressman, E. J., US Patent 5,399,734 1995 to General Electric Co.
16. King, J. A. Jr., Mackenzie, P. D. and Pressman, E. J., US Patent 5,399,734 1995 to General Electric Co.
17. King, J. A. Jr. and Pressman, E. J., US Patent 5,284,964 1994 to General Electric Co.
18. Takagi, M., Miyagi, H., Ohgomori, Y. and Iwane, H., US Patent 5,498,789 1996 to Mitsubishi Chemical Corp.
19. Ableles, B. and Hanak, J. J., *Phys. Lett.*, 34, 165–166, 1971.
20. Hanak, J. J., *J. Appl. Phys.*, 41, 4958, 1970.
21. Hanak, J. J., *J. Mater. Sci.*, 5, 964–971, 1970.
22. Hanak, J. J., *J. Vac. Sci. Technol.*, 8, 172–175, 1971.
23. Hanak, J. J., US Patent 3,803,438 1974 to RCA Corporation.
24. Hanak, J. J., *Jpn. J. Appl. Phys.*, Part 1 (Suppl. 2), 809, 1974.
25. Hanak, J. J., US Patent 3,919,589 1975 to RCA Corporation.
26. Hanak, J. J., US Patent 4,027,192 1975 to RCA Corporation.
27. Hanak, J. J., US Patent 4,167,015 1975 to RCA Corporation.
28. Hanak, J. J., *Le Vide* 175, 11, 1975.
29. Hanak, J. J., US Patent 4,157,215 1979 to RCA Corporation.
30. Hanak, J. J., US Patent 4,162,505 1979 to RCA Corporation.
31. Hanak, J. J., *Solar Energy* 23, 145–147, 1979.
32. Hanak, J. J., US Patent 4,272,641 1981 to RCA Corporation.
33. Hanak, J. J., US Patent 4,292,092 1981 to RCA Corporation.
34. Hanak, J. J., US Patent 4,316,049 1982 to RCA Corporation.
35. Hanak, J. J., Bolker, B. F. T., *J. Appl. Phys.*, 44, 5142–5147, 1973.
36. Hanak, J. J., Friel, R. N. and Goodman, L. A., US Patent 4,042,293 1977 to RCA Corporation.
37. Hanak, J. J. and Gittleman, J. I., *Physica*, 55, 555–561, 1971.
38. Hanak, J. J., Gittleman, J. I. and Cohen, R. W., *Phys. Lett.*, 29A, 56–57, 1969.
39. Hanak, J. J., Gittleman, J. I., Pellicane, J. P. and Bozowski, S., *Phys. Lett.*, 30, 201–202, 1969.
40. Hanak, J. J., Lehmann, H. W. and Wehner, R. K., *J. Appl. Phys.*, 43, 1666, 1972.
41. Thompson, S., *Chemtrek: Small-scale Experiments for General Chemistry*, Allyn and Bacon, Needham Heights MA, USA, 1989, pp. 125–136.
42. Hoogenboom, R., Meier, M. A. R. and Schubert, U. S., *Macromol. Rapid Commun.*, 24, 15–32, 2003.
43. Briceno, G., Chang, H., Sun, X., Schultz, P. G. and Xiang, X.-D., *Science,* 270, 273–275, 1995.
44. Gennari, C., Nestler, H. P., Piarulli, U. and Salom, B., *Liebigs Ann./Recl.*, 4, 637–647, 1997.
45. Burgess, K., Lim, H.-J., Porte, A. M. and Sulikowski, G. A., *Angew Chem. Int. Ed.*, 35, 220–222, 1996.
46. Cargill, J. F. and Maiefski, R. R., *Lab. Rob. Autom.*, 8, 139–148, 1996.
47. Hagemeyer, A., Jandeleit, B., Liu, Y., Poojary, D. M., Turner, H. W. Jr.,Volpe A. F. Jr. and Weinberg, H., *Appl. Catal. A* 221, 23–43, 2001.
48. Spivack, J. L., Cawse, J. N., Whisenhunt, D. W. Jr., *App. Cat. A: Gen.*, 254, 5–25, 2003.
49. Harry, M. J., *The Vision of Six Sigma, A Roadmap for Breakthrough*, Sigma Publishing Company, Phoenix, AZ, 1994.
50. Montgomery, D. C., *Design and Analysis of Experiments*, 3rd ed., Wiley, New York, 1991, pp. 357–374.
51. Soloveichik, G. L., Shalyaev, K. V., Patel, B. P., Gao, Y. and Pressman, E. J., "High throughput experiments in design and optimization of catalytic package for direct synthesis of diphenylcarbonate" in *Catalysis of Organic Reactions*, Sowa, J. Jr., Ed., Chemical Industries Series, Taylor & Francis, Boca Raton, FL, 2005, Vol. 104, pp. 185–194.
52. Patel, B. P., Soloveichik, G. L., Whisenhunt, D. W. Jr. and Shalyaev, K. V., US Patent 6,380,418 2002 to General Electric Co.

53. Patel, B. P., Soloveichik, G. L., Whisenhunt , D. W. Jr. and Shalyaev, K. V., US Patent 6,355,824 2002 to General Electric Co.
54. Patel, B. P., Soloveichik, G. L., Whisenhunt, D.W. Jr. and Shalyaev, K. V., US Patent 6,323,358 2002 to General Electric Co.
55. Patel, B. P., Soloveichik, G. L., Whisenhunt, D.W. Jr. and Shalyaev, K. V., US Patent 6,187,942 2002 to General Electric Co.
56. Shalyaev, K. V., Johnson, B. F., Whisenhunt, D. W. Jr. and Soloveichik, G. L., US Patent 6,440,893 2002 to General Electric Co.
57. Shalyaev, K. V., Soloveichik, G. L., Johnson, B. F. and Whisenhunt, D. W. Jr., US Patent 6,372,683 2002 to General Electric Co.
58. Shalyaev, K. V., Soloveichik, G. L., Whisenhunt, D. W. Jr. and Johnson, B. F., US Patent 6,440,892 2002 to General Electric Co.
59. Spivack, J. L., Cawse, J. N., Whisenhunt, D. W. Jr., Johnson, B. F. and Soloveichik, G. L., US Patent 6,201,146 2001 to General Electric Co.
60. Spivack, J. L., Whisenhunt, D. W. Jr., Cawse, J. N., Johnson, B. F., Soloveichik, G. L., Ofori, J. Y. and Pressman, E. J., US Patent 6,420,587 2002 to General Electric Co.
61. Spivack, J. L., Whisenhunt, D. W. Jr., Cawse, J. N., Johnson, B. F., Soloveichik, G. L., Ofori, J. Y. and Pressman, E. J., US Patent 6,197,991 2002 to General Electric Co.
62. Spivack, J. L., Whisenhunt, D. W. Jr., Cawse, J. N. and Johnson, B. F., US Patent 6,143,913 2000 to General Electric Co.
63. Spivack, J. L., Whisenhunt, D. W. Jr., Cawse, J. N., Johnson, B. F., Grade, M. M., Soloveichik, G. L., Ofori, J. Y. and Pressman, E. J., US Patent 6,160,154 2000 to General Electric Co.
64. Spivack, J. L., Whisenhunt, D. W. Jr., Cawse, J. N., Johnson, B. F. and Shalyaev, K. V., US Patent 6,355,597 2002 to General Electric Co.
65. Spivack, J. L., Whisenhunt, D. W. Jr., Cawse, J. N., Johnson, B. F. and Shalyaev, K. V., US Patent 6,114,563 2000 to General Electric Co.
66. Spivack, J. L., Whisenhunt, D. W. Jr., Cawse, J. N., Johnson, B. F., Soloveichik, G. L., Ofori, J. Y. and Pressman, E. J., US Patent 6,380,417 2002 to General Electric Co.
67. Spivack, J. L., Whisenhunt, D. W. Jr., Cawse, J. N., Johnson, B. F., Soloveichik, G. L., Ofori, J. Y. and Pressman, E. J., US Patent 6,143,914 2000 to General Electric Co.
68. Spivack, J. L., Whisenhunt, D. W. Jr., Cawse, J. N. and Soloveichik, G. L., US Patent 6,160,155 2000 to General Electric Co.

9 Catalyst Preparation for Parallel Testing in Heterogeneous Catalysis

Sabine Schimpf, Martin Lucas, and Peter Claus

CONTENTS

9.1   INTRODUCTION

This chapter will focus on catalyst preparation methods used in high-throughput screening of heterogeneously catalyzed reactions in gas as well as multiphase. The chapter is divided into preparation methods: (i) for the discovery stage, where surface science techniques e.g., chemical vapor deposition or physical vapor deposition are predominant, but more and more solution-based methods come to the fore, (ii) for wall-coated reactors, monolith on the one hand, which are used as multichannel reactors and coated microstructured reactors on the other hand and (iii) for the optimization stage. Here mostly conventionally prepared catalysts are used, often by using automated preparation techniques, e.g., pipette robots for incipient wetness deposition.

9.2 CATALYST PREPARATION FOR PRIMARY SCREENING (DISCOVERY STAGE)

The challenge in discovery stage screening (stage I screening) is to find which active compound or mixture of active compounds, mainly transition metals, catalyzes the desired reaction. This stage is characterized by low information depth and massive degree of parallelization, further scale-up is irrelevant at this stage. Very large libraries with large numbers of chemical compositions have to be created, where very small amounts of each catalyst (typically on the order of less than a milligram) are tested. Known from solid-state chemical synthesis and material discovery, surface science techniques are very well suited to the controlled preparation of large libraries of candidate compositions.[1] In solid-state chemical synthesis, several methods of rapidly creating variations in chemical composition have emerged.[2] In 1997, Schultz et al. created small arrays of high-temperature superconducting and giant-magneto-resistive thin-film materials by sputtering using a binary masking scheme.[2,3] This idea demonstrated the potential for much higher diversity in sample composition than previous work and had been extended to include several novel deposition technologies. Physical vapor deposition, electron-beam evaporation, radio-frequency sputtering, and pulsed-laser ablation can be used as the molecular sources with moving or stationary masks to achieve spatial variations in compositions. All these techniques are suited to creating catalyst libraries for initial screening. However, they cannot produce very realistic supported catalyst samples, because this approach ignores many important issues, such as reliable manufacturing processes of supported catalysts with promising material compositions, particle size issues, metal–support interactions, and transport limitations.[2]

9.2.1 THIN-FILM DEPOSITION-BASED CATALYST PREPARATION METHODS FOR HETEROGENEOUS CATALYSIS

Symyx used radio-frequency (RF) sputtering to create a $15 \times 15 \times 15$ library to test and validate combinatorial synthesis and screening techniques in the catalytic oxidation of CO with O_2 or NO. The library of 120 different catalysts (diameter 1.5 mm, thickness about 100 nm) was prepared by depositing three metals (e.g., Rh, Pd, Pt or Rh, Pt, Cu) using masks onto a quartz wafer (7.5 cm diameter and 1.5 mm thick). The techniques used by these investigators are based on the pioneering work of Hanak[4] and more recently by Xiang et al.[5] where the methodologies for parallel synthesis of spatially addressable solid-state materials libraries were reported.[6] Ten repeated steps were performed, where 10 nm of material per catalyst site was added in each step. The preparation of one library takes about 1 h. Each side at the apex of the triangle contained the pure metal, with its concentration decreasing linearly when going away from the apex and reaching zero at the adjacent side of the triangle. A row of 16 blank elements was added for control of background. After annealing the library at 773 K in a stream of hydrogen (5%) in argon the internal mixing in the various alloy catalysts was controlled by XRD. For comparison an additional library was made by sol–gel-based techniques by using automated liquid deposition robotics, revealing identical results as the RF-sputtered library.[1] A sketch of the ternary library prepared is shown in Figure 9.1.

To speed up synthesis of solid-state materials libraries Hanak[4] proposed simultaneous or co-sputtering of multiple-target materials.[6]

Another example for thin-film generation not in the field of heterogeneous catalysis should be mentioned as it became famous. Symyx developed an automated combinatorial method for synthesizing and characterizing thin-film libraries of up to 25,000 different materials, on a 3-inch-diameter substrate, as candidates for new phosphors aiming in the development of flat-panel displays and lighting.[7]

In addition, other thin-film deposition methods, such as thermal[8,9] and plasma chemical-vapor deposition[10,11] molecular beam epitaxy[12] and pulsed-laser deposition[13,14] are conceivable to create solid-state catalyst libraries.[6]

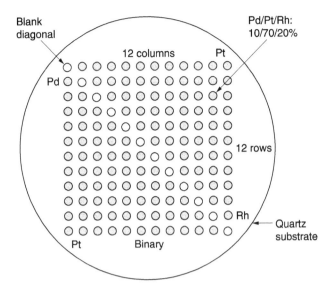

FIGURE 9.1 Catalyst library prepared by sequential sputtering.[6]

The use of thin-film deposition methods enables the creation of very large libraries in a very short time, but these methods imply also a great disadvantage. If larger quantities of each material are required for further testing, it may become difficult to reproduce the preparation by using more technically applicable chemical preparation techniques.[1] Consequently, thin-film deposition-based catalyst preparation methods are normally only suitable for primary screening. That is why there is a general trend to use solution-based methods of synthesis of combinatorial libraries. These methods are adapted from conventional catalyst preparation techniques, therefore up-scaling is much easier than for thin-film deposition methods.

Hence regarding the classification of these preparation methods the transition from stage I to stage II screening is in some cases fluent.

9.2.2 SOLUTION-BASED CATALYST PREPARATION METHODS FOR HETEROGENEOUS CATALYSIS

Solution-dispensing methods, developed for the pharmaceutical industry, have recently been extended to inorganic and organometallic catalyst systems. Solutions containing different amounts of several different metal precursors solutions were dispensed onto the supports in primary screening, mainly by using automated liquid-dispensing systems. Thus thin films of support (e.g., for the discovery stage) up to conventional catalyst support pellets (second stage) can be impregnated by precursor solutions of the active metal salts. This section will focus on the former.

The commercial preparation processes of most catalysts have been developed over decades of trial and error experiments, using solution-based techniques. These can be divided into two primary groups: impregnation and precipitation methods. There are also a number of other techniques that can be considered as variants of these two primary methods; they include complexation, gelation, crystallization, ion exchange, and grafting. Impregnation methods involve the soaking of a porous support (e.g., alumina, silica) with a solution containing the catalytic components, followed sometimes by filtration, drying, and activation. Consequently, the support primarily determines the surface area and mechanical properties of the final catalyst. Precipitation involves the mixing of two or more solutions or suspensions. Co-precipitation implies that the support is precipitated simultaneously if a conventional support is applied, the term used is deposition-precipitation. After (co-)precipitation follows filtration, washing, drying, forming, and activation. The surface area and mechanical properties of the catalytic materials have to be considered as an integral part of the preparation process. Since most commercial catalysts are prepared by techniques that are variations of these two techniques, the miniaturization and automation of impregnation and precipitation methods is likely to be the route to exploit combinatorial techniques in the near future.[6]

9.2.3 IMPREGNATION METHODS

The soaking of a porous carrier (powder or pellets of the support material) with a solution containing the catalytic components is performed by ink-jet print-head technology, manually, using automated liquid dispenser or fully automated preparation robots.

Redington et al. applied solution-based ink-jet print-head technology successfully to prepare a 645-combination Pt-Ru-Os-Ir library for the reforming of methanol.[15] This study led to the discovery of a superior catalyst with composition 44% Pt/41% Ru/10% Os/5% Ir.

Senkan et al. developed an automated, multinozzle liquid-dispensing system to rapidly and precisely prepare microliter-level solution libraries.[16] This system was used for the impregnation synthesis of a 66-combination catalyst library of Pt-Pd-In on γ-Al$_2$O$_3$ which subsequently was used to study the dehydrogenation of cyclohexane to benzene. The catalysts were prepared by mixing precursor solutions into individually addressable test tubes in an array using a computer-controlled x,y,z-translation table and a high-accuracy liquid-delivery system. Pellets were prepared by compacting high-surface γ-Al$_2$O$_3$ powder into cylindrical-shaped pellets and then added to the prepared precursor solutions and allowed to impregnate. The preparation system is shown in Figure 9.2. In the resulting library, the 0.8% Pt/0.1% Pd/0.1% In composition was determined to give the best benzene productivity in the range of conditions investigated.

The same automated liquid dispensation system was also used for the preparation of two 66-member ternary libraries of V-Mo-Li and V-Mo-Rb on γ-Al$_2$O$_3$ for the oxidation, using photoionization detection (PID) as a high-throughput screening tool for the dehydrogenation of ethane and propane as example reactions[17] and 56 quarternary Pt-Pd-In-Na combination catalysts for the selective reduction of NO by propane under both stoichiometric and fuel-lean conditions.[18]

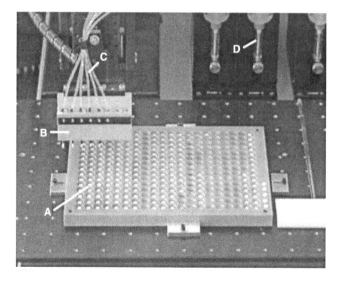

FIGURE 9.2 Micro-jet liquid-dispensation system for the preparation of solution libraries. (a) Wells to hold solution libraries and catalyst carrier pellets, (b) liquid injector nozzles, (c) solution delivery lines, (d) syringe pumps.[6]

A microtiter plate-based reactor for screening of supported catalysts was developed by Claus et al.[19] The reactor plate was built in stainless steel by producing 48 channels. The reactor plate is located between two heated stainless steel plates. The ratio of heated dead volume (0.02 cm^3) and of fill volume (0.69 cm^3) was minimized in order to reduce the vapor phase reaction. For this reason the construction is dedicated to oxidation reactions at temperatures up to 773 K. A multiposition valve with 48 channels (self-construction) is installed before the reactor channels. Only one reactor is flushed with the reaction gas (space velocity 500–20,000 h^{-1}), the others could be overflowed by inert gas. The outlet of the channels ends in a second multiposition valve with 48 channels. The product gas from the "active" channel is analyzed via gas chromatography or mass spectroscopy and therefore the system can be applied to any oxidation reaction with gaseous or evaporable substrates. A scheme of the experimental setup and a photograph are given in Figure 9.3. The catalysts of the experiments were prepared by the incipient-wetness method. Synthesis paths of the catalysts are largely automated by employing a pipetting robot (Miniprep 60, Tecan). For the preparation of the catalysts standard microtiter plates of polystyrene are suitable. Afterwards the material can be dried, calcined, and reduced. The advantage of the design arises from the simultaneous transfer from preparation plate to the reactor plate by transfilling all 48 catalysts in one step. The equipment was successfully tested in the catalytic oxidation of methane over noble-metal catalysts.

Automated liquid-handling robots were also applied by Symyx as a high-throughput primary synthesis method, e.g., to prepare mixed metal oxide catalysts for the gas phase oxidation of ethane to acetic acid.[20] The discovery libraries consisted of arrays of 256 catalysts on quartz wafers, primary screening was performed in parallel for catalytic activity in continuous flow 256-channel parallel microreactors. Acetic acid was detected in parallel using pH-based colorimetric techniques on products adsorbed on silica-coated glass TLC plates. This workflow allows the screening of more than 3000 samples per day. Promising leads were confirmed in focus libraries and are being optimized in secondary screening. MoV was identified as the most active binary of redox metals and was subsequently doped with main group, rare earth, and transition metals to form ternaries. Prior art MoVX (X = Nb, Ni, Sb) catalysts were successfully reproduced and it was shown that

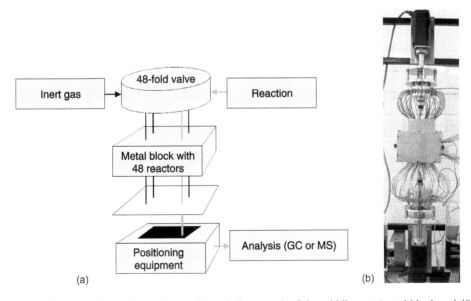

FIGURE 9.3 Scheme of experimental setup (a) and photograph of the middle part (metal block and 48-fold valves) (b).[15]

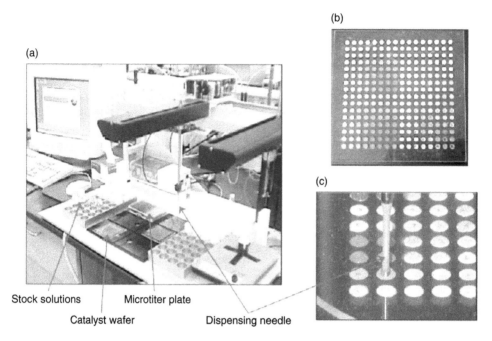

FIGURE 9.4 (a) Synthesis station with liquid handling robot, rack of stock solutions, microtiter plate, and catalyst wafers; (b) quartz wafer with catalyst matrix; (c) liquid dispensing into carrier pre-coated wafer wells (2 mL dispense volume/well; ~0.5 mg catalyst loading/well).[20]

Pd doping significantly increases the catalytic activity of these systems. Wafer-formatted catalyst libraries were designed and synthesized using Symyx Proprietary Library Studio® and Impressionist® Software.[21] Commercial 3 inch × 3 inch quartz wafers were bead blasted through steel masks with α-Al_2O_3 powder to produce arrays of wells which were then precoated with an alumina primer layer by slurry dispensation and dried. The metal precursor solutions were dispensed onto the wafers using automated liquid handling robots (Figure 9.4). The catalysts were prepared by subsequent drying and calcination steps. Aqueous metal nitrates (alkali, earth alkali, rare earth, Group IIIB, IB, IIB metals, Zr, Cr, Mn, Fe, Co, Ni, Al, Ga, In, Pb, Bi, Ru, Rh, Pd, and Pt), oxalates (Ti, V, Nb, Ta, Mo, Sn, and Ge), ammonium salts (vanadate, tungstate, and Sb oxalate) and acids (boric, perrhenic, telluric, selenic, and phosphoric) and in some cases chlorides (Hf, Sn, Ir, and Pt) were used as the standard metal precursors. Carriers used were silica, alumina, titania, and zirconia which were slurry dispensed onto the wafer prior to metal deposition. Initially the wafer was impregnated two to three times with the same library design (from the same microtiter plate) to increase the catalyst loading in the wells (to about 1 mg; 2 ml dispense volume per well per dispense step) to achieve higher acetic acid productivity and thus higher detection signal intensity. A typical post-synthesis treatment involved a drying step at 393 K and calcination in air at 623–673 K for 4 h.

The same equipment was used to prepare libraries consisting of 96 catalysts, which were screened in the selective oxidation of alcohols in liquid phase. Catalyst libraries for primary screening were synthesized in situ also by using Library Studio® and Impressionist® software (Symyx Technologies). The metal precursor solutions were dispensed automatically by Cavro™ robots and the catalysts were prepared by subsequent drying and calcination steps. Wet impregnation techniques of preloaded carriers, freeze-drying methods, and slurry-dispensing of preformed solid catalysts were also employed.[22] Two 96-multi-well reactor systems were developed by Symyx. One, which can tolerate pressures up to 60 bar (Type B), is shown in Figure 9.5.

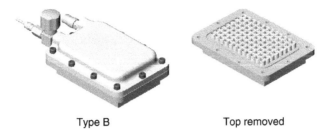

Type B   Top removed

FIGURE 9.5 Ninety-six-well parallel-batch reactor: an 8 × 12 array of 1 mL vials is assembled in a reactor with common headspace. The reactor can tolerate temperatures up to 473 K and pressures up to 60 bar.[22]

Promising hits identified in the high-throughput primary screens were successfully scaled up and optimized in conventional laboratory test units. High-throughput synthesis and screening of polyoxometalate and supported-metal libraries have been developed for the selective aerobic oxidation of alcohols to the corresponding aldehydes/ketones in the liquid phase. Libraries consisting of 96 catalysts were prepared in multi-well reactors and screened for catalytic activity using TLC, GC, and NMR detection methods. Isolated yields confirm high selectivities of more than 90% with quantitative conversions. Substrates tested include primary and secondary alcohols. Specific results are given for hydroxymethyl-substituted heterocycles and bicyclo-octanols.

Another autoclave array high-pressure unit with 24 autoclaves for stage 1 screening was developed by Maschmeyer et al. and successfully tested in the reductive amination of benzaldehyde in the presence of ammonia.[23] In this work the authors used conventional catalyst preparation and support treatment.

Holmes et al. focused on preparation of bulk-mixed oxides without catalytic testing. To explore the utility of liquid-phase automated synthesis for the preparation of bulk-mixed metal oxides they investigated the syntheses of Mo–V–Sb–Nb–O bulk materials by using soluble precursor materials. These catalyst systems are suitable for the selective oxidation of propane to acrolein and acrylic acid. The products were characterized by X-ray powder diffraction and Raman spectroscopic studies. Another objective of this work was the identification of the oxide phases present in the system.[24]

9.2.4 PRECIPITATION METHODS

9.2.4.1 Precipitation and Co-Precipitation

The preparation of gold catalysts, mostly performed by co-precipitation and precipitation, is known to be difficult to reproduce. On this account Schüth et al. addressed this problem by parallelized synthesis and screening the catalysts in the room-temperature oxidation of CO.[25] For the preparation of the gold catalysts a Gilson XL 232 automated dispenser was used.

Au/Co_3O_4 catalysts were prepared by co-precipitation and Au/TiO_2 catalysts by precipitation of gold onto the supports using different concentrations, pH values during precipitation, temperatures of calcination, and support materials. As large amounts of liquid are necessary for precipitation, the preparations were carried out in 50-mL test tubes placed on a shaker tray to avoid settling of the precipitate. The suspensions were transferred by means of a stainless steel tube of 1 mm inner diameter, attached to the dispenser, to a filtration unit to be filtrated and rinsed with water. The whole unit was then placed in an oven for thermal treatment. After that, the catalysts were transferred manually and weighed separately into the sinter plate (46 mg respectively) to guarantee identical space velocity (15.6 mL/min per well). The overall process was highly reproducible, and the data quality is comparable to that of conventional testing in a fixed-bed reactor (200 mg catalyst).

9.2.4.2   Sol–Gel Method

Symyx developed the creation of arrays of thick films prepared from stabilized sol–gel precursors by automated solution deposition techniques.[26,27] Primary screening was performed by using simultaneous MS and photothermal deflection spectroscopy examining the oxidative dehydrogenation within the Mo-V-Nb-O system. Typically, about 150 catalyst compositions (100–200 µg per catalyst) are measured per experiment in the primary screen. Commercially available metal alkoxides in 2-methoxyethanol (0.5 M) after which metal-specific modifiers were added were prepared by refluxing. Libraries of precursor solutions were initially created by using automated liquid-dispensing robots in microtiter plates (11 × 11 × 11 triangular matrix). Along each row of the triangle the metal ratios were decremented by 10% while maintaining constant volume, binary metal alkoxide solutions along the sides, and ternary mixtures within the interior of the triangle. These solutions (3 µL) then are transferred to a chemically and mechanically modified quartz substrate to create two duplicate 11 × 11 × 11 triangular catalyst libraries within a 12 × 12 rectangular array (3 mm in diameter with 4 mm spacing between element centers). The wetting characteristics of the quartz surface were chemically modified by using organosilane reagents. Gelation occurred under ambient conditions followed by annealing under conditions necessary to reproduce X-ray diffraction patterns of bulk samples. Substrates were heated in air at 1 K/min and held at 393 K for 2 h, further heated at 1 K/min and held at 453 K for 2 h, and heated at 2 K/min and held at 673 K for 4 h, after which the sample was allowed to cool naturally.

This setup allows a total throughput of more than 10,000 catalyst compositions per month for which trends are observed under conditions of low reactant conversions. Areas of high product yield may be re-examined in the primary screen or examined in a secondary screen, which was performed under realistic reaction conditions in an array of fixed-bed (50 mg catalyst respectively) reactors with gas chromatography.

In an analogous manner Symxy also prepared and screened V-Al-Nb and Cr-Al-Nb oxide libraries in the ethane oxidative dehydrogenation to ethylene[28] as well as Ni-Nb-Ta-Co mixed oxides. The results were also revised in multichannel fixed bed reactors (50 mg catalyst) and in a conventional bench scale reactor reactors (5 g catalyst).[29]

Maier et al. applied the sol–gel method for the preparation of amorphous microporous mixed oxides (AMM). For the hydrogenation of 1-hexyne 37 combinations of 1–10% Co, Cr, Cu, Fe, Ir, Mn, Ni, Pd, Pt, Rh, Ru, V, Zn on AMM of Si and Ti as catalysts were prepared and screened.[30] The same group also reported the preparation of 33 combinations of 1–6% Ag, Au, Bi, Co, In, Cr, Cu, Fe, Mo, Ni, Re, Rh, Sb, Ta, Te, V, Y on Si, Ti, and Zr AMM as catalysts for the oxidation of propylene[31] and used the same preparation method to investigate binary mixed oxides in the gas-phase oxidation of toluene with air.[32]

9.2.4.3   Hydrothermal Synthesis of Zeolites

One hundred or more solid-state syntheses can be conducted in parallel and employed for the combinatorial hydrothermal syntheses of zeolites at temperatures up to 473 K by using a novel multi-autoclave design, simplest and inexpensive a Teflon block in which 100 reaction chambers are formed by cylindric holes with Teflon-coated septa (see Figure 9.6). These Teflon blocks fit into the compartment of commercial pipette robots for convenient formulation of the synthesis gels. The operation of the multiautoclave was ascertained by the reinvestigation of the complete Na_2O-Al_2O_3-SiO_2 ternary system in a single experiment. After filling in the synthesis gels, they were aged and crystallized, followed by in situ washing and then transferred for X-ray analysis.[33] The same system was used by the same authors for the synthesis of a series of alumo phosphate materials.[34]

Hydrothermal synthesis was also investigated by Maier et al. They developed an 8-µL multichambered microreactor to prepare material libraries through combinatorial hydrothermal synthesis at temperatures up to 573 K. In a model experiment the synthesis of the zeolite TS-1 has been varied

FIGURE 9.6 Multiautoclave showing the mode of stacking of the Teflon blocks and one of the alternative designs using Teflon inserts.[33]

combinatorially. The resulting library is characterized directly (without transfer) by automated microdiffraction.[35]

9.2.4.4 Multinary Oxides by Using Activated Carbon as an Exotemplate

Automated libraries of high surface area multinary oxides are synthesized by using activated carbon as an exotemplate for the generation of small particles. If activated carbon is impregnated with a highly concentrated metal-salt solution (nitrates have been best in most cases) and then dried and calcined, a highly disperse oxide remains. Catalyst impregnation was performed using a commercial pipette system (Gilson XL 232). The carbon was transferred into vials in a 77-well plate. A mixed precursor solution was used which corresponds to incipient-wetness conditions. The whole library was then dried and slowly heated to 773 K and remained there for 2 h, which led to full combustion of the carbon. These catalysts were tested in the oxidative dehydrogenation (ODH) of ethane and in the low-temperature CO oxidation, where catalysts that are free of noble metals have been identified.[36]

9.3 CATALYST PREPARATION FOR CATALYTICALLY COATED REACTORS

This part will deal with monoliths and microreactors, where the catalytically active material is coated as a thin film on the wall of the reactor. In the first part of this subchapter a system for the parallel screening of automobile-exhaust catalysts will be described. In the second part a microreactor consisting of a wafer stack is introduced, where every wafer is coated with a different catalytic material for parallel testing, is presented.

9.3.1 CATALYST PREPARATION IN MONOLITH STRUCTURES

Monolith structures are well known as support of automobile-exhaust catalysts. The catalytic active layer is deposited onto the walls of the monolith mainly by the wash–coat procedure. Claus et al.

used monoliths for high-throughput screening for total oxidation reactions, where the channels are used as single plug-flow reactors containing different catalysts. Therefore we chose as monolith a cordierite based material (Corderit 410, Inocermic GmbH Hermsdorf), which is impermeable for gases and fluids, in contrast to the commonly used monolith materials, which are permeable for gases and fluids. The monolith structure used consists of 10×20 channels with channel width of 2.6 mm (72 cpsi) and a channel length of 75 mm. For the catalytic tests 128 of the 200 channels were used, arranged in an array of eight rows and 18 columns. By a wash–coat procedure[37] the monolith is coated uniformly with the different catalyst support materials (Al_2O_3, SiO_2, TiO_2, etc.). More details about coating with support material, loading with catalytically active compounds and activation of commonly monolithic automobile exhaust gas catalysts and monoliths for other heterogeneously catalyzed reactions can be found in reference 38. Afterwards the coated monolith is sealed at one side and then saturated channel by channel individually with 128 individual aqueous solutions of the metal precursors (mostly the nitrates, but in the case of Ru and Au, the chlorides). After 2–10 s (this time is necessary for saturating the oxide layer completely) the dilutions are removed. The supply of the solutions, the filling of the channels and the removing of the solutions after a defined time are effected by a robot (Tecan Miniprep 60). The desired content of a metal can be adjusted with high accuracy, if the pore volume or the capacity of water absorption of the oxide layer is already known analog to the dry impregnation or incipient-wetness method.[39] The common pretreatment methods, calcination and reduction, are performed in flowing air and hydrogen (3 h at 673 K, flow: 50 l/h), respectively. Alternatively, the immobilization of metals was also achieved by deposition–precipitation using urea. In this case, aqueous solutions of the metal precursors were used, followed by mixing with urea and finally heating to 353 K. At this point, decomposition of urea starts yielding hydroxide ions, which give the corresponding hydroxides. Subsequently, again calcination and reduction steps produce a library of supported oxide or metal catalysts. This monolith-based reactor was used for high-throughput screening of total oxidation of hydrocarbon and CO in the presence of further components like O_2, H_2O, CO_2, NO, SO_2, and inert gas, where the channels are used as single plug-flow reactors containing different catalysts. Water was dosed by a liquid-flow controller into an evaporator and then mixed with the gaseous feed dosed from premixed gas bottles by a mass-flow controller. The reaction gas was conducted to the 3D positioning system and is proportioned channel-by-channel into the monolith. At the same time a gas probe is taken out in 48 mm of depth. The interior of the 3D positioning system is rinsed with Argon. A scheme of the reactor itself, the reaction gas inlet and sample outlet is given in Figure 9.7. Inside this dosing tube there is 1/32″-sample-drawing capillary which juts out 48 mm over the end.

The sample drawing is made by sucking a defined gas amount, which is controlled by a mass-flow controller out of the single channel using a vacuum pump. In the first applications gas chromatography was applied. Then, the gas probe passes a GC-sampling valve, which injects the contents of the sample loop computer-controlled into the GC. Behind the GC-sampling valve and following the flow direction some part of the gas probe taken from the monolith channel is supplied to a mass spectrometer. Alternatively a three-way switching valve can be switched in a way that the gaseous feed mixture can be analyzed. Prompt and exact analyses are achieved by using the mass spectrometer unless there are overlaying spectra (MZ 28), as happened with CO and CO_2, which cannot be separated numerically if there is a large CO_2 excess. In that case a gas chromatograph equipped with a methanizer was used. The time for an analysis cycle can be reduced to 1.5 min, if necessary. More details about the experimental setup, the screening procedure, product analysis, reproducibility, temperature profile and the results can be found in reference 40. For later applications a scanning mass spectrometer was applied consisting of a mass spectrometer having a capillary sample inlet and a sampling device that positions the capillary in the x, y, z-direction at defined positions within the parallel reactor configuration. The gas samples are continuously transferred to the mass spectrometer through the inlet capillary allowing a very fast online analysis. Based on the experience with this type of analysis tool, the scanning MS can be regarded as an almost universal analysis technique for gas-phase reactions in combinatorial catalysis. It can be used with virtually

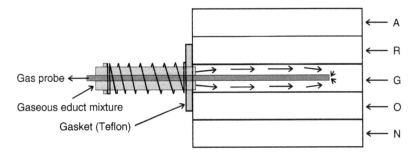

FIGURE 9.7 Sketch showing the principle of gas dosage and sampling in the monolith for the high-throughput screening especially convenient for deactivating catalysts. The gaseous feed (reaction gas) is proportioned channel-by-channel into the monolith. At the same time, a gas probe is taken out in 48 mm of depth.

every reactor configuration and application, where time and spatially resolved sampling is needed and complex gaseous mixtures have to be analyzed. The sample is taken by a replaceable silica capillary in a gas-tight housing. The functioning principle of the sampling device is illustrated in Figure 9.8a[41] and the capillary in front of the product outlets of the monolith reactor is shown in Figure 9.8b. The capillary can be positioned with an accuracy of about 10 μm under hazardous chemical conditions and high thermal demands. During the sampling the capillary moves first in the x-direction and y-direction to the corresponding channel and afterwards in z-direction a certain depth into the product outlet of that channel. The moving parameters and the position of the capillary, adjusted and supervised by a CCD camera equipped with magnifying optics, are freely programmable. The materials of the housing and the sealing were carefully chosen to withstand temperatures up to 723 K. The analysis speed can be determined by choosing the right MS conditions and the application of short and small inlet capillaries. Depending of course on the nature of the products of interest, analysis times of less than 60 s can easily be achieved. Therefore, this scanning mass spectrometry technique is very interesting for primary screening approaches, i.e., screening very large reactor arrays within a short time. By choosing the right capillary and housing configurations it is also possible to apply this technique in high temperatures and high pressures. The system can be run overnight or longer without the need for observing the screening progress. That means that the moving parameters and the position of the capillary are controlled by appropriate software and supervised by a CCD camera while, at the same time, the mass spectrometer analyzes the gas samples according to the capillary position. It is possible to run a predefined program varying the sampling parameters. Furthermore, reaction conditions including residence time, reactant concentration, reactor temperature and pressure are controlled and can be varied during screening.

9.3.2 MICROCHANNEL REACTORS

Another promising approach is the application of microreactor systems. Due to advancements in microfabrication techniques, this type of chemical reactor, namely microchannel reactors, was developed several years ago.[42] Strictly speaking, a microreactor consists of a very large number of parallel microchannels in the range < 1 mm, having a cross-section of, for instance, $500 \times 500\ \mu m^2$. By means of such small channel dimensions, very large heat and mass transfer rates can be achieved. Furthermore, explosive and dangerous reactions can reasonably be carried out in microchannel reactors. As the term microreactor is often used for reactors with small amounts of catalysts, we will here use the more precise expression microchannel reactor or microstructured reactor. Obviously, microchannel reactors for heterogeneously catalyzed reactions are gaining increasing importance because of higher conversion degrees and selectivities as well as process intensification. The

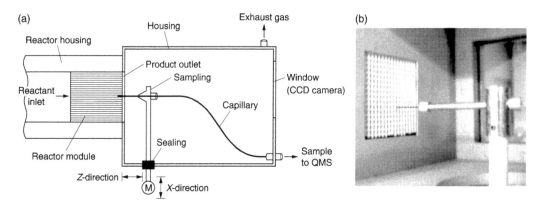

FIGURE 9.8 The principle of how the developed sampling device functions. The sample is taken by the capillary (a) and the capillary in front of the product outlet of the monolith reactor (b).[40]

approach we present here is, therefore, focused on the combinatorial screening of catalysts, where the catalytic active species are immobilized on the channel walls of the microchannels. The key elements of this approach are – as stated above – a suitable reactor module and the analysis technique. Regarding the catalysts on the surface of such microchannels there are of course a number of challenges, such as the adhesion of these catalytic active coatings onto the channel walls, the catalytic activity due to often lower surface areas and the long-term behavior. There is a wide variety of methods to insert the catalytically active wall coating, e.g., sputtering, CVD,[43] anodic oxidation,[44] sol–gel deposition,[45] and deposition of nanoparticles.[46] Also, carbon-based coatings can be achieved.[47] The most widespread method is the use of primers, for example dispersible boehmite alumina systems such as Disperal® (Sasol)[48] to prepare alumina-based coatings, often in combination with organic primers or binders, e.g., polyvinyl alcohol (PVA).[49–51]

The first and simplest generation of the reactor module consists of a stack of metallic frames, as shown in Figure 9.9a.[41] Stacking these metallic frames together, several parallel and independent microreactors are formed. The prototype consisted of 35 microreactors. Each microreactor is then filled with a microstructured inlay containing one catalyst as catalytic active coating on top of its microchannels. It is, therefore, possible to screen 35 different catalysts in parallel. A photograph of the reactor module is shown in Figure 9.9b.

In order to test the catalytic properties of the prepared catalysts, the reactants are uniformly distributed to all parallel microreactors. Then, the reactants flow through the microchannels of the microstructured inlays, react on the catalytic active surface of the microchannels and the products are withdrawn through the product outlets. The catalysts can easily be removed and replaced with a new set of catalysts. The module is heated externally to achieve reaction temperatures up to 723 K. The product outlet at the end of each microreactor has a cross-section of 0.5–2 mm. The reactor module design allows the use of replaceable microstructured catalyst inlays made of different materials such as metals, silicon, ceramics, and glass. This allows many different catalyst synthesis procedures as well as flexible microchannel geometry being essential to the generation of suitable catalyst libraries for different heterogeneously catalyzed reactions. Figure 9.10 presents a typical microstructured support and some examples of catalysts prepared on different substrates.

As the products from each catalyst should be analyzed with low background noise from the other ones, the sample has to be taken inside the product outlet of the corresponding microreactor. Product analysis was performed by the same scanning mass spectrometer already described above (monolith). The catalyst preparation procedure is carried out in parallel for 35 catalysts, but is not yet automated. Therefore, 35 catalysts are produced per day, being in accordance with the time for screening these catalysts under different reaction conditions. An automated liquid-dispensing

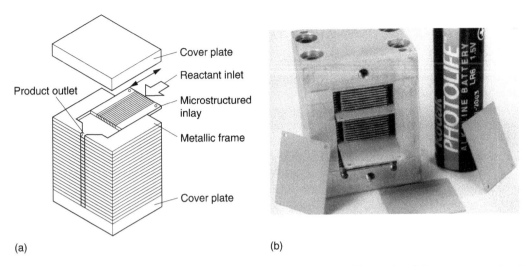

(a)   (b)

FIGURE 9.9 Reactor module consisting of 35 stacked metallic frames. The catalyst inlays are mounted and removed in the direction of the arrow (a). Photograph of the reactor module and the microstructured inlays (b).[41]

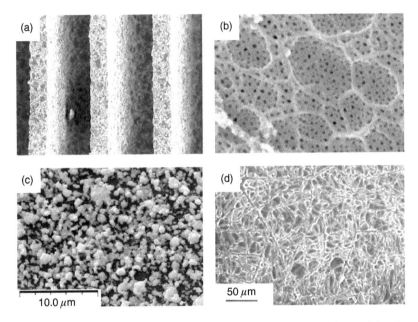

FIGURE 9.10 (a) Typical microstructure with a channel width 300 μm manufactured by electrodischarge micromachining; (b) Al_2O_3 layer prepared by anodic oxidation, pore diameter approximately 40 nm; (c) activated silver catalyst on Al_2O_3 prepared by sputtering; (d) Ru/Al_2O_3 catalyst prepared by sol–gel technique.[62]

robot would facilitate the catalyst preparation procedure. The system was applied to investigate methane oxidation and oxidative dehydrogenation of i-butane to i-butene. The results are summarized in reference 62 and are given in detail in reference 52.

Parallel catalyst preparation by coating and testing was also applied by Pantu and Gavalas.[53] The catalyst samples are prepared in the form of thin films coated on thin quartz rods by dip-coating in solutions of different composition. The system was tested with the reaction of methane reforming with carbon dioxide over $Pt/Ce_{1-x}Gd_xO_{2-0.5x}$ and $Pt/Ce_{1-x}Sm_xO_{2-0.5x}$.

9.4 PREPARATION TECHNIQUES FOR SECONDARY SCREENING (OPTIMIZATION STAGE)

This section deals with the optimization stage of screening. Stage II screening targets the optimization of formulations that exist already using realistic preparation methods under realistic reaction conditions, without necessarily excluding the discovery of new materials.[54] The predominating preparation techniques are solution-based preparation methods, as thin-film deposition methods are not suitable for preparing the larger amounts required. On the other hand, to achieve reliable data for scale-up, the catalysts should be passed through in gas-phase reactions (fixed-bed) or stirred in liquid-phase reactions. On this account mostly conventional preparation methods are used, in some cases automated solution-based impregnation. We will not dwell on conventional preparation methods, as they are sufficiently and accurately described in well-established texts.[55] Thus, in this part the few examples of reactors and fixed-bed reactors, as well as reactors for liquid phase or multiphase reacts are described in detail and a short overview of other examples is given. As not all working groups are equipped sufficiently with highly parallelized fast-screening devices for catalyst discovery, most of the equipment for optimization stage is also used for catalyst discovery. Nevertheless, the configuration of the stage II reactor setups make it possible to vary the reaction conditions (temperature, amount of catalyst, space time, feed concentration, etc.) and the testing should be performed under conditions that are as realistic as possible.

9.4.1 FIXED-BED REACTORS

Gas-phase reactions are mostly performed in fixed-bed reactors, where the catalysts are passed through by the feed. Two examples are described in detail, with increasing focus on scale-up data. The first is a 49-channel reactor developed by Schüth et al. for screening catalysts.[56] The second is a six-flow reactor system used in the group of Kapteijn and Mouljin for kinetic studies.

Figure 9.11 is a photograph of the 49-channel reactor developed by Schuth et al. An example reaction of methane oxidation was chosen.

For product analysis different methods were possible. Here a two-GC setup with a hot and a cold column was used. To prepare the catalysts the same automated system was used as already described in the first part,[25] but the preparation method here was incipient wetness impregnation of the sieved support material. This means that only the amount of precursor solution (Cu, Pt, W, and Mo salts) was added, corresponding to the pore volume of support. The authors checked also the reproducibility and reliability of the preparation procedure for different preparation methods, which was found to be at least as high as for manual preparation. After preparation the catalysts were pressed and crushed to split and had to be transferred to the reactor manually. The results were found to be comparable to the results achieved in a conventional single-tube reactor and with this setup it was shown that it is possible to analyze catalysts under close-to-conventional conditions with a throughput of 150 catalysts per week.

A six-flow reactor for fast catalyst screening and kinetic studies,[57] as well as reaction engineering or scale-up questions such as bed dilution of the catalyst,[58] was developed in the group of Kapteijn and Mouljin and tested for different reactions using conventionally prepared catalysts. The reactor setup is given in Figure 9.12.

A choice of operation modes is depicted in Figure 9.13. The reactors can be filled with five different catalysts (Figure 9.13a) for activity and stability tests. One reactor was always used as a reference. For kinetic studies different amounts of the same catalysts are used (Figure 9.13b).

By varying the molar feed flow (F_A^0) and the amount of catalyst (W) a large range of W/F_A^0 can be obtained. Mass transport limitation was checked by loading the reactors with the same amounts of catalysts of different particle size to indicate intraparticle diffusion limitations (Figure 9.13c) and loading different amounts and feeding with different amounts, such that the space time for each reactor is the same, will reveal absence or presence of external mass transfer (Figure 9.13d). The

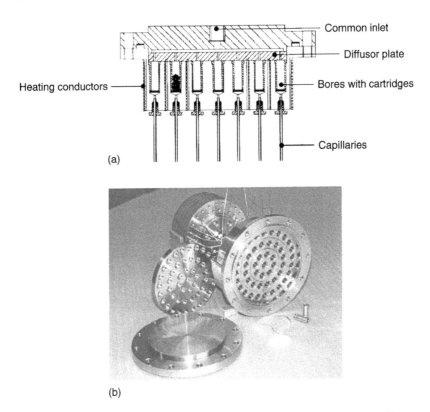

FIGURE 9.11 Schematic drawing of the 49-fold reactor (a). Photograph of the reactor (b).[56]

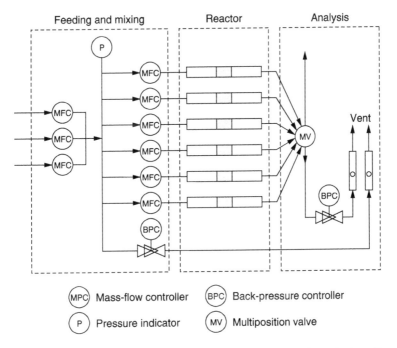

FIGURE 9.12 Scheme of six-flow reactor set-up for environmental catalysis investigations.[57]

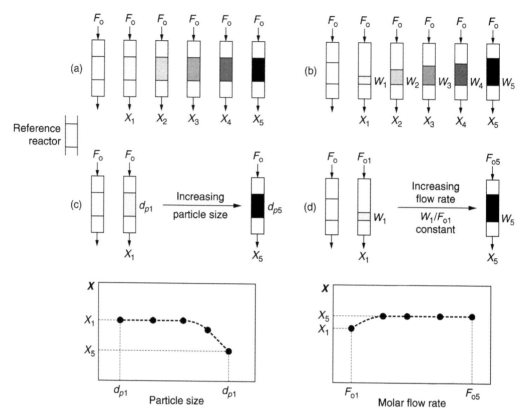

FIGURE 9.13 Applications of parallel reactor systems: (a) comparing different catalysts; (b) kinetic studies; (c) and (d) diagnostic experiments for tests of intraparticle and extraparticle limitations, respectively.[58]

different reactions investigated are summarized in Table 9.1. The product analysis methods depend on the reaction investigated and is also given in Table 9.1.

There are several studies dealing with high-throughput screening under realistic preparation and reaction conditions. An overview of other studies performed in parallelized fixed-bed reactors are summarized in Table 9.2, where the reaction is performed, the catalysts screened, the preparation methods applied, some features and the analysis is given.

9.4.2   REACTORS FOR LIQUID PHASE/MULTIPHASE REACTIONS

Although a number of effective tools to screen catalysts for gas-phase reactions has successfully been developed, only a few developments of parallel and fast-screening methods for heterogeneously catalyzed gas–liquid-phase reactions have been reported, although this type of reaction is widely known in the chemical industry. As the preparation methods do not differ for gas-phase or liquid-phase reactions, this part of the contribution will describe two reactors for screening heterogeneously catalyzed gas–liquid reactions. Multiautoclaves with reactor volumes of 1–2 mL were developed by Symyx[22] and Maschmeyer et al.[23] of about 10 mL by Schüth et al.[53] and of 45 mL by Claus et al.[60]

Schüth et al. developed a high-throughput reactor for the optimization of three-phase reactions specifically for what is termed stage II, the optimization phase.[59] The system allows the catalytic screening of 25 samples simultaneously at a maximum pressure of 50 bar and is shown in Figure 9.14.

TABLE 9.1

Utilization of the Six-Flow Technology in Catalyst Development: Screening and Kinetic Studies[37]

Process	Catalytic System	Product Analysis*
CO oxidation	Cu/Cr/activated carbon	
NO_x reduction with CO	Cu/Cr/activated carbon	GC
	Cu/Cr/γ-Al$_2$O$_3$	NO_x analyzer
Water-gas shift reaction	Cu/Cr/activated carbon	
NO_x reduction with ammonia	Cu/activated carbon	
	Mn oxides/activated carbon	
	Mn oxides/γ-Al$_2$O$_3$	MS
	Modified activated carbons	
	Mn$_2$O$_3$-WO$_3$/γ-Al$_2$O$_3$	
Soot oxidation	Ag/Mn/ZrO$_2$	
	Cu-V/Al$_2$O$_3$	
	Cu/K/Mo/(Cl)/ZrO$_2$	
	Cu/K/Mo/(Cl)/TiO$_2$	
	Cu/ZrO$_2$	NDIR
	Mo/ZrO$_2$	HC analyzer
	V/ZrO$_2$	
	Cs$_2$/MoO$_4$-V$_2$O$_5$	
	CsVO$_3$-MoO$_3$	
	Cs$_2$SO$_4$-V$_2$O$_5$	
Chloroflurocarbons (CFCs)	[Rh, Ru, Re, Pt, Ir]/activated carbon	
hydrogenolysis	Pd/activated carbon	GC
	Pd-zeolite Y	NO_x analyzer
	Pd-mordenite	
N$_2$O decomposition	Calcined hydrotalcites (Co, Pd, La, Rh)	
	[Fe, Co, Rh, Pd]-ZSM-5	GC
NO_x reduction with hydrocarbons	Pt/Al$_2$O$_3$	NO_x analyzer
	Pt-ZSM-5	
Fischer–Tropsch synthesis	Co-zeolite Y	
	Co/SiO$_2$	GC

GC: gas chromatography; NO_x: nitrogen oxides, MS: Mass spectrometer; HC: hydrocarbons; NDIR: non-dispersive infra-red.

The catalytic hydrogenation of crotonaldehyde (CrAld) over bimetallic samples Pt5Mx/AC based on a commercial 5 wt% Pt on activated carbon catalyst was investigated. Automated preparation was performed by using a commercial pipette system (Gilson XL 232) to deliver the nitrate salts of the second metals to pre-weighed Pt/AC (incipient-wetness deposition). Experiments demonstrate the reliability and reproducibility of the reactor. Modification of the mono-metallic Pt sample by the impregnation of aqueous metal salts and various pre-treatments resulted in 140 bimetallic catalysts that were used in the hydrogenation study.

Claus et al. developed a multibatch system consisting of 15 parallel working mini autoclaves (reactor volume: 45 mL), which are arranged in three arrays (size per array: 55 cm × 38 cm × 18 cm). A photograph of one array with five autoclaves is given in Figure 9.15. Each array can be heated and stirred independently from the others. Each autoclave is equipped with a high-precision pressure transducer. Thus, fast on-line monitoring of the catalyst activity via pressure–time dependencies is realized and allows an effective comparison with the results obtained in the other autoclaves already during the experiment. In the present configuration, it is possible to screen 15 different catalysts for gas–liquid reactions, e.g., for selective hydrogenations, in parallel for extended periods of time, or to vary reaction conditions: temperature up to 573 K, pressure up to 117 bar, stirring rate

TABLE 9.2
Overview of Other Studies Performed in Parallelized Fixed-Bed Reactors

Reaction	Catalysts	Preparation Method	Experimental/ Comments	Analysis	Reference
Methane oxidation	36 $Pt/Zr/V/Al_2O_3$ catalyts	Impregnation under incipient wetness conditions using an automatic liquid handler (Gilson)	Catalyst optimization, Al_2O_3 layer onto microstructured Al inlays having flow channel diameters of 250 μm	QMS	62
CO oxidation	Au catalysts	Precipitation, deposition–precipitation, impregnation, sol–gel technique	Catalyst optimization, monolith, where different channels are filled with different catalysts	QMS	62
Oxidative dehydrogenation of cyclohexane to benzene	Ternary Pt, Pd, In library	Impregnation, translation table[16]	Activity as function of time over 24 h, deactivation, Arrhenius diagrams	REMPI + MS	63
Oxidative dehydrogenation of cyclohexane to benzene	α-Al_2O_3 supported multi-metal-oxides	Impregnation, manually preparation (later SOPHAS robot)	Catalyst optimization, 5 generations	GC	64
Low-temperature oxidation of propane	Pt, Pd, Rh, Ru, Au, Ag, Cu, and Mn in groups of 3–5 elements on TiO_2 and α-Fe_2O_3	Impregnation under incipient wetness conditions, single-nozzle liquid-dispensation system	Catalyst optimization	QMS	65
Methane partial oxidation using molecular oxygen	Silica doped with 43 different elements (0.1, 1, 5 mol%)	Impregnation, automated synthesizer, JEOL PTW-100	Different feed gas compositions,	Micro GC, two columns	66
Oxidative coupling of methane	$Mn/Na_2WO_4/SiO_2$ catalysts	Impregnation	Catalyst optimization, different reaction conditions	QMS	67
Low temperature CO oxidation	Au catalysts	Deposition–precipitation, precipitation, impregnation			

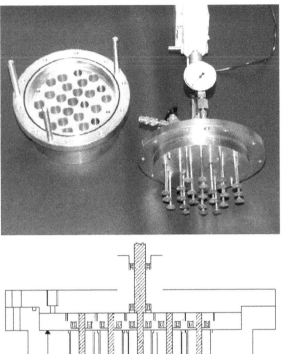

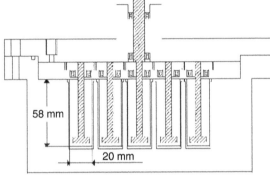

FIGURE 9.14 Photographic and schematic drawing showing the design of the 25-sample HTE reactor.[59]

FIGURE 9.15 Reactor module consisting of five miniautoclaves for fast screening of heterogeneously catalyzed gas–liquid reactions in a multibatch reactor.[60,62]

up to 1200 rpm, type of solvent, concentrations of the organic components and catalysts ($V_{liq.ph}$: 5–20 mL) in a very short time. Enlarging the present system can easily be achieved. The reaction products are analyzed off-line by gas chromatography. The system was successfully tested in the liquid-phase hydrogenation of citral, cinnamaldehyde, and acrolein (to allyl alcohol) at hydrogen pressures up to 70 bar.[60–62]

9.5 SUMMARY

In the beginning of fast parallel testing of heterogeneously catalyzed reactions, the preparation of the catalysts was often the bottleneck regarding the expenditure of time. In the meantime various techniques have been developed for systematic and automatic fast preparation of various types of catalysts for heterogeneously catalyzed reactions. Very large catalyst libraries can be prepared in a very short time by applying fully automated thin-film deposition methods. The disadvantage is the fact that, if higher amounts of catalysts are needed, e.g., for scale-up experiments, it is difficult to reproduce these catalysts with more conventional preparation methods. That is the reason why solution-based preparation methods come to the fore more and more. Nowadays many automated liquid-dispensing systems have been successfully applied for different preparation methods and the reproducibility and ability for scaling up have been demonstrated. Nevertheless, we look forward to further miniaturization and especially automation in catalyst preparation.

REFERENCES

1. Cong, P. C., Doolen, R. D., Fan, Q., Giaquinta, D. M., Guan, S., McFarland, E. W., Poojary, D., Self, K., Turner, H. W. and Weinberg, W. H., High-throughput synthesis and screening of combinatorial heterogeneous catalyst libraries, *Angew. Chem. Int. Ed. Engl.*, 38, 483, 1999.
2. Snively, C. M., Oskarsdottir, G. and Lauterbach, J., Chemically sensitive parallel analysis of combinatorial catalyst libraries, *Catal. Today*, 67, 357, 2001.
3. Sun, X.-D., Wang, K.-A., Yoo, Y., Wallace-Freedman, W. G., Gao, C., Xiang, X.-D. and Schultz, P. G., Solution-phase synthesis of luminescent materials libraries, *Adv. Mater.*, 9, 1046, 1997.
4. Hanak, J., The "Multiple-Sample Concept" in materials research: synthesis, compositional analysis and testing of entire multicomponent systems, *J. Mater. Sci.*, 5, 964, 1970.
5. Xiang, X.-D., Sun, X., Briceno, G., Lou, Y., Wang, K.-A., Chang, H., Wallace-Freedman, W. G., Chen, S.-W. and Schultz, P. G., A combinatorial approach to materials discovery, *Science*, 268, 1738, 1995.
6. Senkan, S., Combinatorial heterogeneous catalysis – a new path in an old field, *Angew. Chem. Int. Ed.*, 40, 312, 2001.
7. Danielson, E., Golden, J. H., McFarland, E. W., Reaves, C. M., Weinberg, W. H. and Wu, X. D., A combinatorial approach to the discovery and optimization of luminescent materials, *Nature*, 389, 944, 1997.
8. Katada, N. and Niwa, M., Silica monolayer solid-acid catalyst prepared by CVD, *Chem. Vapor Deposition*, 2, 125, 1996.
9. Hambrock, M. J., Schröter, M. K., Birkner, A., Wöll, C. and Fischer, R. A., Nano-brass: bimetallic copper/zinc colloids by a nonaqueous organometallic route using [Cu(OCH(Me)CH$_2$NMe$_2$)$_2$] and Et$_2$Zn as precursors, *Chem. Mater.*, 15, 4217, 2003.
10. Kizling, M. B., Jaras, S. G., A review of the use of plasma techniques in catalyst preparation and catalytic reactions, *Appl. Catal. A: General*, 1, 147, 1996.
11. Vossokov, G. P. and Pirgov, P. S., Experimental studies on the plasma-chemical synthesis of a catalyst for natural gas reforming, *Appl. Catal. A: General*, 168, 229, 1998.
12. Kim, Y. J., Gao, K., Chambers, G. A., Selective growth and characterization of pure, epitaxial a-Fe$_2$O$_3$(0001) and Fe$_3$O$_4$(001) films by plasma-assisted molecular beam epitaxy, *Surf. Sci.*, 371, 358, 1997.
13. Gorbunov, A. A., Pompe, W., Sewing, A., Gapanov, S. V., Akhsakhalyan, A. D., Zabrodin, I. G., Kaśkov, I. A., Klyenkov, E. B., Mozorov, A. P., Salashcenko, N. N., Dietsch, R., Mai, H. and Vollmar, S., Ultrathin film deposition by pulsed laser ablation using crossed beams, *Appl. Surf. Sci.*, 96-98, 649, 1996.
14. Mao, X. L., Perry, D. L. and Russo, R. E., Pulse laser deposition of thin films, *CHEMTECH*, 24, 14, 1994.

15. Reddington, E., Sapienza, A., Guraou, B., Viswanathan, R., Sarangapani, S., Smotkin, E. S. and Mallouk, T. E., Combinatorial electrochemistry: a highly parallel, optical screening method for discovery of better electrocatalysts, *Science*, 280, 1735, 1998.

16. Senkan, S. and Ozturk, S., Discovery and optimization of heterogeneous catalysts by using combinatorial chemistry, *Angew. Chem. Int. Ed.*, 38, 791, 1999.

17. Senkan, S., Ozturk, S., Krantz, K. and Onal, I., Photoionization detection (PID) as a high throughput screening tool in catalysis, *Appl. Catal. A: General*, 254, 97, 2003.

18. Krantz, K. W., Ozturk S. and Senkan S. M., Application of combinatorial catalysis to the selective reduction of NO by C_3H_6, *Catal. Today*, 62, 281, 2000.

19. Raif, F., Lucas, M. and Claus, P., 48-fach-Multitube-Reaktor im Mikrotiterplatten-Design zur Untersuchung heterogen katalysierter Oxidationsreaktionen, *Chem.-Ing.-Tech.*, 76, 1333, 2004.

20. Bergh, S., Guan, S., Hagemeyer, A., Lugmair, C., Turner, H., Volpe, A. F. Jr., Weinberg, W. H. and Mott, G., Gas phase oxidation of ethane to acetic acid using high-throughput screening in a massively parallel microfluidic reactor system, *Appl. Catal. A: General*, 254, 67, 2003.

21. Steven, L., McFarland, E. W., Safir, A., Turner, S. J., van Erden, L. and Wang, P. (Symyx Technologies Inc.), Graphical Design of Combinatorial Libraries, EP 1080435, 2001.

22. Desrosiers, P., Guram, A., Hagemeyer, A., Jandeleit, B., Poojary, D. M., Turner, H. and Weinberg, H., Selective oxidation of alcohols by combinatorial catalysis, *Catal. Today*, 67, 397, 2001.

23. Gomez, S., Peters, J. A., van der Waal, J. C. and Maschmeyer, T., High-throughput experimentation as a tool in catalyst design for the reductive amination of benzaldehyde, *Appl. Catal. A: General*, 254, 77, 2003.

24. Holmes, S. A., Al-Saeedi, J., Guliants, V. V., Boolchand, P., Georgiev, D., Hackler, U. and Sobkow, E., Solid state chemistry of bulk mixed metal oxide catalysts for the selective oxidation of propane to acrylic acid, *Catal. Today*, 67, 403, 2001.

25. Hofmann, C., Wolf, A. and Schüth, F., Parallel synthesis and testing of catalysts under nearly conventional testing conditions, *Angew. Chem. Int. Ed. Engl.*, 38, 2800, 1999.

26. Cong, P., Dehestani, A., Doolen, R., Giaquinta, D. M., Guan, S., Markov, V., Poojary, D., Self, K., Turner, H. and Weinberg, W. H., Combinatorial discovery of oxidative dehydrogenation catalysts within the Mo-V-Nb-O system, *Proc. Nat. Acad. Sci.*, 96, 11077, 1999.

27. Giaquinta, D., Devenney, M., Hall, K. and Goldwasser, I. (Symyx Technologies Inc.), Formation of combinatorial arrays of materials using solution-based methodologies, US2003219906, 2003.

28. Liu, Y., Cong, P., Doolen, R. D., Turner, H.W. and Weinberg, W. H., High-throughput synthesis and screening of V–Al–Nb and Cr–Al–Nb oxide libraries for ethane oxidative dehydrogenation to ethylene, *Catal. Today*, 61, 87, 2000.

29. Liu, Y., Cong, P., Doolen, R. D., Guan, S., Markov, V., Woo, L., Zeyß, S. and Dingerdissen, U., Discovery from combinatorial heterogeneous catalysis: A new class of catalyst for ethane oxidative dehydrogenation at low temperatures, *Appl. Catal. A: General*, 254, 59, 2003.

30. Holzwarth, A., Schmidt, H.-W. and Maier, W. F., Detection of catalytic activity in combinatorial libraries of heterogeneous catalysts by IR thermography, *Angew. Chem. Int. Ed.*, 37, 2644, 1998.

31. Orschel, M., Klein, J., Schmidt, H.-W. and Maier, W. F., Detection of reaction selectivity on catalyst libraries by spatially resolved mass spectrometry, *Angew. Chem. Int. Ed.*, 38, 2791, 1999.

32. Holzwarth, A. and Maier, W. F., Catalytic phenomena in combinatorial libraries of heterogeneous catalysts, *Platinum Metals Rev.*, 44, 16, 2000.

33. Akporiaye, D. E., Dahl, I. M., Karlsson, A. and Wendelbo, R., Combinatorial approach to the hydrothermal synthesis of zeolites, *Angew. Chem. Int. Ed.*, 37, 609, 1998.

34. Akporiaye, D. E., Dahl, I. M., Karlsson, A., Plassen, M., Wendelbo, R., Bem, D. S., Broach, R. W., Lewis, G. J., Miller, M. and Moscoso, J., Combinatorial chemistry – The emperor's new clothes?, *Micro. Mesopor. Mat.*, 48, 367, 2001.

35. Klein, J., Lehmann, C. W., Schmidt, H.-W. and Maier, W.F., Combinatorial material libraries on the microgram scale with an example of hydrothermal synthesis, *Angew. Chem. Int. Ed.*, 37, 3369, 1998.

36. Johann, T., Brenner, A., Schwickardi, M., Busch, O., Marlow, F., Schunk, S. and Schüth, F., Real-time photoacoustic parallel detection of products from catalyst libraries, *Angew. Chem. Int. Ed.*, 41, 2966, 2002.

37. Xu, X. D., Vonk, H., Cybulski, A. and Moulijn, J. A., Alumina washcoating and metal deposition of ceramic monoliths, In: *Preparation of Catalysts VI*, Poncelet, G., Martens, J., Delmon, B., Jacobs, P. A. and Grange, P., Eds. *Stud. Surf. Sci. Catal.*, 1995, p. 1069.

38. Scheffler, F., Claus, P., Schimpf, S., Lucas, H. and Scheffler H., Heterogenously Catalyzed Processes with Porous Cellular Ceramic Monoliths, In: *Cellular Ceramics: Structure, Manufacturing, Properties and Applications*, Scheffler, M. and Colombo, B., Eds. Wiley-VCH, 2005.

39. Che, M., Clause, O. and Marcilly, C., Supported catalysts, In: *Handbook of Heterogeneous Catalysis*, Ertl, G., Knözinger, H. and Weitkamp, J., Eds. VCH, Weinheim, 1997, Vol. 1, p. 191.

40. Lucas, M. and Claus, P., High throughput screening in monolith reactors for total oxidation reactions, Special Issue "*Combinatorial Catalysis*," (OD W. F. Maier), *Appl. Catal. A: General*, 254, 35, 2003.

41. Zech, T., Lohf, A., Golbig, K., Richter, T. and Hönicke, D., Simultaneous screening of catalysts in microchannels: methodology and experimental setup, In: *Microreaction Technology: Industrial Prospects, Proceedings of the Third International Conference on Microreaction Technology*, Ehrfeld, W., Ed., Springer, Berlin, 2000, p. 260.

42. Hönicke, D., Microchemical reactors for heterogeneously catalyzed reactions, In: Reaction kinetics and the development of catalytic processes, Froment, G. C. and Waugh, K. C., Eds., *Studies in Surface Science and Catalysis*, Elsevier, Amsterdam, 1999, Vol. 122, p. 47.

43. Janicke, M., Kerstenbaum, H., Hagendorf, U., Schüth, F., Fichtner, M. and Schubert, K., The controlled oxidation of hydrogen from an explosive mixture of gases using a microstructured reactor/heat exchanger and Pt/Al_2O_3 catalyst, *J. Catal.*, 191, 283, 2000.

44. Wießmeier, G. and Hönicke, D., Microfabricated components for heterogeneously catalysed reactions, *J. Micromech. Microeng.*, 6, 285, 1996.

45. Haas-Santo, K., Fichtner, M. and Schubert, K., Preparation of microstructure compatible porous supports by sol–gel synthesis for catalyst coatings, *J. Catal.*, 220, 79, 2001.

46. Pfeifer, P., Fichtner, M., Schubert, K., Liauw, M. and Emig, G., *Microstructured Catalysts for Methanol-Steam Reforming, Proc. IMRET 3*, Ehrfeld, W., Ed. Springer, Frankfurt, April 1999, p. 372.

47. Schimpf, S., Bron, M. and Claus, P., Carbon coated microstructured reactors for heterogeneously catalyzed gas phase reactions: influence of coating procedure on catalytic activity and selectivity, *Chem. Eng. J.*, 101, 11, 2004.

48. Schimpf, S., Lucas, M. and Claus, P., Selective hydrogenation of 1,3-butadiene over supported gold catalysts, *Proc. DGMK-Conference* "Creating Value from Light Olefins – Production and Conversion," Hamburg, Germany, October 10–12, 2001, 97.

49. Müller, A., Drese, K., Gnaser, H., Hampe, M., Hessel, V., Löwe, H., Schmitt, S. and Zapf, R., Fast preparation and testing methods using a microstructured modular reactor for parallel gas phase catalyst screening, *Catal. Today*, 81, 377, 2003.

50. Steinfeld, N., Dropka, N., Wolf, D. and Baerns, M., Application of multichannel microreactors for studying heterogeneous catalysed gas phase reactions, *Trans. IChemE*, 81, 735, 2003.

51. Rouge, A., Spoetzl, B., Gebauer, K., Schenk, R. and Renken, A., Microchannel reactors for fast periodic operation: the catalytic dehydration of isopropanol, *Chem. Eng. Sci.*, 56, 1419, 2001.

52. Zech, T., PhD thesis, *Miniaturisierte Screening-Systeme für die kombinatorische heterogene Katalyse*, Chemnitz University of Technology, Duesseldorf, 2002, Fortschriftberichte/VDI, ISBN 3-18-373203-3.

53. Pantu, P. and Gavalas, G. R., A multiple microreactor system for parallel catalyst preparation and testing, *AIChE J.*, 48, 815, 2002.

54. Holzwarth, A., Denton, P., Zanthoff, H. and Mirodatos, C., Combinatorial approaches to heterogeneous catalysis: strategies and perspectives for academic research, *Catal. Today*, 67, 309, 2001.

55. Preparation of catalysts, Ertl, G., Knözinger, H. and Weitkamp, J., Eds., *Handbook of Heterogeneous Catalysis*, WILEY-VCH, Weinheim, 1997.

56. Hofmann, C., Schmidt, H.-W. and Schüth, F., A multipurpose parallelized 49-channel reactor for the screening of catalysts: methane oxidation as the example reaction, *J. Catal.*, 198, 348, 2001.

57. Pérez-Ramírez, J., Berger, R. J., Mul, G., Kapteijn, F. and Moulijn, J. A., The six-flow reactor technology: A review on fast catalyst screening and kinetic studies, *Catal. Today*, 60, 93, 2000.

58. Moulijn, J. A., Pérez-Ramírez, J., Berger, R. J., Hamminga, G., Mul, G. and Kapteijn, F., High-throughput experimentation in catalyst testing and in kinetic studies for heterogeneous catalysis, *Catal. Today*, 81, 457, 2003.

59. Thomson, S., Hoffmann, C., Ruthe, S., Schmidt, H.-W. and Schüth, F., The development of a high throughput reactor for the catalytic screening of three phase reactions, *Appl. Catal. A: General*, 220, 253, 2001.

60. Lucas, M. and Claus, P., Parallelisiertes Screening von katalysierten Gas-Flüssigphasenreaktionen in einem Multibatch-Reaktor, *Chem.-Ing.-Techn.*, 73, 252, 2001.
61. Lucas, M. and Claus, P., Hydrierungen mit Silber: Hochaktiver und chemoselektiver Ag-In/SiO$_2$-Katalysator für die Direktsynthese von Allylalkohol aus Acrolein, *Chem.-Ing.-Techn.*, 77, 110, 2005.
62. Claus, P., Hönicke, D. and Zech, T., Miniaturization of screening devices for the combinatorial development of heterogeneous catalysts, *Catal. Today*, 67, 319, 2001.
63. Senkan, S., Krantz, K., Ozturk, S., Zengin, V. and Onal, I., High-throughput testing of heterogeneous catalyst libraries using array microreactors and mass spectrometry, *Angew. Chem. Int. Ed.*, 38, 2794, 1999.
64. Buyevskaya, O. V., Brückner, A., Kondratenko, E. V., Wolf, D. and Baerns, M., Fundamental and combinatorial approaches in the search for and optimisation of catalytic materials for the oxidative dehydrogenation of propane to propene, *Catal. Today*, 67, 369, 2001.
65. Rodemerck, U., Wolf, D., Buyevskaya, O. V., Claus, P. Senkan, S. and Baerns, M., High-throughput synthesis and screening of catalytic materials – case study on the search for a low-temperature oxidation catalyst for low-concentration propane, *Chem. Eng. Jour.*, 82, 3, 2001.
66. Yamada, Y., Ueda, A., Shioyama, H. and Kobayashi, T., High-throughput experiments on methane partial oxidation using molecular oxygen over silica doped with various elements, *Appl. Catal. A: General*, 254, 45, 2003.
67. Rodemerck, U., Ignaszewski, P., Lucas, M. and Claus, P., Parallelisierte Synthese und schnelle katalytische Testung von Katalysatorbibliotheken für die Oxidationsreaktionen, *Chem.-Ing.-Techn.*, 71, 873, 1999; Parallel synthesis and fast catalytic testing of catalyst libraries for oxidation reactions, *Chem. Eng. Technol.*, 23, 413, 2000.

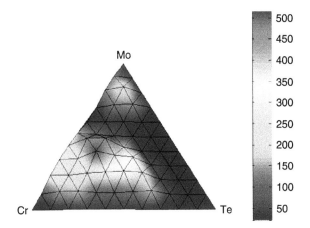

FIGURE 1.5 Catalytic activity of ternary mixed oxides for the oxidation of propene to acroleine as a function of chemical composition. Source: Sieg, S., Stutz, B., Schmidt, T., Hamprecht, F. A., Maier, W. F., *J. Mol. Model.*, in press.

FIGURE 2.15 Single-bead reactor architecture (384-fold and 96-fold).

(a)

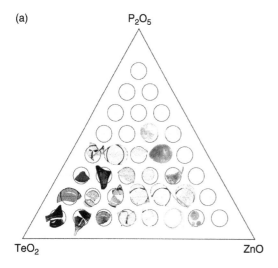

(b)

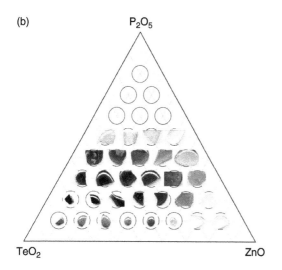

FIGURE 4.14 Appearance pictures of as-quenched P_2O_5-TeO_2-ZnO samples placed on ternary diagram. Batch-melting temperatures were (a) 1000°C and (b) 1100°C.

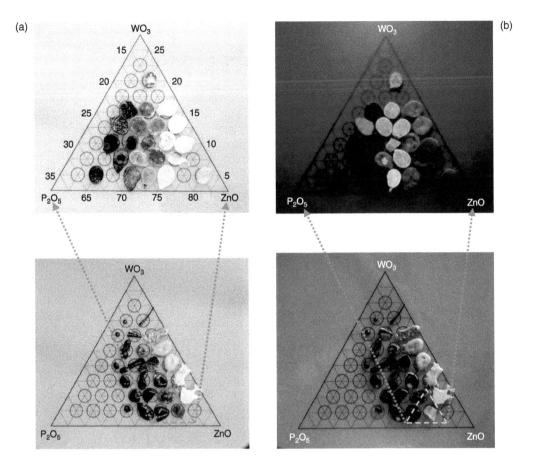

FIGURE 4.26 Photos of samples placed on WO_3-P_2O_5-ZnO ternary composition diagrams taken (a) without irradiation of UV lamp and (b) under irradiation of UV lamp.

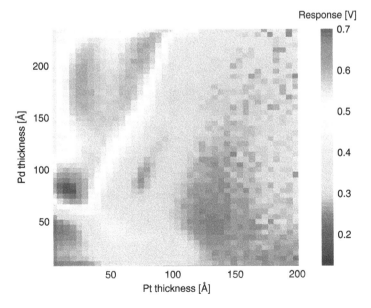

FIGURE 5.11 Response image for a double metal layer, Pt + Pd, with varying metal film thicknesses (Pd on top) and for 100 ppm hydrogen. $T_{sensor} = 140°C$.

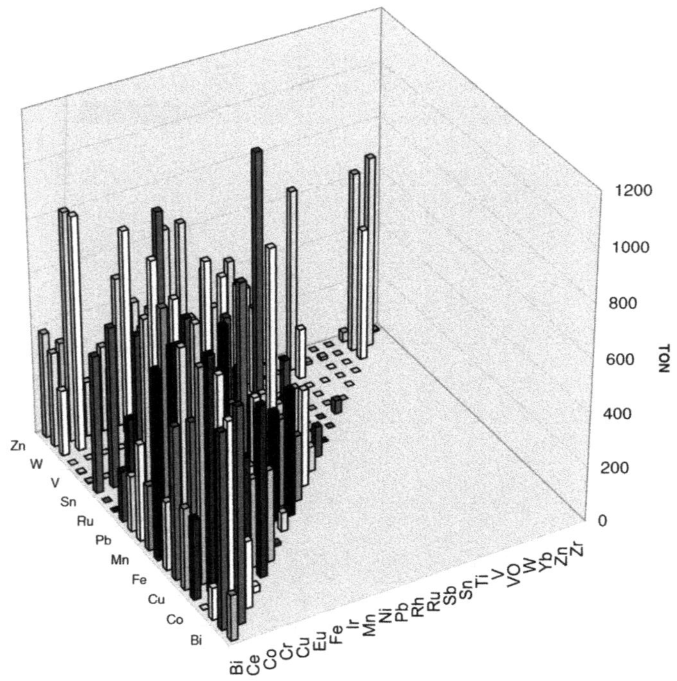

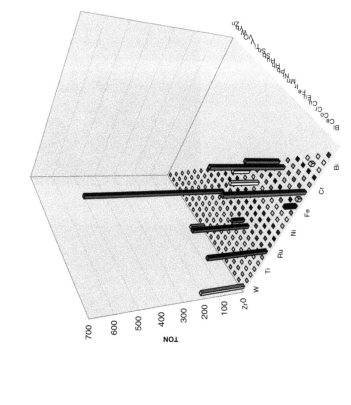

FIGURE 8.8 TON for binary co-catalyst combinations. Single metal co-catalysts are plotted along the diagonal.

FIGURE 8.9 Synergies determined from binary co-catalyst combinations. Synergy is defined as the difference between the TON of the pair of co-catalysts and the sum of the TON for co-catalysts alone.

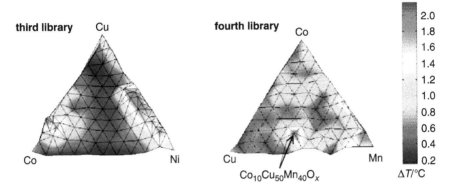

FIGURE 10.3 Representation of the catalytic activity of the third and fourth library screened at 175°C and 165°C, respectively, with 0.3 vol% DMEA in dried air.

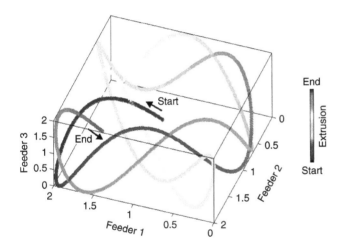

FIGURE 11.3 Computer simulation results of diversity of polymeric formulations produced by the operation of three independent feeders.

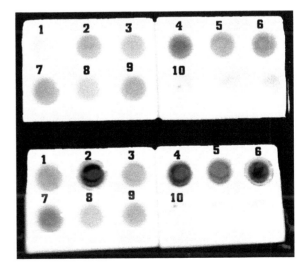

FIGURE 13.1 High-throughput screening of azo dyes in solution. The upper plates were kept in the dark during illumination by masking with a black film.

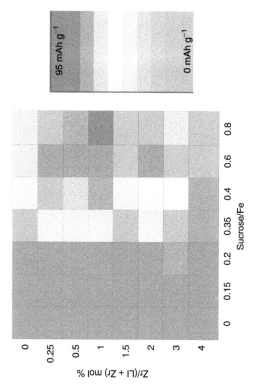

FIGURE 15.15 Color mapping of specific capacity vs. precursor composition variables in array of 56 Zr-doped LiFePO$_4$ electrodes.

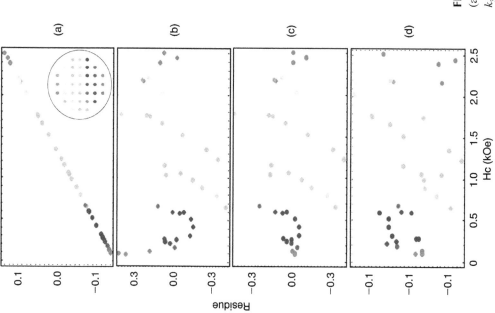

FIGURE 16.11 Graphs showing the residue of all data points sorted according to coercivity value. (a) The fit is to k_0 and the residues range from $+1$ to -1; (b) data fitted to $k_0 + k_1\text{Cr}$; (c) $k_0 + k_1\text{Cr} + k_2\text{Cr}^2 + k_3\text{Cr}^3$; (d) $k_0 + k_1\text{Cr} + k_2\text{Cr}^2 + k_3\text{Cr}^3 + k_4\text{CrTa}$.

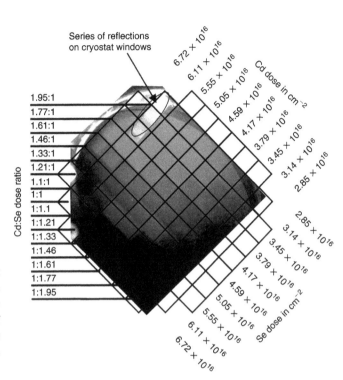

Series of reflections
on cryostat windows

Cd dose in cm⁻²
6.72×10^{16}
6.11×10^{16}
5.55×10^{16}
5.05×10^{16}
4.59×10^{16}
4.17×10^{16}
3.79×10^{16}
3.45×10^{16}
3.14×10^{16}
2.85×10^{16}

1.95:1
1.77:1
1.61:1
1.46:1
1.33:1
1.21:1
1.1:1
1:1
1:1.1
1:1.21
1:1.33
1:1.46
1:1.61
1:1.77
1:1.95

Cd:Se dose ratio

Se dose in cm⁻²
2.85×10^{16}
3.14×10^{16}
3.45×10^{16}
3.79×10^{16}
4.17×10^{16}
4.59×10^{16}
5.05×10^{16}
5.55×10^{16}
6.11×10^{16}
6.72×10^{16}

FIGURE 18.14 A luminescent wafer at 15 K (annealing condition 1000°C, 30 s in pure Ar atmosphere) with implantation pattern. Each of the 88 areas of the dose raster was implanted with a constant and homogeneous Cd:Se dose ratio.

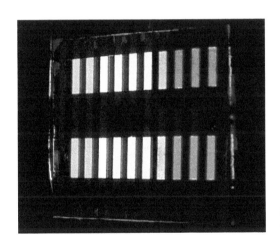

FIGURE 20.9 Color evolution of OLEDs with varying DCM2-doped-layer thickness at $V_{app} = 8$ V.

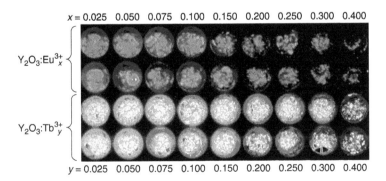

$x = 0.025 \quad 0.050 \quad 0.075 \quad 0.100 \quad 0.150 \quad 0.200 \quad 0.250 \quad 0.300 \quad 0.400$

$Y_2O_3:Eu^{3+}_x$

$Y_2O_3:Tb^{3+}_y$

$y = 0.025 \quad 0.050 \quad 0.075 \quad 0.100 \quad 0.150 \quad 0.200 \quad 0.250 \quad 0.300 \quad 0.400$

FIGURE 21.4 Composition map and photoluminescent photograph of the library under UV excitation.

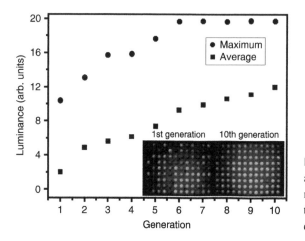

FIGURE 22.12 Experimental maximum and average luminance as a function of generation number. The inset shows libraries of both the first and tenth generation at the 365 nm excitation.

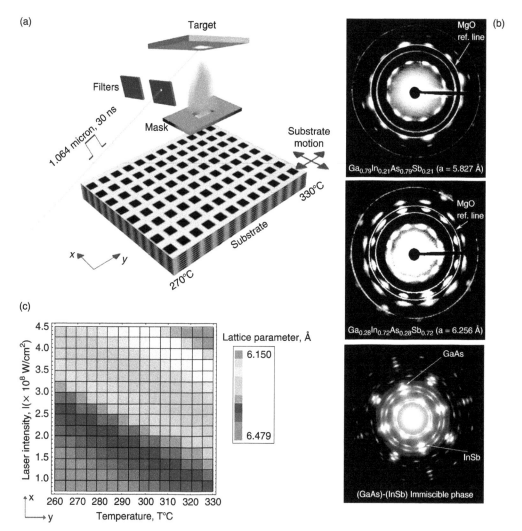

FIGURE 23.3 Synthesis of composition library of single-phase $Ga_{1-x}In_xAs_{1-y}Sb_y$. (a) PLD configuration; (b) electron-diffraction patterns taken from three cells of the library (corresponding to three compositions); (c) dependence of the lattice parameter of synthesized composites on the technology parameters.

10 Tailoring Heterogeneous Catalysts for Pollutant Combustion with High-Throughput Methods

Timm Schmidt, Gerald Frenzer, and Wilhelm F. Maier

CONTENTS

10.1   INTRODUCTION

Industrial processes often use or produce volatile organic compounds (VOCs), such as solvents, reaction intermediates, or by-products, which partially escape into the environment. Because of their more or less aggressive odors and toxic properties, these VOCs are regarded as major problems of air pollution in regions with a high population of industrial production facilities. Several technologies for the end of pipe removal of these VOCs have been developed. Most common are thermal combustion, catalytic (high-temperature) combustion, adsorption on sorbent materials, such as charcoal, evaporation as well as biofiltration.[1–5] For typical high-volume exhaust gas streams with low concentrations of pollutants, such as exhaust gases from large halls or production facilities, direct thermal combustion is not economical, since huge amounts of energy are necessary to bring the exhaust gas to combustion temperature. A very attractive method for such a problem is the pollutant concentration on regenerative sorbents such as charcoal, because the pollutants can be rapidly desorbed by increasing the temperature, thus generating a concentrated pollutant stream, which can much more easily be decontaminated by thermal combustion. Since common desorption temperatures of such concentrated pollutants on sorbents are in the range of 150–250°C and thermal combustion is operated usually at temperatures above 800°C, an attractive alternative in terms of saving energy and thus money is catalytic combustion at the desorption temperature. But state-of-the-art combustion catalysts for the removal of the air pollution usually operate at temperatures between 350 and 600°C, which is much higher than the common desorption temperature. Catalytically active components are Pt, Pd (BASF), Pt-Pd (KataLeuna) or Cu-Cr (Katalizator), in most cases dispersed on γ-Al$_2$O$_3$. A broad review on the state-of-the-art of the catalysts for VOC combustion has been presented by Everaert.[6]

Technically, a combination of pollutant concentration through adsorption/desorption technologies[7–9] with a catalytic converter can be achieved based on present-day technologies. Through clever cycling of exhaust gases and heat exchange most of the heat of combustion can be used to increase the temperature of the stream of desorption.[10,11] Clearly, the operating temperature of the combustion catalyst is essential for the efficiency of such a concept. The lower the working temperature of the catalyst, the lower would be the energy consumption of the exhaust gas cleaning process. Assuming an exhaust gas stream of 10 m³/s, an enrichment factor of 50 and 4 ct per kWh, such a process can reduce cost in the range of 5000–10,000 €/a per 100°C decrease in the working temperature considering only the heating of the exhaust gas. Furthermore, if the operating temperature of the catalyst could be lower than the temperature of the desorption stream, investment and operating costs for the heating of the catalytic converter could be completely saved. A desirable side effect of such a low temperature would most likely be a reduction of carbon monoxide and nitrous oxide formation, commonly only favored at high temperatures. Suitable catalysts, characterized by high catalytic activity at the temperature of desorption, long-time stability, and especially resistance to variable pollutant mixtures, have not been developed yet.

An attractive approach for the tailoring of such catalysts for low-temperature combustion is high-throughput experimentation (HTE). Catalytic combustion is an exothermic reaction, which could be monitored by its heat generation. Since no secondary or parallel reactions have to be considered, the total heat of reaction can be used as a direct measure for catalytic activity. Such an application is ideal for the use of emissivity-corrected IR-thermography (ec-IRT).[12] The use of ec-IRT has the advantage that the effect of elemental variation of the catalyst on catalytic activity can be studied under identical reaction conditions in a parallel manner. With a common library size of 207 samples per library, large parameter spaces can be screened in a reasonable time.[13–15]

Essential for such an approach is the availability of catalyst preparation techniques, which tolerate broad variations of chemical composition as well as additions of a large variety of dopants.

This is very generally the case for catalysts prepared by conventional impregnation, but it is limited by the availability of only a few suitable support materials (thermally and chemically stable porous oxides). Chemical or physical vapor deposition, which is commonly used on gradient or discrete libraries, through moving shutters or in combination with masking techniques in searches for new electronic, magnetic, or other thin-film materials,[16] is not very useful in searches for new catalytic activity, which is a linear function of surface area and the surface area produced on thin films is much too low for detection of any catalytic activity by normal methods. In recent years we have increasingly developed sol–gel procedures for the preparation of mixed oxides.[13–15,17,18] The sol–gel approach has the advantage that extremely flexible synthesis protocols can be developed. The synthesis is carried out under very mild reaction conditions and thus is suitable for use by pipetting robots, which can be used to automate the library preparation. Furthermore, sol–gel methods allow for the preparation of all kinds of mixed oxides and thus have a high potential for discovery of new catalysts through automated, systematic variation of composition.

Another important aspect in the application of high-throughput technologies for such an approach is the variation of the synthesis recipe and the experimental design of the libraries to be investigated. It is clearly counterproductive if changes in a recipe have to be calculated by hand or with the help of an Excel sheet. The transfer of such recipes into the pipetting list of a synthesis or pipetting robot can also be very time-consuming and has to be completely automated. Software specially developed for the experimental design of libraries based on procedures for sol–gel synthesis has been described recently.[19] With this software the scientist can very easily generate a variety of recipes for the sol–gel synthesis of materials. The program allows the automatic generation of recipes for composition spread libraries, which cover the complete composition spread of oxide mixtures of up to five different elements, simply defined by the step size chosen. It also allows the automated generation of recipes for doped materials, which only require the choice of dopant concentration and the selection of doping elements by simple drag and drop from precursor tables. In all cases, the researcher only has to select or provide the suitable synthesis base recipe and handle some other definitions, hence the program calculates all individual recipes and the desired modifications and also generates the associated pipetting lists and sends them to the pipetting robot chosen for the synthesis.

There are several ways to optimize elemental catalyst composition by variation. They all rely on survival of the fittest; the most active catalysts of a library studied are chosen as the basis for the design of the successor library. A more complex approach, which is especially suited to enter new parameter (composition) spaces is the evolutionary approach, whose principal function for materials development has been demonstrated.[20] The evolutionary method among other problems requires a universal synthesis procedure, which in many cases is not available. A more straightforward approach is to start the catalyst search with a highly diverse collection of catalysts, for example binary mixed oxides, and determine the relative catalytic activity. The best catalysts are selected and modified by doping with a large number of oxides. The best modifications are identified and their optimal composition is determined by the step-wise examination of the complete composition space. The optimal catalyst can then be doped again by a large number of oxides, the best will be selected and again optimized for doping composition. This repetitive doping–composition spread approach rapidly leads to new catalysts composed of several metal oxides. This approach has been used successfully for the discovery of new low-temperature CO-combustion catalysts.[14,15] Although more limited than the evolutionary approach, this approach will reach an optimal composition after a rather limited number of search generations and allows to modify and adjust the synthesis recipe from library to library with the increasing complexity of the mixed oxide generated.

In this study the doping–composition spread approach for variation of elemental composition has been applied in the search for noble metal free catalysts for the combustion of VOCs. As model compounds, representing a broad diversity of nitrogen- or sulfur-containing as well as aromatic organic volatiles, dimethylethyl amine (DMEA), dimethyl disulfide (DMDS), and benzene were chosen. For the high-throughput screening of the materials libraries, filled with up to 207 different mixed oxide materials, ec-IRT was applied. The high-throughput experiment itself was controlled by IR-TestRig,[21] which has complete control over all equipment involved with the screening process such as IR camera, mass flow meter, and temperature controllers. Data are automatically treated and deposited in a database. This allowed to automatically screen up to three different temperatures with several hours of reaction time in about 1 day. The course of action of the screening and the verification of the results in a plug-flow reactor is discussed by means of the model compound DMEA, for example.

10.2   EXPERIMENTAL PART

10.2.1   Mixed-Oxide Syntheses

The mixed oxides applied for the high-throughput screening were synthesized in mg quantities using different modified sol–gel recipes. Therefore, suitable alkoxides, nitrates, and chlorides of the desired elements were dissolved in low-boiling alcohols. These solutions served as precursors for the synthesis. Desired aliquots of these liquid precursors have been pipetted by a pipetting robot into the HPLC vials. In most cases the complexing agent 4-hydroxy-4-methyl-2-pentanone and suitable amounts of propionic acid and water have been added. The racks containing the homogenized final sols in the vials were stored for several days at 20–100°C, open to the ambient atmosphere, until clear gels or clear glasses have been obtained. These have been calcined at 300–350°C. After cooling to room temperature, the materials have been crashed into powders by a glass rod in the vials and these powders have been transferred manually into the 207 wells (d ≈ 3 mm, h ≈ 2 mm) of the catalysts library (d ≈ 100 mm, h ≈ 5 mm) made from slate. Design of experiment of the libraries and pipetting of the liquid precursor solutions with the dispensing robot was carried out with the help of the software Plattenbau.[19] The individual synthesis of the catalysts libraries is described in the following.

10.2.1.1   Synthesis 1

This synthesis route was applied for the preparation of the first library, consisting of bulk doped Al_2O_3, TiO_2, and SiO_2. Al(tert)butoxide, Ti(n)propoxide, and tetraethoxysilane were dissolved in isopropanol (Al), ethanol (Ti), and methanol (Si) respectively, each with 0.5 mol/L. Suitable precursors of more than 50 elements (mainly transition and rare earth metals as nitrates or chlorides) were dissolved in methanol (0.1 mol/L). These solutions were mixed in the desired proportions. The sol–gel process was induced by adding 3 mol of 4-hydroxy-4-methyl-2-pentanone, 0.02 mol of propionic acid and 0.02 mol of deionized water per mol of the metal compounds into each vial. The solutions were dried for 5 days at 35°C in the open vials and afterwards calcined at 350°C in air.

10.2.1.2   Synthesis 2

This synthesis route was applied for the preparation of binary compositional spreads of the elements Ce, Co, Cr, Cu, Fe, Mn, Ni, and Zn with an increment of 20 atom% (second library) and the preparation of the eighth and ninth library (consisting of quaternary mixed oxides of Co, Cu, Mn, Ni (eighth library) and Ce-, C-, Cu-, Fe-, Ni-doped Mn oxide (ninth library)). Therefore, the nitrates of the elements were dissolved in ethylene glycol (1 mol/l). These solutions were mixed in the desired proportions. Then 3 mol of 4-hydroxy-4-methyl-2-pentanone and 0.04 mol of propionic acid per mol of the metal compounds were added into each vial. The solutions were dried for 5 days at 90°C in the open vials and afterwards calcined at 350°C in air.

10.2.1.3   Synthesis 3

This synthesis route was applied for the preparation of ternary compositional spreads of the elements Co, Cu, Ni (third library), and Co, Cu, Mn (fourth library) respectively, each with a step size of 10 atom%. The nitrates of the chosen elements were dissolved in methanol, each with 1 mol/L. The solutions were mixed in the desired proportions. Then 3 mol of 4-hydroxy-4-methyl-2-pentanone and 0.04 mol of propionic acid per mol of the metal compounds were added into each vial. The solutions were dried for 5 days at 40°C in the open vials and afterwards calcined at 300°C in air.

10.2.1.4   Synthesis 4

This synthesis route was applied for the preparation of bulk doped optimized compositions (fifth, sixth, seventh library), derived from the third and fourth catalyst libraries. Therefore, the nitrates of the main components were dissolved in methanol according to synthesis 3. The dopant solutions were taken from synthesis 1. These solutions were mixed in the desired proportions. Then 3 mol of 4-hydroxy-4-methyl-2-pentanone and 0.04 mol of propionic acid per mol of the metal compounds was added into each vial. The solutions were dried for 5 days at 40°C in the open vials and afterwards calcined at 300°C in air.

10.2.1.5   Synthesis 5

This synthesis route was applied for the preparation of the tenth catalysts library, consisting of bulk doped TiO_2 and SiO_2 (2–20 atom%). Therefore, Ti(n)propoxide and tetraethoxysilane were dissolved in ethanol (Ti) and methanol (Si) respectively, each with 0.5 mol/L. Proper compounds of more than 50 representative, transition, and rare-earth metals (in most cases nitrates and chlorides) were dissolved in methanol, each with 0.1 mol/L. To each solution Pluronic 123 from BASF was added (10 wt%). These solutions were mixed in the desired proportions. The sol–gel process was induced by adding 0.02 mol propionic acid and 0.02 mol deionized water per mol of the metal compounds into each vial. The solutions were dried for 5 days at 35°C in the open vials and afterwards calcined at 350°C in air.

10.2.1.6   Synthesis of Mixed Metal Oxides Applied for the Conventional Experiments

The same elemental precursors, solvents, complexing agents, and acid sources in the same concentrations and proportions were used according to the high-throughput synthesis. These solutions were mixed in vials (5–20 mL) or small flasks. The solutions were dried for some days in the open vials or flasks and afterwards calcined in air according to the high-throughput synthesis.

10.2.2   HIGH-THROUGHPUT SCREENING

The IR laboratory set-up for high-throughput screening of heat of reactions by ec-IRT and its measuring principles have been described already.[12–15,22,23] Main components of the set-up are a gas-dosing unit, a reactor with sapphire window, in which the library of catalysts is exposed to a flow of the feed gas, an IR camera, and the control units. The flow of events during the high-throughput screening experiment was controlled by the software IR-TestRig.[21] The feed gas was produced by flowing nitrogen through temperature-controlled gas wash bottles containing the desired pollutant. The saturated gas streams were diluted with defined volumes of N_2 and O_2 to reach the desired concentrations of the model compound. The concentrations of the model compounds were verified by leading the gas stream through a cold trap, which was cooled with a mixture of acetone and dried ice, followed by a differential weighing of the cold trap. The concentration of the model compounds was linearly dependent on the nitrogen flux over the working range, which could be verified for example for DMEA with a collection of measurements resulting in a regression coefficient of 0.999. The reaction conditions (model compound, content of the model compound in dried air, temperature, and reaction time) for the high-throughput screening are given in Table 10.1.

10.2.3   EXPERIMENTS IN THE PLUG-FLOW REACTOR

The main components of the plug-flow reactor were a gas dosing unit, a fritted tube in a tempered furnace with temperature controller, and an analysis unit. In contrast to the high-throughput experiments,

TABLE 10.1
Reaction Conditions for the High-Throughput Screening

Library	Model Compound	Content (%)	Temperature (°C)	Time of Reaction (h)
1	DMEA	0.3	350, 400	4
2	DMEA	1	250	1.5
3	DMEA	0.3	175	1.5
4	DMEA	0.3	165	4
5	DMEA	0.3	150	8
6	DMEA	0.3	160	4
7	DMEA	0.3	175	4
1	Benzene	0.5	300, 350	4
2	Benzene	1	250	1.5
8	Benzene	0.5	200	4
9	Benzene	0.5	175	4
10	DMDS	0.5	350	4
Other	DMDS	0.3–1	150–400	1.5–4

the gas dosing unit was used for the dosing and mixing of fixed model pollutants in N_2 and O_2. Defined gas mixtures (1000 ppm for each pollutant in N_2) were used. The catalyst was placed on a frit inside the vertical gas phase flow reactor tube ($d_i = 8$ mm, $l = 39$ mm) made from quartz glass. The analysis unit consisted of a NO_x analyzer equipped with a chemoluminescence sensor, CO- and CO_2-sensitive IR sensors, and a hydrocarbon sensor. The reaction conditions of the experiments are described in the following. The mixed oxides used for the experiments were microporous/mesoporous with surface areas of more than 300 m^2/g (N_2 isotherms) for the Al-based mixed oxides and microporous with surface areas in the range of 20–200 m^2 for the other mixed oxides, determined with the BET method. The mass transfer limitation could typically be excluded for $Co_{20}Mn_{80}O_x$ in the range of 125–250°C for the combustion of DMEA.

10.2.3.1   Combustion of DMEA (Test Series 1a, See Figure 10.9)

The catalyst (200 mg, 100–200 μm) was pretreated in dried air for 30 min at 400°C. Afterwards, the feed gas (800 ppm DMEA, 20 vol% O_2, 80 vol% N_2, LHSV = 30,000 L/(kg h)) was passed through the catalyst. Then the temperature was decreased stepwise. Gas mixtures of the products were analyzed after 20 min for each temperature.

10.2.3.2   Combustion of DMEA (Test Series 1b, See Figure 10.10)

The catalyst (200 mg, 100–200 μm) was pretreated in dried air for 30 min at 300°C. Afterwards, the temperature was decreased to 240°C and the feed gas (800 ppm DMEA, 20 vol% O_2, 80 vol% N_2, LHSV = 30,000 L/(kg h)) was passed through the catalyst-containing tube. Then the temperature was decreased stepwise. Gas mixtures of the products were analyzed after 20 min of reaction time for each temperature.

10.2.3.3   Combustion of DMEA (Test Series 2, See Figure 10.11)

The catalyst (200 mg, 100–200 μm) was pretreated in dried air for 30 min at 275°C. Afterwards, the temperature was decreased to 250°C and the feed gas (800 ppm DMEA, 10 vol% O_2, 90 vol% N_2, LHSV = 30,000 L/(kg h)) was passed through the catalyst-containing tube. Then the temperature was decreased stepwise. Gas mixtures of the products were analyzed after 30 min of reaction time for each temperature.

10.2.3.4 Combustion of Benzene (See Figure 10.12)

The catalyst (200 mg, 100–500 μm) was pretreated in dried air for 30 min at 275°C. Afterwards, the temperature was decreased to 110°C and the feed gas (500 ppm benzene, 10 vol% O_2, 90 vol% N_2, LHSV = 30,000 L/(kg h)) was passed through the catalyst-containing tube. Then the temperature was increased stepwise. Gas mixtures of the products were analyzed after 20 min of reaction time for each temperature.

10.2.3.5 Sequential Combustion of Benzene, DMEA and DMDS (See Table 10.9)

The catalyst (200 mg, 100–200 μm) was pretreated in dried air for 30 min at 275°C. Afterwards, the temperature of the catalysts was decreased to 250°C and 200°C, respectively, and the feed gas for the benzene combustion (800 ppm benzene, 20 vol% O_2, 80 vol% N_2, LHSV = 30,000 L/(kg h)) was passed through the catalyst-containing tube. Then the catalyst was regenerated in dried air for 15 min at 275°C. Afterwards, the temperature of the catalysts was decreased to 250°C and 200°C, respectively, and the feed gas for the DMEA combustion (800 ppm DMEA, 20 vol% O_2, 80 vol% N_2, LHSV = 30,000 L/(kg h)) was passed through the catalyst-containing tube. After the following regeneration in dried air, the feed gas for the DMDS-combustion (800 ppm DMDS, 20 vol% O_2, 80 vol% N_2, LHSV = 30,000 L/(kg h)) was passed through the catalyst-containing tube at 250°C. Then the feed gas for the DMEA combustion was passed through the catalyst-containing tube once again under the same condition as described above. Gas mixtures of the products were analyzed after 30 min of reaction time for each temperature.

10.2.3.6 Simultaneous Combustion of DMEA and DMDS (See Table 10.10)

The catalyst (200 mg, 100–200 μm) was pretreated in dried air for 30 min at 275°C. Afterwards, the temperature of the catalysts was decreased to 250, 225, and 200°C, respectively, and the feed gas for the benzene combustion (750 ppm benzene, 20 vol% O_2, 80 vol% N_2, LHSV = 30,000 L/(kg h)) was passed through the catalyst-containing tube. Then the catalyst was regenerated in dried air for 15 min at 275°C. Afterwards, the temperature of the catalysts was decreased to 250, 225, 200, and 175°C, respectively, and the feed gas for the DMEA combustion (750 ppm DMEA, 20 vol% O_2, 80 vol% N_2, LHSV = 30,000 L/(kg h)) was passed through the catalyst-containing tube. After the following regeneration in dried air, the feed gas for the simultaneous combustion of DMEA and DMDS (750 ppm DMEA, 50 ppm DMDS, 20 vol% O_2, 80 vol% N_2, LHSV = 30,000 L/(kg h)) was passed through the catalyst-containing tube at 250°C. Then the feed gas or the DMEA combustion and the benzene combustion was passed through the catalyst-containing tube once again, under the same conditions as described above. Gas mixtures of the products were analyzed after 30 min during the combustion of DMEA and benzene for each temperature and after 1 h at 250°C and 3.5 h at 225°C (only for $Co_{40}Mn_{60}O_x$) during the combustion of the mixture.

10.3 RESULTS AND DISCUSSION

10.3.1 HIGH-THROUGHPUT SCREENING

The course of action of the high-throughput screening is typically discussed by means of the model compound DMEA. The reaction conditions (model compound, content of the model compound in dried air, reaction temperature and reaction time) for the high-throughput screening of the libraries are given in Table 10.1. The reaction temperatures were gradually decreased during the catalyst development from 350°C for the screening of the first generation to 150–175°C for the screening of some optimization libraries. An adjustment of the reaction temperature to the catalytic activity of the catalyst library is necessary, otherwise a gradient of pollutant concentration would build up across the reactor decrease discriminating against catalysts close to the gas exit of the products. For the model compounds benzene and dimethyl disulfide only selected results are presented.

Since here only mixed oxides are used, and since the oxidation state of the elements may not be constant, especially under reaction conditions, catalysts are identified by their metal ions and their relative content given as atom% in the subscript. The notation $Co_{20}Mn_{80}O_x$ stands for a mixed oxide composed of 20 atom% Co, 80 atom% Mn. In contrast to this notation, commonly used stable oxides like alumina are denoted as usual (e.g., Al_2O_3).

10.3.1.1   Dimethylethyl Amine (DMEA)

In contrast to the extensive literature on the catalytic combustion of hydrocarbons, little is known about the catalytic combustion of nitrogen-containing hydrocarbons: Ono[24] studied ion-exchanged zeoliths for the combustion of trimethyl amine; Haber[25] applied mixed metal oxides of the elements V, Cu, and Cr. Oxides containing noble-metals are less suited for the catalytic combustion of amines, due to their enhanced formation of nitrogen oxides.[26]

The IR image (Figure 10.1) of the diverse first catalyst library (containing the doped base oxides of Al, Si, and Ti) shows the heat of reactions for the combustion of 0.3 vol% DMEA at 350°C in dried air. The change of the surface temperature is illustrated with the gray scale (in °C). Activity was only observed for the doped alumina materials. The most active catalysts are ranked in Table 10.2 by the averaged increase of the surface temperature on the materials at 350 and 400°C. Doping with 2 atom% Fe led to a temperature increase of 2.9°C at 350°C, followed by Cu, Ce, and V. Undoped Al_2O_3 showed only a temperature increase of 1.1°C, while all other dopants deactivated the base oxide.

For the second library the three catalytically most active dopants from the first library, Fe, Cu, and Ce were combined with Cr and Zn, known components of VOC combustion catalysts[6] and Co, Mn, and Ni,[17,18] just discovered as components of catalysts for low-temperature combustion of CO. The second library was composed of binary composition spreads of mixed oxides of these eight elements. At a reaction temperature of 250°C the heat of reaction image (Figure 10.2) shows temperature increases of more than 10°C. Obviously, these mixed oxides are catalytically much more active than the doped Al_2O_3 catalysts. The most active catalysts on this second library were the binary mixed oxides of the elements Co-Cu, Co-Mn and Co-Ni.

Consequently, on the third and fourth library ternary composition spreads (10% spacing) of Cu, Co, Ni and Cu, Co, Mn were investigated. Here, simple display of the IR image of the heat of reactions is not very informative. To facilitate the analysis of the data, it is important to visualize the relative activity in dependence on chemical composition. In Figure 10.3 the relative heat of reaction

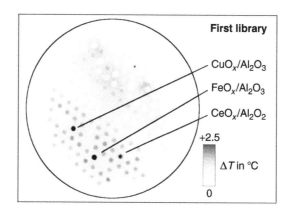

FIGURE 10.1 IR image of the first library screened at 350°C with 0.3 vol% DMEA in dried air.

TABLE 10.2
Catalytic Active Dopants of the First Library (DMEA Combustion)

Composition			Activity (ΔT)		Ranking	
Dopant	Atom%	Matrix	350°C	400°C	350°C	400°C
Fe(III)	2	Al_2O_3	2.9	3.8	1	2
Cu(II)	2	Al_2O_3	2.5	4.2	2	1
Ce(III)	2	Al_2O_3	1.6	2.9	3	3
V(V)	2	Al_2O_3	1.2	2.3	4	4
—	—	Al_2O_3	1.1	1.8	6	8
Ti	2	Al_2O_3	1.1	1.8	7	9
Ga	2	Al_2O_3	1.0	1.9	9	7
Cr	2	Al_2O_3	0.7	2.3	40	5

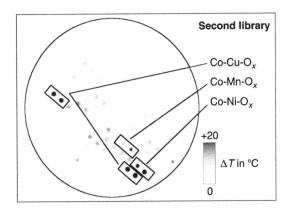

FIGURE 10.2 IR image of the second library screened at 250°C with 1 vol% DMEA in dried air.

of these libraries is presented at reaction temperatures of 175°C (third library) and 165°C (fourth library), respectively (each 0.3 vol% DMEA in dried air). These images not only correlate activity and chemical composition (quantitative composition activity relationship QCAR),[27] they also document a smooth, non-erratic behavior as already observed in other studies.[13–15,27] Two ranges of high catalytic activity can be identified on each library: $Co_{50-90}Cu_{10-50}Ni_{0-20}$, $Co_{0-20}Cu_{0-60}Ni_{40-100}$ and $Co_{0-20}Cu_{30-60}Mn_{30-50}$, $Co_{0-30}Cu_{0-10}Mn_{70-100}$ respectively.

From these ranges of higher activity the mixed metal oxides $Co_{70}Cu_{10}Ni_{20}O_x$, $Co_{20}Cu_{50}Mn_{30}O_x$, and $Co_{20}Mn_{80}O_x$ were doped with more than 50 elements of the periodic table (0.5–5 atom%) and screened for improvement of combustion activity at temperatures as low as 150–175°C. The best dopants are ranked in Tables 10.3–10.5. The doping of $Co_{70}Cu_{10}Ni_{20}O_x$ with 5% of Bi led to an increase in the surface temperature of 0.4°C at 150°C (fifth library, Table 10.3), whereas the undoped $Co_{70}Cu_{10}Ni_{20}O_x$ only showed an increase in temperature of 0.2°C. The Ce and Lu doped $Co_{70}Cu_{10}Ni_{20}O_x$ also displayed a higher increase in temperature than the undoped one. But the temperature differences noticed are very small, thus only a very small increase of the catalytic activity due to the doping with these elements was observed. However, the doping of $Co_{20}Cu_{50}Mn_{30}O_x$ with K led to a comparatively strong increase in the surface temperature of 2.0°C at 160°C (sixth library,

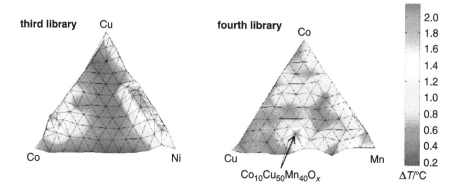

FIGURE 10.3 (See color insert following page 172) Representation of the catalytic activity of the third and fourth library screened at 175°C and 165°C respectively with 0.3 vol% DMEA in dried air.

TABLE 10.3
Catalytic Active Dopants of the Fifth Library (DMEA Combustion)

	Composition		
Dopant	Atom%	Matrix	Activity (ΔT) 150°C
Bi(III)	5	$Co_{70}Cu_{10}Ni_{20}$	0.4
Lu(III)	5	$Co_{70}Cu_{10}Ni_{20}$	0.3
Ce(III)	5	$Co_{70}Cu_{10}Ni_{20}$	0.3
Y(III)	5	$Co_{70}Cu_{10}Ni_{20}$	0.2
Rb(I)	1	$Co_{70}Cu_{10}Ni_{20}$	0.2
Tb(III)	5	$Co_{70}Cu_{10}Ni_{20}$	0.2
Yb(III)	5	$Co_{70}Cu_{10}Ni_{20}$	0.2
Fe(III)	5	$Co_{70}Cu_{10}Ni_{20}$	0.2
—	—	$Co_{70}Cu_{10}Ni_{20}$	0.2

Table 10.4), whereas the undoped $Co_{20}Cu_{50}Mn_{30}O_x$ only displayed an increase in temperature of 1.0°C. The Rb doped $Co_{20}Cu_{50}Mn_{30}O_x$ also displayed a distinct higher increase in temperature than the undoped one. Furthermore, the doping of $Co_{20}Mn_{80}O_x$ with K led to an increase in the surface temperature of 2.2°C at 175°C (seventh library, Figure 10.4, Table 10.5), whereas the undoped $Co_{20}Mn_{80}O_x$ only showed an increase in temperature of 0.7°C. The Rb, Nd, Na, and Mg doped $Co_{20}Mn_{80}O_x$ also displayed a distinct higher increase in temperature than the undoped one. Obviously, the doping of $Co_{20}Cu_{50}Mn_{30}O_x$ and $Co_{20}Mn_{80}O_x$ with alkali metals affects a distinct increase of the catalytic activity. This finding is consistent to the literature knowledge about the promoting effect of alkali metals.[26]

At this point the high-throughput screening with DMEA was terminated due to the lack of significant improvements. The most active mixed oxides were reproduced synthetically on a gram scale and their catalytic activity was confirmed conventionally. Figure 10.5 provides an overview over the whole development of noble-metal-free low-temperature combustion catalysts for DMEA.

TABLE 10.4
Catalytic Active Dopants of the Sixth Library (DMEA Combustion)

Dopant	Composition		Activity (ΔT) 160°C
	Atom%	Matrix	
K(I)	5	$Co_{20}Cu_{50}Mn_{30}$	2.0
Rb(I)	5	$Co_{20}Cu_{50}Mn_{30}$	1.9
Rb(I)	1	$Co_{20}Cu_{50}Mn_{30}$	1.2
—	—	$Co_{20}Cu_{50}Mn_{30}$	1.0
Ag(I)	1	$Co_{20}Cu_{50}Mn_{30}$	1.0
Cs(I)	1	$Co_{20}Cu_{50}Mn_{30}$	0.9

TABLE 10.5
Catalytic Active Dopants of the Seventh Library (DMEA Combustion)

Dopant	Composition		Activity (ΔT) 175°C
	Atom%	Matrix	
K(I)	2	$Co_{20}Mn_{80}$	2.2
Rb(I)	2	$Co_{20}Mn_{80}$	2.1
Rb(I)	5	$Co_{20}Mn_{80}$	1.7
Nd(III)	2	$Co_{20}Mn_{80}$	1.4
Na(I)	2	$Co_{20}Mn_{80}$	1.3
Mg(II)	5	$Co_{20}Mn_{80}$	1.3
—	—	$Co_{20}Mn_{80}$	0.7

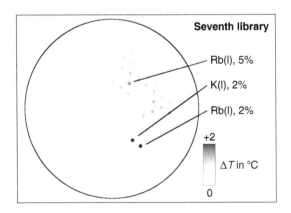

FIGURE 10.4 IR image of the seventh library screened at 175°C with 0.3 vol% DMEA in dried air.

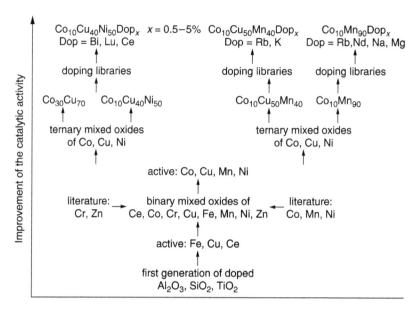

FIGURE 10.5 Development of catalysts for the combustion of DMEA.

10.3.1.2 Benzene

An analogous study as the one described for DMEA has been carried out with benzene. In the literature different mixed oxides can be found for the catalytic combustion of benzene: Papaefthimiou[29] successfully used Pt, Pd, and Co dispersed on Al_2O_3; Hong[30] applied CuO dispersed on TiO_2; Everaerth[6] studied V_2O_5-WO_3/TiO_2.

Again, the first catalyst library was studied for indication of potential catalyst elements. In contrast to DMEA, here only the doped titania catalysts showed significant activity. In Table 10.6 the best dopants are ranked according to increase of surface temperature. Sb, Cu, Fe, Ce, Se, and Te doped TiO_2 significantly increased the combustion activity of titania, while all other dopants had no or negative effects on the catalytic activity. A combination of these elements appeared promising. Due to the significant increase affected by the doping of TiO_2 with the elements Cu, Fe, and Ce, the catalyst second library from the DMEA screening was also screened for benzene combustion. The change in the surface temperature of the library is illustrated in Figure 10.6 for combustion temperature of 350°C. Here, the binary mixed oxides of the elements Fe-Mn, Cu-Co, Cu-Ni, Cu-Zn, and Cu-Cr appeared most promising.

Subsequently, quaternary composition spread of the elements Co, Cu, Mn, and Ni (20% spacing) were investigated at 200°C (eighth library, Figure 10.7). High contents of Mn were obviously most advantageous for the catalytic activity. This could be confirmed by the screening of a library consisting of multiply doped Mn oxides (0–50 atom%), on which catalytic activity could be observed at temperatures as low as 175°C (ninth library, Figure 10.8). The best catalytic activity was found for the combinations of the base oxide Mn in the range of 50–80 atom% and the dopants Fe and Ce in the range of 0–30 atom%. The screening of the seventh library (containing doped $Co_{20}Mn_{80}O_x$) at temperatures in the range of 150–200°C did not result in any new findings. While the dopants Ce, Nd, Nb, and Zn resulted in catalyst activities comparable to the undoped one, all other dopants rather deactivated the base oxide.

The study has not been finished. The effect of Sb, Se, and Te on the catalytic activity, especially on the catalytic activity of the Ce-, Co-, Cu-, Fe-, Mn-, Ni-, and Zn-containing materials is still being investigated.

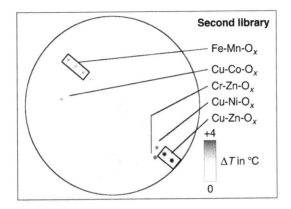

FIGURE 10.6 IR image of the second library screened at 175°C with 0.5 vol% benzene in dried air.

TABLE 10.6
Catalytic Active Dopants of the First Library (Benzene Combustion)

Composition			Activity (ΔT)		Ranking	
Dopant	Atom%	Matrix	350°C	400°C	350°C	400°C
Sb(V)	2	TiO$_2$	3.3	5.9	1	1
Cu(II)	2	TiO$_2$	1.6	4.1	2	2
Fe(III)	2	TiO$_2$	1.3	3.8	3	3
Ce(III)	2	TiO$_2$	1.3	2.6	4	6
Te(IV)	2	TiO$_2$	1.1	3.2	7	4
Se(IV)	2	TiO$_2$	1.1	2.9	6	5
—	—	TiO$_2$	0.4	1.2	26	26

10.3.1.3 Dimethyl Disulfide (DMDS)

The sulfur resistance of catalysts for the combustion of VOCs is often studied by applying DMDS as a model compound: Chu[31,32] applied Pt/Al$_2$O$_3$ and MnO/Fe$_2$O$_3$ for the catalytic combustion of DMDS (with concentrations below 200 ppm) at temperatures above 300°C and 350°C, respectively, Wang[33] successfully used a sulfated CuO-MoO$_3$/Al$_2$O$_3$ at temperatures below 300°C (concentrations below 100 ppm).

When the first catalyst library (containing doped base oxides of Al, Ti, and Si) was studied for activity for DMDS combustion, only little catalytic activity was recorded at temperatures below 400°C. In addition, most catalysts were irreversible deactivated during the time of reaction. Since some weak activities have been detectable with the silica and titania materials, the dopant concentration was increased up to 20% (eighth library). This led to improved catalytic performance; activity was now detectable already at 350°C. Mn, V, Cu, and Ce were the most active dopants, TiO$_2$ as base oxide seems to be more promising than SiO$_2$ (Table 10.7). The deactivation over the time of reaction was negligible. On some other libraries, which have been successfully investigated for the combustion of DMEA and benzene, catalytic activity for the combustion of DMDS was observed even at temperatures below 300°C. But the deactivation behavior was similar to that of the first library. Indeed, the development is still far from a S-stable catalyst for low-temperature combustion. The study is still ongoing.

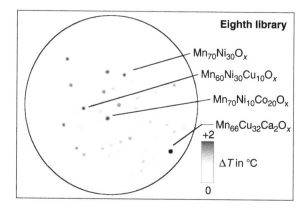

FIGURE 10.7 IR image of the eighth library screened at 175°C with 0.5 vol% benzene in dried air.

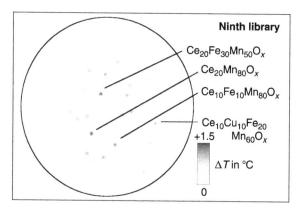

FIGURE 10.8 IR image of the ninth library screened at 175°C with 0.5 vol% benzene in dried air.

10.3.2   CONVENTIONAL CONFIRMATION OF CATALYTIC ACTIVITY

To confirm the catalytic activity of new catalysts discovered by high-throughput experimentation, the catalysts were prepared on a gram scale and their catalytic activity was controlled by gas phase reaction on a conventional plug-flow reactor. To simplify matters, the experiments carried out in the plug-flow reactor were denoted as conventional (to distinguish them from high-throughput). Reported are the T_{50} values, which correspond to the temperature at which 50% conversion of the model compound was observed.

10.3.2.1   Dimethylethyl Amine (DMEA)

The dependence of the conversion on the temperature of selected catalysts studied in a plug-flow reactor is illustrated for the doped Al_2O_3 (test series 1a, Figure 10.9) and mixed metal oxides of the elements Co, Cu, Mn, Ni, and Rb (test series 1b, Figure 10.10). The results of the conventional experiments and the high-throughput screening are compared in Table 10.8.

As expected, the doping of Al_2O_3 with Cu effected a significant decrease of the T_{50} ($T_{50} \approx 280°C$) compared to the undoped or Ga-doped alumina ($T_{50} > 300°C$). The results of the other conventional experiments presented in Table 10.8 correlate also well with the results of the high-throughput

TABLE 10.7
Catalytic Active Dopants of the Tenth Library (DMDS Combustion)

	Composition		
Dopant	Atom%	Matrix	Activity (ΔT) 175°C
Mn(II)	20	TiO_2	2.3
V(V)	20	SiO_2	2.2
V(V)	5	TiO_2	1.9
Cu	20	TiO_2	1.6
Ce	5	TiO_2	1.6
Ce	3	SiO_2	1.4
Mn	5	TiO_2	1.1
Ce	20	TiO_2	1.1
—	—	SiO_2	0.2

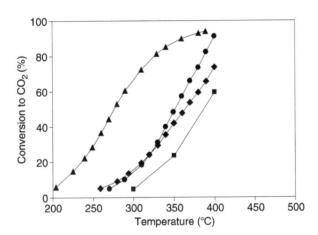

FIGURE 10.9 Test series 1a (800 ppm DMEA, dried air, LHSV = 30,000 L/(kg h), procedure described in the experimental part): blank (■), Al_2O_3 (◆), $Ga_2Al_{98}O_x$ (●), $Cu_2Al_{98}O_x$ (▲).

screening. Even the improvement of the catalytic activity of Rb doped $Co_{20}Cu_{50}Mn_{30}O_x$ could be verified in the conventional experiments.

The formation of nitrogen oxides was studied for selected catalysts (Figure 10.11, test series 2). $Co_{80}Cu_{20}O_x$ and $Co_{10}Cu_{50}Ni_{40}O_x$ were forming 120–160 ppm of nitrous oxides at temperatures of 200–250°C. This corresponds to a selectivity of 24–32%. In contrast, $Co_{20}Cu_{50}Mn_{30}O_x$ was reproducible forming less than 10 ppm of nitrous oxides. A similarly low selectivity for the formation of nitrous oxides during the catalytic combustion of trimethyl amine on Al_2O_3 supported CuO have been reported by Inui.[21] Apparently, not only low temperatures, but also catalyst composition, is effective in reducing the undesired NO_x formation in catalytic combustion.

10.3.2.2 Benzene

To verify the results of the high-throughput screening, the dependence of the conversion on the temperature of selected catalysts was studied in the plug-flow reactor (Figure 10.12). As expected,

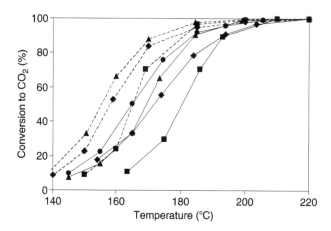

FIGURE 10.10 Test series 1b (800 ppm DMEA, dried air, LHSV = 30,000 L/(kg h), procedure described in the experimental part): $Co_{60}Ni_{40}O_x$ (■), $Co_{10}Ni_{50}Cu_{40}O_x$ (◆), $Co_{30}Cu_{70}O_x$ (▲), $Co_{40}Cu_{30}Mn_{30}O_x$ (●) (solid lines), $Co_{20}Mn_{80}O_x$ (■), $Co_{10}Cu_{50}Mn_{40}O_x$ (◆), $Co_{10}Cu_{48}Mn_{39}Rb_3$ (▲) (dashed lines).

TABLE 10.8
High-Throughput Screening Versus Conventional Experiments

Catalyst	Plug-Flow Reactor T_{50} (°C)	High-Throughput Screening		
		Library	Condition	ΔT (°C)
Al_2O_3	>300	1st	350°C 4 h	1.1
GaO_x/Al_2O_3	>300		0.5%	1.0
CuO_x/Al_2O_3	280		DMEA	2.5
$Co_{60}Ni_{40}O_x$	181	3rd	175°C 1.5 h	<0.5
$Co_{10}Cu_{40}Ni_{50}O_x$	172		0.3%	2.5
$Co_{30}Cu_{70}O_x$	170		DMEA	2.5
$Co_{30}Cu_{70}O_x$	170	4th	165°C 4 h	0.5
$Co_{40}Cu_{30}Mn_{30}O_x$	165		0.3%	1
$Co_{20}Mn_{80}O_x$	165		DMEA	1.5
$Co_{10}Cu_{50}Mn_{30}O_x$	158			2
$Co_{10}Cu_{48}Mn_{39}Rb_3O_x$	155	6th	150°C 4 h 0.3% DMEA	1.9

the catalysts based on Mn showed the highest catalytic activity. Benzene conversion was complete on $Mn_{80}Cu_{20}O_x$, $Fe_{40}Mn_{60}O_x$, and $Mn_{70}Cu_{30}O_x$ respectively, below 250°C. In the high-throughput screening $Mn_{70}Ni_{30}O_x$ showed catalytic activity at 200°C (eighth library, Figure 10.7), $Fe_{40}Mn_{60}O_x$ and $Mn_{80}Cu_{20}O_x$ even at 175°C (ninth library, Figure 10.8). For complete conversion of benzene on the manganese-free $Cu_{80}Zn_{20}O_x$ and $Fe_{60}Ni_{40}O_x$ in the conventional experiment temperatures higher than 300°C were needed. Both mixed oxides were active only above 200°C (second library) in the high-throughput screening. Although the results of the conventional experiments confirm the data from high-throughput screening, for complete conversion higher temperatures or longer residence times are required.

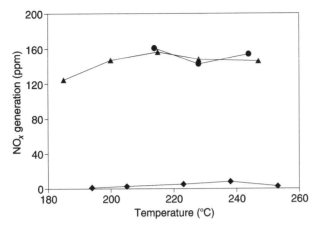

FIGURE 10.11 Test series 2 (800 ppm DMEA, dried air, LHSV = 30 000 L/(kg h), procedure described in the experimental part): $Co_{10}Ni_{50}Cu_{40}O_x$ (●), $Co_{80}Cu_{20}O_x$ (▲), $Co_{20}Cu_{50}Mn_{30}O_x$ (♦).

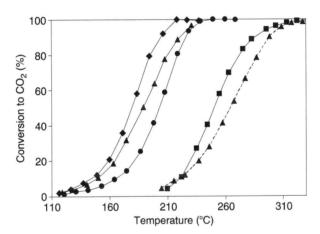

FIGURE 10.12 Benzene combustion (500 ppm benzene, dried air, LHSV = 30,000 L/(kg h), procedure described in the experimental part): $Mn_{80}Cu_{20}O_x$ (♦), $Fe_{40}Mn_{60}O_x$ (▲), $Mn_{70}Ni_{30}O_x$ (●), $Cu_{80}Zn_{20}O_x$ (■), $Fe_{60}Ni_{40}O_x$ (▲, dashed line).

10.3.2.3 Sequential and Simultaneous Combustion of DMEA, DMDS, and Benzene

Besides the development of catalysts for the combustion of single-model compounds, one aim of this study was the development of combustion catalysts for a broad mixture of pollutants at moderate temperatures. Therefore, the most promising catalysts from the above-described developments were subjected to a sequence of pollutants.

All the selected catalysts converted DMEA at 200°C completely (Table 10.9), while benzene was completely converted at 200°C only on $Co_{20}Mn_{80}O_x$. $Ce_{40}Mn_{60}O_x$, $Co_{20}Cu_{50}Mn_{30}O_x$, and $Co_{20}Mn_{80}O_x$ displayed good activity for the catalytic combustion of benzene and DMEA and a reduced formation of nitrous oxides. None of the selected catalysts could maintain a stable catalytic conversion of DMDS over a period of 30 min at 250°C. It is assumed that the catalysts were irreversibly deactivated by DMDS, because the catalytic activity for the repeated combustion of DMEA (see Table 10.9, second run DMEA2) dramatically decreased.

TABLE 10.9
Sequential Combustion of Benzene, DMEA, and DMDS

Catalyst	T (°C)	Benzene Conv. (%)	DMEA Conv. (%)	NO (ppm)	NO$_2$ (ppm)	DMDS Conv. (%)	DMEA2 Conv. (%)
$Co_{20}Mn_{80}O_x$	250	100	100	<10	<10	55	55
	200	100	100	—	—	—	—
$Co_{50}Cu_{50}O_x$	250	100	100	20	190	—	—
	200	70	100	—	—	—	—
$Co_{20}Cu_{50}Mn_{30}O_x$	250	100	100	<10	<10	50	10
	200	60	100	—	—	—	—
$Ce_{40}Mn_{60}O_x$	250	100	100	5	30	30	45
	200	55	100	—	—	—	—
$Cu_{60}Ni_{40}O_x$	250	65	100	65	100	10	20
	200	—	100	—	—	—	—
$Co_{10}Cu_{20}Ni_{70}O_x$	250	75	100	70	30	—	—
	200	—	90	—	—	—	—

Poisoning of the catalysts by means of small amounts sulfur-containing components in the exhaust gas were studied typically with a mixture of DMDS (50 ppm) and DMEA (750 ppm) on $Co_{20}Cu_{50}Mn_{30}O_x$ and $Co_{40}Mn_{60}O_x$ over a period of some hours. The conversion of the catalytic combustion of benzene and DMEA before and after the sulfur poisoning was determined (Table 10.10). Both catalysts displayed a slight decrease in the conversion of benzene after the combustion of the above-mentioned mixture from 90% to 70% at 225°C (Table 10.10 → Benzene 2). As a result, a limited resistance against fast sulfur poisoning by means of small amounts of sulfur-containing components in the exhaust gas can be assumed.

10.4 SUMMARY AND CONCLUSIONS

Within a few generations, the high-throughput screening approach provided several well-defined mixed-oxide catalysts free of noble metals, highly active for the catalytic combustion of benzene and dimethylethyl amine (DMEA) at uncommonly low temperatures. These catalysts appear to be well suited for an application in a catalytic combustor and promise considerable energy savings. Currently, the best catalysts are studied in a plug-flow reactor in g-quantities supported on carriers.

The highest catalytic activity for the combustion of DMEA was found for $Co_{10}Cu_{48}Mn_{39}Rb_3O_x$. The expected reduction of NO_x formation due to the lower temperature of combustion has been confirmed by the experiments. The best catalyst for reduced NO_x formation, $Co_{20}Cu_{50}Mn_{30}O_x$, shows negligible formation of nitrous oxides during DMEA combustion at reaction temperatures below 200°C. Benzene was most successfully combusted on catalysts containing Mn in the range of 60–90% and a combination of Fe and Ce in the range of 0–30%. But combinations of the aforementioned elements with Sb, Se, and Te seem to be also promising. The data also demonstrate that the catalytic activity for the combustion of VOCs is not only dependent on the catalyst's composition, but also very sensitive to the nature of the pollutant. In fact, every pollutant studied leads associated catalysts, whose composition is significantly different to those of the other pollutants. This calls for pollutant-dependent fine tuning.

The catalytic activity of the best catalysts from high-throughput screening has been verified by catalyst synthesis on a laboratory scale and testing in a conventional plug-flow reactor. In several experiments, the feed gas has been enriched with water (20% moisture) to examine the catalyst

TABLE 10.10
Simultaneous Combustion of DMEA and DMDS

Catalyst	T (°C)	Benzene Conv. (%)	DMEA Conv. (%)	Mixture Conv. (%)	Benzene 2 Conv. (%)
$Co_{40}Mn_{60}O_x$	250	100	100	100	—
	225	90	100	100	70
	200	50	100	—	—
	175	—	< 100	—	—
$Co_{20}Cu_{50}Mn_{30}O_x$	250	100	100	100	—
	225	90	100	—	70
	200	50	100	—	—
	175	—	< 100	—	—

stability. In all experiments no deactivation was observed. The effect of moisture has not been studied any further and it can be expected that higher moisture levels may result in catalyst deactivation. However, the catalysts developed for the combustion of benzene and DMEA are rapidly poisoned by the sulfur-containing DMDS. S-stable catalysts for the combustion of DMDS at low temperature have not been found yet, although results of the high-throughput screening indicate that catalysts containing Ti, Mn, V, and Ce show increased S-resistance and might be a suitable basis for sulfur-resistant catalyst.

ACKNOWLEDGMENT

The authors thank the Deutsche Umweltstiftung DBU for support through grant 19594.

REFERENCES

1. Centi, G., Ciambelli, P., Perathoner, S. and Russo, P., Environmental catalysis: trends and outlook, *Catal. Today*, 75, 3, 2002.
2. Navarri, P., Marchal, D. and Ginestet, A., Activated carbon fibre materials for VOC removal, *Filtr. Sep.*, 38, 33, 2001.
3. Qin, Y. J., Sheth, J. P. and Sirkar, K. K., Supported liquid membrane-based pervaporation for VOC removal from water, *Ind. Eng. Chem. Res.*, 41, 3413, 2002.
4. Zhu, X. Q., Alonso, C., Cao, H. W., Kim, B. J. and Kim, B. R., The effect of liquid phase on VOC removal in trickle-bed biofilters, *Water Sci. Technol.*, 38, 315, 1998.
5. Dudukovic, M. P., Larachi, F. and Mills, P. L., Multiphase catalytic reactors: a perspective on current knowledge and future trends, *Catal. Rev.*, 44, 123, 2002.
6. Everaert, K. and Baeyens, J., Catalytic combustion of volatile organic compounds, *J. Hazardous Mat.*, B109, 113, 2004.
7. Schippert, E., Möhner, C., Pannwitt, B. and Chmiel, H., Waste air cleaning using activated carbon fibre cloths regenerable by direct electric heating, *Stud. Surf. Sci. Catal.*, 144, 507, 2002.
8. Schippert, E., Fortschritte bei der thermischen, katalytischen, sorptiven und biologischen Abgasreinigung, VDI-Berichte 1241, VDI-Verlag GmbH, 427, 1996.
9. Bathen, D., Adsorptive Behandlung VOC-haltiger Abluftströme – Stand der Technik. *Chem. Ing. Tech.*, 76, 1631, 2004.
10. Kullavanijaya, E., Cant, N. W. and Trimm, D., The treatment of binary VOC mixtures by adsorption and oxidation using activated carbon and a palladium catalyst, *J. Chem. Technol. Biotechnol.*, 77, 473, 2002.

11. Salden, A. and Eigenberger, G., Multifunctional adsorber/reactor concept for waste-air purification, *Chem. Eng. Sci.*, 56, 1605, 2001.

12. Kirsten, G. and Maier, W. F., Reactive quantification of catalytic activity in combinatorial libraries by emissivity corrected infrared thermography, In: *High Throughput Screening in Chemical Catalysis*, Hagemeyer, A., Strasser, P. and Volpe, A.F., Eds. Wiley-VCH, Weinheim, Germany, 2004, p. 175.

13. Saalfrank, J. W. and Maier, W. F., Doping selection and composition spreads, a discovery of new mixed oxide catalysts for low-temperature CO oxidation, *Comp. Rend. Chim.*, 7, 483, 2004.

14. Saalfrank, J. W. and Maier, W. F., Directed evolution of noble-metal-free catalysts for the oxidation of CO at room temperature, *Angew. Chem. Int. Ed.*, 43, 2028, 2004.

15. Maier, W. F. and Saalfrank, J. W., Discovery, combinatorial chemistry and a new selective CO-oxidation catalyst, *Chem. Eng. Sci.*, 59, 4673, 2004.

16. Xiang, X.-D. and Takeuchi, I., *Combinatorial Materials Synthesis*, Marcel Dekker Inc., New York, 2003.

17. Paul, J. S., Urschey J., Jacobs, P. A. and Maier, W. F., Combinatorial screening and conventional testing of antimony-rich selective oxidation catalysts, *J. Catal.*, 220, 136, 2003.

18. Urschey, J., Kühnle, A. and Maier, W. F., Combinatorial and conventional development of novel dehydrogenation catalysts, *Appl. Catal. A,* 252, 91, 2003.

19. Scheidtmann, J., Saalfrank, J. and Maier, W. F., Plattenbau – automated synthesis of catalysts and materials libraries, *Surf. Sci.*, 145, 13, 2003.

20. Rodemerck, U., Baerns, M., Holena, M. and Wolf, D., Application of a genetic algorithm and a neural network for the discovery and optimization of new solid catalytic materials, *Appl. Surf. Sci.*, 223, 168, 2004.

21. Scheidtmann, J., PhD thesis, Saarbrücken, Germany, 2003.

22. Holzwarth, A., Schmidt, H. W. and Maier, W. F., Detection of catalytic activity in combinatorial libraries of heterogeneous catalysts by infrared thermography, *Angew. Chem. Int. Ed.*, 37, 2644, 1998.

23. Holzwarth, A. and Maier, W. F., Detection of catalytic activation and deactivation phenomena in combinatorial libraries of heterogeneous catalysts with emissivity-corrected IR-thermography, *Platinum Met. Rev.*, 44, 16, 2000.

24. Ono, Y., Fujii, Y., Wakita, H., Kimura, K. and Inui, T., Catalytic combustion of odors in domestic spaces on ion-exchanged zeolites, *Appl. Catal. B*, 16, 227, 1998.

25. Haber, J., Janas, J. and Krysciak-Czerwenka, J., Total oxidation of nitrogen-containing organic compounds to N_2, CO_2 and H_2O, *Appl. Catal. A*, 229, 23, 2002.

26. Haber, J., Machej, T., Sadowska, H. and Janas, J., Catalytic conversion of nitrogen containing organic compounds, In: *Proceeding of the 16th Meeting of NAM Catalysis Society,* Boston, MA, 1999.

27. Scheidtmann, J., Klär, D., Saalfrank, J. W., Schmidt, T. and Maier, W. F., Quantitative composition activity relationships (QCAR) of Co-Ni-Mn-mixed oxide and M1-M2-mixed oxide catalysts, *QSAR Comb. Sci.*, 2, 203, 2005.

28. Muramatsu, G., Abe, A., Yoshida, K. and Takahashi, Y., Exhaust gas cleaner and method of cleaning exhaust gas, US5320999, 1994.

29. Papaefthimiou, P., Ioannides, T. and Verykios, X. E., Combustion of non-halogenated volatile organic compounds over group VIII metal catalysts, *Appl. Catal. B*, 13, 175, 1997.

30. Hong, S.-S., Lee, G.-H. and Lee, G.-D., Catalytic combustion of benzene over supported metal oxides catalysts, *Korean J. Chem. Eng.*, 20, 440, 2003.

31. Chu, H. L. and Lee, W. T., The effect of sulfur poisoning of dimethyl disulfide on the catalytic incineration over a Pt/Al_2O_3 catalyst, *Sci. Total Environ.*, 209, 217, 1998.

32. Chu, H. H. and Tseng, T. K., Laboratory study of poisoning of a MnO/Fe_2O_3 catalyst by dimethyl sulfide and dimethyl disulfide, *J. Hazard Mater.*, 100, 301, 2003.

33. Wang, C.-H., Lee, C.-N. and Weng, H.-S., Effect of acid treatment on the performance of the CuO-MoO_3/Al_2O_3 catalyst for the destructive oxidation of $(CH_3)_2S_2$, *Ind. Eng. Chem. Res.*, 37, 1774, 1998.

Section 3

Development of Functional Polymers

11 One-Dimensional Polymeric Formulated Materials Arrays: Fabrication, High-Throughput Performance Testing, and Applications

Radislav A. Potyrailo and Ronald J. Wroczynski

CONTENTS

11.1 INTRODUCTION

Formulated, additive-based polymeric materials are common in our everyday lives. Additives serve a wide variety of functions in performance polymers during their production and end use. Some important functions of additives include process stabilization, plasticization, lubrication, flame retardation, fluorescent whitening, impact, electrical conductivity, antimicrobial, and antistatic modifications, and many others.[1] Additives are also common in polymer-based sensor materials. Formulated polymeric materials in sensors are used where chemical or biological species of interest are most effectively detected by a polymer-immobilized reagent that changes its property upon exposure to the species of interest.[2] Also, different additives provide improved stability, sensitivity, and selectivity of sensing polymers.[3]

Additives are not only an essential, but also often an expensive, part of many polymer formulations.[4] Certain additives can often be synergistic or antagonistic in their effects.[5,6] Because of the large interactive multitude of additive possibilities and the cumbersome and costly process of evaluating their effects, the introduction of new formulated polymeric materials is extremely slow. Many polymer formulations are simply not optimized because traditional additive packages are often utilized without revalidation for new products or applications, or further development is terminated once a moderate level of performance is achieved.

With the growing acceptance of high-throughput experimentation methodologies,[7–12] it is very attractive to apply these new research tools for fabrication and screening of the best combinations of additives in formulated polymers. Polymeric formulations are typically produced by polymer extrusion,[13,14] solvent casting,[15] and by other methods. Traditional measurement techniques exist for the polymer properties of interest.[16,17] However, many of these methods are not currently suitable for analysis of small samples. Further, many of the existing polymer analysis methods are destructive in nature, limiting the potential for performing multiple measurements on a single test sample. New polymer analysis methods are needed that will improve speed of analysis, reduce sample size 100- to 1000-fold, provide capabilities for analyzing a single sample for multiple properties, and improve precision compared with traditional measurement techniques.

We describe here our strategy for the combinatorial and high-throughput (HT) development of polymer compositions. Our strategy is based on the generation of polymeric formulations as one-dimensional (1-D) materials libraries, their performance testing at predetermined locations, and rapid spatially resolved analysis of the resulting properties. We fabricated these 1-D polymeric arrays using a combinatorial system that included a microextruder and microfeeders.[18,19] Property changes in polymeric formulations were induced along the length of the 1-D arrays by several approaches that included: (1) variation in *composition* of 1-D array during polymer microextrusion; (2) variation in *processing conditions* during polymer microextrusion; and (3) variation in *exposure conditions* of a performance test for different spatial regions along the length of the array. Polymeric 1-D arrays are analyzed either in-line or off-line using spectroscopic and imaging techniques. An application of a small-volume extruder opened several new opportunities in the development of formulated polymeric compositions. First, polymeric formulations were evaluated with up to 200 times less material consumption compared to traditional-scale extruders, thus bringing enormous materials and energy savings. Second, performance tests and characterization of formulated polymeric compositions were performed directly on the coiled polymer arrays, which became possible because of the minute amounts of material produced. Third, the capabilities of the variable composition polymers were explored for the fabrication of chemical sensors.

11.2   CONCEPT OF INTEGRATED MICROEXTRUSION AND HIGH-THROUGHPUT PERFORMANCE TESTING OF ONE-DIMENSIONAL POLYMERIC MATERIALS

Our general concept for the fabrication and high-throughput performance testing of 1-D arrays of polymers is presented in Figure 11.1. According to this concept, a set of polymeric formulations is fabricated in a combinatorial microextruder system that utilizes several feeders that deliver different formulation components into the microextruder to produce an array of polymeric multicomponent formulations. These formulations are extruded and handled as 1-D polymeric arrays. Analysis is done either in-line with polymer extrusion or off-line during or after a performance test. The performance testing processes include exposure of the 1-D library to environments that imitate the end-use applications and alter materials properties in a detectable manner. Similar to testing on a conventional scale, examples of performance testing can include solar radiation, heat, mechanical

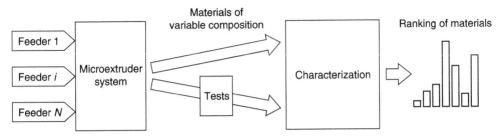

FIGURE 11.1 General concept for the fabrication and high-throughput performance testing of 1-D arrays of polymers.

stress, aggressive fluids, and others. Properties of polymeric materials are further characterized with the goal of ranking materials performance often against a set of worst and best performance controls.[20]

An application of a small-volume extruder opens a possibility of evaluating polymeric formulations with 40–200 times less material consumption compared to traditional-scale extruders used for conventional and HT studies of polymer formulations.[21,18] This is a dramatic improvement in the reduction of generated waste during such experiments, because a traditional-scale extruder typically can generate from 2 to 15 kg of material per minute of operation, making a daily operation quite costly and producing significant amounts of waste. Given the small volume of our microextruder, we typically can generate only 10–20 g of material per minute. Because our microextruder approach for fabrication of formulated polymeric materials requires only a negligible fraction of the material compared to traditional-scale extruders, it provides previously unavailable opportunities for testing and characterization of materials upon extrusion without their pelletization. Instead, we are able to perform a variety of tests with the coiled polymer material (1–2 mm diameter fiber or 5–12 mm wide film) followed by the analysis of resulting polymeric properties. Alternatively, we perform measurements directly on the coiled polymer arrays.

During polymer extrusion, several possibilities exist to produce 1-D arrays. A 1-D array can have an initial variable composition and/or a performance property along its length induced by several approaches. First, variation in *composition* of 1-D polymer array during polymer microextrusion can be induced by applying several feeders in combination with the microextrusion process. Second, variation in *processing conditions* of 1-D polymer array during polymer microextrusion can be induced by changing the extrusion conditions, such as amount of oxygen available in the extrusion barrel, extrusion rate, temperature, and other process parameters. Third, variations in the *conditions of a performance test* (weathering, mechanical, etc.) can be induced for different spatial regions along the length of the 1-D array after material fabrication.

An example of a variation in composition of 1-D polymer array during polymer microextrusion is demonstrated in Figure 11.2 and Figure 11.3. In this computer simulation example, four feeders operate in combination with the microextrusion process, where feeders F_1, F_2, and F_3 are programmed to periodically deliver a known amount of a given polymeric formulation component into the microextruder where a periodic pattern is given by:

$$F_i(t) = 1 + \sin(t/t_i) \qquad (11.1)$$

where t_i are the time constants of the operation of feeders F_1, F_2, and F_3. In our example, three feeders F_1, F_2, and F_3 operate with the time constants 20, 50, and 70 arbitrary units, respectively, producing different sinusoidal patterns of feed rates as shown in Figure 11.2. The fourth feeder operates with the feed rate that is complementary to the sum of the feed rates of feeders F_1, F_2, and F_3, so the total feed rate of four feeders is constant. Thus, the feed rates of feeders F_1, F_2, and F_3 are proportional to the respective concentrations of formulation components delivered into the microextruder.

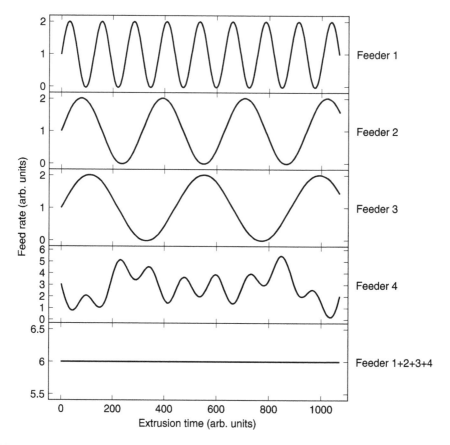

FIGURE 11.2 Composition variation of 1-D polymer array using four feeders operating in combination with the microextrusion process. Each of the feeders is programmed to deliver a known amount of a polymer into the microextruder with a periodic pattern.

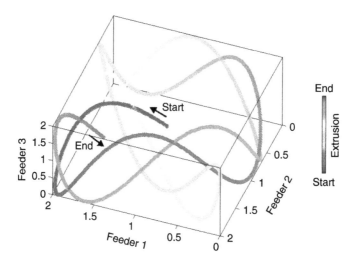

FIGURE 11.3 (See color insert following page 172) Computer simulation results of diversity of polymeric formulations produced by the operation of three independent feeders.

Such multifeeder operation permits generation of different formulated polymeric compositions in a desired number of replicates. Figure 11.3 illustrates the diversity of polymeric formulations produced by the operation of three independent feeders.

The attributes of our characterization methods of 1-D arrays of polymers include reduced sample size, measurements of multiple samples at a time, capability for in situ and/or non-destructive analysis, and often enhanced sensitivity. Depending on the goal of the performance testing, a die attachment to the microextruder can be selected to produce extruded material as a fiber or a film (see Section 4.2). Analysis of the properties of 1-D arrays can be performed in configurations schematically depicted in Figure 11.4 using either serial or parallel measurements with spectroscopic and imaging systems. In-situ analysis of polymer formulations during extrusion (Figure 11.4a) can be performed using luminescence, UV–vis, near-IR spectroscopies for ranking of materials according to their oxidative stability[23] or for determination of the best polymer composition for an end-use application.[24] The 1-D polymeric arrays can be also coiled for performance testing or after performance testing. An extruded fiber can be coiled onto a mandrel to create a cylindrical coil as a helix (see Figure 11.4b) while an extruded film can be coiled as shown in Figure 11.4c.

Application of a small-volume extruder opens several new opportunities in the development of formulated polymeric compositions. For example, small material consumption compared to traditional-scale extruders brings significant materials and energy savings. Also, performance tests and characterization of formulated polymeric compositions can be performed directly on the coiled polymer arrays, previously difficult or impossible to achieve using conventional-scale extrusion. However, there may be some differences in the mixing abilities of the used microextruder when compared to larger lab-scale or production-scale extruders.[18] These differences are expected for complex systems with large quantities of additives or especially for polymer–polymer blends where blend morphology development is critical.[21] However, our experience with low concentrations (1–3 wt%) of small particulate additives did not indicate any noticeable issues with the quality of mixing obtained in the microextruder.

11.3   EXPERIMENTAL METHODOLOGIES

11.3.1   Combinatorial Microextruder System

A microextruder system assembled for our combinatorial applications is shown in Figure 11.5. The basis of our system was a vertical, conical, co-rotating, twin-screw microextruder (DACA Instruments, Goleta, California) operating in a continuous mixing mode. The screw diameter was 14 mm at the

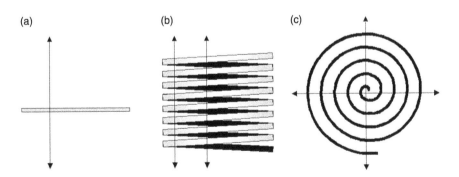

FIGURE 11.4 Analysis of the properties of 1-D arrays of polymers on microscale. (a) Linear 1-D array during in-situ analysis. (b) Coiled fiber array. (c) Coiled film array. Arrows represent regions for performance testing and/or characterization. (Reused with permission from Potyrailo, R. A. et al., *Rev. Sci. Instrum.*, 76, 062222, 2005. Copyright 2005, American Institute of Physics.)

entrance and 5.5 mm at the exit with a total barrel volume of 4.5 cm³ (see Figure 11.5a). For fabrication of 1-D polymer libraries, two independently controlled loss-in-weight feeders (K-Tron International Inc, Pitman, New Jersey) were used in conjunction with the microextruder as shown in Figure 11.5b. The total combined feed from both feeders was kept constant and depending on the experiment was in the range 2–10 g/min.[14] The polymeric arrays were produced as a strand (1–2 mm in diameter) or a film (5–12 mm wide and 0.3–1 mm thick). When needed, polymeric strands and films were coiled using a winder fabricated in-house or manually. In addition, the strand can be pelletized using a small mechanical pelletizer.

11.3.2 High-Throughput Fluorescence Analysis

Spectroscopic analysis of polymers was performed using fluorescence detection. Fluorescence detection has been proven to be useful for materials screening.[25] Other types of analysis from our 1-D materials arrays included colorimetric analysis, melt behavior, torque, volatility, and others as described elsewhere.[14,18,19,23,26–28] Measurements of fluorescence spectra of 1-D arrays of polymers and reference polymer samples were performed using a modular automatic scanning system[18,29,30] that consisted of a light source, a portable spectrofluorometer, and a bifurcated fiber-optic probe. Depending on the application, the light source was either a 266-nm (Nanolase, France), 337-nm N_2 laser (Photon Technology International), 355-nm (Nanolase, France), or 407-nm (Coherent, Auburn, California) laser. In addition, a GaN UV light-emitting diode (LED) source was used with a relatively broad peak emission centered at 375 nm (Nichia Corporation, Tokyo, Japan,). The LED operated in a continuous mode. LED emission was filtered with a 370-nm interference filter, 10-nm full width at half maximum to eliminate visible component of the emission. The spectrofluorometer (Ocean Optics, Inc., Dunedin, Florida) was equipped with a grating blazed at 400 nm and covering the spectral range 250–800 nm, and a linear CCD-array detector. The common end of the bifurcated fiber-optic probe was arranged near an exit slit of the micro-extruder or near a polymer array and provided excitation of the fluorescence of the melted or solid polymer and fluorescence collection. When needed, an in-line long-pass optical filter was used to reject the excitation light from entering the spectrometer. Automated measurements of 1-D arrays after extrusion were performed using this modular automatic scanning system coupled to an X–Y translation stage with a 100-μm step scan.

A system for imaging of 1-D polymeric arrays included a white or laser light source, a beam expander, and a CCD camera.[18,30,31] The white light source was a 450-W Xe arc lamp (SLM Instruments, Inc., Model FP-024). Laser light sources were the same as described above. An assembled

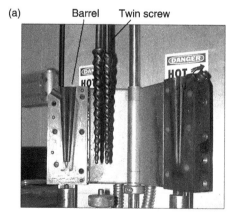
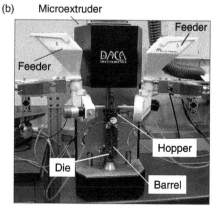

FIGURE 11.5 Microextruder system assembled for fabrication of combinatorial polymer formulations. (a) Open barrel of the microextruder. (b) Two loss-in-weight feeders in conjunction with the microextruder.

in-house beam expander was used for the efficient illumination of the combinatorial arrays. The CCD camera was a scientific-grade, cooled CCD camera (Roper Scientific, Trenton, New Jersey, Model TE/CCD 1100 PF/UV). Reflected-light images were collected with the white light source. Fluorescence images were collected under different excitation conditions through a long-pass or an interference filter. Image analysis was performed using software provided with the CCD camera and Matlab software.

11.3.3 Performance Testing

To demonstrate the applicability of our performance testing, weathering screening was implemented. For this application, an aromatic polymer, bisphenol-A polycarbonate (PC, GE Plastics) was used as a base polymer in combination with a rutile TiO_2 pigment (1 wt%, provided by GE Plastics) and different levels of a UV absorber (Tinuvin 234, T234, Ciba-Geigy, Ardsley, New York). Different concentrations of T234 in polymer were produced by extruding two PC compositions (PC + 1 wt% TiO_2 and PC + 1 wt% TiO_2 + 2 wt% T234) at different ratios of feed rates of the microfeeders while maintaining their total feed rate at 8 g/min and operating the microextruder at 150 rpm screw speed. By changing the feed ratios of the feeders, TiO_2-pigmented PC compositions containing 0, 0.6, 1.0, 1.4, and 2 wt% UV absorber were extruded as a ~1 mm diameter strand.

Multilevel weathering performance testing of 1-D polymeric libraries was done using two systems that have spectral irradiance profiles corresponding to the Sun. The first system was a borosilicate-filtered Xe-arc Weatherometer (Atlas, Model Ci35A) with a 2.7 kJ/h exposure rate and a 0.75 W/m^2 irradiance measured at 340 nm, which corresponds to 1.25 Suns. Exposure dose ranged from 175 to 845 kJ/m^2 (at 340 nm). The second system included a 300-W Xe-arc lamp, focusing optics, and a borosilicate optical filter. The irradiance of this system was measured to be ~18 W/m^2 at 340 nm that corresponded to 30 Suns. For weathering testing and analysis, polymeric libraries were formed by coiling sections of extruded polycarbonate strands with different levels of the UV absorber onto a solid cylindrical mandrel.

11.3.4 Data Analysis and Mining

Data analysis was performed using KaleidaGraph software (Synergy Software, Reading, Pennsylvania), PLS_Toolbox software (Eigenvector Research, Inc., Manson, Washington) operated with Matlab software (The Mathworks Inc., Natick, Massachusetts), and Minitab software (Minitab Inc., State College, Pennsylvania).

11.4 CASE STUDIES

11.4.1 Fluorescence as HT Detection Method of Process Degradation of Polymers

Degradation of thermoplastic polymers during their initial melt processing, end use, and recycling, is typically minimized by using various packages of process stabilizers. Polypropylene (PP), which undergoes chain scission with rapid loss of molecular weight, is rarely used without stabilization.[32] Instead, synergistic combinations of two or more stabilizers are commonly used.[32–34] The efficacy of a process stabilizer system is often evaluated using multiple pass extrusion[1,14,32,33] where the stabilized polymer is subjected to five or more consecutive extruder passes. The stabilizers that maximize retention of performance properties with successive passes are most effective. For PP, melt flow index (MFI) and yellowness index (YI) are the primary evaluated parameters.[1] The multipass extrusion and MFI techniques in their traditional scale consume a relatively large amount of material, are labor-intensive, and slow. Thus, although combinations of stabilizers are often most effective, evaluating different stabilizers and their combinations at different ratios is unmanageably time consuming.

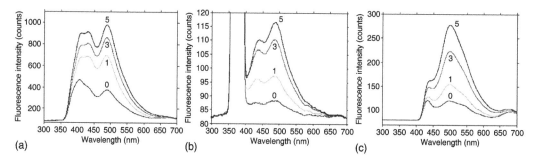

FIGURE 11.6 Typical fluorescence spectra of a stabilized polypropylene formulation as a function of amount of thermal degradation induced by compounding and multiple extrusion process. Excitation light sources and their peak wavelengths: (a) nitrogen laser, 337 nm; (b) GaN diode laser, 407 nm; (c) GaN UV LED, 370 nm. Extrusion passes during multiple extrusions are numbers 1, 3, and 5. Number 0 is compounding. Spectra in (a) and (c) are convoluted with the employed long-pass filters to block an excess of excitation light.

During polymer processing, degradation products constitute one of the main sources for the discoloration of PP.[35] These colored species have luminescence quantum yields which are sufficient for their emission detection.[36,37] Fluorescence is more sensitive than color analysis[38] and is useful for determination of low amounts of degradation products undetectable by color analysis. In our method, fluorescence spectroscopy is coupled with the multivariate analysis to simultaneously quantify color (as yellowness index, YI) and MFI of PP during multiple processing steps.

Fluorescence of unstabilized PP correlates with its thermal degradation[39] and with functional groups that result from and/or promote degradation.[40] With a properly selected excitation wavelength, fluorescence emission can be correlated with the thermal degradation of stabilized PP. The important considerations for the use of this method are the concentrations and fluorescence quantum yields of the stabilizing additives. Under certain conditions, even low concentrations of additives can contribute to the fluorescence of the polymer formulations. In addition, fluorescence can be quenched by contaminant metal ions.

Upon multipass extrusion of PP, the fluorescence intensity was observed to increase with the increase in the amount of degradation products generated during polymer processing. Figure 11.6 illustrates fluorescence spectra of one of extruded formulations (500 ppm of calcium stearate and 1000 ppm of Genox) as a function of the amount of thermal degradation generated by the multiple extrusions and obtained with an excitation with different UV light sources such as a nitrogen laser, diode laser, and UV LED. The trend in the increase of fluorescence intensity from the unprocessed samples to samples after the last extrusion pass was consistent across all formulations. Another important feature of these spectra was the change in the spectral shape.

The fluorescence detection method was further evaluated using a conventional-scale 16-mm twin-screw extruder (see Figure 11.7a). Different amounts of degradation were induced in PP by utilizing three different stabilizer formulations (2000 ppm stabilizer tetrakis [methylene (3,5-di-*tert*-butyl-4-hydroxyhydrocinnamate)] methane, 1000 ppm stabilizer tris(2,4-di-*tert*-butylphenyl) phosphite, or 1000 ppm each of stabilizers tetrakis [methylene (3,5-di-*tert*-butyl-4-hydroxyhydrocinnamate)] methane and tris(2,4-di-*tert*-butylphenyl) phosphate and by exposing these compositions to air during extrusion by opening none, one, or two vent ports to the atmosphere. Figure 11.7b demonstrates an obtained good correlation between the responses of the in-line spectroscopic probe and the MFIs of the extruded compositions.

For the adequate prediction of the melt-flow index and other types of parameters, it is important to understand the variability of measurements of the required parameters using traditional methods. Traditional measurements of MFI were obtained according to the ASTM D1238-95 standard using

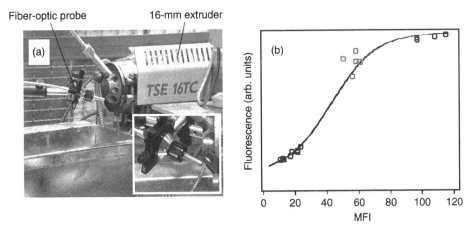

FIGURE 11.7 Fluorescence detection of polypropylene process degradation in a conventional 16-mm twin-screw extruder. (a) Setup for fluorescence measurements, Inset, fiber-optic probe at the die exit. (b) Correlation between the responses of the in-line fluorescence detection and MFI of the extruded compositions.

a Dynisco Kayeness Polymer Test System, Series 4000, Model D4004. The sample conditions were 230°C and 2.16 kg weight with 420 ± 30 s melt time. Five different samples of each several materials with nominal MFI values of about 7, 15, and 21 g/10 min were tested with two measurements obtained for each sample (for a total of ten measurements per material). The total standard deviation across the ten measurements for each material amounted to less than 3% of the mean. This study indicated that 99% of the overall gage variability was captured in a 2 MFI unit (g/10 min) range.

The observed fluorescence spectral features suggested a possibility for the simultaneous determination of color and MFI of stabilized PP with different additives.[27] Experimental validation of the method for simultaneous analysis of color and MFI was performed using a set of diverse polypropylene samples with YI ranging from 2.4 to 10.3 and MFI ranging from 4 to 30. The samples were prepared from six formulations and multiple pass extrusions performed at 260°C in air. Pellets of each formulation were positioned in a 48-well block to provide two measurement replicates of each formulation. Fluorescence measurements of samples arranged in the array format were performed on the modular spectroscopic setup as shown in Figure 11.8. The spectral data were collected from solid samples and arranged as an emission wavelength–fluorescence intensity response matrix. Analysis of data was performed using a multivariate analysis. A multivariate calibration method, partial least-squares (PLS) regression was used to quantify the variations in spectral features as a function of YI and MFI. The predictive performance of the developed PLS model was assessed using the leave-one-out cross-validation (CV) method.[41] The root mean squared error of cross-validation (RMSECV) was the estimator of the quality of the PLS model. Results of the multivariate PLS analysis of the normalized fluorescence spectra of polypropylene pellets to predict YI and MFI values from the spectral features are presented in Figure 11.9. Results of the PLS modeling for prediction of YI and MFI are summarized in Table 11.1. These data demonstrated the applicability of the method for analysis of MFI and YI across diverse types of polymer compositions.

The variability of the method for spectroscopic determinations of MFI was further evaluated using materials produced in the microextruder. For these evaluations, we used multiple replicate individual samples of polymers with different MFI and analyzed their spectroscopic responses. The samples were in the form of pellets produced using a small mechanical pelletizer and loaded into a 48-well block for automatic analysis. To correct for the different geometry of the solid samples and thus different amounts of light collected by the instrument, we used a ratiometric fluorescence measurement approach, where a ratio of peaks in the fluorescence spectrum (see Figure 11.6) was used

TABLE 11.1
Results of the PLS Modeling for Prediction of YI and MFI

Parameter	Number of Latent Variables	Calibration Error	Cross-Validation Error
YI	3	1.16	1.28
	5	0.76	0.91
MFI	3	3.4	3.76
	5	2.2	2.98

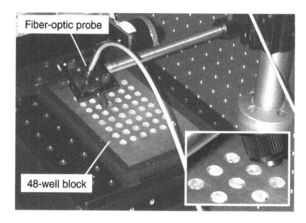

FIGURE 11.8 Automated high-throughput spectroscopic analysis of formulated polypropylene pellets. Multiple formulated polypropylene samples are loaded as a 2-D array. Inset: close-up of the wells with three polymer pellets in each well.

for quantification. The results of the ratiometric determinations are presented in Figure 11.10. This approach provided a significant improvement in the precision of determinations compared to intensity measurements, yet more simple than multivariate spectral analysis. Results of the statistical analysis using Minitab software are presented in Figure 11.10b and Figure 11.10c. A close comparison of the variation of determinations of materials with low and high MFI values indicated that the variability of determinations of high MFI values is more than ten times higher compared to low MFI values. The standard deviations of fluorescence responses of sample with MFI of 4.5 and 19.5 g/10 min were 0.009 and 0.28, respectively. Contributions to this variation have been identified as mostly process sources.

11.4.2 IN-LINE MONITORING

Demonstration of capabilities for real-time analysis of polymers during microextrusion was performed during the generation of step-change and gradient-change polymer compositions.[18] Initially, a base polypropylene polymer (BP Polymers, Fortilene HB9200) was used with two model additives (luminophores, Lumogen F Red 300 and Lumogen Violet 570, BASF). The fiber-optic probe was positioned in the extruder or in close proximity to the exiting polymer in a custom-made die attachment as shown in Figure 11.11. Measurements were performed in fluorescence and white-light reflection modes.

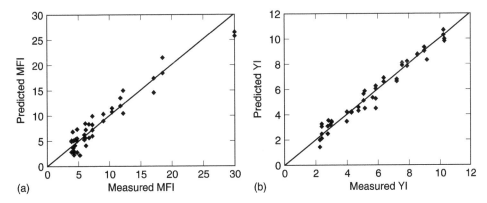

FIGURE 11.9 Prediction of (a) YI and (b) MFI from fluorescence spectral measurements of formulated polypropylene samples using multivariate partial least-squares regression.

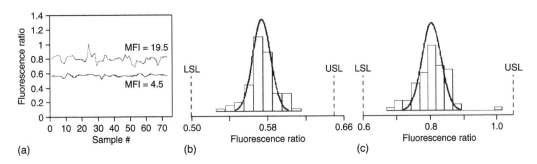

FIGURE 11.10 Statistical analysis of determination of MFI from polypropylene formulations produced in combinatorial microextruder. (a) Fluorescence from two polymer materials with MFI of 4.5 and 19.5 g/10 min measured from multiple ($n = 72$) samples loaded as a 2-D array and measured as shown in Figure 11.8. (b) Histogram of fluorescence measurement capability of the polymeric formulation with MFI = 4.5. (c) Histogram of fluorescence measurement capability of the polymeric formulation with MFI = 19.5. Solid lines in (b) and (c) are Gaussian fits of distributions. LSL and USL are lower and upper spec limits, respectively.

Under the optimized conditions of 10 g/min feed rate and 150 rpm screw speed, we were able to generate step changes in polymer compositions as fast as 30 s and detect these changes as shown in Figure 11.12. Under these operating conditions, polymer formulations could be switched every 45–60 s, requiring only 7.5–10 g of material per formulation. Upon extrusion of this material as a 2-mm diameter strand, the length of a 10-g strand produced over 1 min was about 3 m. A 1 min extrusion of such material as a 1 mm thick and 10 mm wide film yielded a film length of ~1 m.

We were also able to monitor fabrication of gradient polymeric compositions produced by periodically spiking a base polymer formulation with two model additives which were luminophores Lumogen F Red 300 and Lumogen Violet 570. In-line determination of concentrations of these additives in polymer was performed simultaneously at two different wavelengths of fluorescence emission, 580 nm and 420 nm and a single excitation wavelength of 266 nm. The period of these additive introductions was 60 s and the offset between different additives was 30 s. The time required for production of min to max and max to min gradients for the extrusion conditions used (2 g/min feed rate and 150 rpm screw speed) was 20 and 45 s, respectively, as shown in

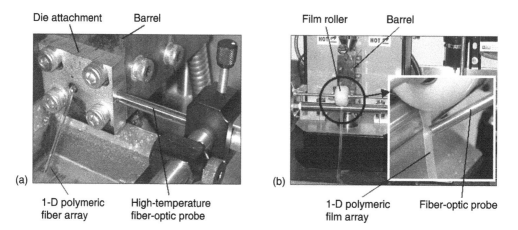

(a) Die attachment   Barrel

1-D polymeric fiber array   High-temperature fiber-optic probe

(b) Film roller   Barrel

1-D polymeric film array   Fiber-optic probe

FIGURE 11.11 In-line spectroscopic monitoring during microextrusion of (a) fiber and (b) film. For scale, diameter of the fiber-optic probe is 1/4 inch.

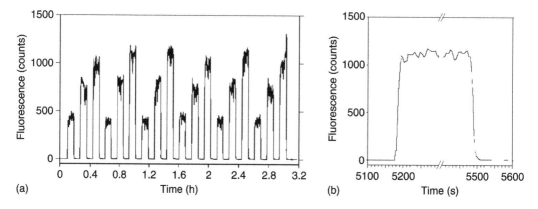

FIGURE 11.12 Capabilities of the combinatorial microextruder system in fabrication of step-changed polymer formulations. (a) Reproducibility of replicate extrusion of step-changed polymer formulations. Three levels of relative amounts of model additive Lumogen F Red 300 are 33.3, 67.3, and 100%. (b) Extrusion throughput of step-changed polymer formulations. (Adapted from Potyrailo, R. A. et al., *Macromol. Rapid Commun.*, 24, 123, 2003. With permission.)

Figure 11.13. This time was dictated by the residence time of material in the extruder. The residence time distribution can be obtained by using a suitable tracer[42–44] and can be described by[45]

$$f(t) = (a^3/2) \, (t - t_d) \exp(-a(t - t_d)) \qquad (11.2)$$

where a is a residence time distribution shape factor, related to the width of the tracer pulse t_w as

$$a = 3/t_w \qquad (11.3)$$

and t_d is the delay time, which is the time interval between when the tracer is added and when the tracer is first detected by a measurement probe, where t_d is related to the mean residence time t_m as

$$t_d = t_m - 3/a \qquad (11.4)$$

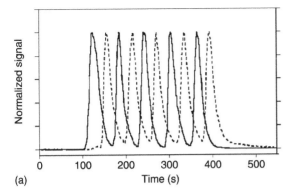
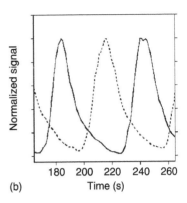

FIGURE 11.13 Capabilities of the combinatorial microextruder system in fabrication of gradient-changed polymer formulations. (a) Reproducibility of generation of replicate gradient compositions with model additives Lumogen F Red 300 (solid line) and Lumogen Violet 570 (dashed line). Introduction period for an additive 60 s, offset between different additives 30 s. (b) Time response in extrusion of gradient changes in polymer formulations.[18]

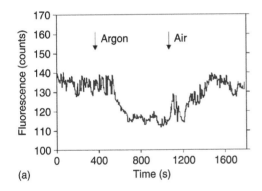
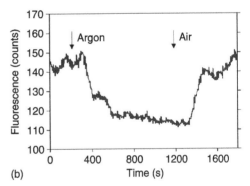

FIGURE 11.14 Results of in-line determination of thermal oxidative stability of polypropylene during microextrusion. Temperature conditions: (a) 230°C and (b) 260°C. Changes from air to argon and back are shown with arrows. Slight offsets are due to the delay in extruder response.

Thus, the microextruder system was capable of producing step or gradient changes in composition in less than a minute per formulation with as little as 2–10 g of polymer.

These in-line measurement capabilities were further applied for the evaluation of thermal oxidative stability of polymers during microextrusion. The variation in the thermal oxidation was induced by changing the extruder atrmosphere from air to argon. Experiments were performed at two temperatures, 230 and 260°C in order to determine the temperature-induced contributions to the spectroscopic signal and the signal change upon variable atmosphere composition. Results of in-line determination of thermal oxidative stability of PP (BP Polymers, Fortilene HB9200, unstabilized homopolymer) during microextrusion experiments are presented in Figure 11.14. These plots demonstrate that under the microextrusion conditions, thermal oxidation of polymer strongly depends on the amount of oxygen in the microextruder. Similarly to the polymer processing in conventional extruders, temperature dependence is also slightly pronounced in the larger initial signal from the degraded polymer and overall larger signal change from air to argon atmospheres. Fluorescence signal was proportional to the amount of thermal degradation products.[19,23]

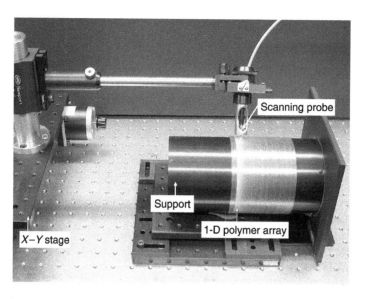

FIGURE 11.15 Automated spectroscopic analysis of 1-D coiled fiber polymeric arrays. Diameter of the coiled fiber is ~ 1 mm. (Reused with permission from Potyrailo, R. A. et al., *Rev. Sci. Instrum.*, 76, 062222, 2005. Copyright 2005, American Institute of Physics.)

11.4.3 ANALYSIS OF COILED ONE-DIMENSIONAL ARRAYS

During extrusion of polymer as a fiber, it can be coiled around a support for further testing and analysis. Such a screening approach was applied to the fabrication of a sensor with a gradient temperature-sensitive plastic optical fiber core.[24] In this sensor, a bisphenol-A polycarbonate (PC, Lexan Grade 100, GE Advanced Materials) fiber was extruded with a spike of a small amount (~ 50 mg) of a temperature-sensitive phosphor ($La_2O_2S:Eu^{3+}$, FluoreScience, Inc., Espanola, New Mexico). This phosphor served as an additive to produce both luminescence under a certain excitation wavelength and scatter and was further employed in gradient temperature sensor.[24] The phosphor concentration exponentially decayed as a function of fiber length as described by Equation 11.2. The extruded gradient-concentration plastic fiber was coiled onto a solid support and arranged for spectroscopic analysis as shown in Figure 11.15. For analysis of scattered and luminescence light, a fiber-optic probe was used. Illumination and collection of light was done through a bifurcated probe end with two 100-μm-diameter optical fibers. The fiber-optic probe was positioned at ~ 0.5 mm to the coiled array and was advanced across the coiled sample with a 100-μm step.

Results of the spectroscopic scanning of the coiled fiber are shown in Figure 11.16. These data illustrated that selection of different emission wavelengths provided a powerful discrimination ability for determination of different properties in 1-D arrays. For example, emission at 495 nm was provided by the base bisphenol-A polycarbonate material only (Figure 11.16a), while emission at 623 nm was due to the phosphorescence of the phosphor (Figure 11.16b). The high spatial resolution of measurements is illustrated in Figure 11.16c, where a small scanned region of the array is shown.

The polycarbonate fiber was further imaged with the imaging system in the scattered light and phosphorescence detection modes. Imaging of scattered light and phosphorescence was done by illuminating the coiled sample with a 400 nm light from the monochromator and collecting scattered light without the filter and phosphorescence through a long-pass optical filter. Figure 11.17 depicts results of scattered light measurements. Similarly to the scanning measurements, the imaging system resolved individual coiled fibers in the array. Results of the phosphorescence imaging and spectroscopy are illustrated in Figure 11.18. From these results, it was possible to select a region of the phosphor-doped fiber with a phosphor concentration adequate for the construction of

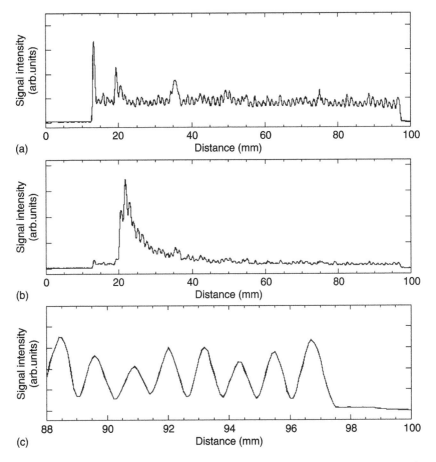

FIGURE 11.16 Results of the automated spectroscopic analysis of the 1-D coiled fiber polymeric array with a gradient concentration of a temperature-sensitive phosphor $La_2O_2S:Eu^{3+}$ along fiber length, laser excitation at 355 nm. (a) Fluorescence emission at 495 nm of polycarbonate. (b) Phosphorescence emission at 623 nm of phosphor. (c) Scattered light detection at 355 nm illustrating high spatial resolution of measurements. Coiled array starts at ~13 and ends at ~ 97 mm. (Reused with permission from Potyrailo, R. A. et al., *Rev. Sci. Instrum.*, 76, 062222, 2005. Copyright 2005, American Institute of Physics.)

a dual-parameter optical sensor.[24] In such a sensor a thin fiber cladding is subsequently added to the temperature-sensitive core and served as a chemically sensitive component. This cladding was made from Nafion®, and was doped with rhodamine 800. Fluorescence at 750 nm from rhodamine 800 was enhanced by the presence of atmospheric moisture, and was used in conjunction with a ratiometric means of measuring temperature provided by the phosphorescence from the fiber core.

During extrusion of polymer as a film, it can be coiled for further testing and analysis. To demonstrate this approach, we fabricated several coiled-film arrays from a polypropylene polymer containing different numbers of additive-containing regions. The polymer formulation was periodically spiked with a small amount (~50 μg) of a model fluorescent additive (500 ppm of nile red, Aldrich) dispersed in polypropylene. For this purpose, nile red was dissolved in chloroform and polypropylene powder was mixed with the solution. The solvent was further evaporated. Dried fluorophore-tagged powder was used for spiking. A general view of this coiled film array is shown in Figure 11.19a. Measurements of this coiled array were performed with a scanning system. Fluorescence scanning was performed with a 407-nm laser as an excitation source. The bifurcated optical fiber tip was scanned across a radius of the coiled array as shown in Figure 11.19b. Results

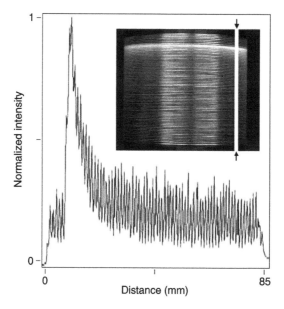

FIGURE 11.17 Quantitative reflected light imaging of the 1-D polymer fiber array with a gradient concentration of a temperature-sensitive phosphor $La_2O_2S:Eu^{3+}$ along its length. Inset, reflected light image of the coiled 1-D fiber array and the cross-section region, indicated with a vertical line.

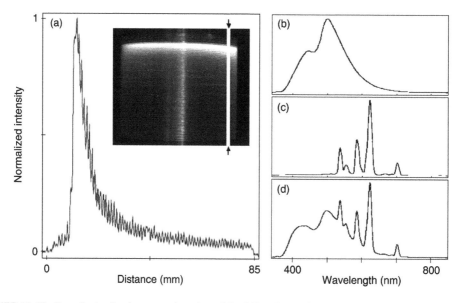

FIGURE 11.18 Quantitative luminescence imaging of the 1-D polymer fiber array with a gradient concentration of a temperature-sensitive phosphor $La_2O_2S:Eu^{3+}$ along its length. (a) Quantitative cross-section of the image. Inset, phosphorescence image of the coiled 1-D fiber array and the cross-section region, indicated with a vertical line. (b) Emission spectrum of a region of the optical fiber without temperature-sensitive phosphor. (c) Emission spectrum of a region of the fiber with high loading of the temperature-sensitive phosphor. (d) Emission spectrum of a region of the fiber with intermediate loading of the temperature-sensitive phosphor.

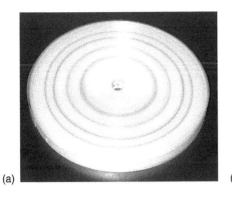
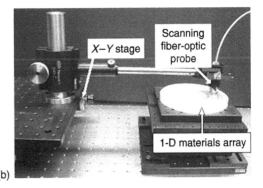

(a)   (b)

FIGURE 11.19 Automated spectroscopic analysis of 1-D coiled film polymeric arrays. (a) General view of a coiled film array with four regions of model (nile red) additive. (b) General view of the automated scanning system. Thickness of the film is ~ 0.5 mm. (Reused with permission from Potyrailo, R. A. et al., *Rev. Sci. Instrum.*, 76, 062222, 2005. Copyright 2005, American Institute of Physics.)

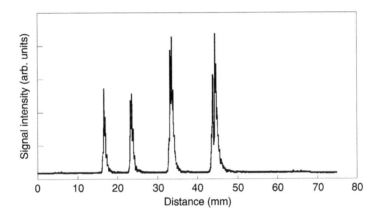

FIGURE 11.20 Results of fluorescence analysis of the coiled 1-D film array with four regions of the model (nile red) additive scanned with a 100-μm step size of the fiber-optic probe.

of the fluorescence analysis of this coiled 1-D array were recorded with a 100-μm step size and demonstrated adequate high spatial resolution of our analysis method as shown in Figure 11.20. For demonstration of imaging capabilities of the coiled film arrays, an array with a single region was used. Figure 11.21a depicts a fluorescence image of the array. A cross-section of the fluorescence image demonstrates the radial distribution of fluorescence intensity (see Figure 11.21b).

We have developed a model to inter-relate the signal obtained from scanning and imaging measurements of coiled arrays to the additive distribution in the 1-D array. In our forward model, we calculated the signal distribution in the coiled array from the signal distribution measured along the length of the 1-D array during the array extrusion. In our backward model, we calculated the signal distribution along the length of the 1-D array from the signal measured from the coiled array. Such modeling of the signal distribution in coiled arrays provided an opportunity to optimize coiling starting diameter in order to achieve the best resolution of measurements. The signal distribution along the length of the 1-D array associated with the change in the property of the array along its length can be either measured with in-line techniques or predicted from the knowledge of extrusion conditions (see Equation 11.2). For a simple demonstration of model performance,

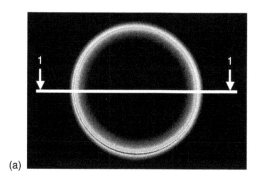 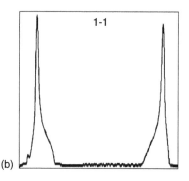

(a)   (b)

FIGURE 11.21 Results of fluorescence imaging of coiled 1-D polymeric film array with a single region of gradient-changed polymer formulations. (a) Fluorescence image under 400-nm excitation. (b) Cross-section of the fluorescence image indicated by a horizontal line. (Reused with permission from Potyrailo, R. A. et al., *Rev. Sci. Instrum.*, 76, 062222, 2005. Copyright 2005, American Institute of Physics.)

we can set four regions along the array with initial Gaussian distributions of concentrations of additives given by:

$$f(t) = \sum_i^4 A_i \exp(-(t - t_i)^2/(\Delta_i)^2) \qquad (11.5)$$

where A_i are initial peak concentrations of additives, t_i are the locations of additives along the 1-D array, and Δ_i are the dispersion widths of additives. An example of additives in four regions of the array as a function of array length is presented in Figure 11.22a. For the visualization of the model performance, we modeled the distribution of all additives as a Gaussian with $A_{1-4} = 1$; $t_i = 200, 400,$ 600, and 800; and $\Delta_i = 20$. When such an array is fabricated as a film and is coiled for analysis, the intensity distribution of fluorescence in the acquired image is presented in Figure 11.22b. Thus, application of our model permits the visualization of the changes of properties as a function of the spatial location along the length of the coiled 1-D array. While Figure 11.22 shows only Gaussian distributions, our developed model can accommodate any additive distribution profiles in the polymer array.

Coiling of 1-D formulated polymeric materials for their testing and characterization became possible due to the application of our microextruder approach that requires ~200 times less material compared to traditional-scale extruders. Our measurement approach of 1-D arrays of polymers adds to the infrastructure of measurement methods applicable to analyze 1-D arrays. Other methods of analysis of 1-D structures include optical time-of-flight techniques,[46–50] frequency domain analysis,[51–53] and cutting extended-length samples to produce discrete regions of sufficiently similar chemical compositions for analysis.[54]

11.4.4 Evaluation of Weathering

Our capabilities of fabrication of 1-D arrays of polymeric formulations were further applied for high-throughput weathering determinations of formulated polycarbonate materials. Weathering is a critical need in the development of new engineering thermoplastic materials for outdoor applications.[55] However, resistance of polymers toward weathering presents a particular challenge for high-throughput screening because exposure times for adequate outdoor weatherability testing are thousands of hours.[55] A determining factor in the outdoor weathering lifetime of a polymer is the received UV radiation dose that leads to material photodegradation.[56] We introduced an approach

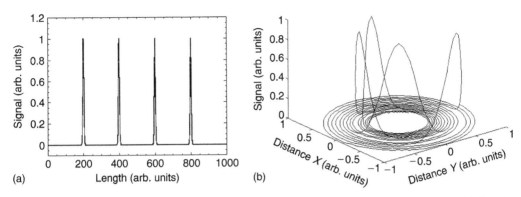

FIGURE 11.22 Example of modeling results of distribution of additives in coiled film 1-D arrays. (a) Gaussian distribution of a model additive in four regions of the film array plotted as a function of array length. (b) Distribution of a model additive in four regions in coiled film array. (Reused with permission from Potyrailo, R. A. et al., *Rev. Sci. Instrum.*, 76, 062222, 2005. Copyright 2005, American Institute of Physics.)

for rapid weathering testing of polymeric compositions that incorporated a locally induced high-irradiance exposure with high-sensitivity detection[31] of polymer photo-oxidative degradation. This acceleration of weathering to levels attractive to HT screening was done by a careful increase of the irradiance while preserving the weathering ranking for UV-stabilized and unstabilized polymer compositions.

Mechanisms of color formation in engineering aromatic thermoplastics upon weathering are complex and still are poorly understood. Suggested mechanisms include oxidation of aromatic rings of polymers, side-chain oxidation, and others.[57] To reduce the rate of color formation, typically UV stabilizers that absorb radiation up to 400 nm are incorporated into the polymer.[1,58] However, the UV absorbers are also subject to photodegradation, and their effectiveness as stabilizers can be lost.[59] Thus, screening of effectiveness of UV absorbers is critical for the evaluation of the long-term performance of plastics in outdoor applications.[60,61]

We use fluorescence spectroscopy and imaging for determination of weathering of polymers[31] by the excitation of fluorescence of degradation products in the weathered polymer at wavelengths within and above the absorbance range of a UV absorber. Thus, the fluorescence intensity is either modulated by both degradation products and the amount of UV absorber (excitation at 300–360 nm) or mostly by the degradation products (excitation above 400 nm). Using fluorescence, the throughput of screening can be increased by 200- to 650-fold compared to determination of color change depending on the type of an engineering thermoplastic.[31] Thus, information about the kinetics of photodegradation should be available much sooner.

Observed fluorescence responses of weathered unstabilized PC/TiO_2 under two excitation wavelengths (355 and 407 nm) correlated well. However, PC/TiO_2 containing different amounts of T234 showed strong initial fluorescence upon excitation at 407 nm that initially decreased upon weathering and then increased as weathering continued. Such initial behavior could be a result of photobleaching of different impurities in the polymer and absorber, followed by the increase of fluorescence due to the photo-oxidation.

In our initial weathering experiments, we evaluated weathering performance of several polymeric formulations fabricated in our combinatorial microextruder system using a conventional weatherometer coupled with fluorescence detection. Several 1-D libraries were produced by coiling strands of PC/TiO_2 compositions with varying levels of T234 UV absorber onto a mandrel as shown in Figure 11.23a. Fluorescence measurements of the coiled 1-D libraries were performed periodically during weathering using the automated fluorescence spectroscopic (Figure 11.23b) and imaging (Figure 11.23c) systems. Fluorescence imaging permits rapid evaluation of all regions of

 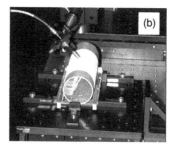 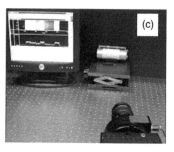

FIGURE 11.23 Demonstration of performance testing of 1-D polymer arrays. (a) General view of a 1-D array with variable amounts of UV absorber additive in a weatherometer. General views of (b) fluorescence spectroscopic scanning system and (c) imaging system for periodic measurements of coiled array after weathering exposures.

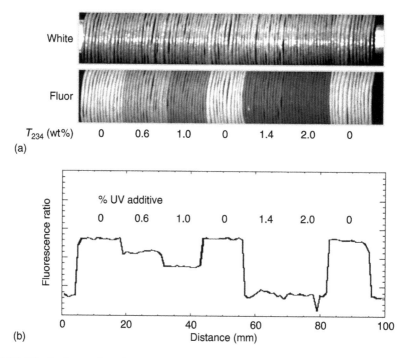

FIGURE 11.24 Weathering testing of 1-D libraries of polymer compositions. (a) White light and fluorescence images of a 1-D library after 845 kJ/m^2 exposure. (b) Fluorescence of a coiled 1-D array containing seven regions with different concentrations of T_{234} UV absorber additive after 845 kJ/m^2 weathering exposure.[18]

1-D libraries with adequate signal-to-noise and spatial resolution. As an example, Figure 11.24a depicts white light and fluorescence images of one of the 1-D libraries after 845 kJ/m^2 exposure. These images illustrate that it is possible to resolve individual coils in the weathered materials array. Also, knowing the locations of the coils from the white-light image, it is possible to correct for any intensity differences in a fluorescence image. To obtain detailed information about weathering, we used an automatic scanning fluorescence spectroscopic system. In scanning measurements, the fiber-optic probe was positioned 1 mm above the polymer library and was moved across the library with 1 mm increments. Fluorescence was recorded as a function of traveled distance. An example of fluorescence signal from the scanned array 845 kJ/m^2 exposure is shown in Figure 11.24b.

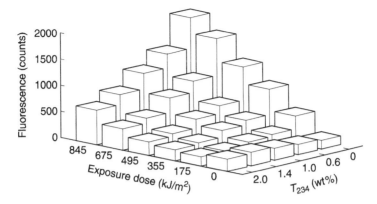

FIGURE 11.25 Weathering response of the 1-D library of polymeric compositions containing increasing amounts of UV absorber T_{234} at increasing levels of weathering exposure.[18]

A comparison of fluorescence for different concentrations of T234 and different weathering conditions is presented in Figure 11.25. Fluorescence intensity was dependent on both the exposure dose and the level of T234 UV absorber in the polymer formulation. This plot also shows that fluorescence intensity of coiled regions before weathering slightly increased as a function of T234 UV absorber. However, upon weathering, the rate of fluorescence increase was reduced by the UV absorber. This rate reduction was proportional to the concentration of UV absorber in the polymer.

When using automated high-throughput analysis systems with high spatial resolution of measurements,[62] performance testing of large materials areas may be not needed if the properties of interest can be reliably evaluated on the small scale. Such an approach may potentially be applied to provide a highly accelerated weathering evaluation. However, it is important to note that, depending on the nature of the polymer matrix, the acceptable increase in irradiance can be different. In general, the increase in the level of environmental stress can be problematic because of loss of correlation with traditional test methods.[17,55,63] However, recently it was shown that certain types of polymeric materials can experience a very high irradiance without the loss of correlation with more traditional weathering.[64] In particular, polycarbonate can withstand irradiance from tightly focused sunlight to up to ~100 Suns without the loss of such correlation.[64] Also, very rapid generation (minutes or seconds) of colored/fluorescent species in PC can be easily produced by lamps emitting below 290 nm and by UV lasers.[65–67] However, these colored products are caused by photodegradation mechanisms different from those caused by terrestrial sunlight and are irrelevant for screening of UV absorbers for end-use applications.

For the HT weathering experiment with a locally induced high-irradiance exposure, we used a light source/optical filter combination similar to that in the weatherometer and provided an additional acceleration in weathering testing by irradiating only the regions of the 1-D library that were further used for weathering analysis. The focused light produced an irradiance of ~18 W/m² at 340 nm that corresponded to ~30 Suns. A 1-D library was produced by coiling strands of PC/TiO₂ formulations with 0 and 2 wt% of the UV absorber onto a mandrel. Small regions of the coiled library were weathered for different amounts of time ranging from 1000 s to 6 h and further were analyzed using fluorescence imaging and scanning systems. A fluorescence image (350 nm excitation) is presented in Figure 11.26a and illustrates the increase in emission intensity upon exposure of small regions of the coiled library to the filtered Xe-arc lamp. This fluorescence image also shows the elevated background intensity from polymer coils containing 2 wt% of T234. By using an excitation wavelength of 310 nm, this fluorescence background was dramatically reduced while preserving the rates of fluorescence increase for UV-stabilized and unstabilized regions. In scanning measurements, fluorescence was recorded as a function of traveled distance across the coiled 1-D library as

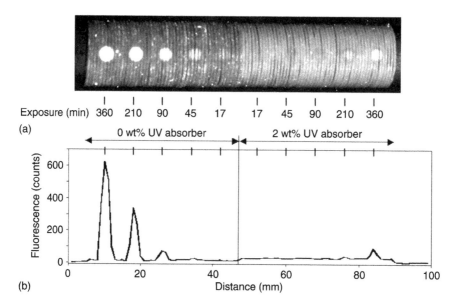

FIGURE 11.26 High-throughput weathering testing of a 1-D polymeric library with locally induced high-irradiance exposure and high-sensitivity detection. (a) Spatially resolved fluorescence profile of the library after exposure of small regions to high-intensity weathering source for different periods of time. (b) Weathering response of the 1-D library of polymeric compositions containing 0 and 2 wt% of T_{234} at increasing levels of weathering exposure.[18]

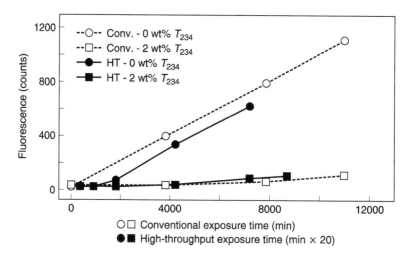

FIGURE 11.27 High-throughput weathering testing of a 1-D polymeric library. Correlation between traditional and locally induced high-irradiance exposure weathering results.[18]

shown in Figure 11.26b. These data conclusively demonstrate the possibility of HT weathering for evaluation of UV additives in polycarbonate.

The comparison between HT and conventional weathering tests is presented in Figure 11.27. This comparison demonstrated a good correlation between these two methods. Similar to experiments on conventional weathering equipment, the rate of fluorescence increase of the UV-stabilized regions was about nine times less than that of the regions without the UV absorber. Thus, the difference between UV-stabilized and unstabilized regions of polymeric libraries was equivalent to Xe-arc data that was generated ~20 times faster. The expected increase in weathering rate was calculated from the irradiance enhancement to be ~24 times. Clearly, such a HT weathering method can be further used to screen different additives in polymers and to generate new knowledge for other types of polymeric materials used for practical applications.

We showed that with our approach it is possible to omit the molding step for sample preparation and to accomplish performance testing directly using the 1-D libraries. Thus, certain performance tests, such as weathering and others, could be accelerated. Such acceleration is provided by testing only local regions, which are further characterized using analytical tools with adequate spatial resolution. Such HT performance testing can generate new knowledge of acceptable high-stress test conditions where the materials still follow the more traditional test methods.

11.5   SUMMARY

The polymer-extrusion system that we have developed incorporates a combination of a microextruder and microfeeders and is applicable for the rapid, material-, and energy-saving fabrication of step- and gradient-level polymeric compositions as 1-D libraries. Depending on the property of interest, we performed the measurements in-line during polymer extrusion or off-line after the material is coiled into a 1-D array. Coiling of 1-D formulated polymeric materials for their testing and characterization became possible due to the application of our microextruder approach that required ~200 times less material compared to traditional-scale extruders. This approach permitted acceleration of performance tests. For example, a weathering evaluation was accelerated by 150- to 800-fold by applying fluorescence compared to color and gloss analysis[31] and by another 20-fold by applying local weathering conditions.[18] Our approach for the fabrication of 1-D polymeric arrays has been applied for the development of dual-parameter optical sensors.[24] While the 1-D libraries that we made in this study were relatively simple, they suggested that it is quite straightforward to increase the number of feeders to three or more. It will enable us to use the microextruder for the generation of more complex step- and gradient-level polymer formulations with multiple components as theoretically demonstrated in this work. Some differences in the mixing abilities between the microextruder and larger lab-scale extruders[18] are expected for complex systems with large quantities of additives and for polymer–polymer blends.[21] However, our experience with low (1–3 wt%) concentrations of additives did not indicate any noticeable issues with the quality of mixing obtained in the microextruder.

ACKNOWLEDGMENT

The work described in this paper was conducted with support of the US Department of Energy (DOE) Award Number DEFC07-01ID14093. However, any opinions, findings, conclusions, or recommendations expressed herein are those of the authors and do not necessarily reflect the views of DOE.

REFERENCES

1. Zweifel, H., *Plastics Additives Handbook*, Hanser, Munich, Germany, 2001.
2. Potyrailo, R. A. and Hieftje, G. M., Use of the original silicone cladding of an optical fiber as a reagent-immobilization medium for intrinsic chemical sensors, *Fresenius' J. Anal. Chem.*, 364, 32, 1999.
3. McQuade, D. T., Pullen, A. E. and Swager, T. M., Conjugated polymer-based chemical sensors, *Chem. Rev.*, 100, 2537, 2000.
4. Seymour, R. B., *Additives for Plastics, State of the Art*, Academic Press, New York, 1978.
5. Ivanov, V. B. and Shlyapnikov, V. Y., Synergism in the photostabilization of polymers, In: *Developments in Polymer Stabilization*, Scott, G., Ed., Elsevier Applied Science, New York, 1987, p. 29.

6. Whisenhunt, D. W. Jr., Carter, R., Shaffer, R., Bulsiewicz, W. and Flanagan, W., High throughput screening of the thermal stability of colorants in a polycarbonate matrix, in *Combinatorial and Artificial Intelligence Methods in Materials Science II. MRS Symposium Proceedings*, Vol. 804. Potyrailo, R. A., Karim, A., Wang, Q. and Chikyow, T., Eds., Materials Research Society, Warrendale, PA, 2004, p. 137.
7. Xiang, X.-D. and Takeuchi, I., *Combinatorial Materials Synthesis*, Marcel Dekker, New York, 2003.
8. Potyrailo, R. A. and Amis, E. J., Editors, *High Throughput Analysis: A Tool for Combinatorial Materials Science*, Kluwer Academic/Plenum Publishers, New York, 2003.
9. Hoogenboom, R., Meier, M. A. R. and Schubert, U. S., Combinatorial methods, automated synthesis and high-throughput screening in polymer research: Past and present, *Macromol. Rapid Commun.*, 24, 15, 2003.
10. Meier, M. A. R., Hoogenboom, R. and Schubert, U. S., Combinatorial methods, automated synthesis and high-throughput screening in polymer research: The evolution continues, *Macromol. Rapid Comm.*, 25, 77–94, 2004.
11. Adams, N. and Schubert, U. S., From science to innovation and from data to knowledge: Escience in the Dutch polymer institute's high-throughput experimentation cluster, *QSAR Comb. Sci.*, 24, 58, 2005.
12. Potyrailo, R. A. and Takeuchi, I., Role of high-throughput characterization tools in combinatorial materials science, *Meas. Sci. Technol.*, 16, 1, 2005.
13. Davis, R. D., Bur, A. J., McBrearty, M., Lee, Y.-H., Gilman, J. W. and Start, P. R., Dielectric spectroscopy during extrusion processing of polymer nanocomposites: A high throughput processing/characterization method to measure layered silicate content and exfoliation, *Polymer,* 45, 6487, 2004.
14. Wroczynski, R. J., Potyrailo, R. A., Rubinsztajn, M. and Enlow, W. P., Micro-scale extrusion for the accelerated development of new polymeric materials, In: *Antec 2002 – Annual Technical Conference (May 5-9, 2002, San Francisco, CA)*, Society of Plastic Engineers, Inc., Brookfield, CT, 2002, paper 693.
15. Wolfbeis, O. S., Editor, *Fiber Optic Chemical Sensors* and *Biosensors*, CRC Press, Boca Raton, FL, 1991.
16. Shah, V., *Handbook of Plastics Testing Technology*, Wiley, New York, 1998.
17. Ryntz, R. A., *Plastics and Coatings. Durability, Stabilization, Testing*, Hanser, Cincinnati, OH, 2001.
18. Potyrailo, R. A., Wroczynski, R. J., Pickett, J. E. and Rubinsztajn, M., High-throughput fabrication, performance testing, and characterization of one-dimensional libraries of polymeric compositions, *Macromol. Rapid Comm.*, 24, 123, 2003.
19. Wroczynski, R. J., Rubinsztajn, M. and Potyrailo, R. A., Evaluation of process degradation of polymer formulations utilizing high-throughput preparation and analysis methods, *Macromol. Rapid Comm.*, 25, 264, 2004.
20. Potyrailo, R. A., Analytical spectroscopic tools for high-throughput screening of combinatorial materials libraries, *Trends Anal. Chem.*, 22, 374, 2003.
21. Potyrailo, Note-2, 2001. Typical output of laboratory extruders is 80–420 g/min. See: http://www.bpprocess.com/twinscrew_lab.htm and http://www.twinscrew.com/product.html.
22. Potyrailo, R. A., Wroczynski, R. J. Spectroscopic and imaging approaches for evaluation of properties of one-dimensional arrays of formulated polymeric materials fabricated in a combinatorial microextruder system, *Rev. Sci. Instrum.*, 76, 062222, 2005.
23. Potyrailo, R. A., Wroczynski, R. J. and Rubinsztajn, M., Spectroscopic sensors for determination of oxidative stability of extruded polymers, In: *Abstracts of Eurosensors xvi, September 15–18, 2002,* Elsevier Science, Prague, Czech Republic, 2002.
24. Potyrailo, R. A., Szumlas, A. W., Danielson, T. L., Johnson, M. and Hieftje, G. M., A dual-parameter optical sensor fabricated by gradient axial doping of an optical fibre, *Meas. Sci. Technol.*, 16, 235, 2005.
25. Schmatloch, S., Bach, H., van Benthem, R. A. T. M. and Schubert, U. S., High-throughput experimentation in organic coating and thin film research: State-of-the-art and future perspectives, *Macromol. Rapid Comm.*, 25, 95, 2004.
26. Potyrailo, R. A., Morris, W. G. and Wroczynski, R. J., Acoustic-wave sensors for high-throughput screening of materials, in *High Throughput Analysis: A Tool for Combinatorial Materials Science*, Potyrailo, R. A. and Amis, E. J., Eds., Kluwer Academic/Plenum Publishers, New York, 2003, ch. 11.
27. Wroczynski, R., Brewer, L., Buckley, D., Burrell, M. and Potyrailo, R., *Accelerated Characterization of Polymer Properties. Final Technical Report doe/id/14093 (July 30, 2003)*, http://www.Osti.Gov/bridge/, US Department of Energy Information Bridge, Washington, DC, 2003.
28. Wroczynski, R. J., Potyrailo, R. A. and Rubinsztajn, M., High-throughput methods for evaluation of process degradation of polymer formulations, in *Antec 2003 – Annual Technical Conference (May 4–8, 2003, Nashville, TN)*, Society of Plastic Engineers, Inc., Brookfield, CT, 2003, p. 2679.

29. Potyrailo, R. A., Wroczynski, R. J., Lemmon, J. P., Flanagan, W. P. and Siclovan, O. P., Fluorescence spectroscopy and multivariate spectral descriptor analysis for high-throughput multiparameter optimization of polymerization conditions of combinatorial 96-microreactor arrays, *J. Comb. Chem.*, 5, 8, 2003.

30. Potyrailo, R. A., Lemmon, J. P. and Leib, T. K., High-throughput screening of selectivity of melt polymerization catalysts using fluorescence spectroscopy and two-wavelength fluorescence imaging, *Anal. Chem.*, 75, 4676, 2003.

31. Potyrailo, R. A. and Pickett, J. E., High-throughput multilevel performance screening of advanced materials, *Angew. Chem. Int. Ed.*, 41, 4230, 2002.

32. Al-Malaika, S., Goodwin, C., Issenhuth, S. and Burdick, D., The antioxidant role of α-tocopherol in polymers II. Melt stabilising effect in polypropylene. *Polymer Degrad. Stab.*, 64, 145, 1999.

33. Fearon, P. K., Marshall, N., Billingham, N. C. and Bigger, S. W., Evaluation of the oxidative stability of multiextruded polypropylene as assessed by physicomechanical testing and simultaneous differential scanning calorimetry-chemoluminescence. *J. Appl. Polym. Sci.*, 79, 733, 2001.

34. Lee, R. E., Papazoglou, E., Johnson, B. and Kim, J. W., The chemistry and service of stabilization, In: Additives 2001, March 18–21, Hilton Hend, SC, USA.

35. Lutz, J. T. Jr., *Thermoplastic Polymer Additives. Theory and Practice*, Marcel Dekker, New York, 1989.

36. Allen, N. S., Homer, J. and McKellar, J. F., Fluorescence and light stability of commercial polypropylene, *Chem. Ind. (London)*, 16, 692, 1976.

37. Allen, N. S., Analysis of polymer systems by luminescence spectroscopy, In: *Analysis of Polymer Systems*, Bark, L. S. and Allen, N. S. Eds., Applied Science Publ. Ltd., Barking, UK, 1982, p. 79.

38. Ingle, J. D. Jr. and Crouch, S. R., *Spectrochemical Analysis,* Prentice Hall, Englewood Cliffs, NJ, 1988.

39. Jacques, P. P. L. and Poller, R. C., Fluorescence of polyolefin. 1. Effect of thermal degradation of fluorescent excitation and emission spectra, *Eur. Polym. J.*, 29, 75, 1993.

40. Jacques, P. P. L. and Poller, R. C., Fluorescence of polyolefin. 2. Use of model compounds to identify fluorescent species in thermally degraded polymers, *Eur. Polym. J.*, 29, 83, 1993.

41. Beebe, K. R., Pell, R. J. and Seasholtz, M. B., *Chemometrics: A Practical Guide*, Wiley, New York, 1998.

42. Hu, G.-H., Kadfu, I. and Picot, C., On-line measurement of the residence time distribution in screw extruders, *Polym. Eng. Sci.*, 39, 930, 1999.

43. Puaux, J. P., Bozga, G. and Ainser, A., Residence time distribution in a corotating twin-screw extruder, *Chem. Eng. Sci.*, 55, 1641, 2000.

44. Carneiro, O. S., Covas, J. A., Ferreira, J. A. and Cerqueira, M. F., On-line monitoring of the residence time distribution along a kneading block of a twin-screw extruder, *Polymer Testing*, 23, 925, 2004.

45. Gao, J., Walsh, G. C., Bigio, D., Briber, R. M. and Wetzel, M. D., Residence-time distribution model for twin-srew extruders, *AIChE J.*, 45, 2541, 1999.

46. Potyrailo, R. A. and Hieftje, G. M., Optical time-of-flight chemical detection: Spatially resolved analyte mapping with extended-length continuous chemically modified optical fibers, *Anal. Chem.*, 70, 1453, 1998.

47. Potyrailo, R. A. and Hieftje, G. M., Spatially resolved analyte mapping with time-of-flight optical sensors, *Trends Anal. Chem.*, 17, 593, 1998.

48. Prince, B. J., Schwabacher, A. W. and Geissinger, P., A readout scheme providing high spatial resolution for distributed fluorescent sensors on optical fibers, *Anal. Chem.*, 73, 1007, 2001.

49. Potyrailo, R. A., *Devices and Methods for Measurements of Barrier Properties of Coating Arrays*, US Patent 6,383,815 B1, 2002.

50. Geissinger, P. and Schwabacher, A. W., Intrinsic fiber optic sensors for spatially resolved combinatorial screening, In: *High Throughput Analysis: A Tool for Combinatorial Materials Science*, Potyrailo, R. A. and Amis, E. J. Eds., Kluwer Academic/Plenum Publishers, New York, 2003, ch. 15.

51. Schwabacher, A. W., Shen, Y. and Johnson, C. W., Fourier transform combinatorial chemistry, *J. Am. Chem. Soc.*, 121, 8669, 1999.

52. Schwabacher, A. W., *One Dimensional Chemical Compound Arrays and Methods for Assaying Them*, World Patent Appl. WO 99/42605, 1999.

53. Schwabacher, A. W., Johnson, C. W. and Geissinger, P., Linear spatially encoded combinatorial chemistry with fourier transform library analysis, In: *High Throughput Analysis: A Tool for Combinatorial Materials Science*, Potyrailo, R. A. and Amis, E. J. Eds., Kluwer Academic/Plenum Publishers, New York, 2003, Ch. 6.

54. Gilman, J. W., Davis, R., Nyden, M., Kashiwagi, T., Shields, J. and Demory, W., Development of high throughput methods for polymer nanocomposite research, In: *High Throughput Analysis: A Tool for Combinatorial Materials Science*, Potyrailo, R. A. and Amis, E. J. Eds., Kluwer Academic/Plenum Publishers, New York, 2003, Ch. 19.

55. Wypych, G., *Handbook of Material Weathering*, ChemTec, Toronto, Canada, 1995.

56. Pickett, J. E. and Webb, K. K., Calculated and measured outdoor uv doses, *Makromol.Chem.*, 252, 217, 1997.

57. Factor, A., Mechanisms of thermal and photodegradations of bisphenol a polycarbonate, In: *Polymer Durability. Degradation, Stabilization, and Lifetime Prediction*, Clough, R. L., Billingham, N. C. and Gillen, K. T., Eds., American Chemical Society, Washington, DC, 1996, p. 59.

58. Davis, A. and Sims, D., *Weathering of Polymers*, Elsevier, London, England, 1986.

59. Pickett, J. E. and Moore, J. E., Photostability of uv screeners in polymers and coatings, In: *Polymer Durability. Degradation, Stabilization, and Lifetime Prediction*, Clough, R. L., Billingham, N. C. and Gillen, K. T., Eds., American Chemical Society, Washington, DC, 1996, p. 287.

60. Pickett, J. E. and Moore, J. E., Photodegradation of uv absorbers: Kinetics and structural effects, *Makromol.Chem.*, 232, 229, 1995.

61. Pickett, J. E., Photostabilization of plastics by additives and coatings, In: *Plastics and Coatings. Durability, Stabilization, Testing*, Ryntz, R. A., Ed., Hanser, Cincinnati, OH, 2001, p. 73.

62. Potyrailo, R. A., Analytical infrastructure for high throughput characterization of combinatorial materials libraries, In: Combi 2002 – The 4th Annual International Symposium on Combinatorial Approaches for New Materials Discovery, January 23–25, San Diego, CA, 2002.

63. Sherbondy, V. D., Accelerated weathering, In: *Paint and Coating Testing Manual*, Koleske, J. V., Ed., Americal Society for Testing and Materials, Philadelphia, PA, 1995, p. 643.

64. Jorgensen, G., Bingham, C., King, D., Lewandowski, A., Netter, J., Terwilliger, K. and Adamsons, K., Use of uniformly distributed concentrated sunlight for highly accelerated testing of coatings, In: *Service Life Prediction. Methodology and Metrologies*, Martin, J. W. and Bauer, D. R., Ed., American Chemical Society, Washington, DC, 2002, p. 100.

65. Chipalkatti, M. H. and Laski, J. J., Investigation of polycarbonate degradation by fluorescence spectroscopy and its impact on final performance, In: *Structure-property Relations in Polymers. Spectroscopy and Performance*, Urban, M. W. and Craver, C. D., Eds., American Chemical Society, Washington, DC, 1993, p. 623.

66. Thompson, T. and Klemchuk, P. P., Light stabilization of bisphenol a polycarbonate, In: *Polymer Durability. Degradation, Stabilization, and Lifetime Prediction*, Clough, R. L., Billingham, N. C. and Gillen, K. T., Eds., American Chemical Society, Washington, DC, 1996, p. 303.

67. Factor, A., Degradation of bisphenol a polycarbonate by light and gamma-ray irradiation, In: *Handbook of Polycarbonate Science and Technology*, LeGrand, D. G. and Bendler, J. T., Eds., Marcel Dekker, New York, 2000, p. 267.

12 A Combinatorial Approach to Rapid Structure–Property Screening of UV-Curable Cycloaliphatic Epoxies

*Fawn M. Uhl, Christine M. Gallagher-Lein,
David A. Christianson, James A. Bahr, Bret J. Chisholm,
Nathan J. Gubbins, and Dean C. Webster*

CONTENTS

12.1   INTRODUCTION

The use of high-throughput methods to synthesize and screen combinatorial libraries of candidate drug compounds has become an established research methodology in the pharmaceutical industry. This approach has also been applied to proteomics and genomics research. The application of combinatorial and high-throughput approaches to the synthesis and screening of materials has been growing rapidly. Combinatorial and high-throughput materials science involves the rapid preparation of libraries of chemically distinct materials, followed by screening of the materials for key performance properties.[1–6] A goal of high-throughput screening is to identify hits or leads: compositions that show promising or extraordinary performance. These leads can either be used to further refine library design in future high-throughput experiments or the leads can serve as candidates for further product development. In addition, data generated in high-throughput experiments can also be used to determine structure–property relationships. This information can provide valuable insight into the effect of key compositional variables on performance properties and guide future research.

There has been a great deal of interest in applying combinatorial and high-throughput techniques to polymer synthesis and screening. Companies such as Dow Chemical have been implementing high-throughput research by the purchase of commercial equipment and creating others as necessary.[7] BASF has developed a complete workflow for polymer synthesis and coating preparation and screening.[4] General Electric has developed a "combinatorial factory" focused on the preparation and screening of coatings for plastics.[8–13] Researchers at the Avery Research Center have shown the applicability of a high-throughput method for the development of pressure-sensitive adhesives.[14] In this particular case there are several variables that must be assessed, such as the amount of tackifier, concentration of crosslinker, and thickness. Using a high-throughput approach to prepare and screen adhesive formulations can significantly speed up the process of discovery and optimization of properties.

Some of the standard methods for monitoring the curing of coatings have been incorporated into combinatorial work flows.[15] de Gans et al. have examined the use of ink-jet printing to create libraries of polymer films which have individually addressable spots of a known composition.[16] They have shown that this technique is applicable to a variety of polymer solutions and would produce easily characterizable libraries. Hoogenboom et al. have shown the effectiveness of coupling gel permeation chromatography (GPC) and gas chromatography (GC) to synthesis robots in combinatorial processes; this allows for the direct monitoring of the polymerization and elucidation of reaction kinetics.[17]

In collaboration with Symyx Technologies, we have developed a high-throughput system for polymer synthesis, coating formulation, and coating screening.[18] When conducting combinatorial and high-throughput research there are a number of variables to be considered and researchers have been working to address exactly how these highly automated techniques would make materials science research valuable.[19–23]

In order to illustrate the use of a high-throughput workflow for the determination of structure–property relationships, an examination of UV curable epoxy coatings was carried out. Libraries were designed and then screened using a combination of high-throughput and conventional analysis techniques (Figure 12.1). Compared to conventional methods, in a relatively few number of experiments, key trends in properties as a function of experimental variables can be observed.

12.2 EXPERIMENTAL

12.2.1 MATERIALS

Cycloaliphatic epoxides used in the study were UVR-6105 (CAE 5), UVR-6110 (CAE 10), and UVR-6128 (CAE 28). These were mixed with several Tone polyols (Tone 0301(PCL 301), Tone 305 (PCL 305), and Tone 0310 (PCL 310)), and a photoinitiator (UVI-6974), a mixed arylsulfonium hexafluoroantimonate salt. All of the above were obtained from Dow Chemical Company and were used without further purification. The structures of the materials used are illustrated in Figure 12.2.

12.2.2 FORMULATIONS

In this study three different cycloaliphatic epoxies and three different polyols were used. The ratio of epoxy equivalents to hydroxyl equivalents (R) was varied from 2 to 5. R can be calculated by the following equation:

$$R = [\text{equivalents epoxide}]/[\text{equivalents polyol}] \qquad (12.1)$$

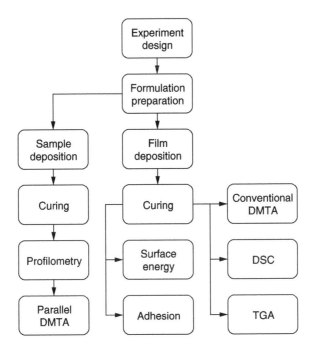

FIGURE 12.1 Workflow used in this study.

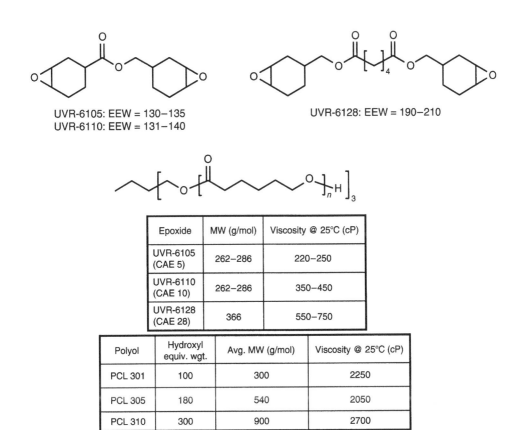

Epoxide	MW (g/mol)	Viscosity @ 25°C (cP)
UVR-6105 (CAE 5)	262–286	220–250
UVR-6110 (CAE 10)	262–286	350–450
UVR-6128 (CAE 28)	366	550–750

Polyol	Hydroxyl equiv. wgt.	Avg. MW (g/mol)	Viscosity @ 25°C (cP)
PCL 301	100	300	2250
PCL 305	180	540	2050
PCL 310	300	900	2700

FIGURE 12.2 Structure and characteristics of epoxies and polyols used.

TABLE 12.1
Sample ID, Polyol, and *R* Value for the Libraries

Sample	Duplicate Sample	Polyol	*R* Value
A1	C1	PCL 301	2
A2	C2	PCL 301	3
A3	C3	PCL 301	4
A4	C4	PCL 301	5
A5	C5	PCL 305	2
A6	C6	PCL 305	3
B1	D1	PCL 305	4
B2	D2	PCL 305	5
B3	D3	PCL 310	2
B4	D4	PCL 310	3
B5	D5	PCL 310	4
B6	D6	PCL 310	5

Three 24-element libraries were created where each library used one epoxide. Library 1 used the CAE 10 cycloaliphatic epoxy, Library 2 used CAE 5, and Library 3 used CAE 28. In each library, the three polycaprolactones were used at four ratios (*R*-values). Table 12.1 lists the sample ID, polyol, and *R*-value for each of the wells in the libraries. Wells A1–B6 are unique and wells C1–D6 are duplicates of the proceeding wells. These designs were created using Symyx Library Studio software. In each case the amount of epoxy was held constant at 3.4 mL and the amount of polyol is varied depending upon the *R*-value. In each formulation 4 wt% photoinitiator was used.

12.2.3 Instrumentation

Formulations were carried out on a Symyx automated formulation station, consisting of a fully automated dual-arm dispensing robot fitted with a disposable pipette tip changer on one robot arm and a viscometer on the second robot arm.[18] Reagents were automatically dispensed into 24 8 mL vials and mixed using a magnetic stir bar. After formulation of the epoxies they were applied to aluminum and glass panels using the Symyx automated film preparation station. This system generates a 24-element array of coatings applied to a pair of $4'' \times 8''$ panels. The casting bar had an 8 mil gap setting, resulting in a cured film thickness of ~0.1–0.2 mm. The formulation and application systems are shown in Figure 12.3. After deposition, the coating arrays were cured for 5 min using a Dymax UV flood lamp having an intensity of 35 mW/cm². Samples were also deposited on the parallel DMTA polyimide substrate and cured under the same conditions.

Differential scanning calorimetry (DSC) was performed on a TA Instruments Q1000 series calorimeter. Samples were subjected to a heat, cool, heat cycle from –50 to 150°C at a ramp rate of 10°C/min. T_g values were determined as the midpoint of the inflection from the first heat cycle. Dynamic mechanical thermal analysis (DMTA) was performed on both a conventional DMTA and a parallel DMTA. The conventional DMTA measurements were carried out using a Rheometric Scientific 3E apparatus in the rectangular tension/compression geometry. T_g is obtained from the maximum peak in the tan delta curves and crosslink densities are calculated from the E' value in the linear portion at least 50°C greater than the T_g. Crosslink density is calculated from the following equation:[24]

$$\nu_e = E/(3RT) \tag{12.2}$$

where ν_e is the crosslink density. Sample sizes for testing were $(10 \times 5 \times (0.1 \text{ to } 0.2))$ mm³. The analysis was carried from –50 to 200°C at a frequency of 10 rad/s and ramp rate of 5°C/min. Symyx

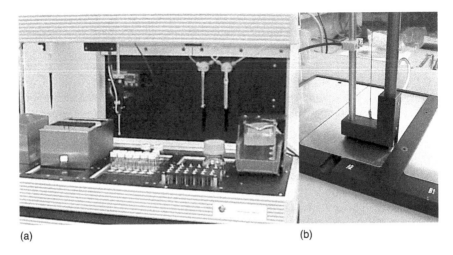

(a)   (b)

FIGURE 12.3 (a) Automated coating formulation system and (b) automated application system. (Reprinted with permission from Webster, D. C., et al., *JCT Coatings Tech* 1 (6), 34–39, 2004.)

Parallel DMTA (pDMTA) was used as the rapid approach to DMTA analysis.[25] Up to 96 samples are analyzed in one experiment using this system. Sample preparation and analysis consists of several steps. The first step is the preparation of the substrate, which is polyimide, ~50 micron Kapton® film. There is an adhesive on one side to allow for attachment to the aluminum pDMTA substrate. It should be noted that there is no adhesive in the testing area. Using an automated pipetting system, liquid samples are deposited in the appropriate area on the Kapton film. After curing, the samples are profiled using an automated laser profilometer. This will allow for the determination of the width and height of the samples, which is needed for the modulus calculations. Finally, the sample array is put into the pDMTA system where the sample array is oscillated against an array of force sensors. The runs were from –25 to 160°C at a heating rate of 1°C/min and a frequency of 10 Hz. The contribution of the Kapton membrane to the stiffness is removed and storage modulus, loss modulus, and loss tangent (tan δ) are calculated.[25]

Symyx coating surface energy system was used to determine the contact angles and interfacial tension (IFT) of the coatings. Contact angles made by water and methylene iodide on the films were measured and a camera was used to image the droplets and calculate the contact angles and interfacial tensions. Three droplets of each liquid are measured. Adhesion testing was performed using a Symyx automated adhesion screening station, which measures the amount of force required to pull off a dollie that is held by an epoxy adhesive on the coating. Analysis of Symyx equipment data was performed using PolyView.

12.3 RESULTS AND DISCUSSION

Determination of structure–property relationships in multicomponent polymer systems can be a tedious process consisting of weighing, mixing, application, curing, and testing of a series of compositions. Often, the sheer magnitude of the effort required limits the experimenter to a relatively small number of formulation variables over a limited range with a small number of increments. Even with the application of statistical experimental design methods, the experimenter tends to be more concerned with how many experiments are required and the time involved than with determining how much information can be gained from the series of experiments.

Radiation-curable polymer systems are a good example of systems where mixtures of oligomeric precursors are mixed at varying ratios in order to achieve an optimum balance of performance properties. These systems are cured with either visible or ultraviolet light or an electron beam. In each of the major classes of radiation-curable polymers (acrylate, epoxy, vinyl ether, donor–acceptor) an incredible number of precursors are available from a number of vendors. It is the formulator's challenge to somehow pick appropriate raw materials and then figure out the optimum composition to arrive at a final product with the desired characteristics. Cycloaliphatic epoxy systems are generally composed of one or more epoxy resins with the addition of a low T_g polyol to improve the flexibility of the coating. These are cured using photoinitiators that generate cations upon exposure to UV radiation.[26] The epoxy resin structure, polyol molecular weight and functionality, and the ratio of the polyol to epoxy will all impact the final properties of the coating film. A structure–property relationship study where a number of these variables are systematically explored can help determine the direction and magnitude of the effects of the formulation variables on key properties of the resulting polymer network.

The cycloaliphatic epoxy formulations were prepared and tested using the workflow shown in Figure 12.1. Three different cycloaliphatic epoxy libraries were prepared using three different epoxies and polyols (Figure 12.2). Library 1 used the CAE 10, Library 2 the CAE 5, and Library 3 used the CAE 28 epoxy. The nominal structures of the CAE 5 and CAE 10 are the same but there are slight differences in the reported epoxide equivalent weight (EEW) and viscosity. The CAE 28 has a different structure in that there are two carboxylate groups and a C_4 chain separating the two, leading to a more flexible structure. In each of the libraries the epoxy was held constant while the polyols were varied. As an example Library 1 used the CAE 10 epoxy in each well and the polyols were varied based on the desired R-value. Each library consisted of 24 samples with the first 12 being distinct samples (A1–B6) while the second half was a replicate of the first 12 (C1–D6). Table 12.1 shows the polyol and corresponding R-value for each of the samples. After formulation, the coatings were applied to aluminum and glass panels using the automated application system and also deposited onto the pDMTA substrate. Coatings were then cured using a UV floodlamp. Coatings on aluminum panels were subjected to surface energy (interfacial tension) and adhesion testing. Glass panels were used to obtain free films for analysis using differential scanning calorimetry (DSC), conventional dynamic mechanical thermal analysis (DMTA), and thermogravimetric analysis (TGA).

12.3.1   Differential Scanning Calorimetry (DSC)

Differential scanning calorimetry was performed on a TA Instruments Q1000 equipped with an auto sampler. Figure 12.4 shows T_g values for each of the libraries. Several key trends are immediately obvious from these data. First, as the R-value increases, there is an increase in the T_g. This is due to the fact that as the R-value increases, less polyol is used in the coating, resulting in both a compositional change and a change in crosslink density. Next, for a given R-value, as the molecular weight of the polyol increases, the T_g decreases. Since the polyols are end-functional, as the molecular weight increases, the molecular weight between crosslinks increases, resulting in a decrease in crosslink density. Third, the differences due to the three cycloaliphatic epoxies are also apparent, with the CAE 28 yielding coatings with a lower T_g than the other two. It is also interesting to note that the T_g values of the CAE 10 and CAE 5 differ considerably in some cases, even though their structures are nominally the same.

12.3.2   Dynamic Mechanical Thermal Analysis (DMTA)

DMTA was performed by both conventional and parallel DMTA. In conventional DMTA, experiments are run one sample at a time but in pDMTA 96 samples are run at the same time. Since our workflow produces 24-sample arrays, we are able to run each sample in quadruplicate using the pDMTA. Figure 12.5 shows four pDMTA plots while Figure 12.6 shows a DMTA for the same sample using conventional DMTA. From Figure 12.5 it is seen that there is some variability in the four parallel DMTA samples. The source of the variability is primarily variations in height and shape of

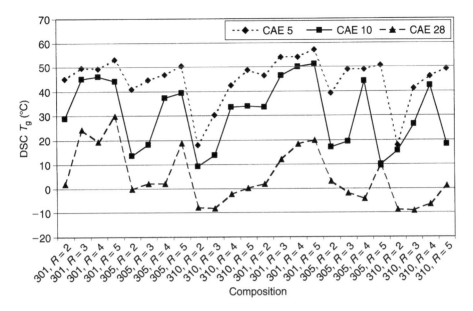

FIGURE 12.4 T_g using DSC for libraries 1, 2, and 3.

the sample on the polyimide membrane. The cause of this arises when a library of formulations having variable viscosity and surface energy are deposited using a liquid-handling robot. Lower viscosity samples have a tendency to flow out on the substrate. We are currently evaluating methods for achieving more uniform sample sizes and shapes. The range in T_g is quite small, only ~9°C, while the other values typically obtained by DMTA have a much broader range. It should be noted that for the conventional DMTA values are not present for all samples. This is because many of the polymer films could not be removed from the glass to yield a suitable film for analysis. Thus, an advantage in using the pDMTA, is that free films do not have to be generated.

Figure 12.7 and Figure 12.8 show the T_g data for all three libraries by conventional and parallel DMTA, respectively. Conventional DMTA is from a single experiment while the pDMTA value is the average of four samples. The same overall trends are observed using DMTA as were seen using DSC. To summarize, as the R-values increase for a particular polyol there is an increase in T_g, as would be expected. Again it is observed that CAE 5 epoxy leads to higher T_g values than CAE 10 and both of these yield higher T_g values than CAE 28. From this we can see several things are easily detected from the T_g data:

1) The effect of R-value on T_g for a particular polyol.
2) The effect of polyol choice on T_g. In these libraries the PCL 305 polyol leads to a higher T_g than the PCL 301, which is higher than the PCL 310 polyol.
3) The effect of epoxy composition on T_g.

A comparison of T_g values obtained from the three methods is shown in Figure 12.9 for a subset of the data. Similar behaviors are observed; that is, an increase in T_g as the R-value increases. Overall, when looking at T_g values for the conventional DMTA, parallel DMTA, and DSC similar trends are observed, while the magnitude of the T_g values is different for the different methods.

In addition to T_g values, DMTA also provides information about the bulk modulus of the coatings at a range of temperatures. Using rubber elasticity theory, the coating crosslink density (XLD) can be calculated from the magnitude of the elastic modulus in the rubbery plateau. Next an examination of crosslink density (XLD) behavior in both conventional and parallel DMTA can be

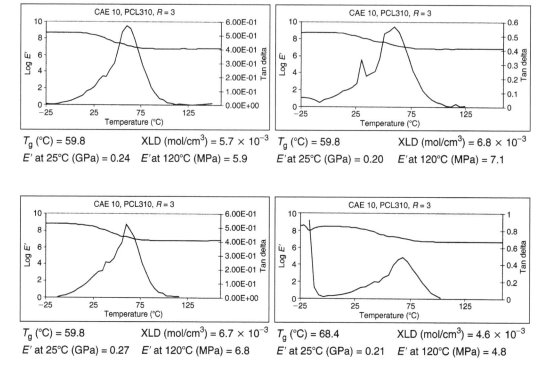

T_g (°C) = 59.8   XLD (mol/cm³) = 5.7 × 10⁻³
E' at 25°C (GPa) = 0.24   E'at 120°C (MPa) = 5.9

T_g (°C) = 59.8   XLD (mol/cm³) = 6.8 × 10⁻³
E'at 25°C (GPa) = 0.20   E'at 120°C (MPa) = 7.1

T_g (°C) = 59.8   XLD (mol/cm³) = 6.7 × 10⁻³
E' at 25°C (GPa) = 0.27   E' at 120°C (MPa) = 6.8

T_g (°C) = 68.4   XLD (mol/cm³) = 4.6 × 10⁻³
E' at 25°C (GPa) = 0.21   E' at 120°C (MPa) = 4.8

FIGURE 12.5 Parallel DMTA plots and data for a set of four replicate samples on one plate.

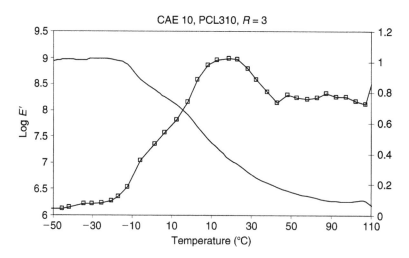

FIGURE 12.6 Conventional DMTA data.

carried out (Figure 12.10 and Figure 12.11). XLD data show some variability between the conventional and parallel DMTA for Library 1 but similar trends are observed between the two. Library 2 shows good consistency between both methods but there are still a couple of points that behave quite differently. Library 3 exhibits similar behaviors as the other two libraries. In some of these we can see an increase in XLD relative to the R-values, where as R increases so does the XLD but this does not hold true for all of the samples. Also it can be observed that XLD appears to be related to the epoxy used, where epoxy CAE 10 shows the highest XLD.

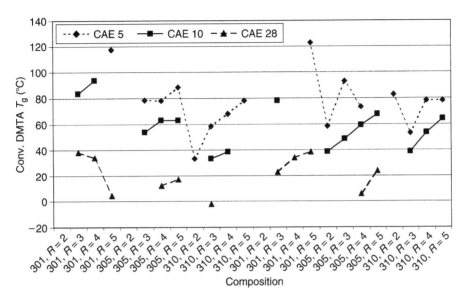

FIGURE 12.7 T_g by conventional DMTA as a function of composition. Missing data points are compositions where free films could not be obtained.

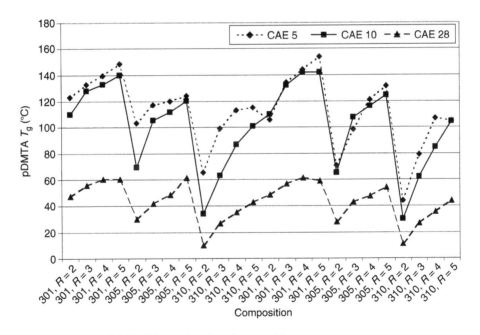

FIGURE 12.8 T_g by parallel DMTA as a function of composition.

Figure 12.12 and Figure 12.13 examine the storage modulus at room temperature. The data show a general increase in modulus as the R-value increases. It should be noted that there is a dependence on polyol used; PCL 301 leads to the highest E' values followed by PCL 305 then PCL 310. Libraries 2 and 3 exhibit similar behaviors as Library 1.

Figure 12.14 and Figure 12.15 examine the E' values above T_g. The storage modulus above T_g for the libraries is quite consistent. Of course there are a few points that do not follow exactly but, in general, behavior in conventional and parallel DMTA is similar. In general, it can also be observed

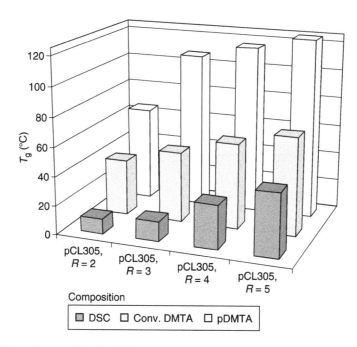

FIGURE 12.9 Comparison of T_g by DMTA (conventional and parallel) and DSC for CAE 10 with PCL 305 epoxy.

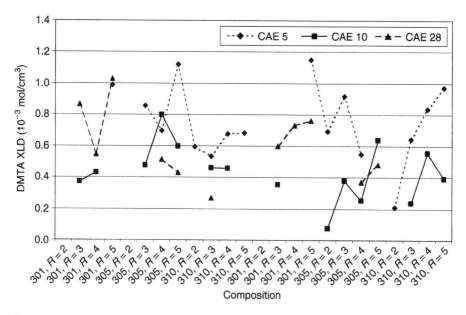

FIGURE 12.10 XLD by conventional DMTA as a function of composition for all three libraries. Missing data points are compositions where free films could not be obtained.

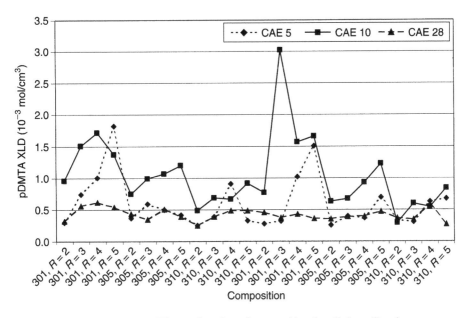

FIGURE 12.11 XLD by parallel DMTA as a function of composition for all three libraries.

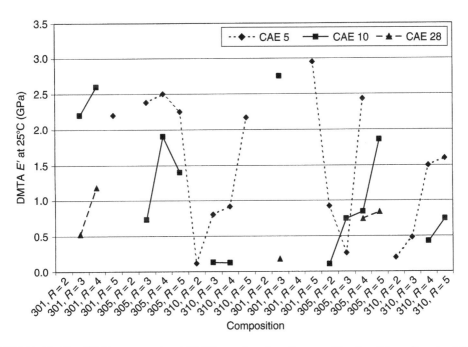

FIGURE 12.12 E' at 25°C by conventional DMTA as a function of composition for all three libraries. Missing data points are compositions where free films could not be obtained.

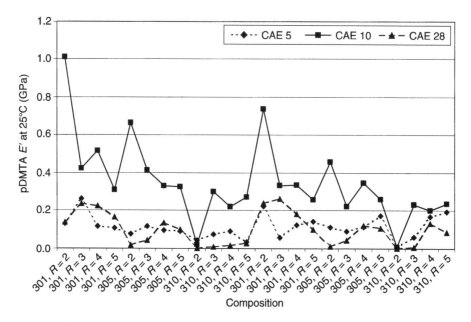

FIGURE 12.13 E' at 25°C by parallel DMTA as a function of composition for all three libraries.

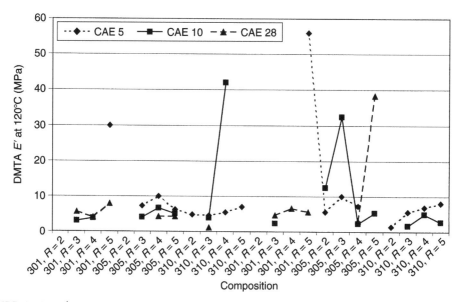

FIGURE 12.14 E' at 120°C by conventional DMTA as a function of composition for all three libraries. Missing data points are compositions where free films could not be obtained.

that as the R-value increases the storage modulus above T_g increases. Also there is a dependence on the polyol used; the storage modulus above T_g increases as follows: PCL 301 > PCL 305 > PCL 310. Epoxy CAE 10 showed the highest E' values both above and below T_g.

12.3.3 THERMOGRAVIMETRIC ANALYSIS (TGA)

TA Instruments Q500 automated TGA was used in this study. Sample films were placed onto an auto sampler tray (15 sample) and then heated from room temperature to 600°C at a rate of

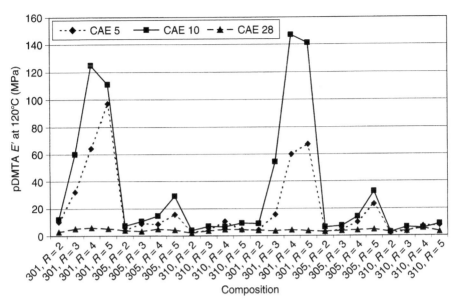

FIGURE 12.15 E' at 120°C by parallel DMTA as a function of composition for all three libraries.

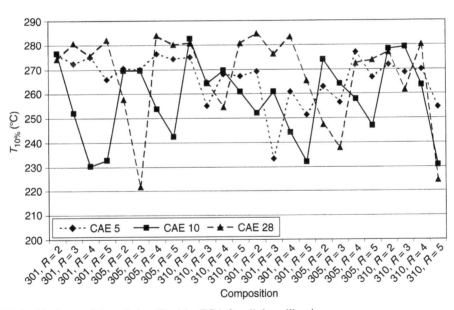

FIGURE 12.16 Onset of degradation ($T_{10\%}$) by TGA for all three libraries.

10°C/min. Onset of degradation ($T_{10\%}$) varied from library to library as seen in Figure 12.16. While all of the values fall within a narrow range, the onset of degradation for these libraries appears to decrease as the R-value is increased. Figure 12.17 shows the behavior of the maximum temperature of degradation (T_{max}) for these polymers. In general, T_{max} decreases with an increase in R-value. Also the CAE 28 tends to have lower values for T_{max}, most likely due to aliphatic ester chain between the cycloaliphatic epoxy groups. Also it can be seen that, in general, CAE 10 formulations have higher T_{max} values compared to the CAE 5 formulations. Figure 12.18 is an examination of the amount of non-volatile residue (char) left at the end of the TGA run. It can be seen that the amount

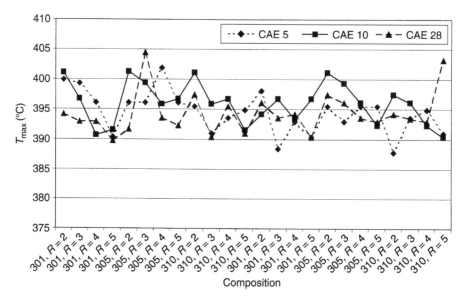

FIGURE 12.17 Maximum decomposition rate temperature (T_{max}) from TGA for all three libraries.

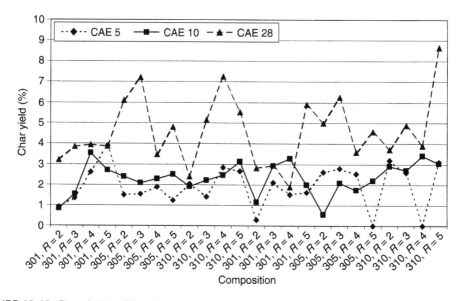

FIGURE 12.18 Char yield by TGA for all three libraries.

of char is dependent upon the type of epoxy used. In this particular case epoxy CAE 28 resulted in the highest amount of char. This is most likely due to the longer chain between the epoxy groups.

12.3.4 CONTACT ANGLE AND SURFACE ENERGY (INTERFACIAL TENSION)

A drop of liquid placed upon a flat surface can spread or remain as a drop.[27–29] Contact angle describes the shape of a liquid drop resting on the solid surface and is the angle between a tangent drawn on a drop's surface at resting (Figure 12.19).[27] The shape of a drop reveals information about the chemical bonding at the surface. Adhesive interactions between a liquid and a solid result in the

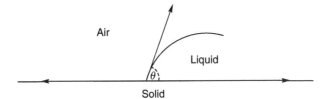

FIGURE 12.19 Schematic of contact angle interface.

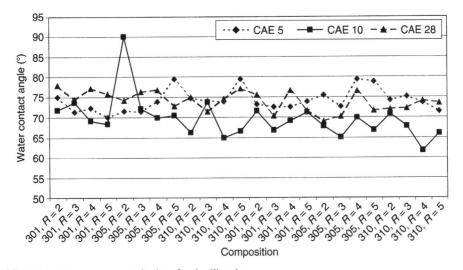

FIGURE 12.20 Water contact angle data for the libraries.

spreading of a liquid (wetting) while liquid cohesive forces result in de-wetting. Contact angles can be used to determine the competition between these two forces and to determine wettability and predict adhesion.[30] Contact angle measurements and semiempirical equations can be used to calculate the surface energy, which is the result of an imbalance of attractive forces between molecules at a surface.[29] In this particular study the test liquids are water and methyl iodide. Water has a large polar component of surface energy while methyl iodide has primarily a dispersion force; both liquids have high surface tensions. The Owens and Wendt equation is used to calculate surface energy or interfacial tension:[29]

$$\gamma_{LV}(1 + \cos \theta) = 2(\gamma_{SV}^{d} \ \gamma_{SV}^{d})^{0.5} + 2(\gamma_{LV}^{p} \ \gamma_{LV}^{p})^{0.5} \tag{12.3}$$

Critical surface tension of compounds with C, O, H at the surface has been shown to be 35–50 dynes/cm.[31] Figure 12.20 and Figure 12.21 show the contact angle and surface energy (IFT) for all three libraries. Contact angles and IFT values do not significantly vary between the libraries. Contact angles were between 63 and 79° while IFT values varied between 48 and 62 mJ/m². It can be seen that CAE 5 leads to a slightly larger contact angle than do the other CAE formulations.

12.3.5 ADHESION

Adhesion is the bonding strength between a polymeric coating and a substrate and it is the reversible separation of the two phases that is expressed by the work of adhesion.[32] In this work adhesion of the epoxy film to an aluminum panel was examined using the Symyx automated

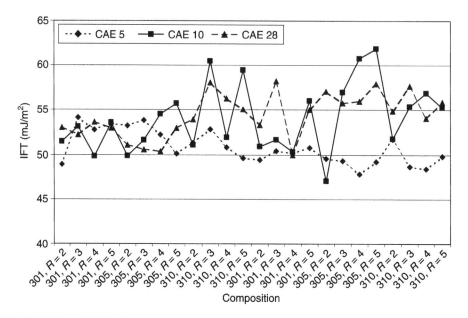

FIGURE 12.21 Interfacial surface tension (IFT) data for the three libraries.

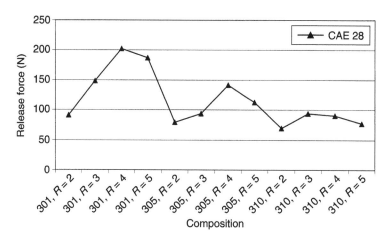

FIGURE 12.22 Force at release for the CAE 28 library.

adhesion screening station. This system mimics ASTM method D4541. An aluminum dollie is attached to the coating surface with a two-part epoxy adhesive and allowed to cure. Three dollies are glued to each coating sample. Next an automated pull-off head grips each dolly and applies a pressure ramp until failure occurs. In this study, the third dolly position was close to the bead of coating left behind from the drawdown blade and did not provide usable data. Thus, the results are presented as an average of four data points: two pull-off tests on two different coating samples.

In these experiments, adhesion of libraries that used CAE 5 and CAE 10 was essentially zero. Figure 12.22 shows data for the CAE 28 library obtained by this method. Key trends observed are that adhesion increases as the R-value increases, i.e., as the amount of epoxy relative to polyol increases. In addition, there is a hint of an optimum value at $R = 4$ which could be explored further. As the molecular weight of the polyol increases, the adhesion decreases also.

12.4   CONCLUSIONS

This work demonstrates the applicability of a combinatorial workflow to help in formulation and analysis of radiation-curable coatings. It shows that large libraries of samples having widely varying composition can be prepared to identify key property trends. Further formulation/characterization can be carried out to see if other desired properties are observed and eventually lead to a final product.

In this work we have shown that T_g by DSC, conventional DMTA, and parallel DMTA behaved similarly for all samples, thus the more efficient parallel DMTA can be used as a rapid screening tool. In these formulations, as the R-value increased there was an increase in T_g. Also, all three analysis methods showed a difference in T_g based on epoxy used where CAE 5 led to the highest T_g and CAE 28 to the lowest. It should also be noted that differences were observed for CAE 5 and CAE 10 formulations. The structures of these two epoxies are similar and one would expect that the properties would generally be the same. But, as can be seen in these results, there are some differences in properties depending upon which epoxy is used. The combinatorial approach has allowed us to see this; if done conventionally we might have only examined one of the two since they were similar in structure. Storage modulus below and above T_g showed a dependence upon the polyol used. TGA showed differences in T_{max} and char depending upon the epoxy used. Some key effects of composition on adhesion to an aluminum substrate were also observed. Thus, this work illustrates that the use of combinatorial and high-throughput techniques for the systematic exploration of the effects of compositional variables on performance properties can provide valuable information that can be used in the further development of materials having an optimum set of properties.

ACKNOWLEDGMENTS

The authors thank the Department of the Navy's Office of Naval Research for sponsorship under grants N00014-03-1-0702 and N00014-04-1-0597.

REFERENCES

1. Cawse, J. N., *Experimental Design for Combinatorial and High Throughput Materials Development*, Wiley-Interscience, New York, 2003.
2. Potyrailo, R. and Amis, E. J., *High-Throughput Analysis: A Tool for Combinatorial Materials Science*, Kluwer Academic/Plenum Publishers, New York, 2004.
3. Wicks, D. A. and Bach, H., The coming revolution for coatings science: high throughput screening for formulations, *Proceedings of the International Waterborne, High-Solids, and Powder Coatings Symposium* 29th, 1–24, 2002, University of Southern Mississippi, Hattiesburg, MS.
4. Iden, R., Schrof, W., Hadeler, J. and Lehmann, S., Combinatorial materials research in the polymer industry: Speed versus flexibility, *Macromol. Rapid Commun.*, 24 (1), 63–72, 2003.
5. Hoogenboom, R. and Schubert, U. S., The fast and the curious: High-throughput experimentation in synthetic polymer chemistry, *J. Polymer Sci., Part A: Polymer Chem.*, 41 (16), 2425–2434, 2003.
6. Schubert, U. S., De Gans, B.-J. and Kazancioglu, E., Combinatorial and high-throughput polymer research: composition of complete workflows, *Polymer. Mat. Sci. Engineering*, 90, 643–644, 2004.
7. Peil, K. P., Neithamer, D. R., Patrick, D. W., Wilson, B. E. and Tucker, C. J., Applications of high throughput research at the Dow Chemical Company, *Macromol. Rapid Commun.*, 25, 119–126, 2004.
8. Cawse, J. N., Olson, D., Chisholm, B. J., Brennan, M., Sun, T., Flanagan, W., Akhave, J., Mehrabi, A. and Saunders, D., Combinatorial chemistry methods for coating development V: generating a combinatorial array of uniform coatings samples, *Prog. Organic Coatings*, 47 (2), 128–135, 2003.
9. Cawse, J. N. and Wroczynski, R. J., Screening using high-dimensional gradient arrays, *Abstracts of Papers, 224th ACS National Meeting*, Boston, MA, United States, August 18–22, PETR-027, 2002.
10. Chisholm, B., Potyrailo, R., Cawse, J., Brennan, M., Molaison, C., Shaffer, R., Whisenhunt, D. and Olson, D., Combinatorial chemistry methods for coating development III. An illustration of an experiment conducted with the combinatorial factory, *Proceedings of the International Waterborne, High-Solids, and Powder Coatings Symposium* 29th, 125–137, 2002, University of Southern Mississippi, Hattiesburg, MS.

11. Chisholm, B., Potyrailo, R., Cawse, J., Shaffer, R., Brennan, M., Molaison, C., Whisenhunt, D., Flanagan, B., Olson, D., Akhave, J., Saunders, D., Mehrabi, A. and Licon, M., The development of combinatorial chemistry methods for coating development I. Overview of the experimental factory, *Prog. Organic Coat.*, 45 (2–3), 313–321, 2002.

12. Chisholm, B. J., Potyrailo, R. A., Cawse, J. N., Shaffer, R. E., Brennan, M. and Molaison, C. A., Combinatorial chemistry methods for coating development: IV. The importance of understanding process capability, *Prog. Organic Coat.*, 47 (2), 120–127, 2003.

13. Potyrailo, R. A., Chisholm, B. J., Morris, W. G., Cawse, J. N., Flanagan, W. P., Hassib, L., Molaison, C. A., Ezbiansky, K., Medford, G. and Reitz, H., Development of combinatorial chemistry methods for coatings: high-throughput adhesion evaluation and scale-up of combinatorial leads, *J. Combin. Chem.*, 5 (4), 472–478, 2003.

14. Grunlan, J. C., Holguin, D. L., Chuang, H.-K., Perez, I., Chavira, A., Quilatan, R. and Akhave, J., Combinatorial development of pressure-sensitive adhesives, *Macromol. Rapid Comm.* 25, 286–291, 2004.

15. Schmatloch, S., Bach, H., van Benthem, R. A. T. M. and Schubert, U. S., High-throughput experimentation in organic coating and thin film research: state-of-the-art and future perspectives, *Macromol. Rapid Commun.*, 25, 95–107, 2004.

16. de Gans, B.-J., Kazancioglu, E., Meyer, W. and Schubert, U. S., Ink-jet printing polymers and polymer libraries using micropipettes, *Macromol. Rapid Commun.*, 25, 292–296, 2004.

17. Hoogenboom, R., Fijten, M. W. M., Abeln, C. H. and Schubert, U. S., High-throughput investigation of polymerization kinetics by online monitoring of GPC and GC, *Macromol. Rapid Comm.*, 25, 237–242, 2004.

18. Webster, D. C., Bennett, J., Kuebler, S., Kossuth, M. B. and Jonasdottir, S., High throughput workflow for the development of coatings, *JCT Coatings Technol.*, 1 (6), 34–39, 2004.

19. Schmatloch, S. and Schubert, U. S., Techniques and instrumentation for combinatorial and high-throughput polymer research: recent developments, *Macromol. Rapid Commun.*, 25, 69–76, 2004.

20. Dar, Y. L., High-throughput experimentation: a powerful enabling technology for the chemicals and materials industry, *Macromol. Rapid Comm.*, 25, 34–47, 2004.

21. Grunlan, J. C., Mehrabi, A. R., Chavira, A. T., Nugent, A. B. and Saunders, D. L., Method for combinatorial screening of moisture vapor transmission rate, *J. Comb. Chem.*, 5, 362–368, 2003.

22. Eidelman, N., Raghavan, D., Forster, A. M., Amis, E. J. and Karim, A., Combinatorial approach to characterizing epoxy curing, *Macromol. Rapid Comm.*, 25, 259–263, 2004.

23. Meier, M. A. R., Hoogenboom, R. and Schubert, U. S., Combinatorial methods, automated synthesis and high-throughput screening in polymer research: The evolution continues, *Macromol. Rapid Commun.*, 25 (1), 21–33, 2004.

24. Wicks, Z. W., Jones, F., N. and Pappas, S. P., *Organic Coatings: Science and Technology, 2nd Edition,* Wiley, New York, 1999.

25. Kossuth, M. B., Hajduk, D. A., Freitag, C. and Varni, J., Parallel dynamic thermomechanical measurements of polymer libraries, *Macromol. Rapid Commun.*, 25 (1), 243–248, 2004.

26. Koleske, J. V., *Radiation Curing of Coatings,* ASTM International, Philadelphia, PA, 2002.

27. Shaw, D. J., *Introduction to Colloid and Surface Chemistry*, 2nd ed. Butterworths, Boston, MA, 1970.

28. www.firsttenangstroms.com.

29. Owen, M. J., Surface energy, In: *Comprehensive Desk Reference of Polymer Characterization and Analysis*, Brady, R. F., Ed., Oxford University Press, New York, 2003, pp. 361–374.

30. Schmidt, D. L. Brady, Jr., R. F., Lam, K., Schmidt, D. C. and Chaudhury, M. K., Contact angle hysterisis, adhesion, and marine biofonling. *Langmuir,* 20 (7), 2830–2836, 2004.

31. Bascom, W. D., The wettability of polymer surfaces and the spreading of polymer liquids, *Adv. Polym. Sci.*, 85, 89–124, 1998.

32. Nelson, G. L., *Paint and Coating Test Manual*, 14 ed. ASTM, Philadelphia, PA, 1995.

13 Combinatorial Synthesis and Screening of Photochromic Dyes and Modified Conducting Polymers

Diego F. Acevedo, María C. Miras, and César A. Barbero

CONTENTS

13.1 INTRODUCTION

Electronic delocalization (conjugation) is for the application of organic materials in electronic, photonic, and optoelectronic devices.[1] The presence of weakly localized electron density affects the electron energy by electric or magnetic fields and/or interaction with photons. In that way, electronic (e.g., transistors), photonic (e.g., photographic plates), or optoelectronic (e.g., light-emitting diodes) devices could be developed. Among conjugated materials, there are transparent polymers, doped with low-molecular-weight photochromic dyes, and conjugated polymers, which exhibit electronic conductivity. In the present chapter, the use of combinatorial chemistry techniques to find photochromic dyes and modified conducting polymers, will be described.

Combinatorial chemistry methods were initially developed to accelerate the discovery of pharmaceutical compounds,[2] and have been extended to the search for other organic co pounds,[3] materials,[4] and polymers.[5–7] Compound libraries are produced by the combinatorial reaction of several reactants through the same reaction. This can be done by parallel reactions in different reactors,[8] or simultaneous ("Split&Pool") reactions.[9,10] The screening of properties is made on the whole set of compounds produced, using simple and fast methods ("high-throughput screening"),[11] and only those compounds having the target property are fully characterized.

The first area of work to be described is the search for photochromic dyes. Photochromism is a result of a reversible change in the molecular structure or charge distribution during irradiation. One way to produce a molecular change involves photoinduced *trans–cis* isomerization, which occurs in azobenzene and similar compounds.[12,13] Indeed, azo compounds have been considered of much interest due to their novel photochromic properties and potential applications in molecular devices such as non-linear optics, molecular switches, and optical data storage.[14–17]

The *trans* (*E*) geometric isomer is thermodynamically more stable while the excited state is more stable in the *cis* (*Z*) isomeric form. Therefore, the molecule is converted from *trans* to *cis* isomer by irradiation. Since the electron distribution is affected by the geometry of the molecule, the optical spectrum will change, producing the photochromic effect. The *cis* isomer will eventually decay to the *trans* isomer either thermally or by irradiation with light. To do the latter, the photon energy has to correspond to the absorption maxima of the *cis* isomer. The rate of decay depends strongly on the aromatic groups attached to the chromophore (–N=N–).[18] The geometric change is also associated with alteration of other properties, such as the molecule size and dipole moment.[19] Therefore, the photoinduced change could also be applied in mechanical or electronic devices. Since azo dyes have been widely used since the 19th century,[20] the photochromic properties of several azo dyes have already been studied. However, to the best of our knowledge, combinatorial techniques have not been used to synthesize and screen photochromic dyes. As a clear color change of the photochromic dye is of interest for technological applications, it is especially amenable to high-throughput screening methods based on optical properties. The decay rate is also a target property. That is, if the system has to retain information for some time, the decay rate should be in the chosen time range (e.g., minutes). The synthetic scheme and high-throughput screening method to obtain photochromic azo dyes will be described.

The other area of work described here involves the synthesis of soluble conductive polymers by post-modification. Since the discovery, by Shirakawa, Heeger, and MacDiarmid, that the conductivity of polyacetylene increases significantly upon doping with electron acceptors,[21–23] much effort has been devoted to synthesize intrinsically conductive polymers (ICPs) and/or improving the properties of those materials.[24,25] Besides conductivity, ICPs have other properties which can be applied in technological applications, ranging from electroluminescence to corrosion inhibition. While an increasing number of ICPs have been synthesized and studied, conducting polymers with valuable properties remain to be discovered. The usual way to produce new ICPs involves the synthesis or acquisition of the monomer, homopolymerization,[26–34] or co-polymerization,[35–38] followed for a detailed study of the polymer properties. Another, less-explored, route to produce materials with varying properties involves post-modification of already synthesized, and well-characterized, conducting polymers. This can be done by covalent bonding to the polymer backbone.[39–44] It has been shown that, when the same modified polymer (e.g., poly(aniline-co-(2-aminosulfonic acid)) is either produced by co-polymerization or post-modification, the latter has higher conductivity.[45]

One (or few related materials) is usually synthesized and its properties thoroughly studied. Such an approach makes the discovery of new conductive polymers rather slow and expensive. A method to overcome those limitations involves the combinatorial synthesis, coupled to high-throughput screening of compounds. Indeed, combinatorial reactions have been used to produce conjugated polymers with success.[46] Combinatorial synthesis could be carried out in solution, but the simplicity of working with immobilized substrates makes it the method of choice. The solid-phase synthesis has been used to produce conjugated molecules of precise length.[47–49] The low solubility of conducting polymers in most solvents makes them ideal to perform combinatorial chemistry reactions on the solid polymer (free standing or supported on inert substrates). In this chapter, the synthesis of novel substituted polyanilines by coupling of combinatorially synthesized diazonium salts with polyaniline is described. Polyaniline (PANI) is one of the most promising materials for technical applications in batteries, supercapacitors, electrochromic displays, sensors, and others, due to its chemical stability and relatively high conductivity.[50–52] The dazonium coupling reaction was chosen for the modification, based on our previous experience of the synthesis of modified conducting polymers.[53–63]

Due to the simplicity of the reaction it is possible to design the combinatorial synthesis of the diazonium salt, using a wide variety of commercial available reactants (aromatic amines). (A substructure search in chemical products catalogues (www.chemexper.com) renders 1186 aromatic amines of type A and 382 of type B. Therefore 46,218 modified polyanilines could be produced using commercially available amines.) Indeed, the reaction fulfills the conditions for "click" chemistry,[64] as it is carried out in water with high yield, allowing easy scale-up from laboratory to industrial production. This is an important factor because soluble conductive polymers could be used in gram to kilogram quantities. The coupling of diazonium salts with PANI was first investigated by Liu and Freund.[65] They carried out the reaction of the diazonium salt with the electrochemically reduced polymer in acid media. In those conditions, they observed substitution on the amine nitrogen by the aryl cation radical, with loss of nitrogen, producing an electroinactive polymer. On the other hand, we obtained a modified polymer with azo-linked groups bonded to the rings of the polymer backbone by the reaction of 4-sulfobenzenediazonium ion with poly(N-methylaniline) at low temperature in basic media. The product is soluble in aqueous basic media and electroactive.[66] The reaction of diazo resins with PANI, with retention of azo groups and electroactivity, has also been observed.[67,68] As will be shown, similar products could be obtained by coupling diazonium ions with PANI in basic media at low temperature.[69,70]

The property we chose to screen is the solubility in common organic solvents and aqueous solutions. PANI base is only soluble in strongly polar amides (e.g., N-methylpyrrolidone) and strong acids.[71] Consequently, making polyaniline soluble by modification would allow processing the polymer from solution a desirable property in technological applications.[72,73] Additionally, the existence of solvent–polymer interactions, necessary to make the polymer soluble, could be correlated to improved miscibility of the polymer with dielectric polymers. Together with solubility, the conductivity of polymer films is evaluated, allowing the materials to be directly applied in conductive coatings. PANI modified with azo groups could also be applied in photochromic systems, photochemically promoted conformational changes, electrochromic devices, pH indicators, drug release polymers, etc. Some of the properties required for such applications are also screened.

13.2 RESULTS AND DISCUSSION

13.2.1 PHOTOCHROMIC DYES

13.2.1.1 Combinatorial Synthesis

Since coupling of diazonium ions with reactive aromatic compounds (amines, phenols, naphthols, etc.) to form azo dyes is a general and high-yield reaction,[74] parallel reaction in solution was chosen as the synthetic method of the dyes. Primary aromatic amines were diazotized by treatment with sodium nitrite and HCl in an ice bath. The diazotized amines were then coupled with reactive

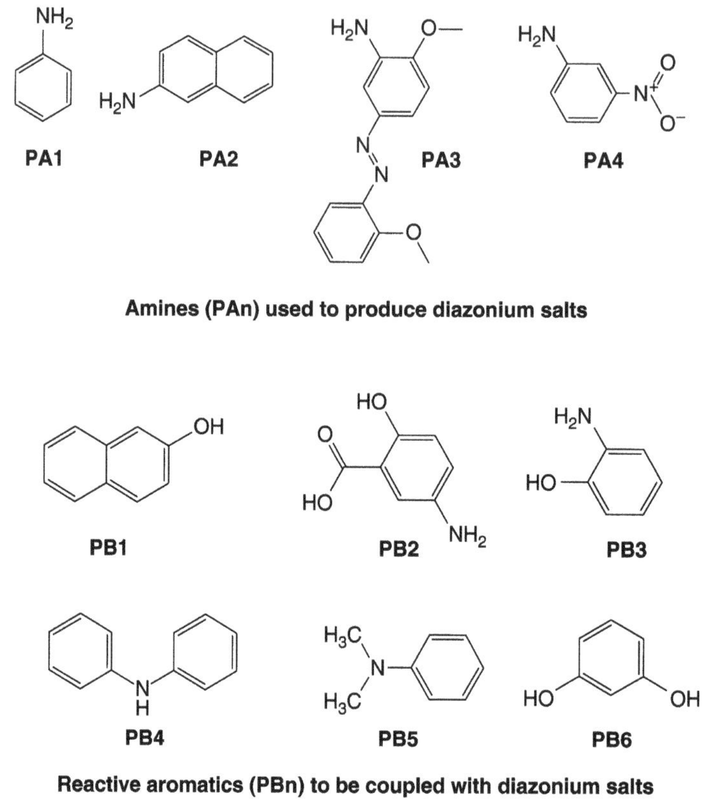

Amines (PAn) used to produce diazonium salts

Reactive aromatics (PBn) to be coupled with diazonium salts

SCHEME 13.1 Compounds used in the synthesis of photochromic dyes.

aromatics, without further purification. The coupling was carried out in an ice bath buffered at pH 8. The dyes were isolated by precipitation and filtration.

Since subsequent coupling is not necessary, as in the modification of conducting polymers, other aromatic compounds could be used such as phenols, naphthols, and non-primary amines. The compounds used to synthesize the dyes are described in Scheme 13.1.

13.2.1.2 High-Throughput Screening (HTS) of Photochromic Dyes

The screening involves detecting spectral changes in the visible range of the spectra upon illumination. It is likely that most azo dyes suffer photochromic changes upon illumination. However, only those that clearly change their color in the visible range of the spectra are of technical interest. Besides, to be useful the photochromic change should be retained for a period of 1–5 min. Therefore, those dyes which either change color in a spectral range outside the visible range, or decay back to the original color in less than 30 s, should not be detected by the HTS procedure.

An easy HTS scheme to detect the properties involves photographing a set of dye solutions with and without illumination. Two sets of solutions were dispensed into identical white porcelain multiwell plates and a digital camera was used to record the colors of the solutions. The plate was then illuminated with an UV lamp (250 nm, 5 min) and photographed 30 s after illumination. The reference plate was kept in the dark.

In Figure 13.1, the photographs of a typical screening test are shown. As can be seen, there are several dyes whose colors change but only some (e.g., numbers 1 and 2) show a clear color difference. Moreover, since contrast is important, a light color in the unmodified state should be chosen.

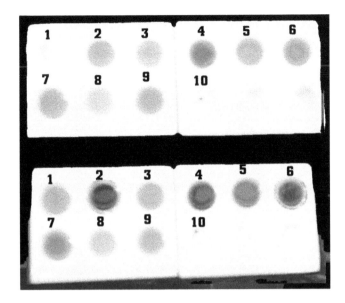

FIGURE 13.1 (See color insert following page 172) High-throughput screening of azo dyes in solution. The upper plates were kept in the dark during illumination by masking with a black film.

Taking all of that into account, dye number 1 (PA1PA1) seems to be a good candidate. Once the photochromic dye was detected, the spectral changes are confirmed by UV–visible differential spectroscopy (Figure 13.2). The method also allows evaluating the reversibility and decay rate of the dye.

As can be seen, the *cis* isomer has a maximum at ca. 490 nm, whereas the *trans* isomer has a maximum at ca. 360 nm. Once the irradiation light is turned off, the *cis* isomer decays thermally into the thermodynamically more stable *trans* isomer. Complete conversion could take up to 30 min. By irradiation at ca. 490 nm, the decay time could be reduced to ca. 5 min. One application of photochromic dyes involves the production of optical memories, setting the dyes into transparent matrices, such as poly(methylmethacrylate) (PMMA) thin films. The doped films could be produced by evaporation of a solution containing both the polymer and the dye. The UV–visible spectra of typical films, taken after illumination at 280 nm (Figure 13.3), reveal that the dye is still able to change color inside the polymer matrix. The spectral maxima for the *cis* and *trans* isomer are present at different wavelength than those measured in the liquid solution (CHCl$_3$). This is likely to be due to the different effective polarity of PMMA and CHCl$_3$. The rate of decay is slightly altered from liquid to a polymeric matrix, as previously found for other dyes.[75]

Using the photochromic dye doped inside a PMMA matrix, it is possible to reversibly transfer images using light (photowriting). The image produced by illumination of a doped film through a dark mask (upper half circle) is shown in Figure 13.4. The illuminated region changes from yellow to red-orange while the masked letters remain yellow.

The image is cleared (lower half circle) by thermal decay or bleached by illumination at the maxima of the colored state. Although the resolution is rather low in the system shown, to produce reversible optical storage using the synthesized photochromic dyes would be possible.

13.2.2 MODIFIED CONDUCTING POLYMERS

To avoid deconvolution steps, parallel synthesis was used to generate modified polyanilines. To synthesize the diazonium salts, a linear synthesis strategy was used.[76] That is, azo dyes are produced

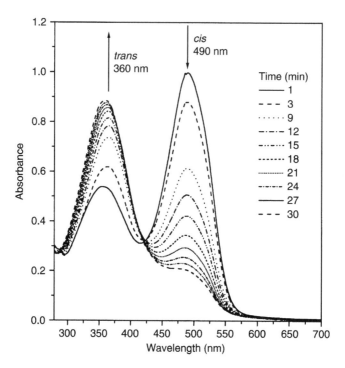

FIGURE 13.2 UV–visible spectra of an azo dye (PA1PA1) in solution (CHCl₃) taken during decay of the *cis* form developed upon illumination (5 min at 250 nm). The spectrum of the solvent is used as reference. Temperature is kept constant at 25°C.

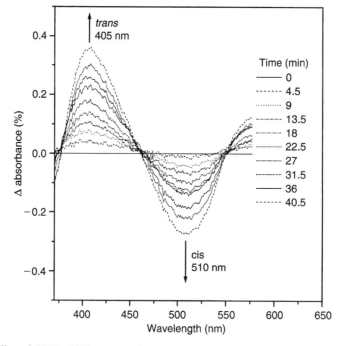

FIGURE 13.3 Differential UV–visible spectra of an azo dye (PA1PA1) in a PMMA matrix during photochromic changes. The spectrum of the film after illumination ($t = 0$) is used as reference.

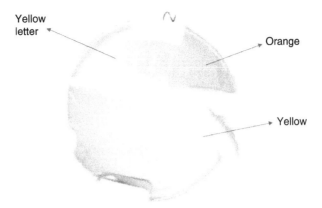

FIGURE 13.4 Photowritten image of a mask on a PMMA film doped with a photochromic azo dye (PA1PA1). Upon illumination, the yellow color of the dye turns orange. The letters are masked and appear yellow on an orange background.

by coupling diazotized aromatic amines with free aromatic amines in solution. Each dye has a free primary amino aromatic group which is then diazotized.

The diazonium ion produced is then coupled with PANI (Scheme 13.2). While this is the simplest synthetic approach, both convergent (reacting PANI sequentially with several diazonium salts) and/or totally combinatorial (reacting several substituted polyanilines with several diazonium salts) strategies have been found to be effective but will not be described here.

13.2.2.1 Combinatorial Synthesis of Diazonium Ions

The azo dyes were synthesized by a combination of three different anilines with three primary aromatic amines (Scheme 13.3). The anilines (type A) bear electron-withdrawing groups while the amines (type B) bear electron-donating (not in the *para* position to the amino group) or are not substituted.

In that way, the produced azo dyes bear a primary aromatic $-NH_2$ group, which could be diazotized to produce a diazonium ion able to react with PANI. Additionally, all the original aromatic amines (type A and B) could be diazotized for coupling with PANI. While no permutation is possible (diazonium salts from type A amines are too strongly deactivated for coupling with diazonium ions), type B amines compounds could give diazonium salts to couple with themselves (The identifying code for the polymers reads: type A type B-P. For example A1B3-P is (*p*-aminosulfonic acid)=>(2-methoxyaniline)=>PANI.). To test if modification has been effected, we determine the FTIR spectra of the films, supported on thin low-density polyethylene (LDPE) films.[77] Infrared spectroscopy is particularly useful in solid-state synthesis as it allows the reaction to be followed from soluble to solid compounds.[78,79]

In most FTIR spectra, a new band (or a shoulder), in the region 1470–1500 cm^{-1}, assigned to the stretching of the azo linkage ($-N=N-$) was found. In those cases where that band was not easily detected, new bands due to the group attached (e.g., a band at 1690 cm^{-1}, assigned to the stretching of the $>C=O$ group in Figure 13.5) were observed.

The UV–visible spectra of the polymers present either new bands in the region 380–480 nm, assigned to the n–π* transition of the azo chromophore or shifts in the bands due to the exciton transition in the PANI backbone. Therefore, it seems that the modification reaction is effective for all the diazonium salts used.

13.2.2.2 Solubility of Conducting Polymers

As was discussed above, the target property is the solubility of the modified conducting polymers. The dissolution of the polymers was tested by coloring of the solvents and discoloring of the

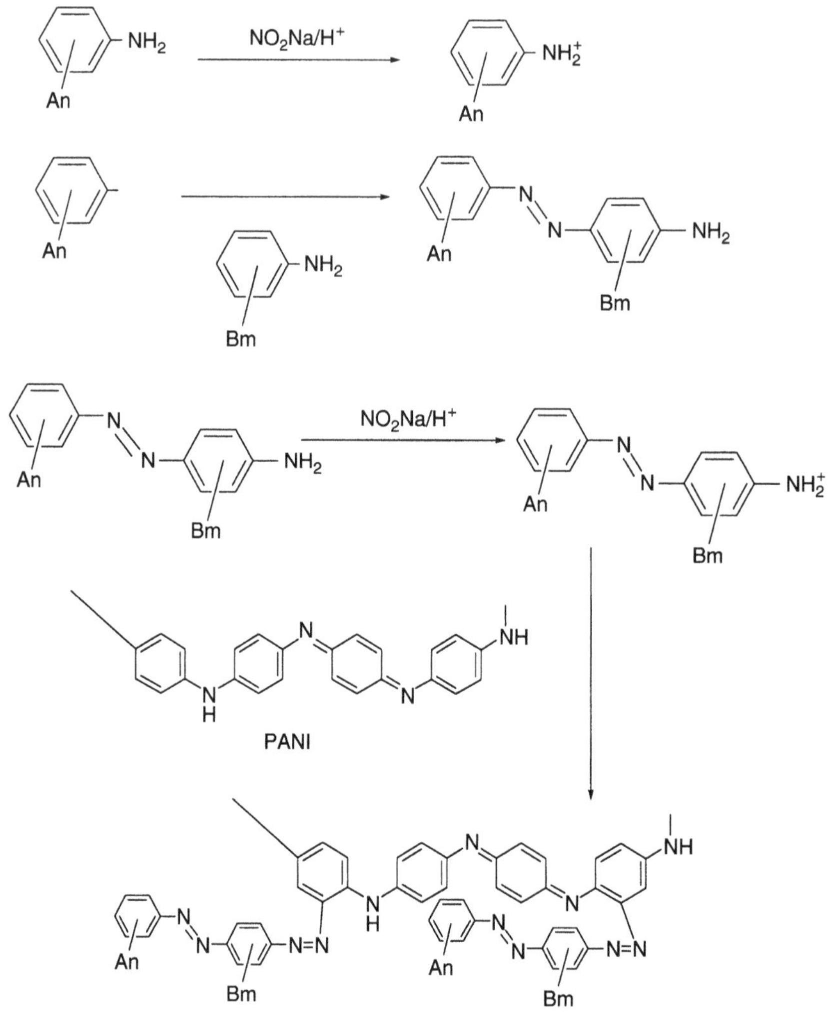

SCHEME 13.2 Modification of polyaniline by coupling with combinatorially synthesized diazonium salts.

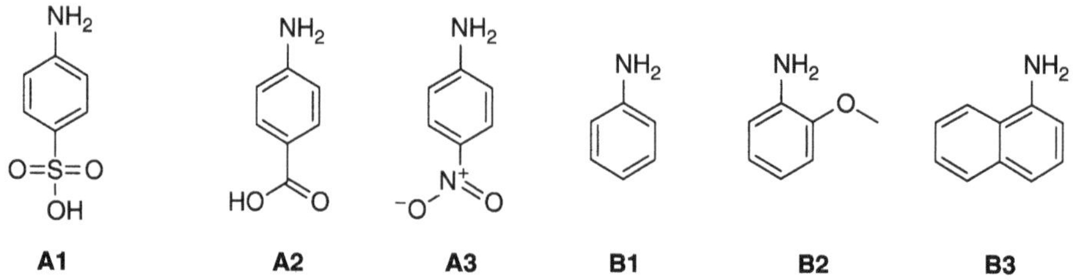

SCHEME 13.3 Anilines used in the combinatorial modification of polyaniline.

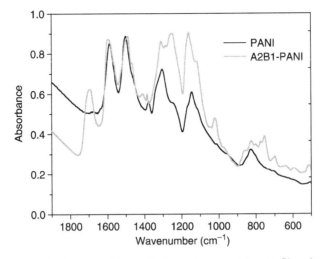

FIGURE 13.5 FTIR of PANI and modified PANI. Both spectra were taken on films deposited onto LDPE.

TABLE 13.1
Solubilities* and UV–Visible Spectral Maxima of Modified Polyanilines

Polymer	Toluene	Acetone	CHCl$_3$	NH$_3$/iPrOH	NH$_3$/H$_2$O
PANI	I	I	I	I	I
A1-P#	—	—	—	—	VS
A2-P	I	I	I	I	VS
A3-P	I	S	I	I	I
B1-P	S	S	I	S	I
B3-P	I	S	I	I	I
B2-P	S	S	VS	S	I
A1B1-P	I	I	I	S	VS
A1B2-P	I	I	I	I	I
A1B3-P	I	S	I	I	VS
A2B1-P	I	S	I	I	S
A2B2-p	I	I	I	I	S
A2B3-P	I	S	I	S	I
A3B1-P	I	S	I	I	I
A3B2-P	I	S	I	I	I
A3B3-P	I	S	I	I	I

*VS = 1% w/v, S = 0.1% w/v, I = insoluble.
This polymer is highly soluble in water, being impossible to isolate.

plastic covering. After the soluble modified polyanilines had been obtained, the solubility was tested by dissolving known weights of modified polymers in the solvents. The results of the solubility tests are shown in Table 13.1.

As can be seen, most materials are not soluble in common solvents. However, some are soluble in organic solvents or water. Those modified polyanilines soluble in acetone and toluene, good solvents for polymer deposition, are of special interest. It seems that phenyl and naphthyl moieties and the nitro group induce solubility in this solvent. The presence of anionic groups induces solubility in basic aqueous solution through dipole interactions with ionic groups in the polyelectrolyte.

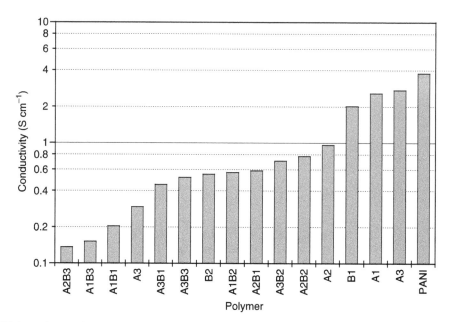

FIGURE 13.6 Conductivity of different modified polyanilines.

13.2.2.3   Measurement of Polymer Conductivity

While solubility is an important property for the polymer processing, modified polyanilines with good solubility but conductivities as low as 10^{-9} S/cm have been previously reported.[80] Therefore, the conductivity of the modified polymers has to be measured. The resistance of the films, deposited onto LDPE, was determined by the four-probe method. The thickness of the films necessary to calculate the conductivity was estimated using the relationship proposed previously.[82]

In Figure 13.6, the conductivities of the modified polymer films are shown. The conductivity of all modified polyanilines is lower than that of PANI. This is to be expected, due to a reduction in π-conjugation of the chain caused by steric effects from the attached groups.[81] Electron-withdrawing substituents in the polymer further decrease the conductivity by lowering the basicity of the imine units.[82] Indeed, it seems that bulky groups (e.g., naphthylamine) and strongly electron-withdrawing groups induce the maximum effect. However, the highest decrease measured is less than one order of magnitude. As the polymers are soluble and possess other valuable properties, the decrease in conductivity seems to be a reasonable trade-off.

13.2.2.4   Structure–Property Relationships in the Polymer Library

Since the chemical structure of the attached functional groups is known, it is possible to evaluate structure–property relationships, such as the effect of electronic density of the substitutents on the polymer conductivity. The electronic density could be evaluated by the σ parameter of Hammet.[83]

In Figure 13.7, the relationship between the Hammet parameter σ and electronic conductivity of some polymer films is shown.[84] It seems clear that a relationship between the structure of the attached functional group and the polymer conductivity exists.

13.2.2.5   Photochromic Conducting Polymers

As discussed above, the compounds containing azo groups could suffer photoisomerization upon illumination. Considering that the functional groups are bonded with the polyaniline backbone by

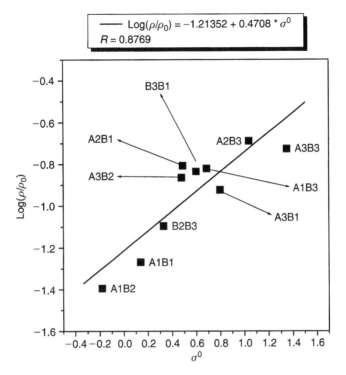

FIGURE 13.7 Relationship between the logarithm of the conductivity and the electronic Hammet parameter. Conductivity was normalized to that of unmodified polyaniline.

azo group, it is likely that modified polyanilines could be photochromic. Photochromic effects have been detected in polyanilines with azo groups attached to the amine nitrogen,[85] present in the counterions,[86] or in the backbone.[87] Since azo groups are attached to the PANI backbone in the soluble polyanilines described above, it is worthwhile looking for photochromic effects. Using the HTS procedure, developed to detect photochromism in dyes only on those polyanilines which are soluble, it was possible to find photochromic polyanilines. Interestingly, the photochromism remains when the polymers are coated onto transparent substrates (Figure 13.8).

In view of the fact that the dipole moment of the attached functional group affects the conductivity (Figure 13.7), it is probable that the change of molecular structure during photoisomerization could have some effect on the polymer conductivity. Nevertheless, it was found that polyanilines modified with strongly electron withdrawing or releasing groups show negligible conductivity changes upon illumination. On the other hand, polymers bearing bulky groups (e.g., naphthylamino) showed clearer effects. It was rationalized that bulky groups interfere with electron hopping between chains, affecting film conductivity. Accordingly, a polyaniline modified with a long chain by coupling with the diazonium salt of 4-dodecylaniline, shows a clear reversible change of film conductivity (Figure 13.9).

It is noteworthy to mention that this material was not created by a purely combinatorial method. However, it was obtained by designing a chemical structure with the knowledge acquired using combinatorially produced chemical libraries. This fact suggests that the best method to produce novel materials is through a proper combination of combinatorial chemistry, physicochemical studies, and chemical structure design.

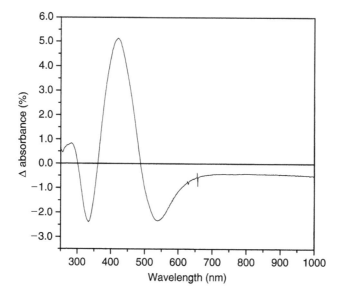

FIGURE 13.8 Differential UV–visible spectrum of a modified polyaniline (A1B1-PANI) film on LDPE upon illumination. The spectrum of the film taken in the dark is used as reference.

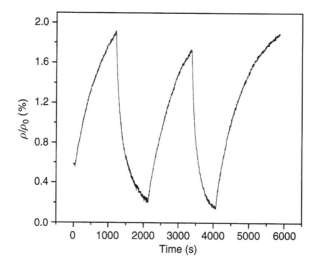

FIGURE 13.9 Film resistance change upon illumination of a modified polyaniline.

13.3 CONCLUSIONS

It has been shown that synthesis of photochromic azo dyes and soluble polyanilines using combinatorially synthesized diazonium salts is possible.

A clear photochromic effect, in the visible range that lasts several minutes, was detected in some of the dyes synthesized. Using such effect, an effective photowriting procedure was devised.

Several members of the combinatorial library originated from six aromatic amines and one polymer (PANI) present solubility in common solvents, including some not previously known to dissolve modified polyanilines. While the conductivity of modified polymers is lower than that of

PANI, in most cases it is only one order of magnitude lower. Therefore, the trade-off of solubility for conductivity seems to pay off. The conductivity is strongly affected by the electron density of the groups attached to the PANI backbone, as shown by the existence of a linear free-energy relationship between the Hammet electronic parameter and polymer conductivity. It was possible to design a conducting polyaniline, which changes electronic conductivity upon illumination, using the knowledge acquired by studying the photochromism of azo-modified polyanilines.

13.4 EXPERIMENTAL

The solvents used in spectroscopy are of HPLC grade (Sintorgan). All other reagents are of analytical quality.

13.4.1 POLYMER SYNTHESIS

Polyaniline (emeraldine form) was synthesized by oxidative polymerization of aniline,[88] by oxidation of aniline (0.1 M) in 1 M HCl with ammonium persulfate (equimolar) at temperatures below 5°C. The temperature is monitored during the polymerization. When the reaction is completed, after the maximum of temperature has occurred, the polymer is filtered out under suction and washed with 1 M HCl solution (500 ml) and water (1 L). The polymer is then converted into its base form by stirring for 24 h in a 0.1 M NH_4OH solution. The emeraldine base form (50% oxidation) used in all reactions has an intrinsic viscosity ($[\eta]$) of 1.12.

Films of polyethylene (LDPE, thickness 2 μm) were immersed in the polymerization solution to produce supported films of PANI. For LDPE films previously rendered hydrophilic by surface oxidation induced by immersing for 1 min in chromic acid solution,[89] it was found that the ICP film adhesion improves significantly after such treatment. The LDPE films were then washed with water and immersed in the polymerization media described above. The reaction was monitored by following the temperature as function of the reaction time.[90] Since the LDPE has few absorption bands in the mid-IR, and none in the UV–visible, the IR and UV–visible spectra of the supported film could be easily measured without cumbersome isolation of the products.

Poly(methylmethacrylate) (MW = 80,000) was purchased from Scientific Polymer Products and dissolved in chloroform without purification.

13.4.2 DYE SYNTHESIS

The azo dyes were synthesized in solution. The amines were diazotized with sodium nitrite and concentrated HCl in an ice bath. Special provisions were made with *p*-aminosulfonic and *p*-aminobenzoic acids which were dissolved initially in sodium carbonate solution. The coupling amine was dissolved in Tris(hydroxymethyl)aminomethane (TRIS®) buffer (pH = 8) and mixed with the diazonium salt solution in an ice bath. The dye precipitated and was washed with distilled water.

13.4.3 POLYANILINE MODIFICATION

All reactive dyes, bearing a primary aromatic amino group, were diazotized with sodium nitrite and concentrated HCl in an ice bath. PANI was suspended in phosphate buffer (pH = 8) and mixed with the diazonium salt solution in an ice bath. The solid was filtered under vacuum and washed first with 1 L of 1 M HCl solution and then with 1 L distilled water. The products were filtered out of the mixture under vacuum and dried (dynamic vacuum for 48 h). In some cases the modified polymer remained suspended in water and could not be filtered out. Therefore, centrifugation was used to separate the polymer. PANI films supported on LDPE were also treated with the diazonium salts, washed with acidic (1 M HCl) solution and water, and dried under vacuum.

13.4.4 SOLUBILITY MEASUREMENTS

One hundred mg of each polymer product were mixed with 10 mL of dissolving solution in two groups of 9 × 5 square set of test tubes. The whole set was subjected to stirring. The absence of remaining solid was considered as proof of solubility at the given concentration.

13.4.5 CONDUCTIVITY MEASUREMENTS

Conductivity measurements were carried out using the Van der Pauw method on polymer films deposited onto LDPE. The samples were first doped by a 1 h immersion in aqueous 1 M HCl and then dried in vacuum during 24 h. Four pads were painted on the films with conductive paint. The current was applied with a galvanostat (Amel 2049) while the voltage was measured with a 6½ digit multimeter (HP).

13.4.6 FTIR SPECTROSCOPY

Fourier transform infrared spectroscopy (FITR) spectra were performed in a Nicolet Impact 400 spectrometer. To obtain FTIR spectra of PANI and PANI modified, the LDPE-polymer films were previously washed with 1 M HCl solution and later dried under vacuum.

13.4.7 UV–VISIBLE ABSORPTION SPECTROSCOPY

UV-vis spectra were performed in a Hewlett Packard 8453 spectrometer. To obtain the spectra of PANI and PANI modified, the LDPE-polymer films were treated as described above.

13.4.8 UV–VISIBLE EMISSION (FLUORESCENCE) SPECTROSCOPY

The fluorescence measurements of polymer solutions, in quartz cells, were made in a Spex Fluoromax Fluorometer.

ACKNOWLEDGMENTS

The authors acknowledge the support of CONICET, FONCYT, Agencia Córdoba Ciencia and SECYT-UNRC. D.F. Acevedo thanks CONICET for a fellowship. C. Barbero is a research fellow of CONICET. The donation of aromatic amines by Vilmax S.A. is gratefully acknowledged.

REFERENCES

1. Shaheen, S. E., Ginley, D. S. and Jabbour, G. E., Toward low–cost power generation, *M.R.S. Bulletin*, 30, 10, 2005.
2. Wilson, R. and Czarnik, A. W., Eds, *Combinatorial Chemistry Synthesis and Application*, 1st ed, J. Wiley & Sons, New York, 1997, Ch. 1.
3. Obrecht, D. and Villalgordo, J. M., *Solid-Supported Combinatorial Synthesis of Small-Molecular-Weight Compound Libraries*, 1st ed, Pergamon, New York, 1998, Ch. 1.
4. Jandeleit, B., Schefer, D. J., Powers, T-S., Turner, H. W. and Weinberg, W. H., Combinatorial materials science and catalysis, *Angew. Chem. Int. Ed.*, 38, 2494, 1999.
5. Brocchini, S., James, K., Tangpasuthadol, V. and Kohn, J., A combinatorial approach for polymer design, *J. Am. Chem. Soc.*, 119, 4553, 1997.
6. Kim, D. Y. and Dordick, J. S., Combinatorial array-based enzymatic polyester synthesis, *Biotechnol. Bioenerg.*, 76, 200, 2001.

7. Schmatloch, S., Meier, M. R. and Schubert, U. S., Instrumentation for combinatorial and high-throughput polymer research: A short overview, *Macromol. Rapid. Commun.*, 24, 33, 2003, and refs therein.

8. Obrecht, D. and Villalgordo, J. M., *Solid-Supported Combinatorial Synthesis of Small-Molecular-Weight Compound Libraries*, 1st ed, Pergamon, New York, 1998, p. 105.

9. Sun, Y., Chan, B. C., Ramnarayanan, R., Leventry, W. M., Mallouk, T. E., Bareand, S. R., Willis, R. R., Split-pool method for synthesis of solid-state material combinatorial libraries, *J. Comb. Chem.*, 4, 569, 2002.

10. Crawse, J. M., Ed., Experimental Design for Combinatorial and High Throughput Materials Development, 1st ed, Wiley, New York, 2002, p. 129.

11. Potyrailo, R. A., Chisholm, B. J., Morris, W. G., Cawse, J. N., Flanagan, W. P., Hassib, L., Molaison, C. A., Ezbiansky, K., Medford, G. and Reitz, H., Development of combinatorial chemistry methods for coatings: High-throughput adhesion evaluation and scale-up of combinatorial leads, *J. Comb. Chem.*, 5, 472, 2003.

12. Moorey, W. F., Photochemistry and photophysics of surfactant trans-stilbenes in supported multilayers and films at the air-water interface, *J. Am. Chem. Soc.*, 106, 5659, 1984.

13. Rau, H., *Photochemistry and Photophysics*; col 2, Rabek, J. K., Ed., CRC Press, Boca Raton, FL, 1990, Vol. 2, p. 119.

14. Kawamura, S., Tsutsui, T., Salto, S., Murao, Y. and Kina, K., Photochromism of salicylideneanilines incorporated in a Langmuir-Blodgett multilayer, *J. Am. Chem. Soc.*, 110, 509, 1988.

15. Ashwell, G. J., Photochromic and non-linear optical properties of $C_{16}H_{33}$-P_3CNQ and $C_{16}H_{33}$-Q_3CNQ Langmuir-Blodgett films, *Thin Solid Films*, 186, 155, 1990.

16. Manda, N. H., Yabe, A. and Kawabata, Y., Photochemical switching in conductive Langmuir-Blodgett films, *J. Am. Chem. Soc.*, 111, 3080, 1989.

17. Liu, Z. F., Hashimoto, K. and Fujishima, A., Photoelectrochemical information storage using an azobenzene derivative, *Nature,* 347, 658, 1990.

18. Barrett, C., Natansohn, A. and Rochon, P., Azo polymers for reversible optical storage the effect of polarity of the azobenzene groups, *Chem. Mater.*, 7, 899, 1995.

19. Natansohn, A. and Rochon, P., Photoinduced motions in azo-containing polymers, *Chem. Rev.*, 102, 4139, 2002.

20. Zollinger, H., *Diazo Chemistry: Aromatic and Heteroaromatic Compounds,* 1st ed., VCH, Weinheim, Germany, 1994, Ch. 1.

21. Shirakawa, H., Louis, E. J., MacDiarmid, A. G., Chiang, C. K. and Heeger, A. J., Synthesis of electrically conducting organic polymers: halogen derivatives of polyacetylene, $(CH)_x$, *J. C. S. Chem. Commun.*, 16, 578, 1977.

22. Heeger, A. J., "Synthetic metals": A novel role for organic polymers (Nobel lecture), *Angewandte Chemie International Edition*, 40, 2591, 2001.

23. MacDiarmid, A. G., Semiconducting and metallic polymers: The fourth generation of polymeric materials (nobel lecture), *Angewandte Chemie International Edition*, 40, 2581, 2001.

24. Bredas, J. L. and Silbey S., *Conjugated Polymers: The Novel Science and Technology of Highly Conducting and Nonlinear Optically Active Materials*, 1st ed., Kluwer Academic Publishers, Amsterdam, 1991.

25. Skotheim, T. A., Elsenbaumer, R. L. and Reynolds, J. R., *Handbook of Conducting Polymers*, 2nd ed., Marcel Dekker, New York, 1998, Ch. 29.

26. Orata, D. and Buttry, D. A., Determination of ion populations and solvent content as functions of redox state and pH in polyaniline, *J. Am. Chem. Soc.*, 109, 3574, 1987.

27. Cui, S. Y. and Park, S., Electrochemistry of conductive polymers XXIII: polyaniline growth studied by electrochemical quartz crystal microbalance measurements, *Synth. Met.*, 105, 91, 1999.

28. Zimmermann, A. and Dunsch, L., Investigation of the electropolymerization of aniline by the in situ techniques of attenuated total reflection (ATR) and external reflection (IRRAS), *J. Mol. Struct.,* 165, 410, 1997.

29. Kpzoel, K., Lapkowski, M. and Genies, E., ESR study of polyaniline doping at various concentrations of anions, *Synth. Met.*, 84, 105, 1997.

30. Nyffenegger, R., Ammann, E., Siegenthaler, H. and Haas, O., In-situ scanning probe microscopy for the measurement of thickness changes in an electroactive polymer, *Electrochim. Acta*, 40, 1411, 1995.

31. Baba, A., Advincula, R. C. and Knoll, W., In situ investigations on the electrochemical polymerization and properties of polyaniline thin films by surface plasmon optical techniques, *J. Phys. Chem. B*, 106, 1581, 2002.

32. Morales, G. M., Miras, M. C. and Barbero, C., Anion effects on aniline polymerisation, *Synth. Met.*, 101, 686, 1999.

33. Sbaite, P., Huerta-Vilca, D., Barbero, C. Miras, M. C. and Motheo, A. J., Effect of electrolyte on the chemical polymerization of aniline, *Eur. Pol. Jour.*, 40, 1445, 2004.

34. Stejskal, V. and Gilbert, R. G., Polyaniline. Preparation of a conducting polymer (IUPAC technical report), *Pure Appl. Chem.*, 74, 857, 2002.

35. Sari, B. and Talu, M., Electrochemical copolymerization of pyrrole and aniline, *Synth. Met.*, 94, 221, 1998.

36. Ram, M. K., Sarkar, N., Ding, H. and Nicolini, C., Synthesis of controlled copolymerisation of aniline and *ortho*-anisidine: a physical insight in its Langmuir–Schaefer films, *Synth. Met.*, 123, 197, 2001.

37. Thiemann, C. and Brett, C. M. A., Electrosynthesis and properties of conducting polymers derived from aminobenzoic acids and from aminobenzoic acids and aniline, *Synth. Met.*, 123, 1, 2001.

38. Salavagione, H., Acevedo, D. F., Miras, M. C., Motheo, A. J. and Barbero, C. A., Comparative study of 2-amino and 3-aminobenzoic acid copolymerization with aniline synthesis and copolymer properties, *J. Pol. Sci.: Part A: Pol. Chem.*, 42, 5587, 2004.

39. Yue, J., Wang, Z. H., Cromack, K. R., Epstein, A. J. and MacDiarmid, A. G., Effect of sulfonic acid group on polyaniline backbone., *J. Am. Chem. Soc.*, 113, 2665, 1991.

40. Wei, X. L., Wang, Y. Z., Long, S. M., Bobeczko, C. and Epstein, A. J., Synthesis and physical properties of highly sulfonated polyaniline, *J. Am. Chem. Soc.*, 118, 2545, 1996.

41. McCoy, C. H., Lorkovic, I. M. and Wrighton, M. S., Potential-dependent nucleophilicity of polyaniline, *J. Am. Chem. Soc.*, 117, 6934, 1995.

42. Han, C.-C., Lu, C.-H., Hong, S.-P. and Yang, K.-F., Highly conductive and thermally stable self-doping propylthiosulfonated polyanilines, *Macromolecules*, 36, 7908, 2003.

43. Han, C.-C., Hong, S.-P., Yang, K.-F; Bai, M.-Y., Lu, C.-H. and Huang, C.-S., Highly conductive new aniline copolymers containing butylthio substituent, *Macromolecules*, 34, 587, 2001.

44. Han, C.-C., Hong, S.-P., Yang, K.-F., Bai, M.-Y., Lu, C.-H. and Huang, C.-S., Combination of electrochemistry with concurrent reduction and substitution chemistry to provide a facile and versatile tool for preparing highly functionalized polyanilines, *Chem. Mater.*, 11, 480, 1999.

45. Han, C.-C. and Jeng, R.-C., Concurrent reduction and modification of polyaniline emeraldine base with pyrrolidine and other nucleophiles, *Chem. Commun.*, 2 (6), 553, 1997.

46. Lavastre, O., Illitchev, I., Jegou, G. and Dixneuf, P. H., Discovery of new fluorescent materials from fast synthesis and screening of conjugated polymers, *J. Am. Chem. Soc.*, 124, 5278, 2004.

47. Huang, S. and Tour, J. M., Rapid solid-phase synthesis of oligo(1,4-phenylene ethynylene)s by a divergent/convergent tripling strategy, *J. Am. Chem. Soc.*, 121, 4908, 1999.

48. Jones, L., Schumm, J. S. and Tour, J. M., Rapid solution and solid phase syntheses of oligo(1,4-phenylene ethynylene)s with thioester termini: molecular scale wires with alligator clips. Derivation of iterative reaction efficiencies on a polymer support, *J. Org. Chem.*, 62, 1388, 1997.

49. Huang, S. and Tour, J. M., Rapid solid-phase syntheses of conjugated homooligomers and [ab] alternating block cooligomers of precise length and constitution, *J. Org. Chem.*, 64, 8898, 1999.

50. Heeger, A. J. and Smith, P., *Conjugated Polymers: The Novel Science and Technology of Highly Conducting and Nonlinear Optically Active Materials*, Bredas, J. L., 2nd ed., Kluwer Academic Publishers, Amsterdam, 1991, p. 141.

51. Evans, G. P., *Advances in Electrochemical Science and Engineering,* Vol. 1, 1st ed., Gerischer, H. and Tobias, C. W., Eds., Weinheim, 1990, Ch. 1.

52. Genies, E. M., Boyle, A., Lapkowski, M. and Tsintavis, C., Polyaniline – A Historical Survey, *Synth. Met.*, 36, 139, 1990.

53. Barbero, C., Miras, M. C., Haas, O. and Kötz, R., Alteration of the ion exchange mechanism of an electroactive polymer by manipulation of the active site: Probe beam deflection and quartz crystal microbalance study of poly(aniline) and poly(N-methylaniline), *J. Electroanal. Chem.*, 310, 437, 1991.

54. Barbero, C. and Kötz, R., Electrochemical formation of a self-doped conductive polymer in the absence of a supporting electrolyte. The copolymerization of o-aminobenzenesulfonic acid and aniline, *Adv. Mater.*, 6, 577, 1994.

55. Barbero, C., Miras, M. C., Schnyder, B. and Kötz, R., Sulfonated polyaniline films as cation insertion electrodes for battery applications. Part 1—Structural and electrochemical characterization, *J. Mater. Chem.*, 4, 1775, 1994.

56. Salavagione, H. J., Morales, G. M., Miras, M. C. and Barbero, C., Comparative study of the ion exchange and electrochemical properties of sulfonated polyaniline (SPAN) and polyaniline (PANI), *Acta Polymerica*, 50, 40, 1999.

57. Barbero, C., Miras, M. C., Haas, O. and Kötz, R., Comparative study of the ion exchange and electrochemical properties of sulfonated polyaniline (SPAN) and polyaniline (PANI), *Synth. Met.*, 55, 1539, 1993.
58. Morales, G. M., Miras, M. C. and Barbero, C., Covalent chlorine incorporation during aniline polymerization, *Synth. Met.*, 101, 678, 1999.
59. Barbero, C., Morales, G. M., Grumelli, D., Planes, G., Salavagione, H., Marengo, C. R. and Miras, M. C., New methods of polyaniline functionalization, *Synth. Met.*, 101, 694, 1999.
60. Salavagione, H. J., Miras, M. C. and Barbero, C., Chemical lithography of a conductive polymer using a traceless removable group, *J. Am. Chem. Soc.*, 125, 5290, 2003.
61. Salavagione, H. J., Acevedo, D. F., Miras, M. C. and Barbero, C., Redox coupled ion exchange in copolymers of aniline with aminobenzoic acids, *Port. Electrochim. Acta*, 21, 245, 2003.
62. Grumelli, D. E., Forzani, E. S., Morales, G. M., Miras, M. C., Barbero, C. A. and Calvo, E. J., Microgravimetric study of electrochemically controlled nucleophilic addition of sulfite to polyaniline, *Langmuir*, 20, 2349, 2004.
63. Barbero, C., Salavagione, H. J., Acevedo, D. F., Grumelli, D. E., Garay, F., Planes, G. A., Morales, G. M. and Miras, M. C., Novel synthetic methods to produce functionalized conducting polymers I. Polyanilines, *Electrochimica Acta*, 49, 3671, 2004.
64. Hartmuth C., Kolb, M. G., Finn, K. and Sharpless, B., Click Chemistry: Diverse Chemical Function from a Few Good Reactions, *Angew. Chem. Int. Ed.*, 40, 2004, 2001.
65. Liu, G. and Freund, M. S., Nucleophilic substitution reactions of polyaniline with substituted benzenediazonium ions: a facile method for controlling the surface chemistry of conducting polymers, *Chem. Mat.*, 8, 1164, 1996.
66. Planes, G., Morales, G. M., Miras, M. C. and Barbero, C., Soluble and electroactive polyaniline obtained by coupling of 4-sulfobenzenediazonium ion and poly(N-methylaniline), *Synth. Met.*, 97, 223, 1998.
67. Sun, J., Cheng, L., Liu, F., Dong, S., Wang, Z., Zhang, X. and Shen, J., Covalently attached multilayer assemblies containing photoreactive diazo-resins and conducting polyaniline, *Colloids and Surfaces. A.*, 169, 209, 2000.
68. Zang, Y., Guan, Y., Liu, J., Xu, J. and Cao, W., Fabrication of covalently attached conducting multilayer self-assembly film of polyaniline by in situ coupling reaction, *Synth. Met.*, 128, 305, 2002.
69. Acevedo, D. F., DellaMea, J. M., Miras, M. C. and Barbero, C., Combinatorial synthesis of environmentally suitable ("green") azo dyes, In: *Green Chemistry Series N° 8*, Cerichelli, G., Tundo, P., Eds, INCAA, Venice, 2003, p. 103.
70. Acevedo, D. F., Miras, M. C., Planes, G. A. and Barbero, C., Síntesis de Polianilinas Modificadas con grupos azoicos, *Argentine Patent*, Nr. P020104980, 2003.
71. Ponzio, E. A., Echevarria, R., Morales, G. M. and Barbero, C., Removal of N-methylpyrrolidone hydrogen-bonded to polyaniline free standing films by protonation-deprotonation cycles or thermal heating, *Polym. Intl.*, 50, 1180, 2001 and refs therein.
72. Yin, W. and Ruckernstein, E., Water-Soluble Self-Doped Conducting Polyaniline Copolymer, *Macromolecules*, 33, 1129, 2000.
73. Hua, F. and Ruckenstein, E., Copolymers of aniline and 3-aminophenol derivatives with oligo(oxyethylene) side chains as novel water-soluble conducting polymers, *Macromolecules*, 37, 6104, 2004.
74. Furniss, B. S., Hannaford, A. J., Smith, P. W. G, Tatchell, A. R., *Vogel's Textbook of Practical Organic Chemistry*, 5th ed., Prentice Hall, Essex, England, 1989, p. 920.
75. Barrett, C., Natansohn, A. and Rochon, P., Thermal cis-trans isomerization rates of azobenzenes bound in the side chain of some copolymers and blends, *Macromolecules*, 27, 4781, 1994.
76. Smith, M. B., *Organic Synthesis*, McGraw-Hill, 1st Ed., New York, 1994, Ch. 5.
77. Acevedo, D. F., Miras, M. C. and Barbero, C., Solid support for high throughput screening of conducting polymers, *J. Comb. Chem.*, 7, 513–516, 2005.
78. Bandel, H., Haap, W. and Jung, J., *Combinatorial Chemistry*, 1st Ed., Jung, G., Ed., Wiley-VCH, Tübingen, The Netherlands, 1999, Ch. 1.
79. Leugers, A., Neithamer, D. R., Sun, L. S., Hetzner, J. E., Hilty, S., Hong, S., Krause, M. and Beyerlein, K., High-throughput analysis in catalysis research using novel approaches to transmission infrared spectroscopy, *J. Comb. Chem.*, 5, 238, 2003.
80. Hany, P., Genies, E. M. and Santier, C., Polyanilines with covalently bonded alkyl sulfonates as doping agent. Synthesis and properties, *Synth. Met.*, 31, 369, 1989.

81. Mav, I. and Zigon, M., Chemical copolymerization of aniline derivatives: Preparation of fully substituted PANI, *Synth. Met.*, 119, 145, 2001.
82. Yue, J., Wang, Z. H., Cromack, K. R., Epstein, A. J. and MacDiarmid, A. G., Effect of sulfonic acid group on polyaniline backbone, *J. Am. Chem. Soc.*, 113, 2665, 1991.
83. Hammett, L. P., *Physical Organic Chemistry*. McGraw Hill, New York, 1940, Ch. 3.
84. Ehrenson, S., Brownlee, R. T. C. and Taft, R. W., *Prog. Phys Org. Chem.* 10, 1, 1973.
85. Huang, K., Qiu, H. and Wan, M., Synthesis of highly conducting polyaniline with photochromic azobenzene side groups, *Macromolecules,* 35, 8653, 2002.
86. Huang, K. and Wan, M., Self-assembled polyaniline nanostructures with photoisomerization function, *Chem. Mater.,* 14, 3486, 2002.
87. Alva, K. S., Lee, T.-S., Kumar, J. and Tripathy, S. K., Enzymatically synthesized photodynamic polyaniline containing azobenzene groups, *Chem. Mater.*, 10, 1270, 1998.
88. MacDiarmid, A. G. and Epstein, A. J., Polyanilines: a novel class of conducting polymers, *Faraday Discuss. Chem. Soc.*, 88, 317, 1989.
89. Briggs, D. and Fairley, N., *Surface and Interface Analysis,* 33, 283, 2002.
90. Morales, G. M., Llusa, M., Miras, M. C. and Barbero C., Effects of high hydrochloric acid concentration on aniline chemical polymerization, *Polymer*, 38, 5247, 1997.

Section 4

Energy-Related Materials Development

14 High-Throughput Screening for Fuel Cell Technology

Jing Hua Liu, Min Ku Jeon, Asif Mahmood, and Seong Ihl Woo

CONTENTS

14.1 INTRODUCTION

Combinatorial chemistry and high-throughput screening have an enormous impact on drug discovery in the pharmaceutical industry and are complementary technologies for the preparation of a large number of formulations and their performance testing.[1–3] In the field of catalysis, combinatorial techniques have been proved to be an effective and efficient approach in screening the most active composition within a very short time. Since the activity of electrocatalysts is one of the most important factors in a fuel cell, the adoption of combinatorial methods and high-throughput screening in the study of fuel cells will definitely be helpful in the discovery of new and active catalyst materials suitable for fuel cell applications.

After nearly three decades of research and optimization, PtRu has become one of the best-known anodic materials[4–8] and Pt the best cathodic material. However, since the first introduction of combinatorial methods to the fuel cell research by Reddington et al.,[9] many kinds of new catalyst materials have been found which possess better properties. Some of these materials are comprised of up to four components. As a result of the technique of high-throughput screening, the period for the optimization of material compositions is greatly reduced, compared to the conventional one-by-one screening methods. Using the high-throughput screening, precise distinction can be made since all samples are measured simultaneously or successively at the same conditions and thus the systematic error is reduced dramatically. Combined with the traditional electrochemical measurements, the high activity of these new catalyst materials discovered by high-throughput screening can be validated.

Since the first introduction of optical screening by Reddington et al., many other screening methods have been adopted. In this review, we have classified the combinatorial study for fuel cell technology based on different screening techniques, i.e., optical screening, electrochemical screening, scanning electrochemical microscopy, and IR thermography screening.

14.2   OPTICAL SCREENING

Reddington et al.[9] first reported combinatorial electrochemistry by using quinine as a fluorescent indicator, which is luminescent in its acidic form but not in its basic form. Figure 14.1[10] shows the screening result, which indicates that Pt(44)Ru(41)Os(10)Ir(5) is more active than commercial Pt(50)Ru(50) black in direct methanol fuel cell (DMFC). The principle of this optical screening can be explained as follows: the most active catalyst compositions produce the greatest local pH change at the lowest overpotential. Thus the compositions that catalyze the anode reaction at a given applied potential generate a more acidic pH in the area immediately surrounding the electrode spot than less-active compositions, and the compositions that catalyze the oxygen reduction reaction create a more alkaline pH. By using an indicator that is fluorescent in the appropriate form, one can determine optically which areas of composition space are the most active for a particular anode or cathode reaction.

Choi et al.[11] suggested a very active and stable quaternary electrocatalyst, Pt(77)Ru(17)Mo(4) W(2), which was found by high-throughput screening after repeated cyclic voltammetry experiments. A calibrated ink-jet printer was used for synthesizing the combinatorial libraries, as shown in Figure 14.2. Figure 14.3 shows the screening results by fluorescence microscopy at low and high potential. This method expanded the combinatorial electrochemistry by using fluorescence indicators into a screening tool to find the catalyst that not only showed an initial excellent performance but also had a stable activity in the long-term operation. However, it was very difficult

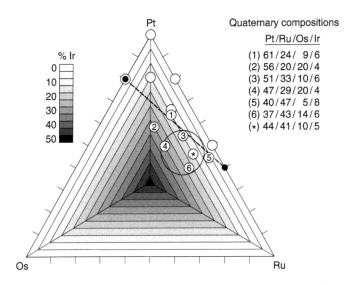

FIGURE 14.1 Pt-Ru-Os-Ir composition map, looking through the Pt-Ru-Os ternary face. Increasing Ir concentration is shown in gray scale. Black dots on the Pt-Ru and Pt-Os axes show the binary solubility limits, and the single-phase ternary region is approximately defined by the area above the dashed line joining these binary limits. Anode catalyst compositions are indicated by open circles, and the region of highest catalytic activity is shown as a circle containing the best catalyst found by combinatorial screening, $Pt_{44}Ru_{41}Os_{10}Ir_5$ (*).

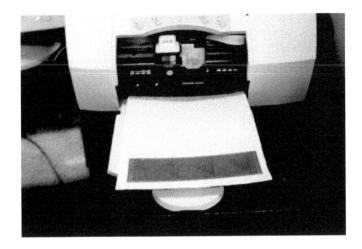

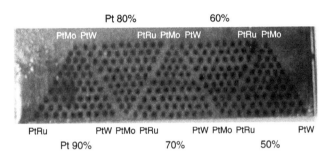

FIGURE 14.2 Combinatorial array of catalysts prepared by a modified ink-jet printer.

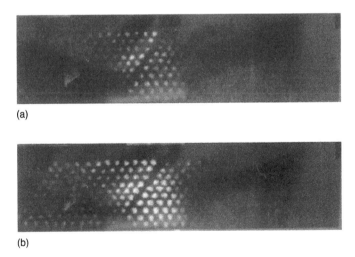

FIGURE 14.3 Combinatorial array and screening results by fluorescence imaging. Active compositions for methanol electro-oxidation are shown as bright spots: fluorescence image (a) at lower overpotential (≈ 0.4 V vs. RHE) and (b) at higher overpotential (≈ 0.46 vs. RHE) after the 20th cyclic voltammetry loop.

(a)

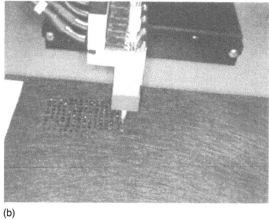

(b)

FIGURE 14.4 Preparation of combinatorial array of catalysts using a micro-liquid dispensing system (Cartesian Technology, Inc.) (a) Preparation of the stock solutions, delivered from four metal precursor solutions, in a 384-well plate. (b) Preparation of combinatorial array on a sheet of carbon paper using the stock solutions.

to prepare a homogeneous phase in a dot of the combinatorial array using an ink-jet printer. Therefore, we used a micro-liquid-dispensing system, as shown in Figure 14.4. We first prepared 645 stock solutions in a 384-well plate and treated them in an ultrasonic bath for obtaining the homogeneous phase. Then the solutions of metal precursor were dispensed on carbon paper. Finally the precursor-loaded carbon paper was heat treated in a H_2/N_2 atmosphere for obtaining metals and metal alloys. The as-prepared array containing 645 compositions is shown in Figure 14.5.

Since the polymer membrane is a solid phase, it does not penetrate deeply into the electrode as does a liquid and the reaction area is limited to the contact surface between the electrode and the membrane. To increase this contact surface area, an ionomer like Nafion was impregnated into the catalytic layer.[12,13] Chu et al.[14] reported a combinatorial screening result that evaluated the influence of the Nafion ionomer on the PtRu alloy for methanol electro-oxidation by using acid indicators. Anodes with different molar ratio of Pt to Ru and Nafion loading were prepared, as shown in Figure 14.6. The fluorescence images revealed that the most active composition of the Pt–Ru alloy electrocatalysts for electro-oxidation of methanol was Pt(54.5)Ru(45.5), and the most active

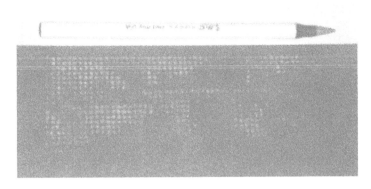

FIGURE 14.5 Combinatorial array (645 compositions) synthesized using micro-liquid dispensing system.

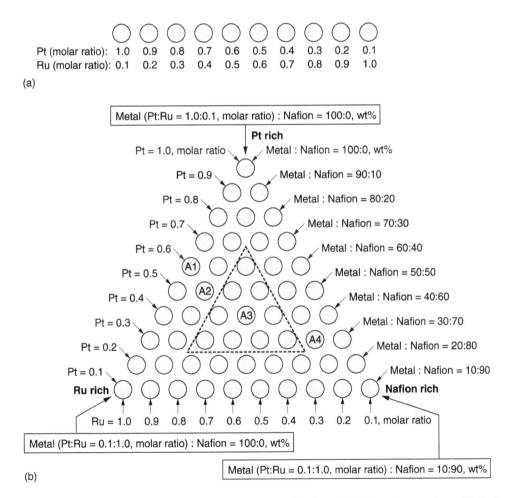

| Pt (molar ratio): | 1.0 | 0.9 | 0.8 | 0.7 | 0.6 | 0.5 | 0.4 | 0.3 | 0.2 | 0.1 |
| Ru (molar ratio): | 0.1 | 0.2 | 0.3 | 0.4 | 0.5 | 0.6 | 0.7 | 0.8 | 0.9 | 1.0 |

(a)

Metal (Pt:Ru = 1.0:0.1, molar ratio) : Nafion = 100:0, wt%

Pt rich

Pt = 1.0, molar ratio   Metal : Nafion = 100:0, wt%
Pt = 0.9   Metal : Nafion = 90:10
Pt = 0.8   Metal : Nafion = 80:20
Pt = 0.7   Metal : Nafion = 70:30
Pt = 0.6   Metal : Nafion = 60:40
Pt = 0.5   Metal : Nafion = 50:50
Pt = 0.4   Metal : Nafion = 40:60
Pt = 0.3   Metal : Nafion = 30:70
Pt = 0.2   Metal : Nafion = 20:80
Pt = 0.1   Metal : Nafion = 10:90

Ru rich   **Nafion rich**

Ru = 1.0   0.9   0.8   0.7   0.6   0.5   0.4   0.3   0.2   0.1, molar ratio

Metal (Pt:Ru = 0.1:1.0, molar ratio) : Nafion = 100:0, wt%

Metal (Pt:Ru = 0.1:1.0, molar ratio) : Nafion = 10:90, wt%

(b)

FIGURE 14.6 Combinatorial array for methanol electro-oxidation; (a) Pt–Ru alloy catalyst, (b) Pt–Ru–Nafion ternary catalyst. Note: inside of dotted area denotes the active compositions for methanol electro-oxidation.

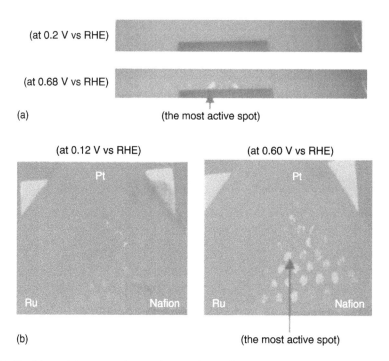

FIGURE 14.7 Combinatorial array and screening results by fluorescence imaging. Active compositions for methanol electro-oxidation are shown as bright spots; (a) Pt–Ru, (b) Pt–Ru–Nafion.

composition of the Pt–Ru–Nafion electrocatalysts for electro-oxidation of methanol was metal (Pt:Ru = 1:1, molar ratio):Nafion = 63.6:36.4, wt.%, as can be seen in Figure 14.7.

Besides the optical screening of anodic catalyst, Chen et al. also used Ni–PTP (Ni^{2+} complexed with 3-pyridine-2-yl-(4,5,6)triazolo-(1,5-a)pyridine) and Phloxine B as proton indicators for the reaction of oxygen evolution and reduction, respectively, in the optical screening of bifunctional catalyst for regenerative fuel cells.[15] Figure 14.8, Figure 14.9 and Figure 14.10 show the optical screening results, which indicate that the ternary catalyst Pt(4.5)Ru(4)Ir(0.5) is superior to the PtIr binary catalyst for both the water oxidation and oxygen reduction reactions.

14.3 ELECTROCHEMICAL SCREENING

While fluorescent screening provides a rapid identification of active zones of a library, experimental complications associated with eliminating background fluorescence, lateral spreading of the fluorescent region, and insensitivity to small differences in activity can limit their applicability. Furthermore, the need to couple fluorescence to an electrochemically generated (or consumed) secondary product (or reactant) limits its versatility. To solve these problems, Sullivan et al.[16] used automated serial electrochemical measurements, which can distinguish small differences in activity. This technique is not limited for the detection of a secondary reactant, and does not require continuous operator input. This suggests a combinatorial approach that combines both coarse fluorescent screening on large arrays to identify active compositional zones, followed by a finer automated electrochemical screen of these zones to identify the optimum electrode materials.

In order to execute a combinatorial approach for discovery of novel electrocatalyst, DuPont has developed rapid half-cell screening techniques to evaluate catalyst activity for methanol oxidation and oxygen reduction.[17] Because these measurements are performed in a liquid electrolyte and

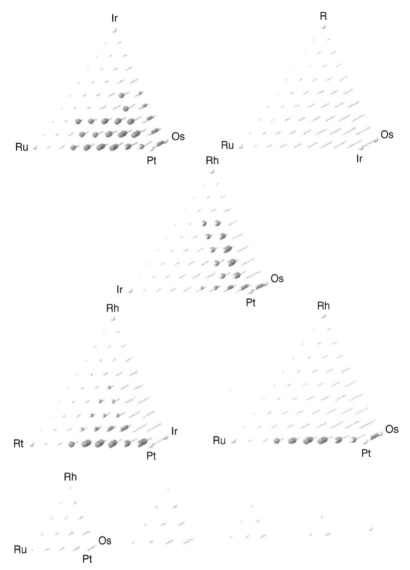

FIGURE 14.8 Catalyst activity map for oxygen reduction. Large tetrahedra represent quaternary regions of the map, with ternary faces, binary edges, and elemental vertices. The five smaller tetrahedra at the bottom of the figure are tetrahedral sections of the pentanary region of the composition map, with Ir content increasing progressively from left to right, from 11 to 55%. Larger gray spheres indicate compositions that gave visible fluorescence from Phloxine B at +550 mV, and smaller gray spheres indicate those that were slightly less active (fluorescence at +500 mV).

32 electrodes can be screened in parallel in a three-electrode configuration, it is possible to generate quantitative electrocatalytic activities by using this experimental setup.

Guerin et al.[18] developed hardware and software for fast sequential measurements of cyclic voltammetric and steady-state currents in 64-element half-cell arrays. Figure 14.11 and Figure 14.12 shows the schematics of the testing system and the cell. The analysis software developed enables the semiautomated processing of the large quantities of data. For example, filters may be applied to define figures of merit relevant to fuel cell catalyst activity and tolerance. This means that peak

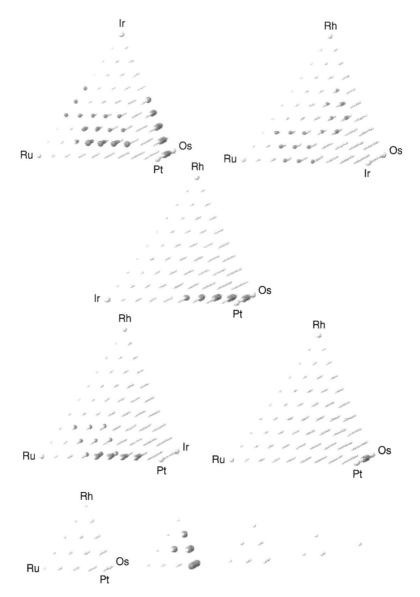

FIGURE 14.9 Catalyst activity map for oxygen evolution. Large and small gray spheres indicate Ni–PTP fluorescence at $+1350$ and $+1400$ mV, respectively.

potentials, currents, and charges can be measured for all of the electrodes in a single step. Using this method, the structure of small particles was found to be associated with edge sites in addition to the closely packed planes, from the results of CO stripping voltammetry. Specific activity for steady-state methanol oxidation and oxygen reduction at room temperature in H_2SO_4 electrolyte was found to be a maximum for the largest particle sizes.

Liu et al.[19] developed a device for high-throughput screening of membrane electrode assembly (MEA), which has no moving parts and state-of-art flow field delivery of reactant streams to the porous electrode surfaces. As shown in Figure 14.13, Figure 14.14 and Figure 14.15, the device accommodates an array of 25 spots permitting repetitive testing of catalysts.

Strasser et al.[20] fabricated an array of 64 individually addressable, circular electrode pads. They were fabricated by using lithographic techniques on an insulating $3''$ quartz wafer, and then

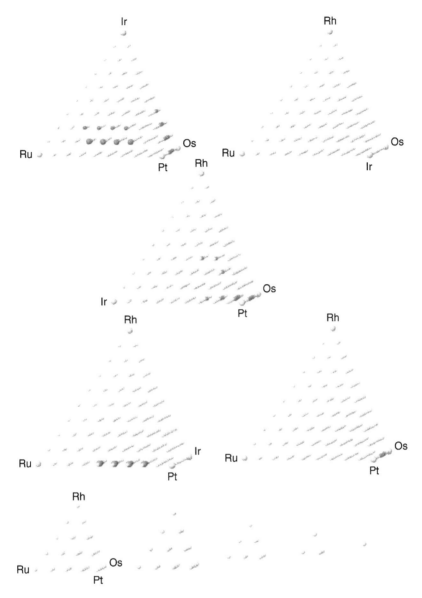

FIGURE 14.10 Consensus map for bifunctional catalyst activity. Larger gray spheres indicate compositions that had the highest level of activity in Figure 14.8 and Figure 14.9. Note that the Pt–Ru–Ir ternary appears on the ternary faces of the two quaternary regions, so there is actually only one region of highest bifunctional activity. Smaller gray spheres indicate combinations of highest and next highest activity rankings from Figure 14.8 and Figure 14.9.

depositing the individual constituents of the synthesized catalyst composition onto the 64-element electrode array through contact masks by using automated rf-magnetron vacuum sputtering (demonstrated in Figure 14.16 and Figure 14.17). Using a sputter-thin-film approach, the influence of unconcerned parameters was neglected and the alloy stoichiometry was precisely controlled. The dense and smooth film makes the geometric area of the electrode equal to the real surface area, which facilitates the comparison. Combined with a cylindrical cell body, the array of electrocatalysts was screened by electrochemical characterization using a multichannel potentiostat/galvanostat.

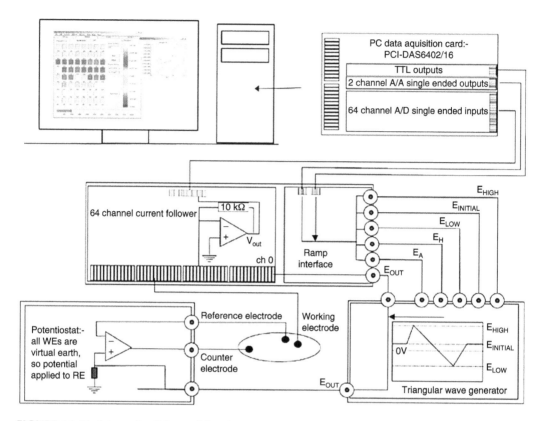

FIGURE 14.11 Schematic of the modular electrochemical control system developed for the combinatorial electrochemical screening experiments. There are several modules depicted above. Among them are the computer for data acquisition (top), a 64-channel current follower (center), a potentiostat (bottom left), a triangular sweep generator (bottom right), and an electrochemical array cell (bottom center).

The study showed that the observed activity of Pt/Ru/Co 1:1:3 was almost eight times as high as the PtRu standard catalysts and about 37 times higher than the Pt standard catalyst.

In order to screen different types and concentrations of electrolyte and reactants, Jiang et al.[21] and Moore et al.[22] demonstrated a movable electrolyte probe (shown in Figure 14.18) that can be conveniently used to form an electrochemical cell with counter, working, and reference electrodes to fast-screen the array of working electrodes for combinatorial analysis. This method is suitable not only for half-cell but also for full-cell research. They investigated the effects of catalyst loading and methanol concentrations and the results showed that 1 mol L^{-1} methanol was suitable for DMFC. This method can be used not only for identifying the best candidate from a large number of new electrochemical materials but also for conducting systematic basic electrochemical research by rapid screening of the arrays of electrodes containing different catalysts, electrolytes, solutions, and reactants for various parallel experiments.

Recently Stevens et al.[23] designed a 64-channel combinatorial infrastructure and measured the performance of compositional arrays of oxygen reduction catalyst for proton exchange membrane fuel cell (PEMFC) under "real world" conditions. Cyclic voltammograms were measured on sputtered film with constant platinum loading at each channel. The results (Figure 14.19) indicated that the data showed excellent channel-to-channel repeatability, with all channels showing CV features typical for nano-particulate platinum. Therefore, high-throughput screening of a real cell can be obtained more rapidly and precisely using a multichannel potentiostat.

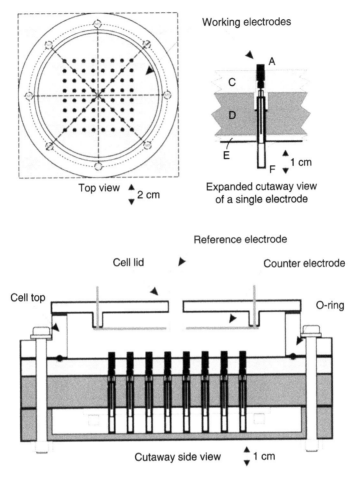

FIGURE 14.12 Schematic of the cell (viewed from the top, the side, and an expanded view of a single electrode) used for electrochemical screening of the 64-element array: vitreous carbon electrode (A), spring-loaded electrical contact, glass filled PTFE array base plate (C), polypropylene contact holder (D), and a printed circuit board (E).

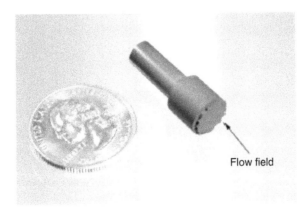

FIGURE 14.13 Sensor electrode.

FIGURE 14.14 Ceramic sensor electrode array flow fields.

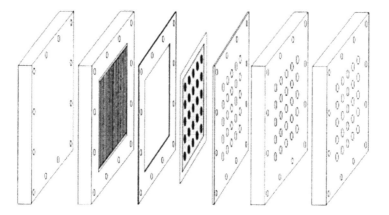

FIGURE 14.15 Schematic of exploded high-throughput device.

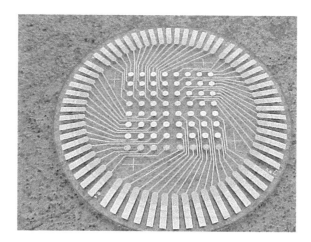

FIGURE 14.16 A 64-element addressable electrode array on a 3″ quartz wafer is used as support for the synthesis of chemically diverse electrocatalyst libraries.

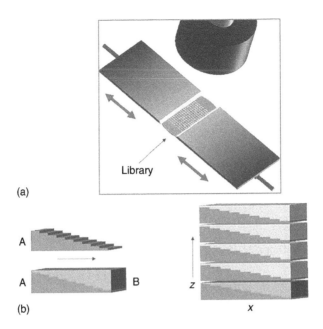

FIGURE 14.17 (a) An Rf magnetron sputter technique with automated target selection as well as automated moving shutters and physical shadow masking is employed for deposition of the electrocatalyst libraries. (b) Sequential gradient sputtering of very thin material slabs lead to in-situ thin-film alloy formation on the contact spots. The width of the slabs on the left extends over the entire width of the electrode array. Superlattice deposition in the z-direction results in the buildup of thicker films of the desired stoichiometric gradient.

14.4   SCANNING ELECTROCHEMICAL MICROSCOPY

Jayaraman et al.[24] used scanning electrochemical microscope (SCEM) as a screening tool for screening of fuel cell catalysts of various compositions. The combinatorial array was fabricated by electrochemical deposition onto multielement electrode band samples (shown in Figure 14.20). The principle of SCEM can be explained as follows (shown in Figure 14.21): the potential of the tip electrode was held at −1.0 V vs. RHE, at which the reaction is the diffusion-limited reduction of protons to hydrogen.

$$2H^+ + 2e^- \rightarrow H_2$$

When the tip electrode is positioned near the catalyst area on the substrate, the tip current increases because hydrogen produced at the tip is oxidized at the catalyst and the generated protons diffuse back to the tip and reduce back to hydrogen, resulting in an increase in the tip current (termed positive feedback), which is a function of the rate constant for hydrogen oxidation at the catalyst. This technique is particularly useful for indirectly measuring the onset of hydrogen oxidation through a catalyst surface that has been poisoned with a strongly bound adsorbate such as carbon monoxide. The results (shown in Figure 14.22) showed the ability of Pt_xRu_y electrodes to oxidize hydrogen in the presence of a CO monolayer at a potential of 0.35 V, which is lower than that of pure Pt. This indicated the ability of Ru to dissociate water at potentials much lower than Pt. The study also demonstrated that Mo-containing electrodes improved the onset potential by an additional 0.2 V.

Jambunathan et al.[25] also used a combination of electrochemical and scanning differential electrochemical mass spectrometry (SDEMS) to characterize methanol oxidation on a series of electrodeposited Pt_xRu_y band electrodes to elucidate information about both the activity and the reaction pathway of methanol oxidation as a function of catalyst composition and temperature in a combinatorial fashion.

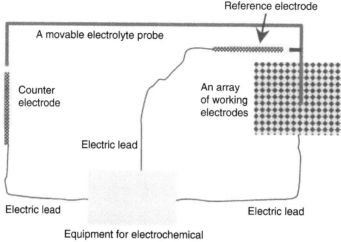

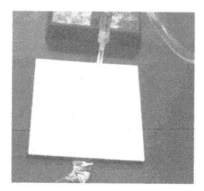

FIGURE 14.18 Upper figure, schematic view of combinatorial electrochemical measurements with an array of working electrodes and a movable electrolyte probe. Lower figure, a picture of an experimental array of working electrodes for combinatorial measurements.

Fernandez et al.[26] used a modified tip generation-substrate collection (TG-SC) mode of SECM technique, as illustrated in Figure 14.23. The electrocatalyst array was prepared by dispensing the corresponding precursor with a micropipette and then reducing with sodium borohydride, as shown in Figure 14.24. When an ultramicroelectrode (UME) tip with a constant oxidation current applied is placed close to the substrate by using the positioning control of the SECM, oxygen generated from the oxidation of water diffuses to the substrate and then reduces at the electrocatalyst spot. One advantage of this technique is that we only measure the oxygen reduction current, which is quantitative and direct, in order to test the catalytic activity. The analysis is fast and requires only a very small quantity of catalysts, which makes the method attractive for combinatorial analysis of multi-component electrode materials.

14.5   IR THERMOGRAPHY SCREENING

Yamada and his co-workers[27-29] used IR thermography for the rapid evaluation of the anode catalyst for PEMFC. The catalysts were placed in an aluminum vessel, which were then arrayed in a

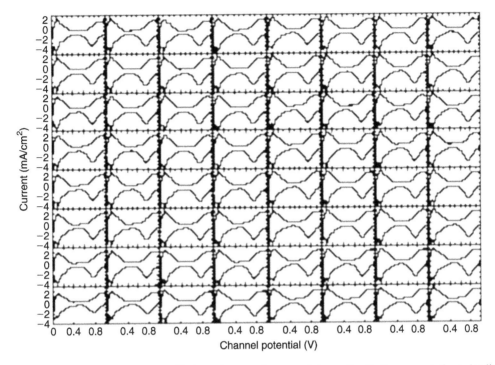

FIGURE 14.19 CVs measured on a 64-channel array of sputtered dots, all with the same platinum loading. The CVs were measured at 100 mV/s at room temperature with humidified argon flow over the cathode array and hydrogen over the anode. The anode is a continuous coating of platinum-loaded carbon in Nafion on carbon paper.

tube with a ZnSe window. The reaction gas was passed over the catalyst and the reaction heat was observed by IR thermography by using a NEC San-ei Instruments Thermo Tracer TS 7302. The experiments revealed that nine metal oxides could improve the CO tolerance of Pt/C in the gas phase (shown in Figure 14.25).

14.6 CONCLUSION

Even though from 1960 to now, the development of DMFC has rapidly progressed, there are many technical problems that hinder its commercialization. Among them, the most challenging is the enhanced electrocatalytic activity of each electrode. According to this point of view, combinatorial chemistry will play an important role in the development of DMFC. Until now, four different screening methods, i.e., optical screening, electrochemical screening, scanning electrochemical microscopy, and IR thermography screening have been used in the high-throughput screening of electrocatalysts for fuel cells. High-throughput screening of a combinatorial array that does not contain noble metals should also be tested.

ACKNOWLEDGMENTS

This research was funded by the Center for Ultramicrochemical Process Systems (CUPS) sponsored by Korea Science and Engineering Foundation (2005) by Korea Institute of Industrial Technology

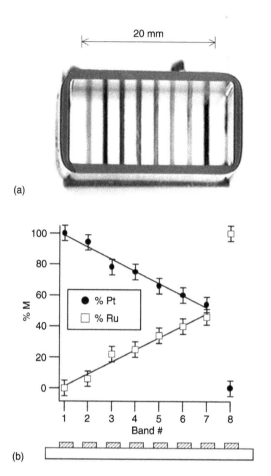

FIGURE 14.20 (a) Optical micrograph of eight-element band electrode. (b) Compositions of Pt$_x$Ru$_y$ band electrodes as determined by Auger electron spectroscopy (AES) with schematic of multielement band electrode (bottom).

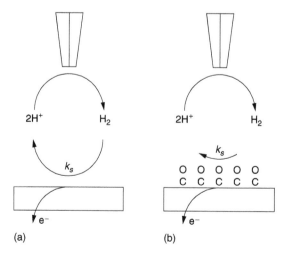

FIGURE 14.21 Schematic of tip–substrate interface during scanning electrochemical microscopy of the hydrogen oxidation reaction: (a) active substrate; (b) CO-covered substrate.

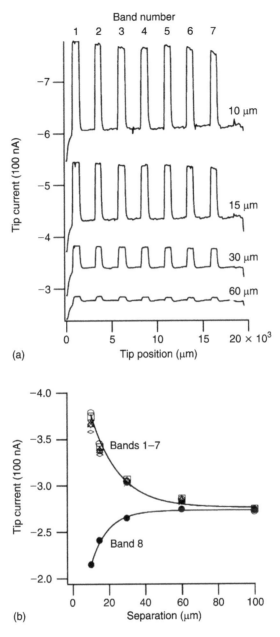

FIGURE 14.22 (a) SECM line scans over CO-coated Pt_xRu_y band electrodes at a tip–substrate separation of 15 mm in nitrogen-purged 0.01 M H_2SO_4/0.1 M Na_2SO_4 with the tip potential held at −1.0 V. Each curve is offset vertically for clarity. (b) Summary of onset potentials for hydrogen oxidation on CO-coated Pt_xRu_y electrodes. The solid line has been added to serve as a guide to the eye.

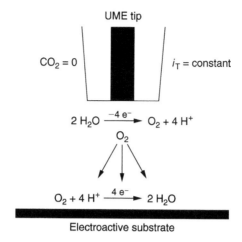

FIGURE 14.23 Scheme of the modified TG-SC mode for the study of the ORR in acidic medium.

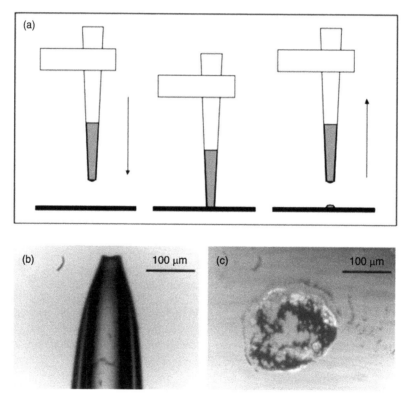

FIGURE 14.24 (a) Method used for the deposition of catalyst spots on the glassy carbon support. (b) Optical micrograph of the bottom end of a glass micropipette used for the preparation of spots. (c) Optical micrograph of a typical Pt spot supported on glassy carbon.

The arrangement of the Pt/C catalysts
with/without additives

Y   Ti   V   Ga
Nb

La   Zr   Ir   In
   Ta

(Pt/C)
Ca   Br   none

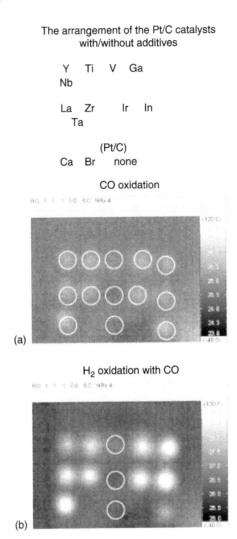

FIGURE 14.25 IR thermographs of (a) 1000 ppm CO in air passed over Pt/Cs with/without additives; and (b) 2000 ppm H_2 and 1000 ppm CO in air.

Evaluation and Planning (ITEP), Samsung Advanced Institute of Technology, LG Chem., Korea Storage Battery, Haesong P&C, and Hankook Tire.

REFERENCES

1. Cong, P., Doolen, R. D., Fan, Q., Giaquinta, D. M., Guan, S., McFarland, E. W., Poojary, D. M., Self, K., Turner, H. W. and Weinberg, W. H., High-throughput synthesis and screening of combinatorial heterogeneous catalyst libraries, *Angew. Chem. Int. Ed.,* 38, 483, 1999.
2. Pascarmona, P. P., van der Waal, J. C., Maxwell, I. E. and Maschmeyer, T., Combinatorial chemistry, high-speed screening and catalysis, *Catal. Lett.*, 63, 1, 1999.
3. Senkan, S. M., High-throughput screening of solid-state catalyst libraries, *Nature*, 394, 350, 1998.
4. Watanabe, M., Uchida, M. and Motoo, S., Preparation of highly dispersed Pt + Ru alloy clusters and the activity for the electrooxidation of methanol, *J. Electroanal. Chem.*, 229, 395, 1987.
5. Rolison, D. R., Hagans, P. L., Swider, K. E. and Long, J. W., Role of hydrous ruthenium oxide in Pt-Ru direct methanol fuel cell anode electrocatalysts: the importance of mixed electron/proton conductivity, *Langmuir*, 15, 774, 1999.

6. Gasteiger, H. A., Markovic, N., Ross, P. N. Jr. and Cairns, E. J., Temperature-dependent methanol electro-oxidation on well-characterized Pt-Ru alloys, *J. Electrochem. Soc.*, 141, 1795, 1994.
7. Liu, L., Pu, C., Viswanathan, R., Fan, Q., Liu, R. and Smotkin, E. S., Carbon supported and unsupported Pt-Ru anodes for liquid feed direct methanol fuel cells, *Electrochim. Acta*, 43, 3657, 1998.
8. Chu, D. and Gilman, S., Methanol electro-oxidation on unsupported Pt-Ru alloys at different temperatures, *J. Electrochem. Soc.,* 143, 1685, 1996.
9. Reddington, E., Sapienza, A., Gurau, B., Viswanathan, R., Sarangapani, S., Smotkin, E. S. and Mallouk, T. E., Combinatorial electrochemistry – a highly parallel, optical screening method for discovery of better electrocatalysts, *Science,* 280, 1735, 1998.
10. Gurau, B., Viswanathan, R., Liu, R., Lafrenz, T. J., Ley, K. L., Smotkin, E. S., Reddington, E., Sapienza, A., Chan, B. C., Mallouk, T. E. and Sarangapani, S., Structural and electrochemical characterization of binary, ternary, and quaternary platinum alloy catalysts for methanol electro-oxidation, *J. Phys. Chem. B,* 102, 9997, 1998.
11. Choi, W. C., Kim, J. D. and Woo, S. I., Quaternary Pt-based electrocatalyst for methanol oxidation by combinatorial electrochemistry, *Catal. Today,* 74, 235, 2002.
12. Ren, X., Wilson, M. S. and Gottesfeld, S., High performance direct methanol polymer electrolyte fuel cells, *J. Electrochem. Soc.,* 143, L12, 1996.
13. Murphy, O. J., Hitchens, G. D. and Manko, D. J., High power density proton-exchange membrane fuel cells, *J. Power Sources*, 47, 353, 1994.
14. Chu, Y.-H., Shul, Y. G., Choi, W. C., Woo, S. I. and Han, H.-S., Evaluation of the Nafion effect on the activity of Pt–Ru electrocatalysts for the electro-oxidation of methanol, *J. Power Sources,* 118, 334, 2003.
15. Chen, G., Delafuente, D. A., Sarangapani, S. and Mallouk, T. E., Combinatorial discovery of bifunctional oxygen reduction – water oxidation electrocatalysts for regenerative fuel cells, *Catal. Today*, 67, 341, 2001.
16. Sullivan, M. G., Utomo, H., Fagan, P. J. and Ward, M. D., Automated electrochemical analysis with combinatorial electrode arrays, *Anal. Chem.,* 71, 4369, 1999.
17. Atanassova, P., *2002 Annual Progress Report,* US DOE, 2002.
18. Guerin, S., Combinatorial electrochemical screening of fuel cell electrocatalysts, *J. Comb. Chem.,* 6, 149, 2004.
19. Liu, R. and Smotkin, E. S., Array membrane electrode assemblies for high throughput screening of direct methanol fuel cell anode catalysts, *J. Electroanal. Chem.*, 535, 49, 2002.
20. Strasser, P., Fan, Q., Devenney, M., Weinberg, W. H., Liu, P. and Nørskov, J. K., High throughput experimental and theoretical predictive screening of materials – a comparative study of search strategies for new fuel cell anode catalysts, *J. Phys. Chem. B*, 107, 11013, 2003.
21. Jiang, R. and Chu, D., A combinatorial approach toward electrochemical analysis, *J. Electroanal. Chem.,* 527, 137, 2002.
22. Moore, J. T., Corn, J. D., Chu, D., Jiang, R., Boxall, D. L., Kenik, E. A. and Lukehart, C. M., Synthesis and characterization of a Pt_3Ru_1/vulcan carbon powder nanocomposite and reactivity as a methanol electrooxidation catalyst, *Chem. Mater.,* 15, 3320, 2003.
23. Stevens, D. A., Domaratzki, R. E. and Dahn, J. R., 64-channel fuel cell for testing sputtered combinatorial arrays of oxygen reduction catalysts, *206th Meeting of The Electrochemical Society,* Hawaii, October 3–8, 2004, p. 1903.
24. Jayaraman, S. and Hillier, A. C., Screening the reactivity of Pt_xRu_y and $Pt_xRu_yMo_z$ catalysts toward the hydrogen oxidation reaction with the scanning electrochemical microscope, *J. Phys. Chem. B,* 107, 5221, 2003.
25. Jambunathan, K., Jayaraman, S. and Hillier, A. C., A multielectrode electrochemical and scanning differential electrochemical mass spectrometry study of methanol oxidation on electrodeposited Pt_xRu_y, *Langmuir,* 20, 1856, 2004.
26. Fernandez, J. L. and Bard, A. J., Scanning electrochemical microscopy. 47. Imaging electrocatalytic activity for oxygen reduction in an acidic medium by the tip generation-substrate collection mode, *Anal.Chem.*, 75, 2967, 2003.
27. Shioyama, H., Yamada, Y., Ueda, A. and Kobayashi, T., Screening of carbon supports for DMFC electrode catalysts by infrared thermography, *Carbon,* 41, 579–607, 2003.
28. Yamada, Y., Ueda, A., Shioyama, H. and Kobayashi, T., High-throughput screening of PEMFC anode catalysts by IR thermography, *Appl. Surf. Sci.*, 223, 220, 2004.
29. Kobayashi, T., Ueda, A., Yamada, Y. and Shioyama, H., A combinatorial study on catalytic synergism in supported metal catalysts for fuel cell technology, *Appl. Surf. Sci.*, 223, 102, 2004.

15 High-Throughput Discovery of Battery Materials

Alan D. Spong, Girts Vitins, and John R. Owen

CONTENTS

15.1 INTRODUCTION: FIGURES OF MERIT

Research into battery materials has continued since Volta's experiments on a variety of different metals to the present day when battery technologies, particularly lithium-ion, are continually updated by the discovery of new and better materials. The scope for discovery is enormous; new materials for these specialist applications need not even be new chemical compositions or crystal phases – even the most minor manipulations of structure or impurity content induced by processing, for example, can have profound effects on the performance and marketability of a new product. The scope of this chapter, by contrast, is limited and therefore we will concentrate on the desired properties of the three major components of a battery, i.e., the positive electrode, negative

TABLE 15.1
Figures of Merit for Materials Used in the Li-Ion Cell

	Negative Electrode	Separator Electrolyte	Positive Electrode
Example	Li_xC_6	1 M $LiPF_6$ in	$Li_{1-x}CoO_2$
	$0 < x < 1$	EC/DMC	$0.5 < x < 1$
Charge capacity (mAh g^{-1})	372	None	150
Potential range (V vs. Li)	0.1–0.5	0.1–4.2	3.8–4.2
Electronic conductivity (S cm^{-1})	10^2	None	10^{-1}
Ionic conductivity (S cm^{-1})	10^{-6}	10^{-2}	10^{-6}
Diffusion constant (cm^2 s^{-1})	10^{-10}	10^{-6}	10^{-10}

electrode, the electrolyte-containing separator. Some figures of merit for these are shown in Table 15.1, which illustrates the state of the art in commercial lithium rechargeable batteries with some data on the materials used in the lithium carbon/cobalt oxide cell.[1–5]

The above figures combine to determine the two most important figures of merit for the complete cell, which are the energy and power per unit mass or volume. Also important are the number of cycles possible and the operational life, which depend not only on the individual materials but also on the interfaces between each electrode and the electrolyte. Thus we can already see a rapid expansion in the number of experiments required to characterize a system, calling for high-throughput techniques.

Whereas high-throughput techniques have been used for decades in pharmaceutical research[6] and for many years for catalysts,[7–13] superconductors,[14–19] magnetic materials,[14,19,20] and optical data storage materials,[21] their application to battery materials is relatively new.[22–34] Therefore, this chapter will first explain the principles of the fabrication and characterization of battery materials before discussing how high-throughput methods have been used to date. Finally, we will conclude with an anticipation of how the field may develop in the future.

15.2   CHARGE CAPACITY AND ITS REVERSIBILITY

The main property required from the electrodes is charge storage and release. The way charge is stored within a battery cell is shown schematically in Figure 15.1, which identifies the stored charge as an increase in the number of electrons in the negative electrode and a decrease in the positive side. Because of the need for electroneutrality of bulk matter (by contrast with thin films or interfaces) to avoid high voltages developing, we must compensate the charge surplus and deficiency by insertion or extraction of ions from the electrolyte.

The operation of the battery then consists of discharging the electrons through the circuit, while the ions move internally to maintain charge neutrality. The *specific capacity*, or charge stored per unit mass of each electrode, is therefore determined by the limiting values of chemical composition of the electrodes according to the *redox* reactions exemplified in Equations 15.1 and 15.2.

Negative electrode discharge:

$$[\text{active material(s)}] \rightarrow [\text{discharge product(s)}]^{n+} + ne^-$$

$$\text{e.g., } Zn \rightarrow Zn^{2+} + 2 e^- \qquad (15.1)$$

$$\text{Specific capacity} = 2F/M_{Zn} = 820 \text{ mAh g}^{-1}$$

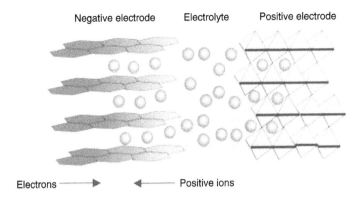

Negative electrode   Electrolyte   Positive electrode

Electrons ⟶   ⟵ Positive ions

FIGURE 15.1 Schematic diagram of a lithium-ion cell. The balls represent lithium ions that compensate the electrostatic charge due to electrons passing in and out of the electrodes while current passes in the external circuit.

Positive electrode discharge:

$$[\text{active material(s)}]^{m+} + me^- \rightarrow [\text{discharge product(s)}]$$

$$\text{e.g., } MnO_2 + H^+ + e^- \rightarrow MnOOH \qquad (15.2)$$

$$\text{Specific capacity} = F/M_{MnO_2} = 308 \text{ mAh g}^{-1}$$

Obviously, light atoms are preferred from the point of view of maximizing the charge/mass quotient (or the specific capacity); however, given that the electrodes are solid (see below), we must consider the effects of the redox reaction on the solid structure. In Equation 15.1 above, the zinc is dissolved, and can be re-plated on charging. In Equation 15.2, however, we find that manganese dioxide, in common with most other positive electrode materials, undergoes a solid-state transformation during discharge. The *reversibility* of such a reaction is essential to the operation of a rechargeable battery and it is this factor that is probably the most difficult to predict.

Recently, batteries have used the rocking-chair principle shown in Figure 15.1, where the metallic negative material is replaced by a host structure that stores the electroactive ions.[1–5,35,36] An example is given in the lithium–graphite intercalate LiC_6, which is formed on storing one lithium ion and one electron per C_6 ring of graphite:

$$C_6 + Li^+ + e^- \rightarrow LiC_6 \qquad (15.3)$$

$$\text{Charge per unit mass} = 372 \text{ mAh g}^{-1}$$

Obviously the specific capacity is much less than what could be obtained with pure lithium as the negative electrode (3861 mAh g^{-1}). However LiC_6 is preferable to lithium metal because it shows better reversibility. The latter transforms on cycling from a smooth metal surface to a finely divided form[5,35,37–42] that has been known to cause fires and even explosions.

15.3 ACTIVE POTENTIAL RANGE

The cell potential can be expressed as the difference between the electrode potentials of the two electrodes,

$$E_{cell} = E_{Pos} - E_{Neg} \qquad (15.4)$$

Clearly, E_{cell} should be maximized by choosing positive and negative electrodes with high and low potentials respectively when fully charged. The individual potentials should not, however, be so extreme as to cause electrolyte decomposition. These two factors that contribute to optimization of the cell potential are discussed below.

The individual electrode potentials can be measured in an electrochemical cell versus a standard reference electrode. Each electrode potential depends on the electrode composition at each stage in the discharge process. A reasonably constant cell potential during discharge requires reasonably constant values of the individual electrode potentials as shown in Figure 15.2, which shows how the two potentials subtract to give the cell potential as a function of charge passed. In this example it is the negative electrode that contributes most to a sloping discharge profile for the cell. Most importantly, both electrodes show sharp potential excursions at the extremities, where the electron charges can no longer be neutralized by ion insertion. These points define the potential ranges of the electrodes.

According to Faraday's law, the fundamental principle of electrochemical charge storage, the external current should correspond to the rate of formation of products of the electrode reactions. In efficient electrochemical charge storage device the reactions should be totally reversible on reversing the current. In practice, parasitic currents can arise through undesired reactions; electrolysis of water, for example, produces hydrogen and oxygen gases that escape from the cell and, in the case of non-aqueous electrolytes, more complex side reactions detract from reversibility of the intended cell reaction. Parasitic currents increase as the cell is charged because the electrodes become increasingly reactive, reducing or oxidizing the electrolyte as in the example of Equation 15.5[43] which is contrasted against the intended reaction of Equation 15.3.

$$2C_2H_4CO_3 + 2e^- + 2\,Li^+ \longrightarrow C_2H_4 \qquad + Li^+_2(CH_2CO_3^-)_2 \qquad (15.5)$$

$$\text{ethene carbonate (EC)} \qquad\qquad \text{ethene gas + lithium succinate}$$

Such reactions contribute to a loss of charge/discharge efficiency, and the term "irreversible charge capacity" refers to the shortfall in the charge passed during a normal discharge with respect to the charge passed during charge. More serious consequences are gas evolution and the precipitation of solid products that may foul the electrode surfaces. Sometimes, however, the formation of

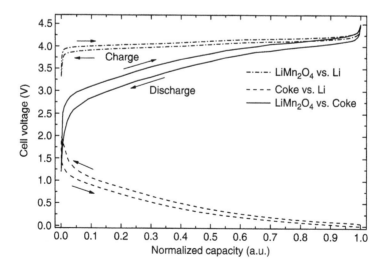

FIGURE 15.2 Charge discharge plots of a Li-ion cell calculated from second cycle data of two cells: $LiMn_2O_4$ vs. Li cycled at C/5; 2) Carbon (Coke) vs. Li cycled at C/5.

thin passivating films is welcome as it increases apparent stability of the cell materials. For instance, AlF_3 films in $LiPF_6$- or $LiBF_4$-based electrolytes passivate and stabilize the Al current collector against dissolution.[44–49] Therefore Al is normally used as a positive electrode current collector in lithium batteries.

The necessity to minimize electrolysis places limits on the maximum operating voltage of a cell and hence the total charge and energy stored.

15.4 SPECIFIC ENERGY AND POWER

The theoretical specific energy may be calculated from the measurements of cell charge and potential as described above according to Equation 15.6.

$$\text{Theoretical specific energy, } G = \frac{\int_0^{q_{max}} E_{cell} \, dq}{m} \tag{15.6}$$

where E_{cell} = equilibrium cell potential, dq = charge increment, q_{max} = maximum charge capacity, m = total cell mass.

The specific pulse power (i.e., the power obtained during a very short period of discharge) is mainly determined by the total resistance of the electrolyte and electrode materials, which is often called the equivalent series resistance. Under the optimum load matching condition this is given by Equation 15.7

$$\text{Specific pulse power, } P_{pulse} = \frac{E^2}{4Rm} \tag{15.7}$$

where R is the equivalent series resistance of the cell.

During longer discharge periods, the discharge current, and hence the power, is further limited by the requirement for diffusion to transport ions from deep inside the electrode particles to the electrolyte. A rough estimate of the specific continuous power given in Equation 15.8 highlights the importance of knowing the diffusion coefficient of the electrode materials.

$$\text{Specific continuous power, } P_{cont} \approx \frac{G}{\tau} \approx \frac{GD}{r^2} \tag{15.8}$$

where τ = shortest discharge time, r = average electrode particle radius or total electrode thickness and D = effective diffusion coefficient.

The actual rate of discharge of a battery in a given situation is often expressed as the "C-rate." This is the reciprocal of the time, in hours, for a fully charged battery to discharge completely at the given rate. The expression can also be used to describe the maximum rate capability of a battery, in which case we can estimate the C-rate to be $1/\tau$, with τ given in hours. For example, a shortest discharge time of 10 min would correspond to a rate of $6C$.

$$\text{Maximum power discharge rate}/C \sim \tau^{-1}/h^{-1} \tag{15.9}$$

15.5 PERFORMANCE, CYCLABILITY, AND LIFE

It was seen above that operational figures of merit such as the specific energy and power may be related to the properties of the constituent materials and some assumptions regarding the cell module design and the packaging materials that will inevitably increase the battery mass. Two more figures of merit are worthy of mention here — the cyclability and life. These can only be tested by multiple cycling of every sample in a test rig over a long period of time. This fact is perhaps the best illustration of the importance of high-throughput techniques in the study of battery materials and it also allows us to define some targets for the development of equipment and testing procedures. The equipment should satisfy at least some of the following criteria:

- Measurement of a property of a single material that affects the cell performance
- Measurement of the properties of material combinations (e.g., electrode/electrolyte interface) that affect the cell performance
- A large number of parallel measurements per unit cost, including space labor
- Reproduction of properties as manifested in a cell environment with minimum error

15.6   COMBINATORIAL SYNTHESIS OF CELL MATERIALS

The general problem facing the high-throughput materials scientist is the selection of a suitable method for the simultaneous, or automated sequential, fabrication of a large number of samples to span the desired range of synthesis variables, such as composition, thermal history, etc. Samples can be produced either as an array of individual samples or as a single sample with a large enough area to allow small areas to be distinguished and probed individually.

Battery electrodes are normally produced as composite thick films in which the electroactive compounds are dispersed in a polymeric binder with carbon black added to improve the electronic conductivity. Research into the materials themselves can be done on compact thin films of the pure material or on composite films, which resemble the state of the material in its application. Both types of sample can be produced combinatorially, as shown below.

Liquid electrolytes are generally contained in porous polymer separators, or immobilized as plasticizers in compatible polymers. Solid or dry polymer electrolytes are also used as thick films. All of these forms are amenable to combinatorial study, although there have been few documented examples to date of high-throughput experiments.

15.6.1   PHYSICAL VAPOR DEPOSITION (PVD)

Ultra-high vacuum (UHV) techniques, from simple evaporation by magnetron sputtering of Ga-Te-Sb thin films[21] to molecular beam epitaxy (MBE), have been favored in catalyst, superconductor, and ferroelectrics studies because of the high degree of purity, and thickness control offered for the deposition of (submicrometer) thin films. A combinatorial molecular beam epitaxy (COMBE), for synthesis of thin films of complex oxides, including high-temperature superconductors, ferroelectrics, ferromagnets, has been reported by Bozovic et al.[17,50] However, thicker films, and therefore faster deposition rates, are generally preferred for battery material testing and consequently sputtering has been preferred to UHV deposition. There have been studies of thin-film battery electrodes prepared by radio-frequency magnetron sputtering, e.g., of $LiMn_2O_4$[51] or $LiCoO_2$.[52]

One problem with sputtering, compared with MBE, is that the variation of flux with distance and tilt from the source, are not easily arranged within a standard sputtering chamber. Early experiments on the deposition of films using multiple sputtering targets achieved the atomic mixing and diversity of flux to produce composition spreads, but these were difficult to predict and control. Xiang et al.[11,18,19,53,54] described a moving shutter method of providing composition control and sequential depositions to combine several component elements together. However, interdiffusion of the layers required heating. An elegant variation of the successive layer deposition method has been demonstrated by Dahn et al.[22,27,55,56] in which several targets were placed on a fast rotating carousel behind tapered masks designed to change the average flux as a predetermined function of the radius. An ingenious feature of the latter design was a cam-driven mechanism for rotating the

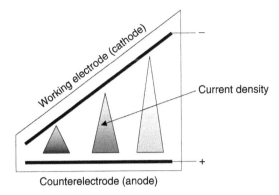

FIGURE 15.3 Design of a Hull cell used for combinatorial electrodeposition.

sample table through 90° as it passed from one target to the next in order to produce a two-dimensional composition spread.

Sputtered arrays or composition spreads have been particularly successful in the case of metal alloys of the type commonly used as negative electrodes in lithium and metal hydride batteries. This is because all components have low vapor pressures at the substrate temperature and therefore are easily incorporated into the solid product. Volatile elements can be lost to the vacuum, however, and difficulties may arise in the preparation of transition metal oxides, particularly positive electrode materials, which are, by definition, highly oxidizing and susceptible to oxygen loss.

15.6.2 ELECTRODEPOSITION

The Hull cell (Figure 15.3) has long been used in electroplating as a means of examining the effects of changing the current density and potential on the composition of an electroplated alloy by introducing a variable resistance path. The potential range can span from the threshold for plating the least-active metal alone at a low current density to the point where both metals are co-deposited as fast as diffusion allows. Under the diffusion limit, the deposition rate is proportional to the concentration, so that the metal ratio is the same in the alloy as the plating bath. The above cell has been used to deposit Cu-Sn alloys[57] of interest as negative electrode materials in lithium batteries.

Electrodeposition of compounds is well known in the world of batteries where lead dioxide has been formed by anodization (electrochemical oxidation) of both lead metal and of lead salts. Electrolytic manganese dioxide (EMD) for alkaline zinc batteries is produced by anodization of manganese sulfate. A combinatorial study has been made of the effects of the composition of the plating solution and the anodization current density on the cathode properties of EMD in KOH.[25] Another study used cathodic electrodeposition, i.e., reduction, of a mixture of soluble peroxo complexes of molybdenum and tungsten to produce WO_3/MoO_3 films.[58,59] Similarly automated systems for electrochemical synthesis and high-throughput screening of photoelectrochemical materials were developed and used to prepare tungsten-based mixed-metal oxides, $W_nO_mM_x$ (M = Ni, Co, Cu, Zn, Pt, Ru, Rh, Pd, and Ag), specifically for hydrogen production by photoelectrolysis of water.[60] Although the goal was a photocatalyst in this case, the method of array synthesis is equally applicable to battery applications.

Figure 15.4 places metal plating and anodic electrodeposition into a general context with two more ways in which compounds may be electrodeposited, by using electrolysis to change the local pH near the electrode causing precipitation of a metal oxide. An interesting example of the combinatorial deposition of a metal oxide by local pH change has been reported in the context of mesoporous ZnO for photocatalytic hydrogen generation.[61] An example library of 56 ZnO samples

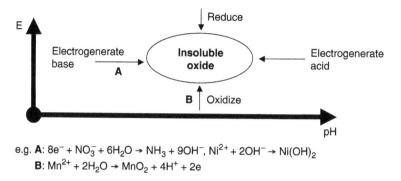

e.g. **A**: $8e^- + NO_3^- + 6H_2O \rightarrow NH_3 + 9OH^-$, $Ni^{2+} + 2OH^- \rightarrow Ni(OH)_2$
   B: $Mn^{2+} + 2H_2O \rightarrow MnO_2 + 4H^+ + 2e$

FIGURE 15.4 Some routes for electrodeposition of metal hydroxides or oxides.

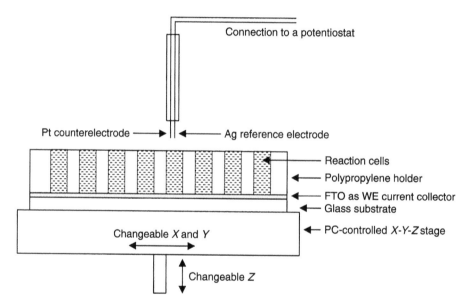

FIGURE 15.5 A combinatorial cell of multiple working electrode deposition. PC-controlled X-Y-Z stage positions sequentially the cells at a Pt counterelectrode and a Ag reference electrode. This concept can be used for various electrochemical depositions.

was synthesized, in the presence of varying concentrations (0–15 wt%) of poly(ethylene oxide)-block-poly(propylene oxide)-block-poly(ethylene oxide) – $EO_{20}PO_{70}EO_{20}$ as a structure-directing agent. High-throughput photoelectrochemical screening for the measurement of water-splitting photocatalysis identified peak performance at 3 wt% of the additive. Although the figure of merit in this case was the photocurrent intensity, the method of array synthesis (Figure 15.5), where the compositional variable was the concentration of a structure-directing agent (surfactant), is equally applicable to many battery materials.

Conducting polymers have also been produced combinatorially by anodic electrodeposition. In this way Barbero et al. made co-polymers based on polyaniline.[62] Yudin et al. have proposed combinatorial electrodeposition of polythiophene derivatives.[63]

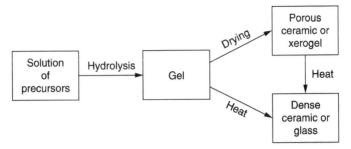

FIGURE 15.6 The sol–gel process. When the gel is dried at high temperature dense ceramics or glasses are formed. Drying at low temperatures above the critical pressure of water produces microporous solids known as xero-gels or aero-gels.

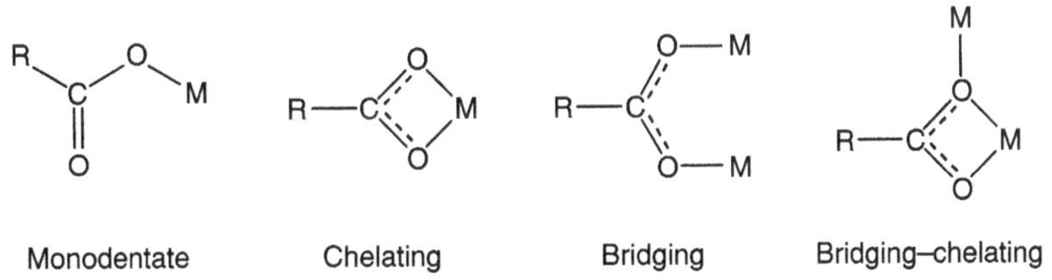

FIGURE 15.7 Types of ligand bonding for carboxylate groups with metal ions.

15.6.3 Sol–Gel Synthesis

Historically, sol–gel chemistry has been utilized for the preparation of many different ceramics, glasses, and inorganic materials.[64] Conventional ceramic powder processing normally requires several successive mixing and high temperature reaction in order to ensure effective interdiffusion of the constituent elements. Such processes are cumbersome for high-throughput experiments because of the difficulties of mixing, grinding, and thermal treatment of powders on multiple samples.

Solution and sol–gel routes were first developed to prepare glasses and ceramic materials,[65] but many solution and sol–gel routes to other inorganic materials including battery materials have reported. Sol–gel routes exist for the preparation of $LiCoO_2$,[66] $LiCo_{1-x}Ni_xO_2$,[67] $LiMn_2O_4$[68] and recently a sol–gel route to $LiFePO_4$ has been reported using a citric acid route.[69–71] Very recently sol–gel preparation has been used in combinatorial studies of optically functional oxides, e.g., $Ca_{1-x}Sr_xZr_{1-y}Cr_yO_3$ and $SrZr_{1-x}Ta_xO_3$.[72]

In the sol–gel process (Figure 15.6) the materials are prepared from solution precursors and compositional variation is easily achieved by mixing in the solution phase.[64] A chelating agent is introduced to the precursors such as a molecule with two or more carboxylic acid functionalities (Figure 15.7). These acid functions form bonds with the precursors in a type of polymerization reaction that leads to the formation of a xerogel as the solvent is removed. This gel is then calcined by heating in a tube furnace to burn off the additives and water yielding a product that should require little further heat treatment. The diffusion step is much more rapid than in a powder synthesis because the precursor elements are already within a few atoms distance of each other. Hence only one step is required and the reaction temperature may be lower.

Homogeneous mixtures can be prepared from solutions of reactive ions dispersed on the atomic scale. However, it is important to ensure that when the solvent is removed, components do not crystallize individually and thereby separate from one another. Inhibition of crystallzsation can be achieved by adding a gelling or fixing agent, such as a long-chain polymer. Studies by

Nguyen et al. have produced a route to the synthesis of oxide powders by polymeric steric entrapment in which transition metal precursors are confined in the correct stoichiometry within polymer chains of PVA.[73]

The sol–gel process can easily be adapted for high-throughput synthesis, where many compositions can be prepared rapidly by automated preparation of various combinations of different components from stock solutions using a commercial liquid handling system. Another advantage of sol–gel for combinatorial synthesis is the one-step reaction, simplifying the process for rapid preparation in an array.

Recent work in our laboratory has shown that the sol–gel route can be used to prepare electrode materials directly on an array of carbon current collectors for high-throughput electrochemical characterization.[74,75] An electrode material of current interest, $LiFePO_4$, was prepared from aqueous solutions of $LiCH_3COO$, $Fe(NO_3)_3$, and H_3PO_4 in stoichiometric proportions with added sucrose in a molar ratio of 0.25 to Fe. The sucrose had three functions: first to allow the $LiFePO_4$ precursors to dry into a homogeneous glass, which implied very fine mixing of reactants; second, it facilitated the reduction of Fe^{3+} to Fe^{2+}; and third, it provided, on pyrolysis, a carbon coating on the $LiFePO_4$ particles, enhancing the electronic conductivity in the calcined samples.

This preparation method was adapted[74,75] for the preparation of other olivine solid solutions, e.g., $LiFe_{1-x}M_xPO_4$ (M = Co, Ni, Mn, Mg) and $Li_{1-x}M_xPO_4$ (M = Zr, Nb, Ti) which are of interest in order to increase both the charge transport and the electrode potential of the material.[76–82] Cobalt, nickel, and zirconyl nitrates were correspondingly added in the above preparation from liquid solutions with sucrose.

15.6.4   PASTE DEPOSITION OF THICK ELECTRODE FILMS

Arrays of thick-film elements resembling real battery electrodes were produced in this laboratory using an automated robotic liquid handler. In this work, the objectives were:

a)   To evaluate the quality of composite electrodes produced automatically by a robotic liquid handler on an 8 × 8 array of current collectors
b)   To compare the results of electrochemical characterization of the array with the results of individual experiments

The following components were used:

1.   Active material: $LiMn_2O_4$ *Carus* (specific capacity 106–107 mAh g^{-1} at 4.0 V vs. Li plateau): 90, 89, 88, 87, 85, 83, 80, 70 wt%
2.   Conductive additive: Acetylene black (AB 100% compressed, *Chevron*)
     Eight compositions: 0, 1, 2, 3, 5, 7, 10, 20 wt%
3.   Binder: PVDF-HFP *Aldrich*; 10 wt% in all the composite films

Two inks were made manually, one containing $LiMn_2O_4$ dispersed at 30 wt% in a solution of the binder in cyclopentanone and another containing a 4 wt% dispersion of acetylene black. (The given solid contents were optimized to give a good balance between a long dispersion stability and an acceptably low viscosity.)

The inks were stirred while the robot extracted the requisite aliquots for mixing in individual 10 mL wells to give the desired carbon contents. This procedure was found to define the concentrations within an accuracy of 1%. Aspiration by robot was used to homogenize the samples before automatic dispensation of 3 μL samples onto the 3-mm-diameter current collectors. Problems of segregation and viscosity were noted for this stage and the automated array was replicated by a manually dispensed array from automatically mixed samples.

Quantification of the errors in the dispensation of dispersions of electrode materials is an important ongoing study. The interim conclusions are that the critical factors for automated electrode fabrication are:

a)   The settling time constant must be long compared with the array deposition time
b)   The volume of dispersion in each element, cf. the dispensing accuracy

Results for the electrochemical characterization of these arrays are presented in Section 15.8 of this chapter.

15.6.5   ELECTROLYTE SYNTHESIS

Solid electrolytes can, in principle, be made by any of the techniques outlined above and there are many reports of the preparation of thin films of solid electrolytes for specialized batteries.[83–86] Crystalline and glassy materials containing mobile ions of hydrogen, lithium, and silver, in particular, were investigated over large composition ranges in the search for fast ion-conducting solids to replace the more traditional liquid electrolytes used in the battery industry. However, the recent development of rechargeable lithium technology eventually evolved with a preference for liquid electrolytes based on solutions of lithium salts in organic solvents. This was largely because of the problems of fabricating and maintaining a good contact between two solids at the interface between the electrolyte and an electrode undergoing repeated cycles of expansion and contraction during battery operation.

Liquid electrolytes are ideally suited to combinatorial synthesis because they are simply mixtures of salts and solvents. The main benefit of a mixed solvent is an enhanced operational temperature range and the scope for optimizing the conductivity, e.g., by having one component for decreasing the viscosity and another for suppressing ion pair formation. Polymer electrolytes can be included as liquids here — indeed it is their liquid-like mechanical compliance that makes them preferable to hard solids. Again we have mixtures of polymer solvent, lithium salts, and liquid co-solvents or plasticizers that can be mixed mechanically and extruded as films or dissolved in a volatile solvent which is subsequently evaporated off to give thin films.

15.7   HIGH-THROUGHPUT ELECTROCHEMICAL CHARACTERIZATION

This section will outline the types of measurement technique that are used for the characterization of battery materials, and how they can be made applicable to the simultaneous, or fast sequential, screening of large arrays of material. Several commercial instruments are available, and the following notes may serve as a guide to match such products to the research programme.

15.7.1   GALVANOSTATIC CYCLING

Galvanostatic (constant current) cycling is a method in which a sample electrode material under test is used as one of the electrodes in a cell containing a reference material, e.g., lithium metal, as the other electrode. The cell potential thus represents the sample electrode potential relative to the constant reference. The cell is subjected to constant current discharges and charges, reversing the current when the cell potential reaches predetermined positive and negative voltage limits. The cell voltage in Figure 15.8(a) changes gradually with the electrode potential as described in Section 3, while the electrolyte resistance adds a constant *IR* potential that changes sign between charge and discharge.

The differential capacity, *dQ/dE*, is plotted against potential in Figure 15.8(b). The peaks indicate the potentials at which most of the charge is stored and released. These correspond to plateaus in the plot of potential versus specific capacity in Figure 15.8(a). The *IR* contributions add

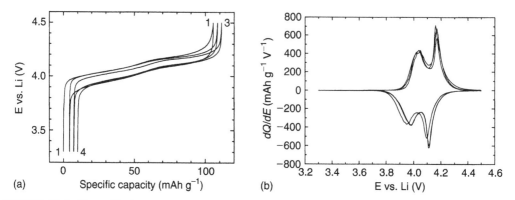

FIGURE 15.8 Charge/discharge plots (a) and differential capacity (b) for $Li_xMn_2O_{4+\delta}$ (Carus) obtained in galvanostatic cycling: rate C/8, 18 mA g^{-1}, 0.33 mA cm^2.

and subtract respectively to the peak potentials, so that the peaks themselves are shifted to the right and left from an almost symmetrical pattern. The *IR* complication is therefore effectively removed from the peak shape. Differential capacity profiles are good characteristics of thermodynamic properties of the electrode materials. Another advantage of the galvanostatic cycling is in its variable voltage sweep, which depends on charge capacity of the electrode materials. The sweep is slow where capacity is large and gives plateaus in time (charge) profiles while it is fast at potentials with little capacity, which are of little interest. So basically the voltage sweep (the response) is controlled by capacitive properties of the electrodes. In result, the cycling appears to be faster and more efficient compared to CV. Yet it is easier to observe charge capacity in galvanostatic cycling than in CV. The charge capacity is directly proportional to the charging time where the cell polarizes from lowest potential limit to the highest limit. In CV capacity can be obtained by integration of current–time profiles. All these features just mentioned are the reason why galvanostatic cycling is a technique most used in testing of slow and highly capacitive cells, e.g., lithium batteries. CV, however, is more preferable in order to observe the rate of electrode reactions and compare the rate capability of different electrode materials. In CV the peak currents correspond to the pseudo-capacity as well as ohmic and/or diffusion limitations in the cell.

Parallel galvanostatic measurements on battery cells have been developed over several decades in the guise of custom-built racks of galvanostat/potentiostats connected to multichannel programming and data collection hardware. However, true high-throughput materials characterization requires the ability to deal with arrays of electrodes rather than complete cells. This is theoretically possible, using a single counterelectrode, a common electrolyte, and an individual current supply with a potential measurement for each array element. However, no instrumentation specifically designed for array characterization, rather than a rack of individual potentiostats/galvanostats, has come to the attention of the authors and is clearly an area for future development.

15.7.2 Cyclic Voltammetry

Cyclic voltammetry (CV) is a method where the voltage is continuously swept between two voltage limits at a given sweep rate. Ideally, the current response should resemble Figure 15.8(b) because the current is the time derivative of the charge and the rate of change of potential is constant:

$$I = dQ/dt \qquad (15.10)$$

$dE/dt = v$, where v is the sweep rate.

$$I = v(dQ/dE) \qquad (15.11)$$

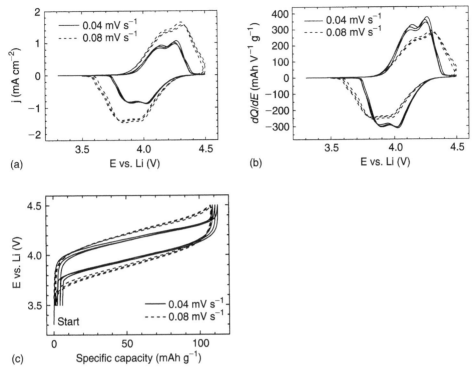

FIGURE 15.9 Cyclic voltammogram (a) differential capacity vs. potential (b) and potential/charge plots (c) for $Li_xMn_2O_{4+\delta}$ (Carus) obtained at two cycling rates: 0.04 and 0.08 mV s^{-1} corresponding to C/6.9, and C/3.5.

In this case, however, the current is not constant and therefore neither is the added *IR* potential. The latter causes a significant distortion of the relation between *dQ/dE* and *E* as shown in Figure 1.9(b). Figure 15.9(c) shows the result of integration of the cyclic voltammogram, which is a good approximation of the galvanostatic discharge curve (Figure 15.8(a)). The distortion is notably diminished at the lower sweep rate, and theoretically becomes insignificant at very slow rates.

The cyclic voltammogram therefore combines information regarding the capacity of the electrode and resistive contributions due to the electrolyte and the electronic resistance of the electrode itself. At fast sweep rates it can be regarded as a useful screening technique that evaluates the overall electrode performance and at low sweep rates it can provide the same information as galvanostatic cycling.

The main advantage of CV in high-throughput measurements is that it can easily be applied to an array in which all cell potentials change together. Provided that the sweep rate is slow enough to avoid excessive *IR* potentials, there should be no potential differences to cause cross talk between elements of an array of electrodes placed in the same electrolyte and using the same reference/counter electrode.

Figure 15.10 shows a cell design for testing an array of sample electrodes. It consists of:

a) An array of test electrodes, each connected to the measurement circuitry via a printed circuit boards
b) A common electrolyte-soaked separator designed to have a small resistance between each electrode and the counterelectrode but a relatively large lateral resistance between the test elements
c) A common counterelectrode pressed onto the separator to minimize the electrolyte thickness and encourage good contacts with the array

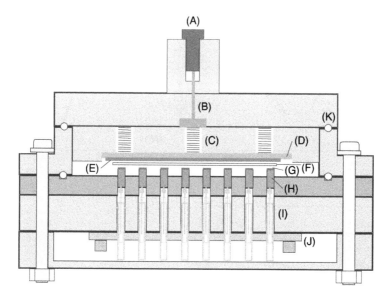

FIGURE 15.10 Combinatorial cell used for potential controlled experiments: A –negative electrode connector, B – stainless steel contact, C – spring contact, D – stainless steel current collector, E – lithium foil, F – glass separator, G – electrode material, H – aluminum working electrode, I – spring contact, J – printed circuit board.

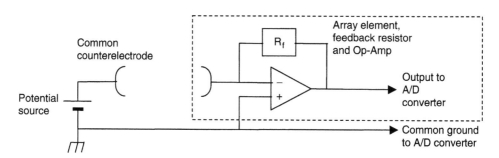

FIGURE 15.11 A simplified circuit diagram of the hardware used to control the potential and measure the current in a single array element.

A basic multichannel system for current monitoring of an array of electrodes under potential control is shown in Figure 15.11.The essential electronic component required here is a multichannel analog-to-digital converter, or a combination of a fast voltmeter with a multiplexer that can rapidly switch between inputs from different array elements. At the time of writing, 16-bit, 64-channel A/D converters can be obtained at a similar price to a desktop personal computer.

The advantages of the above system are:

1. It requires only one connection per array element.
2. It uses a single counter electrode and a single separator.

15.7.3 Conductivity Measurements and Complex Impedance Spectroscopy

Commercial instruments exist for the automated application of a four-point probe for the determination of electronic conductance of thin or thick films on insulating substrates.

This method can be applied to an array by simple translation between sequential conductance measurements. The measurements themselves can be made using direct current or, preferably, using alternating current for an improvement of the signal-to-noise ratio. In the latter case, an impedance spectrometer is the most convenient instrument to use.

An alternative approach to the four-point apparatus for measurement of impedance within the plane of the sample has been developed by Simon et al.[87,88] Arrays of capacitive gas sensor materials on Al_2O_3 were probed with an array of screen-printed platinum interdigitated connectors. A similar arrangement of a "gap" electrode has also been suggested by the principal author for use in the measurement of ionic conductivity in solid electrolyte films. A theoretical simulation determined how the impedance result may be interpreted directly to give the conductivity without independent knowledge of the film thickness.[89]

The four-point probe measures the lateral conductance, parallel to the film surface. This may be appropriate for isotropic materials, but complications may arise if there is surface conduction, or if orientation effects in composite films give rise to sample anisotropy. In such cases, a measurement of the conductance perpendicular to the film surface should be considered. A high-throughput method for examining electronic conductivity across an electrode array has been developed in our laboratory by modifying the equipment described in Figure 15.10. The modification consisted of removing the lithium foil counter electrode from the current collector and substituting a large conducting rubber disc for the electrolyte-soaked separator. The rubber is a carbon-filled electrically conductive silicone elastomer with an electronic conductivity of 0.3 S/cm and thickness 1.5 mm. The array was subjected to, for example, 25 and 50 mV potential steps during the instantaneous current response. The conductance of each array element was measured by dividing the applied potential by the current response over a short time period in each case and subtracting the resistance due to the carbon-filled elastomer.

Impedance spectroscopy is a common technique in the study of ohmic resistance in cells. In the most favorable circumstances the technique can deconvolute the resistances of the electrodes, electrolyte, and the interfaces between them. In that case the impedance is measured across the whole cell. Commercial instruments are available for impedance measurement on many cells, using switching circuitry to perform automated sequential measurements.

15.8 RESULTS

15.8.1 Thin Films of Metal Alloys and Oxides

Results of thin film arrays of metal alloys as negative electrodes have been given in a number of publications from Dahn's group.[22,24,27,55,56] The work showed systematic variations of capacity with composition in the Si/Al/Sn system.[24] Cyclic voltammetry of some amorphous alloys gave good combinations of high capacity and cycle life, e.g., 1500 mAh g^{-1} reversible over ten cycles.

Whitacre et al.[32] have shown sputter depositions of thin-film cells containing an array of $LiMn_xNi_{2-x}O_4$ as the positive electrodes, sputtered LiPON as the (solid) electrolyte, and lithium metal as the negative. They highlighted the composition $LiMn_{1.4}Ni_{0.6}O_4$ as one which had the highest voltage plateau, at 4.7 V vs. Li and a primary discharge capacity of 115 mAh g^{-1}.

15.8.2 Percolation Effects in Thick Films of Composite-Positive Electrodes

A selection of cyclic voltammograms taken from a 64-electrode array is shown in Figure 15.12. The liquid handler was used to prepare inks for the deposition of composite electrode elements

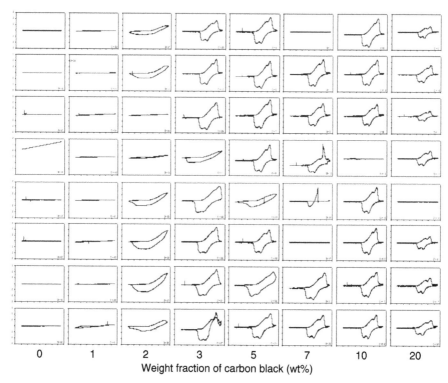

0   1   2   3   5   7   10   20

Weight fraction of carbon black (wt%)

FIGURE 15.12 Raw CVs taken at 0.05 mV s^{-1} from 3.2 to 4.5 V on an array of 64 LiMn$_2$O$_4$ electrodes. The current scales are $\pm$ 60 μA in each case.

containing powdered positive electrode material LiMn$_2$O$_4$, PVdF binder, and carbon black. Results of electrochemical characterization of automatically deposited arrays, as previously described,[30] combined errors due to automated sample deposition with those due to the electrochemical cell array. For the purposes of validating the electrochemical characterization, we present here the results for the manually deposited array. Samples were nominally identical down the columns of the array while the carbon content varied systematically across the array. The array was characterized as described in Section 15.3 using 1 M LiPF$_6$ in propylene carbonate as the electrolyte. The following observations can be made on the raw voltammograms:

a) There is much scatter in the results, presumably due to the sampling errors that are inherent in the dispensation of small volumes of inks.
b) A systematic trend is apparent, but partially obscured by the noise.
c) The voltammograms with a large amount of carbon show peaks, whereas those with small amounts only show loops at the discharged end.

(The small areas at high carbon content reflect the fact that the amount of active material decreased across the series because of the need to control viscosity.)

The amount of data contained in the voltammograms is impressive but daunting. Therefore the extraction numeric parameters forming each voltammogram are a very important part of the overall procedure. The simplest parameters are the total charge (time-integrated current) per half-cycle and the maximum current during a cycle. Provided that clear peaks are shown, the charge may be related to a specific capacity according to Equation 15.12. This is done in Figure 15.13.

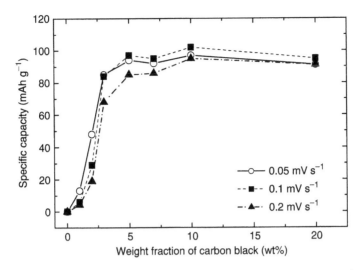

FIGURE 15.13 Capacity vs. carbon loading at different sweep rates: 0.05, 0.1, 0.2 mV s^{-1}.

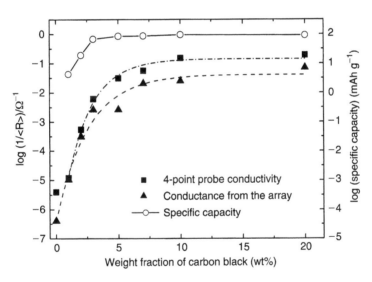

FIGURE 15.14 Electronic conductivity vs. carbon loading measured by two different methods compared with the specific capacity of carbon-loaded LiMn$_2$O$_4$ composite electrodes measured at a sweep rate of 0.2 mV s^{-1} in voltage range 3.2–4.5 V vs. Li.

$$\text{Specific Capacity} = \left. \int_{V=\text{discharged}}^{V=\text{charged}} I\,dt \middle/ m \right. \tag{15.12}$$

Figure 15.14 shows the observed specific capacity at the 0.1 mV s^{-1} scan rate (0.4 C) as a function of the carbon content. It is clear that the theoretical capacity is indeed reached at high carbon loadings but that the capacity is very low below 3% carbon. The result concurs well with the predictions of percolation theory, which shows a sharp drop in the effective conductivity of the matrix below the percolation threshold. It also shows that the characterization procedure succeeds in measuring the specific capacity to an accuracy of about 10%, which is useful, at least as a screening process. Further processing of the result could be made to predict the energy density for a cell, assuming a given potential-charge profile for the associated negative electrode.

Another interesting result is that the capacities are dependent on the cycling rate, particularly near the percolation threshold. At low cycling rates, the measured capacities are less dependent on the electronic conductivity and the percolation effects. In this case the measurement includes the full redox storage capability of the structure, which may either be related to the electronic band structure or the ion site limitations of the material. This is the maximum capacity that can be obtained, and should be constant at all sweep rates below the one used in the measurement.

At high cycling rates, the measured capacity becomes increasingly limited by electronic conductivity and diffusional limitations. Therefore, fast cycles are a way of probing the power density during continuous operation. Here the observed capacity is a strong function of the sweep rate as shown in Figure 15.13.

15.8.3 ELECTRONIC CONDUCTIVITY

Given that in both the examples above the currents and charges were related to the electronic conductivity, a correlation was sought with direct measurements of the electronic conductivity of the composite materials. Using replicate arrays of $LiMn_2O_4$, PVdF, and acetylene black to the one used in the capacity tests, the conductivity was measured both by the four-point probe method and the potential steps. Figure 15.14 compares the results of both methods.

Both methods give the same shape for the conductance–composition relation, and both are consistent with percolation theory, showing a typical increase of 10^5 between zero and 5% carbon.

Figure 15.14 also shows the rate-limited capacity values extracted from cyclic voltammetry. The shape is again similar, but the variation is only a factor 10^2. This illustrates the relation between the specific capacity and the conductivity as estimated in Equation 15.13 which applies only in the case where the capacity is limited by the conductivity:

$$\text{Observed capacity,} \quad Q \approx \int_E^{V_1} \frac{V - E}{Rv} dV \tag{15.13}$$

where E is the thermodynamic potential for insertion, V_1 is the potential sweep limit, R is the effective resistance and v is the sweep rate.

The relation applies in the region of low carbon content, but the limiting capacity is, of course, reached much earlier because of the theoretical limitations of the electrode capacity according to the reaction stoichiometry. In the former case, the voltammogram peaks are very broad, showing an almost constant current, whereas in the latter case sharp peaks indicate the absence of kinetic control.

15.8.4 SOL–GEL SYNTHESES

An array of $LiFePO_4$-based electrodes was produced by the sol–gel method described in Section 15.4, using sucrose as a glass former, reducing agent, and source of conductive carbon.[74,75] The latter was considered very important because the electrode material, nominally $LiFePO_4$, is insufficiently conductive without an additive to increase intergranular conductivity. Two compositional parameters were varied simultaneously, namely the sucrose content and a zirconium dopant which was of topical interest due to the recent announcement that the dopant could give a dramatic conductivity increase.

The result, shown in Figure 15.15, is obviously very noisy. This has been a characteristic of all our experiments using the sol–gel method with 3 μL liquid samples. Generally, the specific capacities

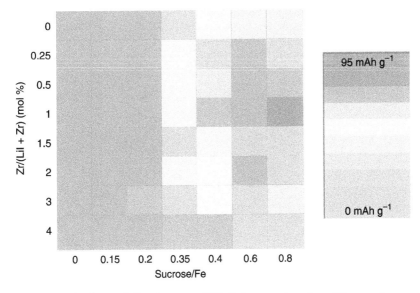

FIGURE 15.15 (See color insert following page 172) Color mapping of specific capacity versus precursor composition variables in array of 56 Zr-doped LiFePO$_4$ electrodes.

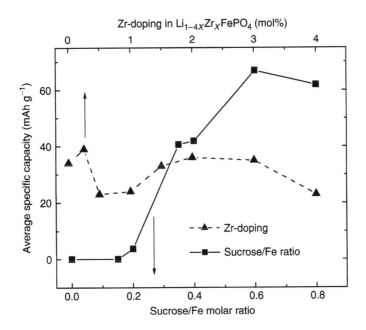

FIGURE 15.16 An analysis of the data shown in Figure 15.15.

are below theoretical and this is attributed to electronic contact problems between the electrodes and the current collectors, which were, in turn, caused by gas evolution during the calcination step and poor cohesion of the thick film products. However, despite the noise we can observe a systematic increase in the peak current with sucrose content, similar to that obtained with the carbon additive above. The averaged plot of Figure 15.16 shows that the capacity rises almost linearly from zero to about 60% of the theoretical value between 0.2 and 0.6 mol sucrose per Fe. This is consistent with

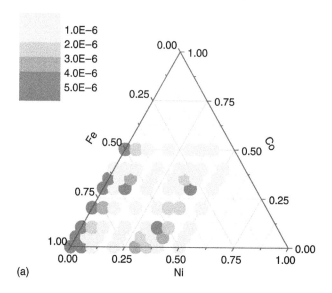

(a)

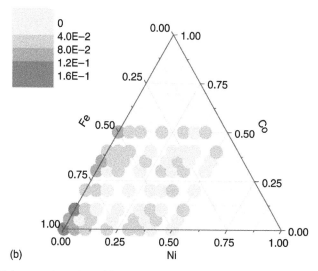

(b)

FIGURE 15.17 (a) Color mapping of specific capacity and (b) peak current at the C/6 rate as a function of Ni and Co substitution for Fe in LiFePO$_4$.

the pyrolysis of sucrose, part of which was burnt off when oxidized by the nitrogen oxides evolved. However, by contrast with the literature report, Figure 15.16 shows no systematic increase with increasing zirconium content, i.e., no amount of zirconium was seen to compensate for poor electronic conductivity at low levels of sucrose in the precursors.

One of the greatest goals of high-throughput measurement is the discovery of a valuable material buried in a small region of a complex composition diagram. Therefore, we have begun the search for alternative compositions in the LiFePO$_4$ system by doping with other metals. This study has been complicated by difficulties in quantification of the amount of carbon present. An example result is shown in Figure 15.17, which shows "hot spots" in the Li(Fe$_x$Co$_y$Ni$_z$)PO$_4$ system. It was tempting to interpret the increased currents for these samples in terms of substitutional defects. We have found similar "hot spots" in the Li(Fe$_x$Co$_y$Ni$_z$)PO$_4$ system. It was tempting to interpret the increased currents for these samples in terms of substitutional defects. However, the other explanation is that the lowered nitrate addition during doping with divalent metals would leave more residual sucrose, which, according to our previous experiments, is a much more plausible explanation for the results. Clearly more combinatorial experiments on the same lines as the zirconium investigation are needed to confirm which mechanism is correct.

Figure 15.17 shows examples of raw data and a ternary diagram of the results of substituting iron with cobalt and nickel. The initial result is that most of the doped compounds show lower peak currents than $LiFePO_4$ itself; overall, cobalt substitution has little effect on the capacity, whereas nickel is detrimental. However, it is interesting to note that around 30% cobalt improves the peak current, irrespective of the Fe/Ni ratio.

15.9 CONCLUSIONS AND FUTURE PROSPECTS

It has been shown that combinatorial synthesis methods can be devised for several types of battery material, including electrolytes and electrodes. The methods chosen to date have generally been compatible with those used industrially. Particular attention has been drawn to combinatorial synthesis of composite electrodes where the electronic conductivity is enhanced by addition of a conductivity additive.

High-throughput electrochemical characterization has been demonstrated in many ways, with array measurements of capacity, current capability and electronic conductivity made by cyclic voltammetry and other methods where the whole array is controlled by the same potential signal.

Overall results thus far have served to validate the techniques, verifying what is already known. Much work is still to be done in improving the signal-to-noise ratio of these experiments by optimizing array dimensions, deposition, and measurement techniques.

The most impressive result of these measurements is the sheer volume of data produced in a relatively short time. This emphasizes the need for computerized analysis techniques to reduce the data to the relevant parameters of capacity, conductivity, etc., as a function of the experimental variables. In the next few years we can look forward to real discoveries that would not have been possible without combinatorial methods.

REFERENCES

1. Dahn, J. R., VonStacken, U. and Fong, R., In: *The Electrochemical Society Extended Abstracts*, ECS Meeting, ECS, Seattle, WA, U.S., 1990, p. 66.
2. Nishi, Y., Azuma, H. and Omaru, A., United States Patent, 4,959,281, 1990.
3. Ratnakumar, B.V., Smart, M. C., Huang, C. K., Perrone, D., Surampudi, S. and Greenbaum, S. G., Lithium ion batteries for Mars exploration missions, *Electrochim. Acta*, 45, 1513, 2000.
4. Ratnakumar, B.V., Smart, M.C., Kindler, A., Frank, H., Ewell, R. and Surampudi, S., Lithium batteries for aerospace applications: 2003 Mars Exploration Rover, *J. Power Sources*, 119, 906, 2003.
5. Guyomard, D. and Tarascon, J. M., Li metal-free rechargeable $LiMn_2O_4$/carbon cells: Their understanding and optimisation, *J. Electrochem. Soc.*, 139, 937, 1992.
6. Czarnik, A. W. and DeWitt, S. H., *A Practical Guide to Combinatorial Chemistry*, American Chemical Society, Washington, DC, 1997, p. 450.
7. Maier, W. F., Kirsten, G., Orschel, M., Weiss, P.A., Holzwarth, A. and Klein, J., Combinatorial chemistry of materials, polymers and catalysts, *Macromol. Symp.*, 165, 1, 2001.
8. Reddington, E., Sapienza, A., Gurau, B., Viswanathan, R., Sarangapani, S., Smotkin, E.S. and Mallouk, T.E., Combinatorial electrochemistry: A highly parallel, optical screening method for discovery of better electrocatalysts, *Science*, 280, 1735, 1998.
9. Yajima, K., Ueda, Y., Tsuruya, H., Kanougi, T., Oumi, Y., Ammal, S.S.C., Takami, S., Kubo, M. and Miyamoto, A., Combinatorial computational chemistry approach to the design of $deNO_{(x)}$ catalysts, *Appl. Catal. A-Gen.*, 194, 183, 2000.
10. Tuinstra, H. E. and Cummins, C. L., Combinatorial materials and catalyst research, *Adv. Mater.*, 12, 1819, 2000.

11. Xiang, X. D., Combinatorial materials synthesis and high-throughput screening: An integrated materials chip approach to mapping phase diagrams and discovery and optimization of functional materials, *Biotechnol. Bioeng.*, 61, 227, 1999.

12. Wang, Z., Yu, Z. and You, X. Z., Solid-state combinatorial chemistry and its applications in material science, *Chin. J. Inorg. Chem.*, 16, 167, 2000.

13. Francis, M. B., Jamison, T. F. and Jacobsen, E. N., Combinatorial libraries of transition-metal complexes, catalysts and materials, *Curr. Opin. Chem. Biol.*, 2, 422, 1998.

14. Hullinger, J. and Awan, M. A., A "single-sample concept" (SSC): A new approach to the combinatorial chemistry of inorganic materials, *Chem.-Eur. J.*, 10, 4694, 2004.

15. Hanak, J. J., A quantum leap in the development of new materials and devices, *Appl. Surf. Sci.*, 223, 1, 2004.

16. Boyle, T. J., Rodriguez, M. A., Ingersoll, D., Headley, T. J., Bunge, S. D., Pedrotty, D. M., De'Angeli, S. M., Vick, S. C. and Fan, H. Y., A novel family of structurally characterized lithium cobalt double aryloxides and the nanoparticles and thin films generated therefrom, *Chem. Mat.*, 15, 3903, 2003.

17. Bozovic, I., Atomic-layer engineering of superconducting oxides: yesterday, today, tomorrow, *IEEE Trans. Appl. Supercond.*, 11, 2686, 2001.

18. Xiang, X. D., Sun, X. D., Briceno, G., Lou, Y. L., Wang, K. A., Chang, H. Y., Wallacefreedman, W. G., Chen, S. W. and Schultz, P. G., A combinatorial approach to materials discovery, *Science*, 268, 1738, 1995.

19. Xiang, X. D., Combinatorial synthesis and high throughput evaluation of functional oxides – An integrated materials chip approach, *Mater. Sci. Eng. B-Solid State Mater. Adv. Technol.*, 56, 246, 1998.

20. Yoo, Y. K. and Xiang, X. D., Combinatorial material preparation, *J. Phys.-Condes. Matter*, 14, R49, 2002.

21. Kyrsta, S., Cremer, R., Neuschutz, D., Laurenzis, M., Bolivar, P. H. and Kurz, H., Characterization of Ge-Sb-Te thin films deposited using a composition-spread approach, *Thin Solid Films*, 398, 379, 2001.

22. Fleischauer, M. D. and Dahn, J. R., Combinatorial investigations of the Si-Al-Mn system for Li-ion battery applications, *J. Electrochem. Soc.*, 151, A1216, 2004.

23. Fleischauer, M. D., Hatchard, T. D., Rockwell, G. P., Topple, J. M., Trussler, S., Jericho, S. K., Jericho, M. H. and Dahn, J. R., Design and testing of a 64-channel combinatorial electrochemical cell, *J. Electrochem. Soc.*, 150, A1465, 2003.

24. Hatchard, T. D., Topple, J. M., Fleischauer, M. D. and Dahn, J. R., Electrochemical performance of SiAlSn films prepared by combinatorial sputtering, *Electrochem. Solid State Lett.*, 6, A129, 2003.

25. Devenney, M., Donne, S. W. and Gorer, S., Application of combinatorial methodologies to the synthesis and characterization of electrolytic manganese dioxide, *J. Appl. Electrochem.*, 34, 643, 2004.

26. Takada, K., Fujimoto, K., Sasaki, T. and Watanabe, M., Combinatorial electrode array for high-throughput evaluation of combinatorial library for electrode materials, *Appl. Surf. Sci.*, 223, 210, 2004.

27. Fleischauer, M. D., Hatchard, T. D., Bonakdarpour, A. and Dahn, J. R., Combinatorial investigations of advanced Li-ion rechargeable battery electrode materials, *Meas. Sci. Technol.*, 16, 212, 2005.

28. Bonakdarpour, A., Hewitt, K. C., Hatchard, T. D., Fleischauer, M. D. and Dahn, J. R., Combinatorial synthesis and rapid characterization of $Mo_{1-x}Sn_x$ ($0 <= x <= 1$) thin films, *Thin Solid Films*, 440, 11, 2003.

29. Cumyn, V. K., Fleischauer, M. D., Hatchard, T. D. and Dahn, J. R., Design and testing of a low-cost multichannel pseudopotentiostat for quantitative combinatorial electrochemical measurements on large electrode arrays, *Electrochem. Solid State Lett.*, 6, E15, 2003.

30. Spong, A. D., Vitins, G., Guerin, S., Hayden, B. E., Russell, A. E. and Owen, J. R., Combinatorial arrays and parallel screening for positive electrode discovery, *J. Power Sources*, 119, 778, 2003.

31. Suzuki, K., Kuroiwa, Y., Takami, S., Kubo, M. and Miyamoto, A., Combinatorial computational chemistry approach to the design of cathode materials for a lithium secondary battery, *Appl. Surf. Sci.*, 189, 313, 2002.

32. Whitacre, J. F., West, W. C. and Ratnakumar, B. V., A combinatorial study of $Li_yMn_xNi_{2-x}O_4$ cathode materials using microfabricated solid-state electrochemical cells, *J. Electrochem. Soc.*, 150, A1676, 2003.

33. Yanase, I., Ohtaki, T. and Watanabe, M., Application of combinatorial process to $LiCo_{1-x}Mn_xO_2$ ($0 <= X <= 0.2$) powder synthesis, *Solid State Ionics*, 151, 189, 2002.

34. Spong, A. D., Vitins, G. and Owen, J. R., in *Combinatorial Arrays and Parallel Screening for Positive Electrode Discovery*, Abstracts of IMLB 11, Monterey, CA, U.S., 2002, Abstract No: 267.

35. Owen, J. R., Rechargeable lithium batteries, *Chem. Soc. Rev.*, 26, 259, 1997.

36. Brousse, T., Fragnaud, P., Marchand, R., Schleich, D. M., Bohnke, O. and West, K., All oxide solid-state lithium-ion cells, *J. Power Sources*, 68, 412, 1997.

37. Monroe, C. and Newman, J., Dendrite growth in lithium/polymer systems – A propagation model for liquid electrolytes under galvanostatic conditions, *J. Electrochem. Soc.*, 150, A1377, 2003.

38. Aurbach, D., Zinigrad, E., Teller, H. and Dan, P., Factors which limit the cycle life of rechargeable lithium (metal) batteries, *J. Electrochem. Soc.*, 147, 1274, 2000.

39. Brissot, C., Rosso, M., Chazalviel, J. N. and Lascaud, S., Dendritic growth mechanisms in lithium/polymer cells, *J. Power Sources,* 82, 925, 1999.
40. Brissot, C., Rosso, M., Chazalviel, J. N., Baudry, P. and Lascaud, S., In situ study of dendritic growth in lithium/PEO-salt/lithium cells, *Electrochim. Acta,* 43, 1569, 1998.
41. Takehara, Z., Future prospects of the lithium metal anode, *J. Power Sources,* 68, 82, 1997.
42. Ribes, A. T., Beaunier, P., Willmann, P. and Lemordant, D., Correlation between cycling efficiency and surface morphology of electrodeposited lithium. Effect of fluorinated surface active additives, *J. Power Sources,* 58, 189, 1996.
43. Aurbach, D., Markovsky, B., Weissman, I., Levi, E. and Ein-Eli, Y., On the correlation between surface chemistry and performance of graphite negative electrodes for Li-ion batteries, *Electrochim. Acta,* 45, 67, 1999.
44. Tachibana, K., Sato, Y., Nishina, T., Endo, T., Matsuki, K. and Ono, S., Passivity of aluminum in organic electrolytes for lithium batteries (1) film growing mechanism, *Electrochemistry,* 69, 670, 2001.
45. Wang, X. M., Yasukawa, E. and Mori, S., Inhibition of anodic corrosion of aluminum cathode current collector on recharging in lithium imide electrolytes, *Electrochim. Acta,* 45, 2677, 2000.
46. Zhang, S. S., Ding, M. S. and Jow, T. R., Self-discharge of $Li/Li_xMn_2O_4$ batteries in relation to corrosion of aluminum cathode substrates, *J. Power Sources,* 102, 16, 2001.
47. Zhang, S. S. and Jow, T. R., Aluminum corrosion in electrolyte of Li-ion battery, *J. Power Sources,* 109, 458, 2002.
48. Song, S. W., Richardson, T. J., Zhuang, G. V., Devine, T. M. and Evans, J. W., Effect on aluminum corrosion of $LiBF_4$ addition into lithium imide electrolyte; a study using the EQCM, *Electrochim. Acta,* 49, 1483, 2004.
49. Kanamura, K., Okagawa, T. and Takehara, Z., Electrochemical oxidation of propylene carbonate (containing various salts) on aluminium electrodes, *J. Power Sources,* 57, 119, 1995.
50. Bozovic, I. and Matijasevic, V., COMBE: A powerful new tool for materials science, *Materials Science Forum,* 352, 1, 2000.
51. Yamamura, S., Koshika, H., Nishizawa, M., Matsue, T. and Uchida, I., In situ conductivity measurements of $LiMn_2O_4$ thin films during lithium insertion/extraction by using interdigitated microarray electrodes, *J. Solid State Electrochem.,* 2, 211, 1998.
52. Liao, C. L. and Fung, K. Z., Lithium cobalt oxide cathode film prepared by rf sputtering, *J. Power Sources,* 128, 263, 2004.
53. Xiang, X. D., Combinatorial materials synthesis and screening: An integrated materials chip approach to discovery and optimization of functional materials, *Annu. Rev. Mater. Sci.,* 29, 149, 1999.
54. Schultz, P. G. and Xiang, X. D., Combinatorial approaches to materials science, *Curr. Opin. Solid State Mat. Sci.,* 3, 153, 1998.
55. Dahn, J. R., Trussler, S., Hatchard, T. D., Bonakdarpour, A., Mueller-Neuhaus, J. R., Hewitt, K. C. and Fleischauer, M., Economical sputtering system to produce large-size composition-spread libraries having linear and orthogonal stoichiometry variations, *Chem. Mat.,* 14, 3519, 2002.
56. Barkhouse, D. A. R., Bonakdarpour, A., Fleischauer, M., Hatchard, T. D. and Dahn J. R., A combinatorial sputtering method to prepare a wide range of A/B artificial superlattice structures on a single substrate, *J. Magn. Magn. Mater.,* 261, 399, 2003.
57. Beattie, S. D. and Dahn, J. R., Single-bath electrodeposition of a combinatorial library of binary $Cu_{1-x}Sn_x$ alloys, *J. Electrochem. Soc.,* 150, C457, 2003.
58. Baeck, S. H. and McFarland, E. W., Combinatorial electrochemical synthesis and characterization of tungsten-molybdenum mixed metal oxides, *Korean J. Chem. Eng.,* 19, 593, 2002.
59. Baeck, S. H., Jaramillo, T. F., Choi, K. S., Stucky, G. D. and McFarland, E. W., Photocatalytic and electrochromic properties of mesoporous tungsten-molybdenum mixed oxides synthesized by combinatorial electrochemistry, *Abstr. Pap. Am. Chem. Soc.,* 225, 910, 2003.
60. Baeck, S. H., Jaramillo, T. F., Brandli, C. and McFarland, E. W., Combinatorial electrochemical synthesis and characterization of tungsten-based mixed-metal oxides, *J. Comb. Chem.,* 4, 563, 2002.
61. Jaramillo, T. F., Baeck, S. H., Kleiman-Shwarsctein, A. and McFarland, E. W., Combinatorial electrochemical synthesis and screening of mesoporous ZnO for photocatalysis, *Macromol. Rapid Commun.,* 25, 297, 2004.
62. Barbero, C., Salavagione, H. J., Acevedo, D. F., Grumelli, D. E., Garay, F., Planes, G. A., Marales, G. M. and Miras, M. C., Novel synthetic methods to produce functionalized conducting polymers I. Polyanilines, *Electrochim. Acta,* 49, 3671, 2004.
63. Yudin, A. K. and Siu, T., Combinatorial electrochemistry, *Curr. Opin. Chem. Biol.,* 5, 269, 2001.
64. Shriver, D. F., Atkins, P. W. and Langford, C. H., *Inorganic Chemistry: Second Edition,* Oxford University Press, Oxford, 1998, p. 771.

65. Wen, J. Y. and Wilkes, G. L., Organic/inorganic hybrid network materials by the sol-gel approach, *Chem. Mat.*, 8, 1667, 1996.

66. Sun, Y. K., Oh, I. H. and Hong, S. A., Synthesis of ultrafine LiCoO2 powders by the sol-gel method, *J. Mater. Sci.*, 31, 3617, 1996.

67. Chang, S. H., Kang, S. G. and Jang, K. H., Electrochemical properties of $Li_xCo_yNi_{1-y}O_2$ prepared by citrate sol-gel method, *Bull. Korean Chem. Soc.*, 18, 61, 1997.

68. Huang, H. T. and Bruce, P. G., A 3 Volt Lithium Manganese Oxide Cathode for Rechargeable Lithium Batteries, *J. Electrochem. Soc.*, 141, L76, 1994.

69. Dominko, R., Bele, M., Gaberscek, M., Remskar, M., Hanzel, D. and Jannik, J., Porous olivine composites synthesised by sol-gel technique, 125, 2004.

70. Arcon, D., Zorko, A., Dominko, R. and Jaglicic, Z., A comparative study of magnetic properties of $LiFePO_4$ and $LiMnPO_4$, *J. Phys.-Condes. Matter*, 16, 5531, 2004.

71. Hsu, K. F., Tsay, S. Y. and Hwang, B. J., Synthesis and characterization of nano-sized $LiFePO_4$ cathode materials prepared by a citric acid-based sol-gel route, *J. Mater. Chem.*, 14, 2690, 2004.

72. Henderson, S. J., Armstrong, J. A., Hector, A. L. and Weller, M. T., High-throughput methods to optically functional oxide and oxide-nitride materials, *J. Mater. Chem.*, 15, 1528, 2005.

73. Nguyen, M. H., Lee, S. J. and Kriven, W. M., Synthesis of oxide powders by way of a polymeric steric entrapment precursor route, *VAJ. Mater. Res.*, 14, 3417, 1999.

74. Vitins, G., Spong, A. D. and Owen, J. R., in *Combinatorial Discovery of Electrode Materials*, Extended Abstracts of Battery and Fuel Cell Materials Symposium, Besenhard, O. J., Moler, K.-C. and Winter, M., Eds., International Battery Materials Association, Graz, Austria, 2004, p. 15.

75. Spong, A. D., Vitins, G. and Owen, J. R., in *Combinatorial screening of lithium battery materials*, Battery and Fuel Cell Materials Symposium, Besenhard, O. J., Moler, K.-C. and Winter, M., Eds., International Battery Materials Association, Graz, Austria, 2004, p. 185.

76. Chung, S. Y. and Chiang, Y. M., Microscale measurements of the electrical conductivity of doped $LiFePO_4$, *Electrochem. Solid State Lett.*, 6, A278, 2003.

77. Chung, S. Y., Bloking, J. T. and Chiang, Y. M., Electronically conductive phospho-olivines as lithium storage electrodes, *Nat. Mater.*, 1, 123, 2002.

78. Yamada, A., Kudo, Y. and Liu, K. Y., Reaction mechanism of the olivine-type $Li_{1-x}(Mn_{0.6}Fe_{0.4})PO_4$ $(0 <= x <= 1)$, *J. Electrochem. Soc.*, 148, A747, 2001.

79. Yamada, A., Hosoya, M., Chung, S. C., Kudo, Y., Hinokuma, K., Liu, K. Y. and Nishi, Y., Olivine-type cathodes achievements and problems, *J. Power Sources*, 119, 232, 2003.

80. Yamada, A. and Chung, S. C., Crystal chemistry of the olivine-type $Li(Mn_yFe_{1-y})PO_4$ and $(Mn_yFe_{1-y})PO_4$ as possible 4 V cathode materials for lithium batteries, *J. Electrochem. Soc.*, 148, A960, 2001.

81. Wolfenstine, J. and Allen, J., $LiNiPO_4$-$LiCoPO_4$ solid solutions as cathodes, *J. Power Sources*, 136, 150, 2004.

82. Amine, K., Yasuda, H. and Yamachi, M., Olivine $LiCoPO_4$ as 4.8 V electrode material for lithium batteries, *Electrochem. Solid State Lett.*, 3, 178, 2000.

83. Kuwata, N., Kawamura, J., Toribami, K., Hattori, T. and Sata, N., Thin-film lithium-ion battery with amorphous solid electrolyte fabricated by pulsed laser deposition, *Electrochem. Commun.*, 6, 417, 2004.

84. Takada, K., Inada, T., Kajiyama, A., Kouguchi, M., Sasaki, H., Kondo, S., Michiue, Y., Nakano, S., Tabuchi, M. and Watanabe, M., Solid state batteries with sulfide-based solid electrolytes, *Solid State Ionics*, 172, 25, 2004.

85. Takada, K., Inada, T., Kajiyama, A., Sasaki, H., Kondo, S., Watanabe, M., Murayama, M. and Kanno, R., Solid-state lithium battery with graphite anode, *Solid State Ionics*, 158, 269, 2003.

86. Takada, K., Inada, T., Kajiyama, A., Kouguchi, M., Kondo, S. and Watanabe, M., Research on highly reliable solid-state lithium batteries in NIRIM, *J. Power Sources*, 97, 762, 2001.

87. Simon, U., Sanders, D., Jockel, J., Heppel, C. and Brinz, T., Design strategies for multielectrode arrays applicable for high-throughput impedance spectroscopy on novel gas sensor materials, *J. Comb. Chem.*, 4, 511, 2002.

88. Frantzen, A., Scheidtmann, J., Frenzer, G., Maier, W. E., Jockel, J., Brinz, T., Sanders, D. and Simon, U., High-throughput method for the impedance spectroscopic characterization of resistive gas sensors, *Angew. Chem.-Int. Edit.*, 43, 752, 2004.

89. Fitt, A. D. and Owen, J. R., Simultaneous measurements of conductivity and thickness for polymer electrolyte films: a simulation study, *J. Electroanal. Chem.*, 538, 13, 2002.

Section 5

Electronic Materials Development

16 Innovation in Magnetic Data Storage Using Physical Deposition and Combinatorial Methods

Erik B. Svedberg

CONTENTS

16.1 INTRODUCTION

Advances in data storage technologies and especially in disk drives (Figure 16.1(a)), are driven by discoveries and improvements in thin-film technologies. To produce a disk, numerous layers with different functions have to be deposited (see Figure 16.1(b)).[1] Some layers have the purpose of storing the actual data "bit," others are used to prime the correct crystallographic texture of the subsequent layers[2] and yet others are used for corrosion protection or lubrication.[3,4] Similarly the sensor or "head" that reads and writes the data is a complex thin-film structure as well.[5–7] In order to further fine-tune these structures, it is important to be able to rapidly screen many different material sets and also to create transfer functions between different levels of parameters, so that it is possible to calculate the response and optimize the systems at a higher level. For example, it is important to be able to predict the coercivity as a function of alloy composition. At yet another level, it is important to predict the signal decay rate (Figure 16.1(c)), based on coercivity as well as other parameters.[8,9] Hence, transfer functions and rapid screening of suitable thin-film combinations are important for the development of tomorrow's disk drives.

16.2 METHODOLOGY

Experiments depositing a thin film with changing thickness are often performed by gradually removing a shutter surface that shadows the substrate from the active source either continuously[10–12] or in steps.[13] Thus, a gradient or wedge is created that can be incorporated in a thin-film structure. Another traditional deposition system consists of a moving shutter where a slot is made

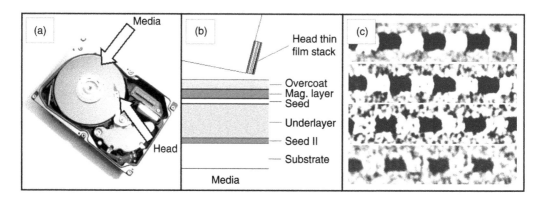

FIGURE 16.1 (a) Photo of an opened disk drive showing the head and the media. (b) Schematic of the head–media region, showing the multilayered structure of the head and the media. (c) Magnetic force microscopy image of written bits with different signal to noise ratio (SNR); from top-to-bottom, decreasing SNR due to aging.

(Figure 16.2(a)). This system can be used for producing a thickness gradient across a wafer if the speed at which the shutter is moved is gradually changed. Similarly the flux from the deposition source can also be changed while keeping the shutter at a constant speed or stepped.[14] If the shutter is moved while the sample is stationary, the angle of incidence is changing for the incoming species, which might affect the properties of the film. If the shutter is fixed and instead the sample is moved behind the shutter, the direction of the incoming flux will be constant across the sample.[15–19] Again, moving the shutter at a constant speed and instead changing a parameter other than its speed, such as temperature, or deposition–source combination, one can achieve gradients other than thickness. Shown in Figure 16.2(b) is a system with dual shutters; in this way any point or position on a wafer (a round disk) can be accessed in any order by, in this case, the three deposition sources below (where one is active). This system in Figure 16.2(b) can be used to produce combinatorial libraries on a wafer.[20,21] The advantage of this approach is that any change in parameters can be carried out between each "point" on the wafer. However, the disadvantage is that each point is deposited in turn. Thus, a library or wafer would take almost as long to produce as the individual data points on separate wafers. The speed can be increased somewhat if samples can be produced with only one shutter in place at a time, so that samples with orthogonal gradients in subsequent layers are produced.[22,23] Another way of masking the substrate during deposition makes use of accurately positioned masks and just a few deposition steps generating a multitude of unique samples. For example, using a quarternary masking scheme,[24–26] it is possible to deposit 256 unique samples in 16 deposition steps. In this way the time spent using individual shutters, shown in Figure 16.2(b), can be significantly reduced. Figure 16.2(c) shows the deposition scenario using a mask with one of its quadrants open for the deposition flux to reach the sample. After each 90° rotation of the mask, an additional material can be deposited, producing up to a total of four materials on different parts of the disk. Through subsequent use of masks with self-similar patterns (Figure 16.2(c) shows masks 1–3 of 4) it is possible to add material to the substrate up to a maximum of 256 unique positions. Figure 16.2(c) also shows the four positions for mask 3 where material has been deposited in 16 regions on the wafer for each rotation of the mask. This approach is useful when the material is later being "mixed," for example, by a heat treatment,[27] so that homogeneous alloys are being produced. Mixing can also occur if the layers are so thin that the energy from the deposition process itself produces an alloy. Other masking schemes[28–30] can also be used to manipulate the sample in order to, for example, make selective ion-implantation across the wafer in different doses. Yet more exotic schemes of combinatorial depositions have been made by vapor jet printing of molecular organic semiconductors for direct mask-free patterning[31] and compositional changes along fibers.[32]

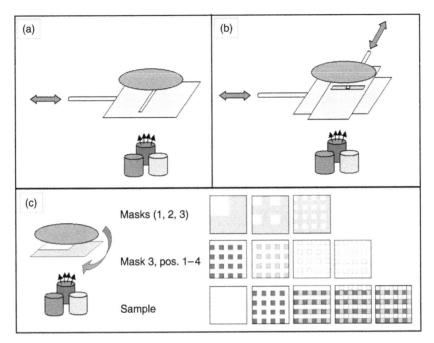

FIGURE 16.2 (a) Traditional deposition system with a moving slot for producing a gradient across a wafer. Shown are a circular wafer and three deposition sources (where one is active). (b) Dual orthogonal moving slots to produce combinatorial systems. (c) Quarternary masking scheme where up to four masks rotated 90° between each deposition can generate a total of 256 unique library positions in only 16 depositions.

One way to avoid the drawback of sequential deposition is to use parallel deposition of all the data points.[2,33-35] This approach can be made if the gradient is produced simultaneously across the wafer by tilting or producing an offset between the sample and the directional source, so that there are different deposition rates at different points across the sample/wafer. This approach should not be used if the microstructure or the properties of the film are heavily dependent on the angle of incidence. It will be difficult to separate the dependence on thickness or composition with the angle since they will be completely dependent on each other.

Figure 16.3 describes two different systems that use faster ways of depositing the gradients than the moving-shutter approach. In Figure 16.3(a) and Figure 16.3(b) a planetary system with an offset between the magnetron deposition source–center and the wafer–center is shown. In Figure 16.3(c) and Figure 16.3(d) a system with the magnetrons tilted relative to the sample surface normal is shown. The advantage of a planetary system is that several different substrates or seed layers can be deposited simultaneously in one run, creating a set of wafers with identical libraries on them for different post-treatments. The benefits with the other system shown in Figure 16.3(c) and Figure 16.3(d) is that more complex compositions of the film can be deposited, since there is a possibility of having more than two magnetrons in this configuration. In this configuration (Figure 16.3(c) and Figure 16.3(d)), there are six magnetrons placed in a triangular pattern with two sets of three magnetrons each at two different angles relative to the sample surface normal. Another distinct advantage of this system is the possibility to rotate the wafer to produce intermediate layers that have a uniform composition and thickness, making it possible to optimize individual layers in a complex multilayer device, while other layers are kept at a constant thickness and composition.

In Figure 16.3(a), an image of a planetary deposition system is shown. This system is designed so that it can alternately deposit two gradients that are orthogonal to each other across a wafer. The four-planet planetary system consists of a table that rotates counter-clockwise around its mid-point.

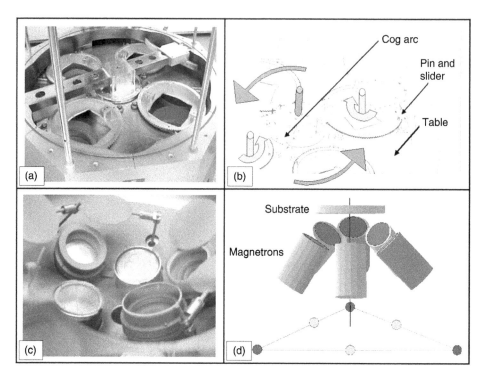

FIGURE 16.3 (a) A photograph of a planetary deposition system with four planets on a rotating table that is set up so that two orthogonal gradients can be deposited. (b) Schematic diagram showing the cog arc and the pin/slider that is used to rotate the sample 90° between each deposition position. (c) Photograph showing five of six magnetrons in a triangular pattern with their deposition direction tilted in relation to the substrate surface normal. (d) Drawing of the setup in (c) which illustrates the relationship of the magnetrons to the substrate.

The central non-rotating arm has a cog arc (also shown in Figure 16.3(b)) that can engage the rim of each planet as it moves past. This arc has the proper length to rotate the passing planet 90° counter-clockwise around the planet's center. Thereafter, for approximately 25% of one rotation of the table, that planet does not rotate around its own mid-point, the only motion is the non-stop movement of the table around its own mid-point. During this part of the rotation, the planet will pass over a magnetron sputter gun depositing one of the material gradients onto the sample that has been fixed to the planet. On the opposite side to the cog arc, the table will pass a stationary slider arm attached to the outer part of the deposition chamber. This slider will make contact with a pin on the rim of the planet and thus turn the planet 90° clockwise around the planet's mid-point as the planet passes by. The planet has now resumed its original position on the table and, during the final quarter turn of the table, the planet will pass the second magnetron and deposit the second and final material gradient. As mentioned, the system will alternately deposit two gradients that are 90° orthogonal to each other across a wafer for each revolution of the table. In this way, multilayers can be investigated. By carefully tailoring the magnetron deposition rate and the rotational speed to each other, one can, in each pass over a magnetron, deposit either a layer thin enough to ensure total mixing with the next layer by the energy from the process, or deposit a layer thick enough to form a unique layer. In this way, alloys, multilayers, and the transition region between layered and mixed materials can be investigated.

The highest accuracy or independency between the effects caused by the two gradients occur if the gradients are at right angles to each other, or in other words orthogonal. However, in order to introduce compositions with more than two materials, other angles have to be chosen. Figure 16.4(a) and

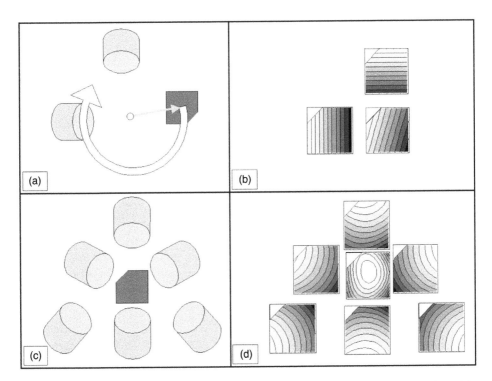

FIGURE 16.4 (a) Two magnetrons and the moving sample are depicted together with the direction of the main motion of the sample. (b) The film thickness/deposition rate across the sample from each magnetron is shown to the left and at the top, while in the bottom right corner is the resulting total thickness gradient for the film, for a given rate ratio between the magnetrons. (c) Six tilted magnetrons and the stationary sample are depicted together with the indication of a possible rotational motion to create uniform samples. (d) The six outer patterns show the thickness profile/deposition rate for each magnetron position while the center image shows the total for a case where all six magnetrons have almost similar rates.

Figure 16.4(b) show an orthogonal two-magnetron system in a simplified form. In Figure 16.4(a), the two magnetrons and the sample are depicted together with the main motion of the sample. In Figure 16.4(b), the effect of the deposition from each magnetron is shown to the left and at the top, while in the bottom right corner is the resulting total thickness gradient for the film. The linear thickness gradients produced by the individual magnetrons are constant across the wafer in the direction of the movement, and completely orthogonal to each other, while the angle of the total thickness gradient is dependent on the ratio between the two fluxes. This can be contrasted to the setup where the sample is stationary (Figure 16.4(c)). In the stationary case the radial spread from the magnetrons will give a pattern similar to those shown in Figure 16.4(d). The six outer patterns show the thickness profile or deposition rate for each magnetron position, while the center image shows the total for a case where all six magnetrons have similar rates.

When using a setup with a stationary sample, where the radial spread from the magnetrons will give patterns similar to those shown in Figure 16.4(d), there is still a possibility to produce fairly linear thickness gradients in the direction of the magnetron. Figure 16.5 shows a case with a Pt film that is 17 nm thick closest to the magnetron and 7 nm furthest away; the deviation from a linear slope is less than 5% in the worst case.

The linearity of a gradient can be further improved if a mask is placed a short distance from the source. The opening in the mask would then be tailored to the necessary changes in the total flux needed to make the gradient linear.

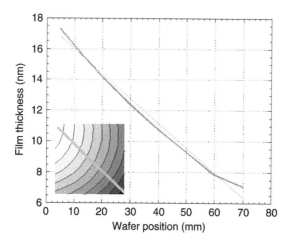

FIGURE 16.5 Measurement of linearity of a Pt film that is 17 nm thick closest to the magnetron and 7 nm furthest away; the deviation from a linear slope is less than 5% where the deviation is worst. The inset shows the complete thickness gradient across the wafer determined by X-ray diffraction in 36 data points.

However, this improvement in linearity is easily achieved by the other system as well (Figure 16.4(a) and Figure 16.4(b)) by placing similar masks over the sources. Figure 16.6(a) shows another type of mask that can be used to limit the deposition across a wafer to discrete points, Figure 16.6(b) shows a wafer deposited with this type of mask. Masking is useful in order to avoid different areas of the wafer to affect each other. One example where it can be important to separate a film into different areas is when the whole wafer is exposed to a magnetic field in order to measure the magnetic hysteresis loop, but each position on the wafer is measured "in turn." If the magnetic film is not separated in such a case, a magnetic domain wall can nucleate in one region of the wafer and sweep across the wafer and the region being measured thus giving rise to results that would differ from experiments made with uniform non-combinatorial samples. Figure 16.6(c) shows a drawing of the improved linear gradients across the wafer for one combinatorial layer while Figure 16.6(d) shows a possible complete multilayer deposited on a substrate.

16.2.1 EXAMPLE

Co/Pt multilayers with perpendicular anisotropy have been proposed as perpendicular recording media.[36,37] Several important factors such as magnetization direction and exchange bias are dependent on the thickness of the layers and the interface roughness between the layers. It has previously been found that below a certain Co layer thickness, ~0.8 nm, these materials exhibit a strong perpendicular anisotropy. This interfacial anisotropy can be exploited in Co/Pt multilayers and a strong out-of-plane anisotropy can be maintained for thicker films, making perpendicular recording media experiments possible.[38] By choosing the power to Co and Cr targets in the magnetron setup shown in Figure 16.3(c) and Figure 16.3(d) properly, it is possible to achieve almost orthogonal gradients of composition and layer thickness across a wafer surface. Figure 16.7 shows the thickness gradient from the Co and Cr magnetrons across the wafer surface, as determined by X-ray diffraction (XRD), as well as the expected total thickness and resulting Cr concentration in the co-deposited film. The compositional gradient is almost orthogonal to the thickness gradient as indicated by the orthogonal height curves in Figure 16.7(e). The final bilayer period was again determined by XRD and the period corresponds quite well to the thickness contours deviating less than 6% over the wafer surface. This CoCr layer can now be repeatedly deposited between uniform Pt layers. The setup in Figure 16.3 is suitable to produce alternating layers of uniform thin films with layers having a

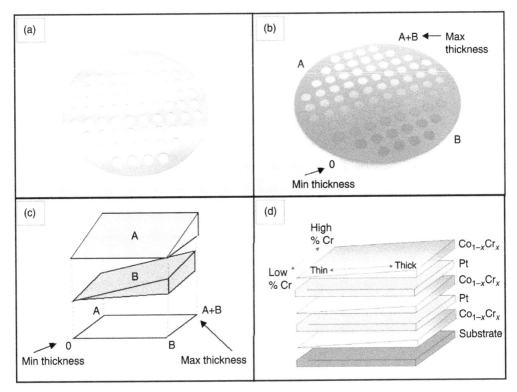

FIGURE 16.6 (a) Photograph of a mask for 75-mm-diameter wafers to produce separate "points" on a thin film. (b) Photograph of a wafer where material "A" and "B" have been deposited in a planetary system. (c) Schematic of the two orthogonal gradients that were generated across the wafer in (b). (d) Multilayer with alternating layers with a thickness and compositional ($Co_{1-x}Cr_x$) gradient and Pt layers of uniform thickness.

compositional as well as thickness gradient across the surface. Since the wafer is not translated in relation to the magnetron positions, it is possible to achieve the gradients without rotating the wafer around its center position. However, if the sample wafer is rotated around its center position and the focus of the magnetron is aimed slightly off the sample center, a relatively uniform film can be deposited.

A CoCr/Pt multilayer consisting of ten bi-layers was deposited by magnetron sputtering at an Ar pressure of 5 mTorr. The deposition rate was in the range of 0.02–0.07 nm/s. The thickness of each Pt layer was kept constant over the surface of the entire wafer, while the thickness of the CoCr layer was varied. The seed layer, as well as the Pt spacers, was deposited during a 10 rpm rotation of the substrate, thus creating a uniform film thickness within ±2%, while the CoCr layer was deposited without rotation.

In Figure 16.8, the correlation of the coercivity to the thickness and Cr concentration can be seen. A maximum of 1.0 kOe was achieved for 24% Cr content and a CoCr layer thickness of 0.60–0.65 nm. The hysteresis loops for all values have a high squareness equal or close to 1. Some shearing of the loop was present for the thickest CoCr layer, close to 0.8 nm; this finding might be indicative of a higher decoupling between the grains. However, as observed for this example, it is possible to relatively quickly find the maximum coercivity or the multilayer structure that gives the desired properties. Furthermore, since the measurements were made with an automated magneto-optic Kerr effect (MOKE) instrument, the whole library can be loaded in one run and the number of data points, and spacing between them, easily set from the program running the MOKE. The way

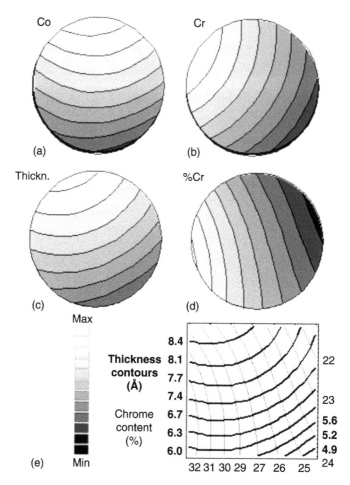

FIGURE 16.7 Deposition from a system of the type in Figure 16.3(c) and Figure16.3(d) where there are several magnetrons focused on a stationary disk. (a) Uniformity of Co thickness across the sample and (b) uniformity of Cr thickness. The thickest region, which has the lightest color, was closest to the magnetron. (c) Total thickness when the Co and Cr magnetrons co-deposit a CoCr layer and (d) the percentage of Cr in the co-deposited CoCr layer. (e) Gray scale and the thickness contours and composition gradient for the wafer surface.

data were illustrated here by a surface function, makes it easy to rapidly find the maxima or minima of the data and in which direction the data are changing. However, there is sometimes a risk involved with this way of depicting the data and a multiple "plot" approach is recommended. In Figure 16.9 three different ways of showing the same set of data across a wafer are depicted. In Figure 16.9(a) with raw numbers plotted at the correct position, Figure 16.9(b) with a two-dimensional second-degree polynom (in this case including the following components: constant, x, y, x^2y, xy^2, x^2, y^2) and as a gray scale, with the data interval of the image (0.270–0.985) translated to a gray scale ranging from black to white. In the first case it is difficult to see the trends, but it is possible to get some appreciation for the real absolute values. In the second case one can quickly obtain the trends, but it is possible to miss data points that are unusual, so-called "outliers." The final plot easily shows data points that are outliers and the trends at the same time, but not what the absolute values are. Using several of these multiple ways of plotting the data is recommended in order to minimize the risk of misinterpretation.

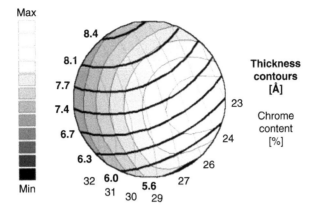

FIGURE 16.8 Coercivity across a wafer as a function of position on the wafer. The coercivity ranged from a minimum of 0.4 kOe to a maximum of 1.0 kOe across the wafer, as indicated by the gray scale. On the wafer at the right hand side, the gradients of thickness iso-lines (thick lines) and chrome content iso-lines (thin lines) are also shown.

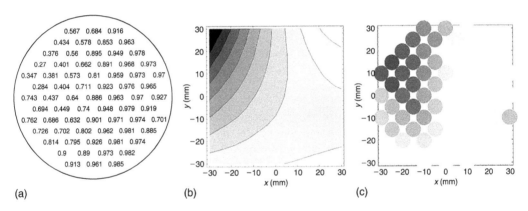

FIGURE 16.9 Three ways of showing the same set of data across a wafer: from left to right, (a) raw numbers, (b) with a fitted polynom (here the terms were: constant, x, y, x^2y, xy^2, x^2, y^2) and finally (c) with the data interval of the image (0.270–0.985) translated to a gray scale for each point.

16.3 MODELS

To make appropriate models of the dependencies and connecting composition and thickness to parameters such as coercivity or any other measured property, computer programs are needed and can be used to establish linear or non-linear models. The steps for determining a suitable model can be achieved by an iterative process, where model parameters or basis functions are cycled through a large number of permutations. In such a run, all the permutations or possible combinations of basis functions that are reasonable are also tested for validity as they are fitted. Within the model there are the basis functions fj specified, i.e., the predictors as functions of the independent variables. The resulting models for the response variable will be of the form: $y_i = \beta_1 f_1 i + \beta_2 f_2 i + \cdots + \beta_p f_p i + e_i$, where y_i is the ith response, f_{ji} is the jth basis function evaluated at the ith observation, and ei is the ith statistical error. In a combinatorial material case of the type described here, the

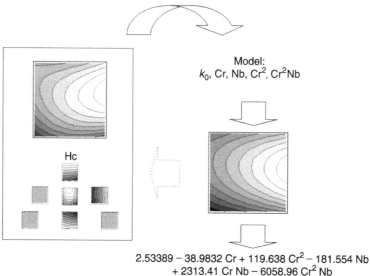

$$2.53389 - 38.9832 \ Cr + 119.638 \ Cr^2 - 181.554 \ Nb$$
$$+ 2313.41 \ Cr \ Nb - 6058.96 \ Cr^2 \ Nb$$

FIGURE 16.10 A "picture" of the iterative process, where model parameters (also called, basis functions) are cycled through a large number of permutations, in this case, constant k_0, Cr, Nb, CrNb, Cr^2, and Cr^2Nb are fitted to the input data in the rectangular box to the left. In a complete run all the permutations or possible model combinations that are reasonable are tested for validity and minimization of the statistical error is performed. For non-optimal solutions a new iteration is performed, the dotted arrow, or the final model is reached, bottom of the image.

coercivity, Hc, could be related to the alloy composition ratio between material A and B, and the film thickness, t, as follows: $H_c = \beta_1 A + \beta_2 B + \beta_3 AB + \beta_4 t + e$. In this equation, the base functions are linear combinations of the percentages of material A and B and the film thickness. Normally the terms of the type AB are called cross-terms and do not easily reveal themselves in typical experiments where only one variable is varied at a time.[39] Estimates of the coefficients β_1, β_2 ..., β_p can then be calculated to minimize $\Sigma_i e_i^2$, the error or residual sum of squares.

As an example, one can look at a case where Cr and Nb are used as additives to the Co used as a magnetic thin film for data storage. Parameters A and B in this case are thus the percentage of Cr and Nb concentrations in the total film as measured in each data point composing the set of samples for that test wafer.

In Figure 16.10 the iterative process is illustrated. The input parameters are found in the rectangular box to the left; they are the two-dimensional thickness gradients of Co, Cr, and Nb across the wafer. These gradients are shown as the three of the six plots that are non-gray in the "triangular" magnetron configuration. Since only three of the six magnetrons were in operation the unused ones are shaded uniform gray in the figure. Another input parameter is the total thickness, which is shown in the center of the six individual thickness gradients from the magnetrons. The final input parameter is the measured variable, in this case the coercivity (the large plot in the top of the box). In the iterative process the individual thickness gradients are used to calculate the composition concentrations across the wafer. As the possible models are being stepped through, in this iteration the basis functions were a constant k_0, and the value of the concentrations of Cr, Nb, CrNb, Cr^2, and Cr^2Nb, we can plot the model as seen in the large plot to the right, and compare it to the measured coercivity. In a complete run, all the permutations or possible combinations of terms making up a model that are reasonable, are tested for validity and minimization of the statistical error is done until a suitable model is found. In this case the final model with the best fit is shown in the lower part of the figure.

After the fit to a particular model under test has been calculated, the different basis functions have to be evaluated to assess if they correctly describe the measured data, or if any of the functions are redundant. To evaluate the importance of each basis function, estimates of the standard errors, and t-statistics can be performed by analysis of variance.[40–43] P-values can be calculated by comparing the obtained statistics to the t-distribution with $n - p$ degrees of freedom, where n is the sample size or number of data points and p is the number of basis functions used. This iterative process of studying the error or the remaining residues after the model is subtracted is continued until there are no more obvious dependencies remaining. The statistical P-values calculated indicate the validity of the different basis functions or input factors and hence can be used to validate their presence in the model, i.e., should a specific β_i be zero or not. The P-value can be regarded as the smallest α level at which the data are significant. Here, α is the probability of a type I error. (A type I error is to have the hypothesis rejected when it is true.) The values of P are normally accepted when they are below 0.05, a value that indicates the level of "risk" one is willing to take in order to include that base function in the model. Our t test is defined as a verification of the hypothesis. The hypothesis typically tests if there is a difference in the average between two sets of random samples from two different populations of samples. X_{an} is here a random sample n from population α. If $\bar{X}$ is the mean and σ the variance we can define Z as:

$$Z = \frac{\bar{X}_1 - \bar{X}_2 - (\mu_1 - \mu_2)}{\sqrt{\dfrac{\sigma_1^2}{n_1} + \dfrac{\sigma_2^2}{n_2}}} \qquad (16.1)$$

Now, the hypothesis that $\mu_1 - \mu_2 = 0$ can be tested. If Equation 16.1 falls in the interval from $-z_{\alpha/2}$ to $z_{\alpha/2}$, we will have to reject the hypothesis, with α being the desired level of confidence. However, in our case the hypothesis is if the β_i is zero or not. So again, the P-value is the smallest level of confidence that would lead to rejection of the hypothesis. A method to test the significance of this regression is called analysis of variance. The procedure partitions the total variability in the response variable into meaningful components as the basis of the test. The procedure has an identity as follows:

$$\sum_{i=1}^{n} (y_i - \bar{y})^2 = \sum_{i=1}^{n} (\hat{y}_i - \bar{y})^2 + \sum_{i=1}^{n} (y_i - \hat{y})^2 \qquad (16.2)$$

Equation 16.2 is in essence our t test where the two components of the right-hand side of Equation 16.2 measure the amount of variability in y_i accounted for by the regression line (the error sum of squares) and the residual variation left unexplained by the regression line (the regression sum of squares). For more information see references 40–43. Hence, the sum-of-squares calculation shows what the total portion of the data that is explained or predicted by the model and remaining error. The square of the multiple correlation coefficient is called the coefficient of determination R^2. Here R^2 is given by the ratio of the model sum of squares to the total sum of squares and it is a summary statistic that describes the relationship between the predictors and the response variable. The adjusted R^2, R^2_{adj}, is defined as:

$$R^2_{adj} = 1((n - 1)/(n - p))(1 - R^2). \qquad (16.3)$$

It provides a value that can be used to compare subsequent subsets of models. In the process of determining the underlying model, it is helpful to view the output parameters in relation to all the input parameters. Again, it is important to point out that, following these procedures, as described herein, will give models that are very accurate in representing the dependencies of all the variables used and the validity of incorporating each of the variables, it is not just one fit out of numerous possible fits.

The following example shows typical snapshots out of the iterative process of finding a suitable model for the data. The data set is from a wafer with a CoCrTa alloy where there is a composition and thickness gradient on the wafer, i.e., the percentage of Cr and Ta in the Cr alloy is changing. If one tries to fit the data set with simply one constant: $Hc = k_0$ the result is a large error (Table 16.1). In Table 16.1 the parameter included is listed under the parameter table together with the estimated value (from the optimization) and the P-value. Also listed is an ANOVA table showing the analysis of variance, as well as the degrees of freedom (DF) and the sum of squares for the model and the error as well as the total.

The sum of squares error indicates that this model, in Table 16.1, does not explain the behavior of the data. A much more realistic model is to use a dependency on the Cr concentration and fit the data to: $Hc = k_0 + k_1Cr$ to obtain values shown in Table 16.2

Now the sum of squares from the model is more than 95% of the total sum of squares (Table 16.2); i.e., the error represents less than 5%. Following the optimization of the Cr dependency the following fit to the data brings the process along: $Hc = k_0 + k_1Cr + k_2Cr^2 + k_3Cr^3$ with values shown in Table 16.3.

Now in Table 16.3 the sum of squares from the model is better than 97% of the total sum of squares; the error represents less than 3%. Larger P-values are obtained, but still within the 0.05 range. In the sequence of trials, the following fit to the data: $Hc = k_0 + k_1Cr + k_2Cr^2 + k_3Cr^3$ gives values shown in Table 16.4.

Even though the sum of squares in Table 16.4 for the model is higher than in the previous step, the P-values for the parameters are too high. This means that the addition of the Cr^4 term was incorrect, since it does not contribute in any useful way to explain the data. However, by dropping that parameter and continuing the process with the addition of Ta terms, eventually the following fit to the data will be tested: $Hc = k_0 + k_1Cr + k_2Cr^2 + k_3Cr^3 + k_4CrTa + k_5Ta$ with values shown in Table 16.5.

In this case (Table 16.5) the P-value for the Ta term is really high and should be dropped in the next run, and eventually the following and final equation will be obtained: $Hc = k_0 + k_1Cr + k_2Cr^2 + k_3Cr^3 + k_4CrTa$.

TABLE 16.1

Parameter Table	Estimate	P value
k_0	1.107	0.0

ANOVA Table	DF	Sum of
Model	0	0
Error	36	22.257
Total	36	22.257

TABLE 16.2

Parameter Table	Estimate	P value
k_0	−2.107	0.0
Cr	26.60	0.0

ANOVA Table	DF	Sum of
Model	1	21.234
Error	35	1.0237
Total	36	22.257

TABLE 16.3

Parameter Table	Estimate	P value
k_0	4.17	0.01185
Cr	−126.3	0.00397
Cr^2	1186.6	0.00153
Cr^3	−2952.2	0.00342

ANOVA Table	DF	Sum of
Model	3	21.677
Error	33	0.5798
Total	36	22.257

TABLE 16.4

Parameter Table	Estimate	P value
k_0	−3.529	0.59571
Cr	150.96	0.52344
Cr^2	−2441.8	0.42694
Cr^3	17518.1	0.31154
Cr^4	−42109.8	0.23767

ANOVA Table	DF	Sum of
Model	4	21.703
Error	32	0.5547
Total	36	22.257

TABLE 16.5

Parameter Table	Estimate	P value
k_0	4.216	0.00003
Cr	−129.87	0.0
Cr^2	1147.68	0.0
Cr^3	−2815.2	0.0
CrTa	59.774	0.29442
Ta	2.3631	0.73755

ANOVA Table	DF	Sum of
Model	5	22.0938
Error	31	0.16369
Total	36	22.257

One can immediately see that all the P-values in Table 16.6 are well below 0.05 and that the model explains 99.3% of the data by using the following model $Hc = 4.223 − 127.8Cr + 1125Cr^2 − 2767Cr^3 + 78.49CrTa$. As the fitting proceeds there are no additional models that explain the data. The model can now be used to predict further behavior within the same parameter space and to find optimal tradeoffs between coercivity and other parameters. As an example, the surface roughness of the film is usually proportional to the film thickness and ultimately linked to the corrosion stability of the film. If in this case the film thickness had affected the coercivity, a well-justified

TABLE 16.6

Parameter Table	Estimate	P value
k_0	4.223	0.00002
Cr	−127.83	0.0
Cr^2	1124.94	0.0
Cr^3	−2767.11	0.0
CrTa	78.4866	0.0

ANOVA Table	DF	Sum of
Model	4	22.0932
Error	32	0.16430
Total	36	22.257

tradeoff between coercivity and corrosion stability can be made. In the following section, it is possible to get an indication of the underlying physics based on the fast analysis achieved by the combinatorial approach.

As shown in Figure 16.11, one of the important tools to use in determining the appropriateness of a model is the examination of residues. The residue of each data point is the difference between the actual value in the data set and the predicted value by the model currently chosen. In Figure 16.11(a) the fit has been made to the simple model of a constant $k_0 = 1.1$ (the average of the data set) and the residues range from +1.40 to −1.01 as plotted against the coercivity range of the sample. In Figure 16.11(b) the data are now fitted to the model $k_0 + k_1Cr$, here one can see that this is still not the right answer to a complete description of the dependencies of the coercivity even though the residues now span a smaller range. In Figure 16.11(b) one can also make out "lines" of data points, in this case representing a direction across the wafer where the Ta concentration is changing. In Figure 16.11(c) the more extensive model of $H_c = k_0 + k_1Cr + k_2Cr^2 + k_3Cr^3$, does fit to the data only slightly better, the model is still lacking a Ta component and the "line pattern" is still present in the residues. With the final description in Figure 16.11(d) using the model $k_0 + k_1Cr + k_2Cr^2 + k_3Cr^3 + k_4CrTa$ the "lines" in the residue plot are gone and only a "random" pattern of residues in the +0.1 to −0.1 range is left. At this stage, where all the P-values for the parameters are well below 0.05 and there is no discernible pattern in the residue, plot one has a model that is ready for an interpretation attempt. In this case the following model $Hc = 4.223 − 127.8Cr + 1125Cr^2 − 2767Cr^3 + 78.49CrTa$ seems to best explain our data from a pure linear fit. The model does contain terms describing the coercivities dependency on Cr content. Hc is also dependent on the Ta concentration but not directly, instead it is the Cr Ta cross term CrTa that is the only Ta-containing term making sense during the linear fit.

Before looking at the scientific implications of the above result it is necessary to look at the immediate technical implications. Firstly, in this case the coercivity is not affected by the thickness of the film at all. Secondly, if one inputs some specific values of Cr and Ta concentration into the alloy it is possible to determine how much the coercivity is affected by these components. As an example 14% Cr and 8% Ta will give a coercivity of 4.2 − 3.43 + 0.87 kOe, given the 3.43 kOe is coming from the Cr containing base functions 0.87 kOe from the cross term containing both Cr and Ta and 4.2 is the constant. This way one can see that the biggest contribution comes from the Cr and that the cross term is 25% of the solitary Cr term. Hence, in a production process it is more important to keep the Cr addition within limits and that the deposition time, i.e., the thickness, is not important for the final coercivity of the film. (However, the total thickness will determine the total moment and is thus still important. But, in our case it is independently adjustable.) From a

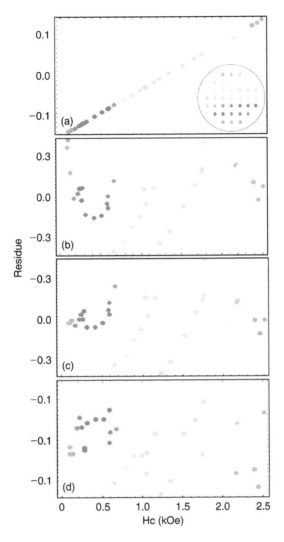

FIGURE 16.11 (See color insert following page 172) Graphs showing the residue of all data points sorted according to coercivity value. In image (a) the fit is to k_0 and the residues range from $+1$ to -1. (b) Data fitted to $k_0 + k_1 Cr$, (c) $k_0 + k_1 Cr + k_2 Cr^2 + k_3 Cr^3$, (d) $k_0 + k_1 Cr + k_2 Cr^2 + k_3 Cr^3 + k_4 CrTa$.

scientific point of view what has been observed when Cr is added to Co at elevated deposition temperatures is that the Cr segregates to the grain boundaries, where it aids in decoupling the grains and thus enhances the coercivity. Inaba et al.[44] found from transmission electron microscopy analysis that an alloy with an average of 15% Cr had a concentration of 23% Cr within 1 nm of the grain boundary, while the rest of the grain was at half the average concentration. Furthermore, Kemner et al.[45] have shown, by extended X-ray absorption analysis, that the Ta atoms are preferentially distributed to the Cr-enriched regions in a CoCr film and that the Ta has a direct effect on the average local environment of the Cr atoms by introducing a large amount of local structural disorder. Thus it can be concluded that the Ta enhances the effect of Cr and that the effect of Ta itself is negligible in increasing the coercivity. In light of this physical explanation of the CoCrTa alloy, it is understandable that there will be base functions describing the dependency of Cr and a cross-term having both Cr and Ta, while there is no single Ta term.

TABLE 16.7

Physical Vapor Deposition	Input Parameters	Responses	Under Study
Sputtering	Temperature	Coercivity	Seed layers
Evaporation	Thickness	Squareness	Corrosion
Electron beam	Chemical composition	Total magnetic moment	Lubrication
Spraying		Surface roughness	Additives
		Decay rates	Stability
			Topography

16.4   SUMMARY

In general, combinatorial methods can be employed in two ways: (i) for discovery of new materials, i.e., to find so-called "sweet spots" or (ii) for optimization of structures or specific layers to reach optimal performance across a set of, often contradictory, properties. Although combinatorial approaches are well known in the chemistry arena for discovery/testing of drugs, physical deposition methods (Table 16.7), lend themselves to this approach after the modifications discussed. For magnetic applications, composition, temperature, and thickness are the most often varied input parameters (see Table 16.7). There is no limit to the number of responses that can be measured. However, on a practical note, only the most important properties that can be assessed rapidly are assessed in the initial screening stages. For more on combinatorial screening see for example references 46–50.

REFERENCES

1. O'Grady, K. and Laidler, H., The limits to magnetic recording – media considerations, *J. Magn. Magn. Mater.*, 200(1–3), 616, 1999.
2. Svedberg, E. B., Van de Veerdonk, R., Howard, K. J. and Madsen, L. D., Method for seed and underlayer optimization of perpendicular magnetic recording media, *J. Vac. Sci. Technol. A*, 20(4), 1341, 2002.
3. Shukla, N., Svedberg, E. B., van der Veerdonk, R. J. M., Ma, X., Gui, J. and Gellman, A. J., Water adsorption on lubricated a-CH$_x$ in humid environments, *Tribology Lett.*, 15(1), 9, 2003.
4. Svedberg, E. B. and Shukla, N., Adsorption of water on lubricated and non lubricated TiC surfaces for data storage applications, *Tribology Lett.*, 17(4), 973, 2004.
5. Khizroev, S. and Litvinov, D., Perpendicular magnetic recording: Writing process, *J. Appl. Phys.*, 95(9), 4521, 2004.
6. Svedberg, E. B., Howard, K. J., Bønsager, M. C., Pant, B. B., Roy, A.G. and Laughlin, D. E., Inter-diffusion in CoFe/Cu multilayers and its application to spin-valve structures for data storage, *J. Appl. Phys.*, 94(2), 1001, 2003.
7. Svedberg, E. B., Howard, K. J., Bønsager, M. C., Pant, B. B., Roy, A. G. and Laughlin, D. E., Diffusion in Co$_{90}$Fe$_{10}$/Ru multilayers, *J. Appl. Phys.*, 94(2), 993, 2003.
8. Svedberg, E. B., Litvinov, D., Gustafson, R., Chang, C. H. and Khizroev, S., Magnetic force microscopy of skew angle dependencies in perpendicular magnetic recording, *J. Appl. Phys.*, 93(5), 2828, 2003.
9. Svedberg, E. B., Khizroev, S. and Litvinov, D., Signal-to-noise deterioration in perpendicular storage media by thermal and magnetic field aging as determined by magnetic force microscopy, *J. Appl. Phys.*, 92(11), 6714, 2002.
10. McGee, N. W. E., Johnson, M. T., de Vries, J. J. and aan de Stegge, J., Localized Kerr study of the magnetic properties of an ultrathin epitaxial Co wedge grown on Pt(111), *J. Appl. Phys.*, 73(7), 3418, 1993.
11. Katayama, T., Suzuki, Y., Hayashi, M. and Geerts, W., Change of magneto-optical Kerr rotation due to interlayer thickness in magnetically coupled films with noble-metal wedge, *J. Appl. Phys.*, 75(10), 6360, 1994.

12. Krishnan, K. M., Nelson, C., Echer, C. J., Farrow, R. F. C., Marks R. F. and Kellock, A. J., Exchange biasing of permalloy films by Mn_xPt_{1-x}: Role of composition and microstructure, *J. Appl. Phys.*, 83(11), 6810, 1998.
13. Lee, J., Lauhoff, G., Hope, S., Daboo, C., Bland, J. A. C., Schillé, J. Ph., van der Laan, G. and Penfold, J., Variation of the magnetic moment and strain in epitaxial Cu/Ni/Cu sandwiches, *J. Appl. Phys.*, 81(8), 3893, 1997.
14. Hayashi, H., Ishizaka, A., Haemori, M. and Koinuma, H., Bright blue phosphors in ZnO–WO_3 binary system discovered through combinatorial methodology, *Appl. Phys. Lett.*, 82(9), 1365, 2003.
15. Wu, Y., Parkin, S. S. P., Stohr, J., Samant, M. G., Hermsmeier, B. D., Koranda, S., Dunham, B. and Tonner, B. P., Direct observation of oscillatory interlayer exchange coupling in sputtered wedges using circularly polarized x rays, *Appl. Phys. Lett.*, 63(2), 263, 1993.
16. Li, D., Freitag, M., Pearson, J., Qiu, Z. Q. and Bader, S. D., Magnetic and structural instabilities of ferromagnetic and antiferromagnetic Fe/Cu(100), *J. Appl. Phys.*, 76(10), 6425, 1994.
17. Ives, J. R., Hicken, R. J., Bland, J. A. C., Daboo, C., Gester, M. and Gray, S. J., High-field polar MOKE magnetometry as a probe of interlayer exchange coupling in MBE-grown Co/Cu/Co(111) and Fe/Cr/Fe(001) wedged trilayers, *J. Appl. Phys.*, 75(10), 6458, 1994.
18. Bader, S. D., Li, Dongqi and Qiu, Z. Q., Magnetic and structural instabilities of ultrathin Fe(100) wedges, *J. Appl. Phys.*, 76(10), 6419, 1994.
19. Qiu, Z. Q., Pearson, J. and Bader, S. D., Magnetic coupling of Fe/Mo/Fe and Co/Cu/Co sandwiches across wedged spacer layers, *J. Appl. Phys.*, 73(10), 5765, 1993.
20. Takeuchi, I., Chang, H., Gao, C., Schultz, P. G., Xiang, X.-D., Sharma, R. P., Downes, M. J. and Venkatesan, T., Combinatorial synthesis and evaluation of epitaxial ferroelectric device libraries, *Appl. Phys. Lett.*, 73(7), 894, 1998.
21. Aronova, M. A., Chang, K. S., Takeuchi, I., Jabs, H., Westerheim, D., Gonzalez-Martin, A., Kim, J. and Lewis, B., Combinatorial libraries of semiconductor gas sensors as inorganic electronic noses, *Appl. Phys. Lett.*, 83(6), 1255, 2003.
22. Bloemen, P. J. H., van Dalen, F. L., de Jonge, W. J. M., Johnson, M. T. and aan de Stegge, J., Short period oscillation of the interlayer exchange coupling in the ferromagnetic regime in Co/Cu/Co(100), *J. Appl. Phys.,* 73(10), 5972, 1993.
23. Bloemen, P. J. H., van de Vorst, M. T. H., Johnson, M. T., Coehoorn, R. and de Jonge, W. J. M., Magnetic layer thickness dependence of the interlayer exchange coupling in (001) Co/Cu/Co, *J. Appl. Phys.*, 76(10), 7081, 1994.
24. Chang, H., Gao, C., Takeuchi, I., Yoo, Y., Wang, J., Schultz, P. G., Xiang, X.-D., Sharma, R. P., Downes, M. and Venkatesan, T., Combinatorial synthesis and high throughput evaluation of ferroelectric/dielectric thin-film libraries for microwave applications, *Appl. Phys. Lett.*, 72(17), 2185, 1998.
25. Cheng, H.-W., Zhang, X.-J., Zhang, S.-T., Feng, Y., Chen, Y.-F., Liu, Z.-G. and Cheng, G.-X., Combinatorial studies of $(1-x)Na_{0.5}Bi_{0.5}TiO_3-xBaTiO_3$ thin-film chips, *Appl. Phys. Lett.*, 85(12), 2319, 2004.
26. Xiang, X.-D. and Schultz, P. G., The combinatorial synthesis and evaluation of functional materials, *Physica C*, 282–287, 428, 1997.
27. Takeuchi, I., Chang, K., Sharma, R. P., Bendersky, L. A., Chang, H., Xiang, X.-D., Stach, E. A. and Song, C.-Y., Microstructural properties of (Ba, Sr)TiO_3 films fabricated from $BaF_2/SrF_2/TiO_2$ amorphous multilayers using the combinatorial precursor method, *J. Appl. Phys.*, 90(5), 2474, 2001.
28. Vossmeyer, T., Jia, S., DeIonno, E., Diehl, M. R., Kim, S.-H., Peng, X., Alivisatos, A. P. and Heath, J. R., Combinatorial approaches toward patterning nanocrystals, *J. Appl. Phys.*, 84(7), 3664, 1998.
29. Liu, X. Q., Li, Z. F., Lu, W., Shen, S. C., Chen, C. M., Zhu, D. Z., Hu, J. and Li, M.-Q., Application of the combinatorial approach to the fabrication of a quantum well multiwavelength emitting chip, *Appl. Phys. Lett.,* 75(17), 2611, 1999.
30. Litvinov, D., Wolfe, J. C., Svedberg, E. B., Ambrose, T., Howard, K., Chen, F., Schlesinger, T. E. and Khizroev, S., Ion implantation of magnetic thin films and nanostructures, *J. Magn. Magn. Mater.*, 283, 128, 2004.
31. Shtein, M., Peumans, P., Benziger, J. B. and Forrest, S. R., Direct mask-free patterning of molecular organic semiconductors using organic vapor jet printing, *J. Appl. Phys.*, 96(8), 4500, 2004.
32. Yoshikawa, A., Boulon, G., Laversenne, L., Canibano, H., Lebbou, K., Collombet, A., Guyot, Y. and Fukuda, T., Growth and spectroscopic analysis of Yb^{3+}-doped $Y_3Al_5O_{12}$ fiber single crystals, *J. Appl. Phys.*, 94(9), 5479, 2003.

33. Vandervelde, T. E., Kumar, P., Kobayashi, T., Gray, J. L., Pernell, T., Floro, J. A., Hull, R. and Bean, J. C., Growth of quantum fortress structures in $Si_{1-x}Ge_x$/Si via combinatorial deposition, *Appl. Phys. Lett.*, 83(25), 5205, 2003.

34. Svedberg, E. B., CoCr/Pt multilayers with perpendicular anisotropy and texture-controlled coercivity, *J. Appl. Phys.*, 92(2), 1024, 2002.

35. Svedberg, E. B., Van de Veerdonk, R. J. M., Howard, K. J. and Madsen, L. D., Quantifiable combinatorial materials science approach applied to perpendicular magnetic recording media, *J. Appl. Phys.*, 93(9), 5519, 2003.

36. Hashimoto, S., Ochiai, Y. and Aso, K., Perpendicular magnetic anisotropy and magnetostriction of sputtered Co/Pd and Co/Pt multilayered films, *J. Appl. Phys.*, 66, 4909, 1989.

37. Hashimoto, S., Ochiai, Y. and Aso, K., Ultrathin Co/Pt and Co/Pd multilayered films as magneto-optical recording materials, *J. Appl. Phys.*, 67, 2136, 1990.

38. Maat, S., Takano, K., Parkin, S. S. P. and Fullerton, E. E., Perpendicular exchange bias of Co/Pt multilayers, *Phys. Rev. Lett.*, 87, 087202, 2001.

39. Madsen, L. D. and Weaver, L., In situ doping of silicon films prepared by low pressure chemical vapour deposition using disilane and phosphine, *J. Electrochem. Soc.*, 137, 2246, 1990.

40. Montgomery, D. C., *Design and Analysis of Experiments*, Wiley, New York, 1997.

41. Montgomery, D. C. and Runger, G. C., *Applied Statistics and Probability for Engineers*, Wiley, New York, 1999.

42. Myers, R. H. and Montgomery, D. C., *Response Surface Methodology*, Wiley, New York, 1995.

43. Box, G. E. P. and Draper, N. R., *Empirical Model Building and Response Surfaces*, Wiley, New York, 1987.

44. Inaba, N. and Futamoto, M., Effects of Pt and Ta addition on compositional microstructure of CoCr-alloy thin film media, *J. Appl. Phys.*, 87, 6863, 2000.

45. Kemner, K. M., Harris, V. G., Elam, W. T., Feng, Y. C., Laughlin, D. E., Woicik, J. C. and Lodder, J. C., Preferential site distribution of dilute Pt and Ta in CoCr-based films: An extended x-ray absorption fine structure study, *J. Appl. Phys.*, 82, 2912, 1997.

46. Potyrailo, R. A. and Morris, W. G., Parallel high-throughput microanalysis of materials using microfabricated full bridge device arrays, *Appl. Phys. Lett.*, 84(4), 634, 2004.

47. Potyrailo, R. A., Enhancement in screening throughput and density of combinatorial libraries using wavelet analysis, *Appl. Phys. Lett.*, 84(25), 5103, 2004.

48. Olk, C. H., Tibbetts, G. G., Simon, D. and Moleski, J. J., Combinatorial preparation and infrared screening of hydrogen sorbing metal alloys, *J. Appl. Phys.*, 94(11), 720, 2003.

49. Isaacs, E. D., Marcus, M., Aeppli, G., Xiang, X.-D., Sun, X.-D., Schultz, P., Kao, H.-K., Cargill, G. S. III and Haushalter, R., Synchrotron x-ray microbeam diagnostics of combinatorial synthesis, *Appl. Phys. Lett.*, 73(13), 1820, 1998.

50. Drolet, F. and Fredrickson, G. H., Combinatorial screening of complex block copolymer assembly with self-consistent field theory, *Phys. Rev. Lett.*, 83(21), 4317, 1999.

17 High-Throughput Screening of Next Generation Memory Materials

Chang Hwa Jung, Eun Jung Sun, and Seong Ihl Woo

CONTENTS

17.1 INTRODUCTION

Combinatorial chemistry is defined as high-speed R&D.[1] The combinatorial chemistry and high-throughput experimentation have emerged during the last decade as a response to the challenges of materials development in these increasingly complex experimental spaces.[2] Although its antecedents can be traced to the beginning of the 20th century, combinatorial experimentation really began to take off around 1990 in the pharmaceutical industry and 1995 in materials development.[3–5] This resulted from a convergence of technologies in robotics, semiconductor processing, computer software, and analytical capabilities, and from the simple realization that such productivity was possible.[6–9] The rapid growth of this technology brought the revolution in the field of materials science research. With this technique, we can synthesize a thin-film array with various chemical compositions on a wafer and then carry out the parallel characterization of the physical and chemical properties of the samples rapidly.[10] This combinatorial method helps us to find suitable materials for each application and finally design and predict the structure or chemical composition with desired properties. In this chapter we focus on the application of combinatorial methods in non-volatile memory chips.

17.2 DYNAMIC/FERROELECTRIC RANDOM ACCESS MEMORY (DRAM/FRAM)

17.2.1 DRAM

For DRAM, many capacitor materials are suggested. Among these materials, $(Ba,Sr)TiO_3$ (BST) is one of the most widely researched materials. Composition study of the dielectric materials is one of the main focuses of the recent researches because the properties of these materials are sensitive to composition. That is why combinatorial method and high-throughput screening technique are useful. In the earlier stages of combinatorial methods, researchers manufactured the combinatorial chips by using masks (Figure 17.1). But in this case, too many masks would be needed to obtain the combi-chips that have diverse compositions. To overcome this problem, two methods are suggested: one is to use the moving shutter (Figure 17.1) and the other is the co-deposition method (Figure 17.2).

Takeuchi et al.[11] made the $(Ba,Sr)TiO_3$ combi-chip using pulsed-laser deposition. The moving shutter could make the composition spread on the substrate. But when the film is fabricated from the two metal precursors, $BaTiO_3$ and $SrTiO_3$, the film may retain the amorphous state. To overcome this problem, the $(Ba,Sr)TiO_3$ combi-chip is deposited at above 600°C using pulsed-laser deposition with the moving shutter.[12–14] After the annealing process, it was ensured that the film became crystalline in character and showed a well-defined diffraction pattern. After the fabrication the dielectric constant of the fabricated BST combi-chip was measured by using scanning evanescent microwave microscopy (SEMM), which measures the impedance characteristic very quickly. On measuring the dielectric constant, this device applied the microwave energy with high resonant frequency (above 1 GHz) into the sample to occur with the resonant frequency shift. Such a change in the frequency can be converted into the dielectric constant. On the theoretical basis, the evanescent microwave microscope is a powerful device for measuring the independent characteristics including the dielectric constant: it is non-destructive and has excellent spatial resolution. As a result, it is revealed from SEMM that $Ba_{0.65}Sr_{0.35}TiO_3$ film exhibited the highest dielectric constant and the composition of $Ba_{0.65}Sr_{0.35}TiO_3$ is the transition point from paraelectric phase to ferroelectric phase (Figure 17.3).

17.2.2 FRAM

Ferroelectric materials have been studied worldwide because they are used as the capacitor of ferroelectric random access memory (FRAM).[15,16] The memory has non-volatile characteristics as well as the merits of DRAM including high-speed read/write and high integration. This is due to the fact that FRAM has the same structure (1T-1C) as DRAM. For commercial applications of FRAM, it is very important to develop ferroelectric material with the desired properties.[17,18] The ability of ferroelectric materials to switch their polarization direction between two stable polarized states provides the basis for the binary code-based non-volatile ferroelectric random access memories (FRAM).

In general, uniform alignment of electric dipoles only occurs in certain regions of a crystal, while in other regions of the crystal spontaneous polarization may be in the reverse direction. Such regions with uniform polarization are called ferroelectric domains. The ferroelectric domains are generated in ferroelectric materials because the system with polydomain is in a state of minimum free energy. The interface between two domains is called the domain wall.[16] Figure 17.4 shows P–E hysteresis loop of typical ferroelectric materials. If we first apply a small electrical field, we will have only a linear relationship between polarization and electric field because the field is not large enough to switch any domains and the crystal will behave as a normal dielectric material (paraelectrics). This case corresponds to the segment OA of the curve. As the electrical field strength increases, a number of the negative domains (which have a polarization opposite to the direction of the field) will be switched over in the positive direction (along the field direction) and the polarization will increase rapidly (segment AB) until all the domains are aligned in the positive direction

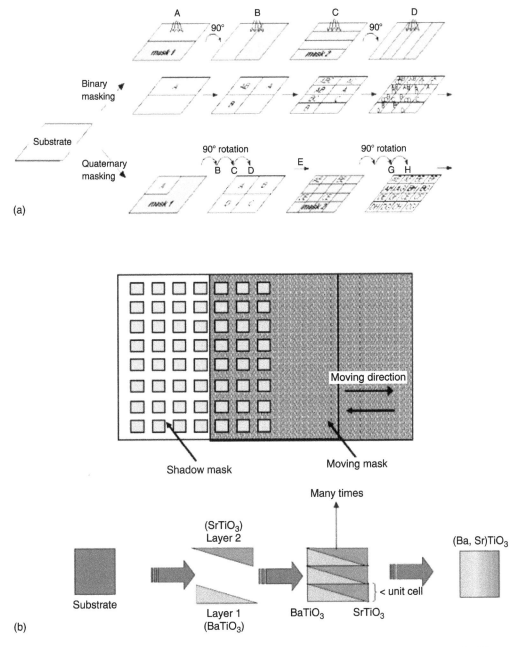

FIGURE 17.1 Masking schemes for combinatorial thin-film library; (a) binary or quaternary mask, (b) movable mask (moving shutter).

(segment BC). As the field strength decreases, the polarization will generally decrease (at the point D) but never return to zero. When the field is reduced to zero, some of the domains will remain aligned in the positive direction and the crystal will exhibit P_r. The extrapolation of the linear segment BC of the curve back to the polarization axis (at the point E on the vertical axis) represents the value of the P_s. The P_r in a crystal cannot be removed until the applied field in the opposite (negative) direction reaches a certain value (at the point F). The strength of the field required to reduce the polarization to zero is called the coercive field (E_c). Further, increase of the field in the negative

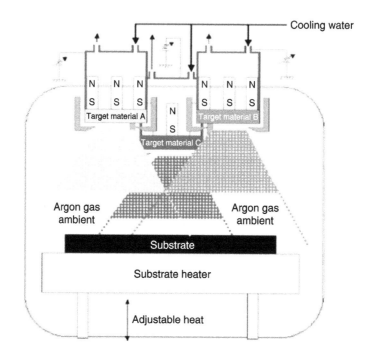

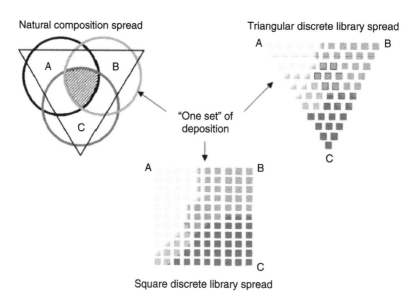

FIGURE 17.2 Combinatorial thin-film library synthesis by co-sputtering.

direction will cause a complete alignment of the dipoles in the direction and the cycle can be completed by reversing the field direction once again.[16] The fact that ferroelectric layers can maintain an induced polarization, even in the absence of an external electric field, provides the unique non-volatility of FRAM. Among various ferroelectric materials, Pb(Zr,Ti)O$_3$ (PZT) has been widely investigated because PZT has the advantages of large remanent polarization (P$_r$) (20–70 μC/cm^2) and low processing temperature (500–600°C). But PZT has the critical problem in fatigue endurance. Therefore other materials are suggested, such as SrBi$_2$Ta$_2$O$_9$ (SBT), La-substituted BTO(Bi$_4$Ti$_4$O$_{12}$),

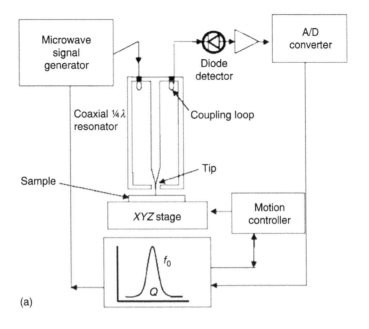

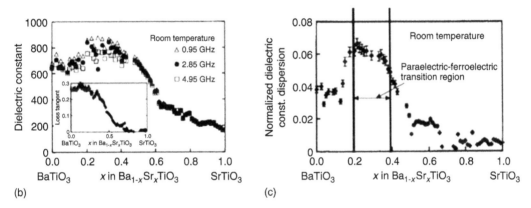

FIGURE 17.3 (a) Schematic drawing of a SEMM, (b) dielectric constant vs composition on a $(Ba,Sr)TiO_3$ spread at 0.95, 2.85, and 4.95 GHz at room temperature, (c) normalized dielectric dispersion.

Nd, Sm, and Ce-substituted BTO. In these substitutions, the ferroelectric properties of these materials are very dependent on the chemical composition. Therefore, combinatorial method and high-throughput screening technique can be useful in optimizing the chemical compositions.

17.3 PHASE CHANGE RANDOM ACCESS MEMORY (PRAM)

17.3.1 INTRODUCTION

The phase change memory is a reasonable substitute for all kinds of memory such as DRAM, SRAM, and flash memory because of simple cell structure with high scalability, non-volatility, high read/write operation speed and long cycle life.[19] Also, it has radiation tolerance and can be used for space-based applications[20] as it uses the distinction of the resistance difference between amorphous and crystalline states.

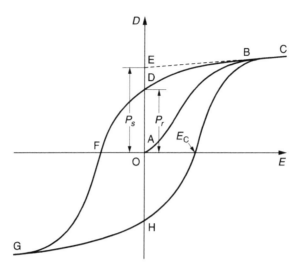

FIGURE 17.4 A typical P–E hysteresis loop in ferroelectric.

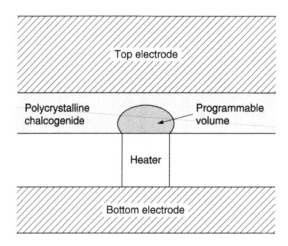

FIGURE 17.5 Basic cross-section of the phase change memory.

In phase change memory, the electric current is applied for the transition of two states. When the electric current heats the phase change material, the programmable volume around the contact region changes (Figure 17.5).[21] For the phase change from the amorphous state to the crystalline state, a temperature between crystallizing temperature and melting point is needed. Therefore, a medium current for a longer pulse time is used to crystallize the phase change material. This crystalline state has lower resistance than that of the amorphous state. This kind of phase change is called a set operation. On the other hand, for the phase change from the crystalline state to the amorphous state, the temperature above the melting point is needed. Therefore, higher current and fast quenching are used to make the material amorphize. This kind of phase change is called a reset operation. The reset operation is the write process and the set operation is the erase process. Figure 17.6 shows the pulse profile for writing and erasing processes.[22] Figure 17.7 shows the I–V curve of the phase change random access memory.[21] We considered the set operation (crystallizing process) firstly. When the voltage reaches the threshold value, the phase change material shifts to the high

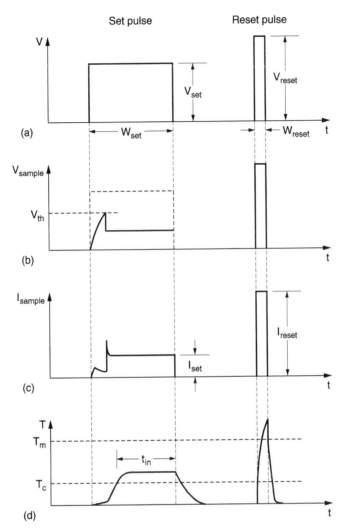

FIGURE 17.6 Schematic voltage, current waveforms, and temperature for the set and reset operations; (a) waveforms of the driving voltage, (b) voltage across the sample, (c) current flowing through the sample, (d) temperature of the sample due to joule heat.

conductive state (low resistance) and we can observe the negative resistance. Then, the current exceeds a minimum level and the I–V curve is identical to the set state. This gives the writing ability characteristics of the memory.

17.3.2 Combinatorial Method for PRAM

The attempts for the improvement of phase change memory are classified into three groups, cell physics study, cell program current reduction, and high-density array manufacturing development.[21] Among these issues, we focus on cell program current reduction because this problem inhibits the low-power consumption operation. Two types of the innovative ideas have been suggested to solve this problem. One is the edge contact method and other is the development of high-quality phase change material.[21] A high-current pulse is required for the reset operation because the phase change material must be heated above the melting point (Figure 17.6). To reduce the power consumption,

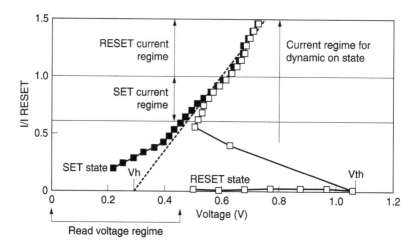

FIGURE 17.7 I–V curve of basic memory element.

the materials of relatively low melting point are needed. Many phase-change materials such as GeSbTe, InSbTe, and GeTe were also suggested because they can be deposited easily on the amorphous phase.[22–33] Among these materials, GeSbTe alloy, which is the same family of material used in optical re-writable CD/DVD RW disks, is the most widely studied material. Composition study of the phase-change materials is the main focus of recent research because the properties of chalcogenide-based phase-change materials are sensitive to composition. That is why the combinatorial method and high-throughput screening techniques are useful in the study.

Krysta et al. fabricated a Ge-Sb-Te alloy using a four-target multi-magnetron sputtering.[34] The films were deposited from separate Ge, Sb, Te, targets on the Si(111) wafers. Figure 17.8 shows the arrangement of the target guns. Each target gun was fixed with a tilting angle of 26°. This tilting enabled to obtain the GeSbTe alloy with various compositions in a wafer. Figure 17.9 shows these various compositions in a wafer using the electron probe mapping analysis (EPMA). A laser diode beam heated the GeSbTe film to change the states between the amorphous and the crystalline. Figure 17.10 shows the mappings of the crystallization time and the writing time of GeSbTe film on the Si wafer. From these two figures, we can obtain the information about the relations between the composition and the crystallization time and between the composition and the writing time. Besides, there are some reports using the combinatorial method in researching the GeSbTe alloy, but these reports have many points in common.[35–37] These reports opened up the possibility of a combinatorial approach on the study of the phase change materials.

17.4 MAGNETORESISTIVE RANDOM ACCESS MEMORY (MRAM)

17.4.1 INTRODUCTION

In a MRAM device, the phenomenon of magnetism is used to store the data. Ferromagnetic material is strongly affected by the external magnetic field. If there is no magnetic field, spins in the ferromagnetic materials exist randomly formed. Therefore, their net magnetic field vector is zero. But in the magnetic atmosphere, they are aligned in the direction of magnetic field and their magnetic properties are retained after the external field has been removed. Because of their excess magnetic properties, ferromagnetic materials are able to be used for non-volatile memory.

A MRAM cell structure includes both bit line and digit lines, which are crossed to each other on a chip. To store bits of data, the MRAM device demands only a small amount of electricity. If a small amount of electricity flows in the bit line and digit line, the magnetic field is formed. In this

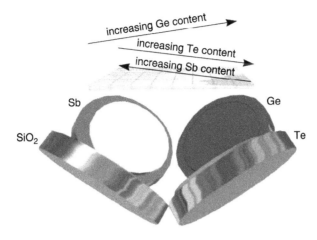

FIGURE 17.8 Arrangement of the cathodes and resulting compositional gradient of the films.

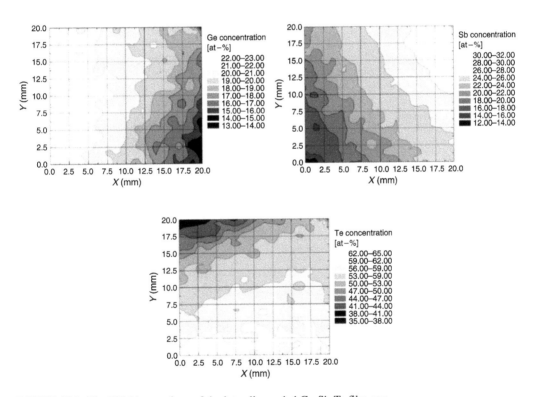

FIGURE 17.9 The EPMA mappings of the laterally graded Ge-Sb-Te film one.

way, we can write the "0" or "1". We can write bits of data using magnetic hysteresis and we can read the data using magnetoresistance of the same cell in which the data are stored (Figure 17.11).[38]

The giant-magnetoresistive (GMR) materials are able to read stored data. Their structure is composed of magnetic film sandwiching ferromagnetic and non-magnetic metal layers such as a non-magnetic copper layer sandwiched with a ferromagnetic cobalt layer. There is little difference between up-spin and down-spin electrons in the copper layer, which is in contrast to the cobalt layer where quite a gap between up-spin and down-spin electrons exists. In the parallel-magnetized

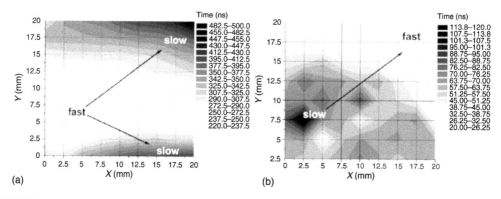

FIGURE 17.10 (a) Mapping of the crystallization time on the laterally film one. (b) Mapping of the writing time on the laterally graded film two with an equivalent composition as compared to film one but an additional Al/SiO$_2$ layer as a heat sink.

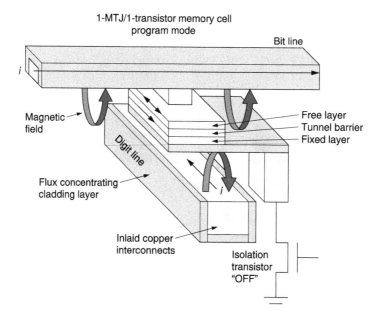

FIGURE 17.11 MRAM cell.

cobalt layer atmosphere, up-spin electrons pass through the copper layer and reach the cobalt layer without scattering. Therefore, the up-spin electrons move freely all over the layer. But in the antiparalled magnetized cobalt layer atmosphere, up-spin electrons are unable to reach the cobalt layer. They only move in the one side of the cobalt layer and copper layer. In the case of parallel magnetized cobalt layer, there is low resistance and we can read its state to "1". In the other case of antiparallel magnetized cobalt layer, there is higher resistivity and we can also read its state to "0".

17.4.2 COMBINATORIAL METHOD FOR MRAM

Fukumura et al.[39] fabricated colossal magnetoresistive oxide La$_{1-x}$Sr$_x$MnO$_3$ film by using combinatorial laser molecular beam epitaxy (laser MBE). The film was fabricated by two target materials (LaMnO$_3$ and SrMnO$_3$) and the resultant La$_{1-x}$Sr$_x$MnO$_3$ film was characterized by using concurrent XRD, scanning superconducting quantum interference device (SQUID) microscopy. The structure of the film was analyzed by XRD and the ferromagnetic property was measured by SQUID.

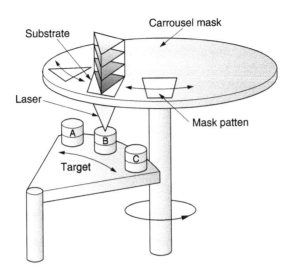

FIGURE 17.12 Schematic illustration of combinatorial pulsed laser deposition system.

Especially, SQUID is a very powerful tool to evaluate the ferromagnetic property – even a very weak magnetic field of a thin film. Fukumura et al.[39] measured the reflectivity and then found that the reflectivity of $La_{1-x}Sr_xMnO_3$ was maximum at $x = 0.47$.

Takahashi et al.[40] recommended the improved combinatorial deposition system with a characteristically designed mask. The important feature of this process is that the apparatus can control the ternary phase diagrams directly between library location and composition. The schematic illustration is shown in Figure 17.12.

In this method, we can obtain the higher screening rate of ternary composition materials and it would be extended to complex multicomponent combinatorial systems.

Svedberg et al.[41] fabricated perpendicular magnetic recording materials using Pt, Ta, and Ti-doped CoCr alloy. The ideal perpendicular magnetic recording materials should have the comparatively square hysteresis loops structure, the remanent magnetization (M_r) of at least several hundred kA/m, appropriate coercivity, stability against stress, as well as temperature and corrosive effect. The composition and thickness of the samples were changed to optimize the ferromagnetic property. In the case of CoCrPtTa layer, the film thickness was changed from 47.5 nm to 55.9 nm and the composition of CoCrPtTa thin film was changed. As a result, the concentration of Pt significantly affected the magnetic properties.

Koinuma et al.[42] investigated some features and magnetic properties of $La_{1-x}Ca_xMnO_3$ (LCMO) and $Nd_{1-x}Sr_xMnO_3$ (NSMO) films using SQUID. They studied the magnetic properties of $Nd_{1-x}Ca_xMnO_3$ (NCMO) film at various temperatures and finally concluded that magnetic properties of NCMO films were not affected by temperature below 100 K. Local magnetic properties of $Nd_{1-x}Sr_xMnO_3$ films were investigated using SQUID at 3 K and observed that the magnitude of magnetic field of $Nd_{1-x}Sr_xMnO_3$ in the ferromagnetic region ($x < 0.5$) was significantly increased by photo-irradiation, almost linearly against exposure time. The enhanced magnetic field was retained for several hours without laser beam.[43]

17.5 CONCLUSION

We reviewed the basic principles of some non-volatile memories and the applications of the combinatorial technology for searching the materials of these memory devices. However, this combinatorial technique has several drawbacks caused by limitation of the characterization tools; even though combinatorial method has dramatically increased the efficiency of obtaining the material library with various compositions. The speed of characterization of these materials is still restricted currently. High-throughput characterization tools are also needed for high-throughput screening and characterization. Much effort is being directed to the development of equipment for combinatorial research, such as microbeam XRD, the scanning superconducting quantum interference device (SQUID), scanning evanescent microwave microscopy (SEMM). If such effort continues, then could well combinatorial methods be considered as state-of-art for research into new materials.

REFERENCES

1. Seong I. W., Ki W. K., Hyun Y. C., Kwang S. O., Min K. J., Naresh H. T., Tai S. K. and Asif M. Current status of combinatorial and high-throughput methods for discovering new materials and catalysis, *QSAR Comb. Sci.*, 24, 138, 2005.

2. Cawse, J. N., *Experimental Design for Combinatorial and High Throughput Materials Development*, John Wiley & Sons, Inc., New Jersey, 2002.

3. Schultz, P. G., Xiang X. and Goldwasser, I. Combinatorial synthesis of novel materials, US patent, 6004617, 2000.

4. Wu X. D., Wang Y. and Goldwasser, I., Systems and methods for the combinatorial synthesis of novel materials, US patent, 6045671, 2000.

5. Schultz, P. G., Xiang X. and Goldwasser, I. Combinatorial synthesis of novel materials, US patent, 6326090, 2001.

6. Maxwell, I. E., Connecting with catalysis, *Nature,* 394, 325, 1998.

7. Koinuma, H., Combinatorial materials research projects in Japan, *Appl. Surf. Sci.,* 189(3–4), 179, 2002.

8. Hewes, J. D. and Bendersky, L. A., High throughput materials research and development: a growing effort at NIST, *Appl. Surf. Sci.,* 189(3–4), 196, 2002.

9. Kimura, S., Prospect of materials research methodology by combinatorial approach, *Appl. Surf. Sci.,* 189(3–4), 177, 2002.

10. Takeuchi, I., Experimental issues in combinatorial investigation of electronic thin-film materials, *Appl. Surf. Sci.* 189(3–4), 353, 2002.

11. Takeuchi, I., Chang, K., Sharma, R. P., Bendersky, L. A., Chang, H., Xiang, X.-D., Stach, E. A. and Song, C.-Y., Microstructural properties of (Ba,Sr)TiO$_3$ films fabricated from BaF$_2$/SrF$_2$/TiO$_2$ amorphous multilayers using the combinatorial precursor method, *J. Appl. Phys.,* 90(5), 2474, 2001.

12. Chang, K. S., Aronova, M., Famodu, O., Takeuchi, I., Lofland, S. E., Hattrick-Simpers, J. and Chang, H., Multimode quantitative scanning microwave microscopy of in situ grown epitaxial Ba$_{1-x}$Sr$_x$TiO$_3$ composition spreads, *Appl. Phys. Lett.,* 79(26), 4411, 2001.

13. Li, J. W., Duewer, F., Gao, C., Chang, H. Y., Xiang, X. D. and Lu, Y. L., Electro-optic measurements of the ferroelectric-paraelectric boundary in Ba$_{1-x}$Sr$_x$TiO$_3$ materials chips, *Appl. Phys. Lett.,* 76(6), 769, 2000.

14. Terai, K., Lippmaa, M., Ahmet, P., Chikyow, T., Koinuma, H., Ohtani, M. and Kawasaki, M., Fabrication of lattice-tunable Ba$_{1-x}$Sr$_x$TiO$_3$ buffers on a SrTiO$_3$ substrate, *Appl. Surf. Sci.,* 223, 183, 2004.

15. Scott, J. F. and Araujo, C. A., Ferroelectric memories, *Science,* 246, 1400, 1989.

16. Xu, Y., *Ferroelectric Materials and Their Applications*, Elsevier Science Publication, Los Angeles, 1991, Ch. 1.

17. Chen, H. D., Udayakumar, K. R., Gaskey, C. J. and Cross, L. E., Electrical properties' maxima in thin films of the lead zirconate-lead titanate solid solution system, *Appl. Phys. Lett.,* 67, 3411, 1995.

18. Park, B. H., Kang, B. S., Bu, S. D., Noh, T. W., Lee, J. and Jo, W., Lanthanum-substituted bismuth titanate for use in non-volatile memories, *Nature,* 401, 682, 1999.

19. Wicker, G. C., Nonvolatile high-density high-performance phase-change memory, *Proc. SPIE,* 3891, 2, 1999.

20. Bernacki, S., Hunt, K., Tyson, S., Hudgens, S., Pashmakov, B. and Czubatyj, W., Total dose radiation response and high temperature imprint characteristics of chalcogenide based RAM resistor elements, *IEEE Trans. Nucl. Sci.,* 47, 2528, 2000.

21. Lai, S., Current status of the phase change memory and its future, in IEDM'03 Technical Digest in IEEE international, Intel Corporation, RN2-05, 2003.

22. Nakayama, K., Kojima, K., Imai, Y., Kasai, T., Fukushima, S., Kitagawa, A., Kumeda, M., Kakimoto, Y. and Suzuki, M., Nonvolatile memory based on phase change in Se-Sb-Te glass, *Jpn. J. Appl. Phys.,* 42, 404, 2003.

23. Bunton, G. V. and Quilliam, R. M., Switching and memory effects in amorphous chalcogenide thin films, *IEEE Trans. Electron Devices,* 20, 140, 1973.

24. Dharam P. G., Minoru N., Tomoyasu S., Masakuni S. and Shuichi O., Nonvolatile memory based on reversible phase transition phenomena in telluride glasses, *Jpn. J. Appl. Phys.,* 28, 1013, 1989.

25. Kang, D. H., Ahn, D. H., Kim, K. B., Webb, J. F. and Yi, K. W., One dimensional heat conduction model for an electrical phase change random access memory device with an $8F^2$ memory cell (F = 0.15 μm), *J. Appl. Phys.,* 94, 3536, 2003.

26. Bhatia, K. L., Singh, M., Katagawa, T., Kishore, N. and Suzuki, M., Optical and electronic properties of Bi-modified amorphous thin films of $Ge_{20}Te_{80-x}Bi_x$, *Semicond. Sci. Technol.,* 10, 65, 1995.

27. Bhatia, K. L., Kishore, N., Malik, J., Singh, M., Kundu, R. S., Sharma, A. and Srivastav, B. K., Study of the effect of thermal annealing on the optical and electrical properties of vacuum evaporated amorphous thin films in the system $Ge_{20}Te_{80-x}Bi_x$, *Semicond. Sci. Technol.,* 17, 189, 2002.

28. Ramesh, K., Asokan, S. and Sangunni, K. S., Electrical switching in germanium telluride glasses doped with Cu and Ag, *Appl. Phys. A: Mater. Sci. Process.,* 69(4), 421, 1999.

29. Tsendin, K. D., Lebedev, E. A., Kim, Y.H., Yoo, I. J. and Kim, E. G., Characteristics of information recording on chalcogenide glassy semiconductors, *Semicond. Sci. Technol.,* 16, 394, 2001.

30. Rajesh, R. and Philip, J., Memory switching in In-Te glasses: results of heat-transport measurements, *Semicond. Sci. Technol.,* 18, 133, 2003.

31. Nakayama, K., Kitagawa, T., Ohmura, M. and Suzuki, M., Nonvolatile memory based on phase transition in chalcogenide thin film, *Jpn. J. Appl. Phys.,* 32, 564, 1993.

32. Nakayama, K., Kojima, K., Hayakawa, F., Imai, Y., Kitagawa, A. and Suzuki, M., Submicron nonvolatile memory cell based on reversible phase transition in chalcogenide glasses, *Jpn. J. Appl. Phys.,* 39, 6157, 2000.

33. Saheb, P. Z., Asokan, S. and Gowda, K. A., Electrical switching studies of lead-doped germanium telluride glasses, *Appl. Phys. A: Mater. Sci. Process.,* 77(5), 665, 2003.

34. Kyrsta, S., Cremer, R., Neuschutz, D., Laurenzis, M., Bolivar, P. H. and Kurz, H., Characterization of Ge-Sb-Te thin films deposited using a composition-spread approach, *Thin Solid Films,* 398–399, 379, 2001.

35. Laurenzis, M., Heinrici, A., Bolivar, P. H., Kurz, H., Krysta, S. and Schneider, J. M., Composition spread analysis of phase change dynamics in $Ge_xSb_yTe_{1-x-y}$ films embedded in an optical multilayer stack, *IEE Proc. Sci. Meas. Technol.,* 151(6), 394, 2004.

36. Kato, N., Takeda, Y., Fukano, T., Motohiro, T., Kawai, S. and Kuno, H., Compositional dependence of optical constants and microstructures of GeSbTe thin films for compact-disc-rewritable (CD-RW) readable with conventional CD-ROM drives, *Jpn. J. Appl. Phys.,* 38, 1707, 1999.

37. Kyrsta, S., Cremer, R., Neuschutz, D., Laurenzis, M., Bolivar, P.H. and Kurz, H., Deposition and characterization of Ge-Sb-Te layers for applications in optical data storage, *Appl. Surf. Sci.,* 179, 56, 2001.

38. Bez, R. and Pirovano, A., Non volatile memory technologies: emerging concepts and new materials, *Mater. Sci. Semicond. Process,* 7, 349, 2004.

39. Fukumura, T., Okimoto, Y., Ohtani, A., Kageyama, T., Koida, T., Kawasaki, A., Hasegawa, T., Tokura, Y. and Koinuma, H., A composition-spread approach to investigate band-filling dependence on magnetic and electronic phases for Perovskite manganite, *Appl. Surf. Sci.,* 189, 339, 2002.

40. Takahashi, R., Kubota, H., Tanigawa, T., Murakami, M., Yamamoto, Y., Matsumoto, Y. and Koinuma, H., Development of a new combinatorial mask for addressable ternary phase diagramming: application to rare earth doped phosphors, *Appl. Surf. Sci.,* 223, 249, 2004.

41. Svedberg, E.B., van de Veerdonk, R. J. M., Howard, K. J. and Madsen, L. D., Quantifiable combinatorial materials science approach applied to perpendicular magnetic recording media, *J. Appl. Phys.,* 93(9), 5519, 2003.

42. Hasegawa, T., Kageyama, T., Fukumura, T., Okazaki, N., Kawasaki, M., Koinuma, H., Yoo, Y. K., Duewer, F. and Xiang, X. D., High-throughput characterization of composition-spread manganese oxide films with a scanning SQUID microscope, *Appl. Surf. Sci.,* 189, 201, 2002.

43. Sugaya, H., Okazaki, S., Hasegawa, T., Okazaki, N., Nishimura, J., Fukumura, T., Kawasaki, M. and Koinuma, H., Photo-induced magnetism in perovskite-type Mn oxides investigated by using combinatorial methodology, *Appl. Surf. Sci.,* 223, 68, 2004.

18 Combinatorial Ion Beam Synthesis of II–VI Compound Semiconductor Nanoclusters

Helmut Karl

CONTENTS

18.1   INTRODUCTION

The interest in nanostructured materials and their unique properties has increased enormously in recent years. The typical spatial dimension of a nanoparticle lies in the range of a few nanometers and their properties are between those of molecules and the corresponding bulk material. Quantum dots (QDs) are therefore often called "artificial atoms."

The physical properties of nanoclusters are essentially determined by two effects. First, there is the large surface-to-volume ratio of nanoparticles and the associated surface-energy contribution and surface states. Second, there is spatial confinement of charge carriers attainable which manifests in a drastic change in the distribution of density of states and energy levels.[1,2] Both effects and their interaction might be the basis of improved or even novel electrical, optoelectronic, magnetic, and mechanical properties.

In addition to the surface and volume effects, interaction between nanocrystallites can be exploited. QDs assembled in close proximity forming dense-packed QD solids can interact with one another.[3,4] This interaction leads to a delocalization of the electron states of the single QDs and to next neighbor and even long-range energy transfer.[5] In these materials a reduction of non-radiative losses[6] and a red shift of the optical transitions[7] can often be observed.

On the basis of the tunable interplay of the size-dependent properties of nanocrystals and the resulting collective physical phenomena of dense-packed QD ensembles, important applications like integrateable QD lasers, light energy harvesting structures, tunneling devices, planar waveguide amplifiers, and efficient micro-LEDs may result.

In some of these applications high-energy densities are involved, a typical issue of highly integrated miniaturized device architectures and laser active materials. This requires the use of chemically and physically stable materials for both the host materials and QDs.

There have been several techniques developed for producing nanoclusters of very small size-distribution.[2] Even core-shell structures of different multielemental semiconducting materials can be grown. In particular the chemical route is effective for producing particles in large quantities and homogeneous size distribution, which can be used to build regular particle assemblies for opto-electronic devices like lasers[4] or as photocatalytic materials. Even single particles can be attached to biomolecules as fluorescent markers. These are only some examples of a much longer list of potential fields of applications.

A promising technique to synthesize buried compounds is ion implantation followed or accompanied by a thermal annealing process. To date, the central focus in the field has been to synthesize either buried elemental semiconductor nanocrystals[8] and metal nanoparticles[9] or to take advantage of chemical reaction of the host material with the implanted species to synthesize buried layers of SiC^{10} or $CoSi_2^{11}$ in a silicon host. Nanocomposites formed by this technique inside a stable host can be quite durable since they are protected from the surrounding environment, a critical issue due to the high reactivity of nanoparticles.

Much less investigated are buried compound semiconductor nanocrystals synthesized by means of ion implantation into $SiO_2^{12,13}$ and other technologically appropriate host materials like Al_2O_3, Si_3N_4, and others. In Figure 18.1 ion beam synthesized CdSe- and ZnTe-nanocrystals are shown. Here, ion implantation offers a very promising and unique tool for the generation of planar, shallowly embedded layers of densely packed nanocrystals.

Ternary II–VI compound semiconductors like nanocrystals of CdSSe embedded in SiO_2 offer a flexible materials system. Similar to pure CdSe, CdSSe has proven to have high oscillator strength. Furthermore its properties can be tuned not only by quantum confinement and particle interaction, but also by variation of the S content. However, as morphology is controlled by complex processes (diffusion and chemical reaction in a highly distorted matrix), it is difficult to analyze and to understand structural evolution of the system as a function of synthesis parameters. In order to explore the multiparameter processing space a combinatorial synthesis approach has been envisaged to systematically investigate the synthesis of buried CdSSe by ion implantation.

In Figure 18.2 the nanoparticle synthesis by sequential ion implantation of the elements is illustrated. When the implantation sequence is complete diffusion and chemical reaction is initiated and controlled by a thermal annealing procedure. The implanted dose ratios of the elements determine the stoichiometry and thus the compound to be formed.

The implanted concentration profiles in an amorphous target is nearly Gaussian shaped and can be approximated by

$$c(depth) = c_p\, e^{\frac{(x-R_p)^2}{2\Delta R_p}}$$

where R_p is the range of implanted ions, ΔR_p represents the straggling, i.e., the width of the concentration profile and c_p is the peak concentration. This depth distribution is an approximation and corrections have to be taken into consideration in the case of high-dose ion implantation. In order to obtain a realistic simulation of the resulting implantation profiles either TRIM[14] or TRIDYN[15] has to be utilized. Moreover the individual energies in cases of multielement high-dose ion implantation have to be adapted in order to achieve the desired overlap of the concentration profiles. This is exemplified at a sample with an implanted Cd:Se dose ratio of 1:1 (implantation energies: Cd 190 keV

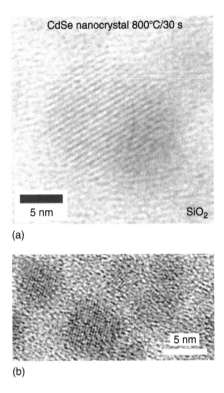

(a)

(b)

FIGURE 18.1 Cross-sectional transmission electron microscopy images (XTEM) of compound semiconductor (a) CdSe and (b) ZnTe nanocrystals synthesized by sequential implantation of the elements into SiO_2 thermally grown on Si.

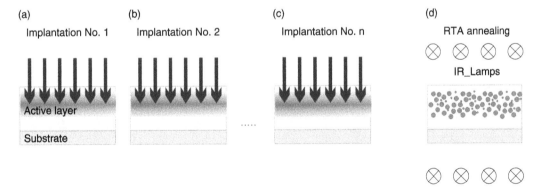

FIGURE 18.2 Appropriate doses of the implanted elements (a)–(c) are sequentially implanted before rapid thermal annealing (RTA) (d).

and Se 130 keV) with Cd implanted first (see Figure 18.3). The concentration profile was simulated by TRIM with additional corrections on account for sputtering and changes of the stopping power. The implanted concentration profile of both elements was measured by SIMS (secondary ion mass spectrometry) depth profiling. The result is shown in Figure 18.3 and indicates that the simulation coincides well with the measured concentration profiles. The concentration maximum is located at approximately 80 nm and only for depths larger than 150 nm deviations between simulation and

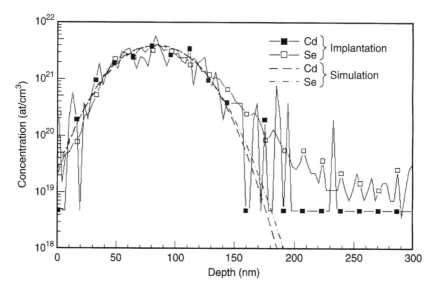

FIGURE 18.3 SIMS and simulated (TRIM) depth profiles of Cd and Se of an as-implanted sample in SiO_2 (Cd 190 keV and Se 130 keV)[41]. The targeted dose of both elements was 2.4×10^{16} at/cm^2.

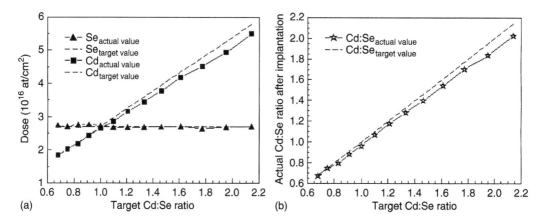

FIGURE 18.4 RBS analysis of a stoichiometry series with a Se dose of 2.7×10^{16} at/cm^2, (a) shows the target and actual dose values of the Cd and Se implantation and (b) the actual Cd:Se ratio versus the nominal implanted Cd:Se dose ratio.[41]

measurement occurs. The actual remaining doses (i.e., nominal implantation doses determined by the ion beam current and loss due to ion sputtering during implantation and out-diffusion during thermal treatment) of Cd and Se were quantitatively determined by RBS (Rutherford backscattering spectroscopy).[40] Details of this procedure applied to In and As in a Si matrix are described in reference 16. In Figure 18.4 the results of both implanted elements Cd and Se versus the nominal (targeted) Cd:Se dose ratios are shown. A reduction of the actual remaining dose compared to the nominally implanted Cd was found for increasing Cd doses. This finding can be explained by ion sputtering during implantation of Se, which removes Cd together with SiO_2 from the substrate surface. Sputter coefficients for this situation have been determined from the shift of an implanted C marker profile to be 3.5 for Cd at 190 keV and 2.1 for Se at 128–138 keV.

FIGURE 18.5 (a) Side and (b) front view of the target holder with shadow masks and scanning device.

18.2 COMBINATORIAL ION IMPLANTATION

When applying the concept of combinatorial materials science (CMS) to ion implantation techniques, the creation of implantation dose and ion sort variations across individual substrate areas has to be realized. This technical problem can be solved in different ways.[16–20]

The use of moving shadow masks allows an implantation scheme where accumulation of implanted ion dose in different areas of the sample at the same time enables partial parallel processing.[19] The spatially selective shadow masks are constructed as a set of moveable shields inserted into the beam line of a commercial-type mid-energy implanter (maximum acceleration voltage 200 kV) (Figure 18.5). The computer-controlled shields stepwise cover, successively or continuously, parts of a wafer so that a lateral pattern of well-defined ion-implanted dose ratio combinations of different sequentially implanted ion species are obtained.

The lateral homogeneity of the ion beam current distribution is monitored in real time during the whole implantation process by measuring it with four Faraday cups (Figure 18.5(b)) at the edges of the scanning area. The ion beam current is used to control the movement of the shields in order to obtain the predefined dose distribution pattern. The size of the homogeneously implanted regions on the wafer is chosen so that they are large enough to match the requirements of the intended analysis and characterization techniques.[19] In the course of this work different types of dose patterns (see Figure 18.6 (a)–(d)) were used for binary and ternary material combinations.

The use of shadow masks enables an enormous increase of throughput. In the case of a parallel series of stripes with increasing implantation doses d_0, d_1, d_2, ..., d_n the ratio D/d_0 of the total implantation dose D and that of the first stripe d_0 can be calculated. In the case of n stripes with the doses d_i scaling in succession like $d_{i+1} = a * d_i$ this ratio is determined by

$$\frac{D}{d_0} = 1 + \sum_{i=1}^{n-1}(a^i - a^{i-1}).$$

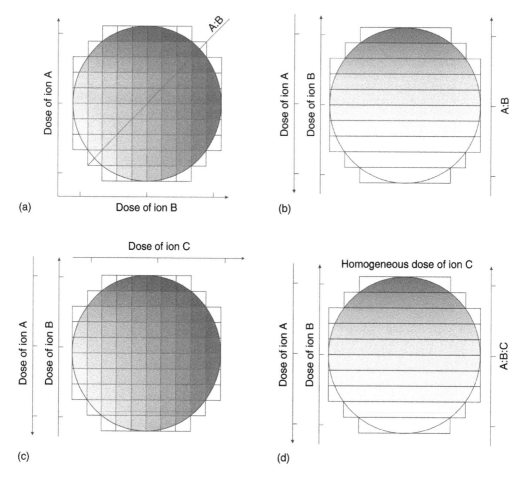

FIGURE 18.6 (a)–(d) Schematics of dose ratio pattern used for the investigation of structural and optical properties of compounds consisting of the two or three elements A, B, and C. Wafer implanted with pattern of (a) and (c) where annealed as a whole, whereas wafer implanted with dose stripes as shown in (b) and (d) were cut into pieces for annealing at different temperatures and annealing atmospheres.

Now a comparison between the generation of a series of ten dose stripes by use of one shadow mask (i.e., either x- or y-shadow mask) and a series of samples implanted one after the other will be made. For that the ratio D/d_0 for ten stripes of successively increasing doses by a factor $a = 1.1$ will be calculated. The result is illustrated in Figure 18.7(a) and Figure 18.7(b). The total implantation dose D for preparing the sample series in a sample by sample approach (Figure 18.7(b)) is approximately 6.6 times larger. The gain in throughput is even higher for two-dimensional multiple element dose pattern (using both x- and y-shadow mask) where, in addition to dose variations, also systematically varied dose ratios are obtained.

18.3   HIGH-DOSE ION IMPLANTATION

18.3.1   COMPOUND FORMATION AND PHASE SEPARATION

In the case of implantation into single crystalline semiconductors the implanted impurities occupy interstitial or substitutional lattice sites after thermal treatment. In contrast, implantation of doses

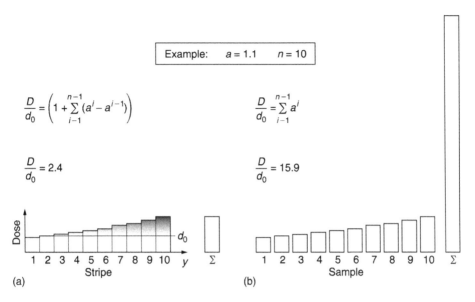

FIGURE 18.7 (a) Illustrates a one-dimensional dose pattern of ten stripes with increasing doses. The dose increases by a factor of 1.1 for successive stripes, (b) same set of implanted samples prepared individually in succession.

creating impurity concentrations above the solubility limit can lead to phase separation processes and finally to the formation of precipitates. With increasing implantation dose the density of displaced atoms becomes higher and higher so that the implanted target material volume becomes completely dissociated. The recovery of the target is initiated either during the ion implantation process or by post-implantation thermal treatment. The size and location of the precipitates are subjected to a temporal and spatial evolution where coarsening, element redistribution and pattern formation during thermal treatment exhibits complex interdependencies.

The driving force for chemical restructuring is the minimization of the Gibbs free energy of the system. Thermal activation initiates chemical reactions directed towards decreasing Gibbs free energies of the reaction products. In this connection the annealing gas atmosphere, i.e, reactive versus inert gas, has a strong influence on the resultant chemical compounds.[21–23] This was shown for the formation of ZnTe nanocrystals which was inhibited by addition of oxygen to the annealing gas atmosphere, whereas Zn and Te concentration and particles density oscillations developed across the SiO_2 layer for pure Ar atmosphere.[21]

According to the law of mass action and the Gibbs free energies the thermodynamically most stable compounds are formed. In case of ZnTe with a stoichiometric amount of oxygen for the recovering SiO_2 ZnTe will be formed. However, overstoichiometric oxygen partially or even completely suppresses the formation of the targeted compound.[21] Table 18.1 gives an overview of Gibbs free energies of II–VI compounds and some reaction products encountered when silicon and oxygen are presented.

In classical nucleation theory it is assumed that a precipitate possesses a homogeneous core, which is surrounded by the matrix representing a heterogeneous materials system. Both phases, the nucleus and the supersaturated matrix, are in a metastable state.

In the case of spherical precipitates of radius r the energy of formation is given by:

$$W = -\frac{3}{4}\pi r^3 \frac{\Delta G}{V_{\text{Mol}}} + 4\pi r^2 \sigma + \frac{3}{4}\pi r^3 \varepsilon$$

TABLE 18.1
Gibbs Free Energies for II–VI Compounds and Reaction Products of the Components with Silicon and Oxygen[38]

Compound	ΔG [KJ/mol]
SiO_2	−900
CdS	−61.9
CdTe	−92.5
ZnS	−192.6
CdSe	−163
SeO_2	−225.4
TeO_2	−322.6
$ZnSiO_4$	−1636.7
$CdSiO_3$	−530
CdO	−260

where ΔG is the Gibbs free energy, V_{mol} the mole volume, σ the specific surface/interface energy and ε the specific energy of deformation. From this equation a critical precipitate radius r_k can be calculated:

$$r_k = \frac{2\sigma}{\dfrac{\Delta G}{V_{Mol}} - \varepsilon}.$$

The criteria for the critical radius is a result of a critical precipitate energy of formation of the form:

$$W_k = \frac{16}{3}\pi\frac{\sigma^3}{\left(\dfrac{\Delta G}{V_{Mol}} - \varepsilon\right)}.$$

This energy W_k has to be overcome for the formation of a precipitate. This is directly associated with the probability of precipitate formation by thermal fluctuations and described by the Boltzmann factor $\exp(-W_k/kT)$. It should be noted that there is considerable debate about the applicability of this classical nucleation theory, since there is no correction for nuclei containing a very small number of monomers, where the thermodynamic and macroscopically mechanical approach of the equations given above ceases to be valid. Once phase separation has set in and the nucleation process has finished Ostwald ripening plays an important role in the temporal and spatial evolution of the nanocrystallite size distribution. The growth of precipitates larger than the critical radius r_k is primarily controlled by an interplay of diffusional material transport to and chemical reaction rates at the particles' surfaces. A precipitate has an equilibrium concentration at its surface which can be described by the Gibbs-Thomson equation:

$$c_G(R) = c_\infty e^{\frac{\alpha}{R}} \cong c_\infty\left(1 + \frac{\alpha}{R}\right) \quad \text{with} \quad \alpha = \frac{2\sigma V_m}{k_B T}$$

where c_∞ is the solidus, α is the capillary length, σ is the surface tension and V_m the molar volume. From this equation it can be concluded that the equilibrium concentration at the surface of a particle with radius R increases with decreasing R. For adjacent particles of different size this leads to diffusional mass transfer from smaller dissolving to larger growing ones and finally to an increase of the average particle size. This particle coarsening is known as Ostwald ripening. Lifshitz, Slyozov, and Wagner formulated the LSW theory[24,25] for systems with macroscopically homogeneous concentration and particle distribution, i.e., the length scale of the initial spatial variation of the concentration is large compared to the microscopic reaction–diffusion and nucleation regions.

The above-described process of phase separation by nucleation and growth takes place when the curvature of the Gibbs free energy versus composition $\partial^2 G^{mixture}/\partial X^2 > 0$ is positive. Such a materials mixture is stable against infinitesimal composition fluctuations. In other words the formation of a nuclei must initiate a large change in composition for precipitation. Aside from the nucleation and growth mechanism there is the so-called spinodal decomposition, which occurs when the curvature of the Gibbs free energy versus composition $\partial^2 G^{mixture}/\partial X^2 < 0$ is negative. In this case infinitesimally small composition fluctuations are sufficient to initiate phase separation and the mixture is thus inherently unstable and separates into two phases without a nucleation barrier.

In contrast to the case of a homogeneous concentration distribution, ion-implanted concentration profiles are highly inhomogeneous; consequently the dynamics and mechanism of phase separation varies locally. In the following, the two decomposition mechanisms of sequentially implanted Cd and Se into SiO_2 on silicon at liquid nitrogen temperatures will be discussed in this context. The 300 nm thin SiO_2 layer was thermally grown on silicon. The temporal and spatial evolution of the concentration distribution profiled by dynamic SIMS measurements of individually and sequentially implanted Cd and Se after rapid thermal annealing reveals completely different diffusion behavior for the single implanted elements Cd and Se.[26] There is a clear correlation to the corresponding diffusion coefficients of both elements in SiO_2. By contrast, when both elements were implanted in overlapping profiles, congruent diffusion of Cd and Se was observed. This behavior can only be explained by assuming that the formation of CdSe takes place in the implanted region at the very beginning of the postimplantation annealing procedure. It is very probable that the phase separation mechanism at these high CdSe concentrations (up to 10^{22} at/cm^3) is spinodal and that after chemical reaction diffusion of electrically neutral CdSe dimers takes place.

It should be noted once more that in the implanted region the SiO_2 network is heavily damaged. However, in case of sufficiently long annealing times the concentration pattern extends far into non-damaged regions of the SiO_2 layer. The structure formation process can thus be modeled in the following way. At the beginning of thermal processing the implanted region acts as a source of material. In the course of particle coarsening it turns from a material source to a material sink. Material reaching the SiO_2/Si interface preferentially precipitates there and forms a second layer of particles close to the interface. Both particle layers represent boundary conditions. This constellation holds potential for self-organization and inherently results in inhomogeneous particle distributions.

18.3.2 Particle Distribution: Self-Organization

High-dose ion-implanted concentration profiles are nearly Gaussian-shaped and easily reach the supersaturation threshold at their maximum. Under those initial conditions the dynamics of the nucleation and growth of precipitations varies with depth accompanied by diffusion of material and chemical reactions, according to the concentration gradients on the flanks of the implanted concentration profile.

It is well known that nucleation and precipitate growth in reaction–diffusion systems holds potential for self-organization phenomena, also known as the Liesegang phenomenon.[32–37] The patterns of rings or layers of precipitates observed depend strongly on the geometry of the system and can be studied in systems where one chemical component (typically dissolved in an appropriate solvent) diffuses into a gel containing the reactant. When reaction takes place a highly insoluble compound (for example $Ag^+ + Cl^- \rightarrow AgCl$) is formed and turns into precipitate above a certain concentration threshold (saturation concentration). The dimensions of the pattern typically found are macroscopic in contrast to the pattern generated in films of submicton thicknesses as described in this chapter.

The conditions for self-organizing phenomena to be observed are non-linear couplings between different dynamic variables in systems far from thermal equilibrium. These prerequisites are fulfilled for conditions obtained by high-dose ion implantation. The position and width of the precipitate layers undergoes a temporal and spatial evolution[39] and for certain systems empirical time, spacing, and width laws can be deduced from experiment and explained by theories.

For two reactants A and B two different scenarios can be considered. Firstly, the ion-product supersaturation mechanism in which A + B forms the precipitate directly when supersaturation is exceeded (A + B → phase separation). Secondly, the intermediate-compound mechanism in which it is assumed that an intermediate compound C is formed and precipitation sets in when the concentration of C exceeds its solubility limit (A + B → AB = C → phase separation).

In the case of a spinodal decomposition where phase separation of the compound AB is accompanied by infinitesimally small concentration fluctuations, the ion product supersaturation mechanism can be applied since the local composition changes only little and, as a consequence, the existence of an intermediate compound C which diffuses before precipitation can be neglected in many cases. Therefore, this model might be applicable in the high-dose region of the implanted concentration profile where spinodal decomposition most probably dominates the phase separation process.

However, the observed congruent diffusion of A = Cd and B = Se towards greater depths of the SiO_2 layer must be modeled by the intermediate compound mechanism with C = CdSe as the diffusing dimer, which nucleates when their concentration exceeds the solubility limit, followed by growth forming CdSe nanocrystalline precipitates. According to Ostwald the growing precipitates deplete their vicinity by consuming material. As a result the concentration C dissolved in the surrounding matrix reduces to levels below supersaturation ceasing further nucleation. The growth of precipitates at greater depths during extended post-implantation thermal treatment after high-dose multiple ion implantation can be modeled by the following set of differential equations:

$$\frac{\partial c_A}{\partial t} = D_A \frac{\partial^2 c_A}{\partial x^2} - k c_A c_B$$

$$\frac{\partial c_B}{\partial t} = D_B \frac{\partial^2 c_B}{\partial x^2} - k c_A c_B$$

$$\frac{\partial c_C}{\partial t} = D_C \frac{\partial^2 c_C}{\partial x^2} + k c_A c_B - s$$

The symbols c_A, c_B and c_C are the concentrations of the implanted materials A, B, and of the reaction product C, k is the reaction rate between the reactants A and B. It is assumed that there is no chemical reaction of A, B, and C with the constituents of the host material, so that either they are dissolved in the host or form separate phases.

The term s in the last equation is a sink term and describes the depletion of the product C due to nucleation and precipitate growth. This sink term u is non-linear and includes nucleation, growth, and dissolution of precipitates formulated according to LSW theory or microscopic nucleation and growth models. Boundary conditions generated by precipitates in the implanted region and those formed at the internal SiO_2–Si interface can cause regular patterning of the precipitate density as a function of depth for certain process parameters. The process of pattern formation is illustrated in Figure 18.8(a)–(c). Due to local concentrations of more than 1 at % of the implanted ions Cd and Se, the chemical reaction product CdSe forms at the very beginning of the thermal processing step in the range of implanted ions R_p. In parallel the ion beam damaged SiO_2 network recovers by consuming stoichiometric amounts of oxygen and silicon. At the flanks of the implantation profile diffusion towards the surface and the SiO_2–Si interface of CdSe dimers and overstoichiometric Cd or

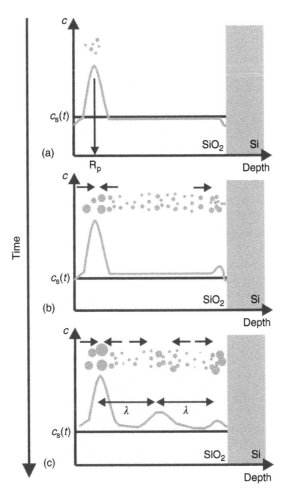

FIGURE 18.8 Schematics of the mechanism of pattern formation. (a) Precipitate formation in the range of the implanted ions R_p begins during post-implantation thermal treatment, in parallel material diffusion occurs; (b) material reaching the SiO_2–Si interface nucleates and precipitates grow close to it; (c) adjacent small precipitates formed in between by fluctuations become dissolved (Ostwald ripening) causing an instability and a 3rd precipitate layer develops.

Se takes place (Figure 18.6(a)). The implantation profile and the SiO_2–Si interface comprise preferred nucleation sites and layers of larger precipitates form around R_p and close the SiO_2–Si interface (Figure 18.6(b)). During the annealing procedure the concentration c of the CdSe dimers approaches the saturation concentration c_s and thermal fluctuations lead to nucleation of CdSe dimers. Larger particles already present in the implanted region and at the SiO_2–Si interface (representing boundary conditions acting now as sinks for diffusing dimers) deplete their surrounding by dissolving smaller particles (Ostwald ripening) and precipitate denuded zones form (Figure 18.6(b)).

Depending on the diffusion length of the CdSe dimers and the SiO_2 layer thickness there is no complete dissolution of the particles formed. The mean size of the remaining particles becomes larger, causing an instability which leads to a depletion, as far as they are concerned, of the adjacent regions. This Ostwald ripening process gives rise to the growth of one or more particle layers inside the SiO_2 layer. A regular pattern may develop and manifests itself as concentration modulations detectable by Rutherford backscattering spectroscopy (RBS) and secondary-ion mass spectroscopy

(SIMS) depth profiling or as a regular variation of the particle densities clearly visible in cross-sectional TEM images.[16]

It was found that Cd surplus increases the diffusivity of both Cd and Se, whereas equal implanted doses of Cd and Se result in a minimum of diffusivity. The amount of material accumulated at the interface can be controlled by implanted dose, dose ratio, annealing duration, and temperature.[26] Regular pattern formation was achieved only for certain processing conditions and is rather an exception. Regular pattern formation was also found for the ZnTe system.[21] There is an indication that the pattern formation is favored by overstoichiometric Cd in the CdSe system.[26]

The nanoparticle size, interparticle distance and distribution of an ensemble of nanoparticles determine the physical properties and provide important information for their interpretation. The size distribution of a QD ensemble can in principle be optimized for a particular application. Ion implantation allows control of the concentration profile of implanted ions and with it the initial supersaturation condition: a prerequisite for controlling the position, size-distribution and density of the precipitated phase. The extraordinary high density together with a complex concentration and size distribution comprises a chance for new applications of ion-beam synthesized semiconductor QDs.

There are some measurement techniques available which allow a quantitative determination of size distribution, interparticle distance, and crystalline structure of buried nanoclusters. Their pros and cons will be discussed briefly in the following.

18.3.3 MEASUREMENT OF PARTICLE SIZE, DISTRIBUTION, AND INTERPARTICLE DISTANCE

18.3.3.1 Cross-Sectional Transmission Electron Microscopy (XTEM)

Cross-sectional transmission electron microscopy (XTEM) images provide a real space projection of the size of nanoclusters within a surrounding matrix. The problem which arises using this technique is the determination of the exact location of the interface between particle and surrounding host material. This problem becomes even more crucial if there are such high nanocluster densities, that the projections of precipitates overlap, conditions most often found in ion beam synthesized dense-packed nanocluster ensembles. Additionally, the highly energetic electron beam used in XTEM measurement can alter the morphology during the period of observation. However this technique is unsuitable for high-throughput analysis of the materials library due to lengthy and complicated preparation of XTEM specimen.

18.3.3.2 X-Ray Diffraction (XRD)

Thin-film small-angle X-ray diffraction measurements can provide information on both nanocluster phase and size. A general problem is the small amount of volume of the ion beam synthesized and randomly oriented nanoclusters, corresponding to film thickness of only a few nanometers, making long integration times necessary.

The Guinier thin-film XRD configuration is the best suited for the investigation of planar and nanocluster ensembles near the surface. In this configuration an X-ray beam enters the substrate in small-angle geometry so that the X-ray intensity is absorbed near the surface where the nanoclusters are located. From the full-width half maximum (FWHM) of the diffraction peaks the crystallite size $<S>$ can be calculated by evaluating Debye–Scherrer's formula:

$$< S > = \frac{0.9\lambda}{B \cos \Theta_B}$$

where B is the 2ϑ FWHM of a diffraction peak at the position Θ_B and λ is the X-ray wavelength, typically Cu Kα of $\lambda = 0.15406$ nm. For evaluation of the FWHM, the broadening of the diffraction peak caused by the instrument itself has to be considered:

$$B^2 = B^2_{\text{Instrument}} + B^2_{\text{Particle}}.$$

Moreover, the form of the particles has influence on the broadening of the diffraction peaks. In the case of spherical-shaped nanoclusters the size $<S>$ evaluated by applying the Debye–Scherrer formula has to be multiplied by 4/3 in order to obtain the correct mean diameter D of the nanocluster ensemble:

$$< D > = \frac{4}{3} < S >.$$

This technique is well suited for automation, since there is no specimen preparation necessary, though long integration times have be taken into consideration.

18.3.3.3 Grazing Incidence Small-Angle X-Ray Scattering (GISAXS)

Synchrotron radiation has been applied successfully in small-angle incidence scattering geometry to study shape, size, and spatial correlations of buried nanoclusters.[27] For evaluating GISAXS spectra different models have been developed.

In the Guinier approximation, or so-called Guinier plot, the average cluster diameter D along the scattering direction can be calculated from the slope of the linear part ln $I(q)$ versus q^2 of a GISAXS profile of the one-dimensional scattering intensity, where I is the scattered X-ray intensity for a wave vector q. If the particle shape differs from that of a sphere a correction factor is needed. Moreover the average distance L between the clusters can be determined from a GISAXS measurement. This interparticle distance L is correlated to the position of the interference peak q_m:

$$L = \frac{2\pi}{q_m}.$$

The interparticle distance L only obtainable from GISAXS measurements, is a very important parameter for interpreting effects originating from QD–QD interactions. Another method to exploit a GISAXS profile is the local monodisperse approximation (LMA). It is refined compared to the Guinier approximation in the way that it considers an ensemble of weighted size distributions. The scattered intensity is expressed in the distorted-wave Born approximation:

$$I(q) = |T(\alpha_i)|^2 \, |T(\alpha_f)|^2 \int_0^\infty P(q, D)S(q, D_{hs}, \eta_{hs})N(D)dD,$$

where $P(q, D)$ is the form factor of a homogeneous sphere of diameter D, $N(D)$ represents the size distribution, $S(q, D_{hs}, \eta_{hs})$ is the structure factor as a function of the volume fraction η_{hs} of the hard spheres (hs). Finally $|T(\alpha_i)|^2$ and $|T(\alpha_f)|^2$ are Fresnel transmission coefficients for the angle of incidence α_i and exit α_f. The use of LMA requires the introduction of several different size distribution functions with appropriate weighting factors. GISAXS might be the most promising analysis technique, since it can be automated and the availability of intensive synchrotron X-ray beams enables a short integration time.

18.3.4 PHASE ANALYSIS

For studying the interdependencies of dose, dose ratio, and annealing conditions, materials, libraries with constant low (2.7×10^{16} at/cm²) and high (4.7×10^{16} at/cm²) Se dose and varying Cd dose were produced. Already after 30 s annealing in pure Ar atmosphere at 700°C CdSe precipitates with hexagonal crystal structure form for all Cd:Se dose ratios (Figure 18.9). In the case of dose ratios smaller than 1.95 the series of diffraction peaks situated between 23° and 27° merge to a broad structure most probably due to line width broadening by size effects. For these samples an evaluation of

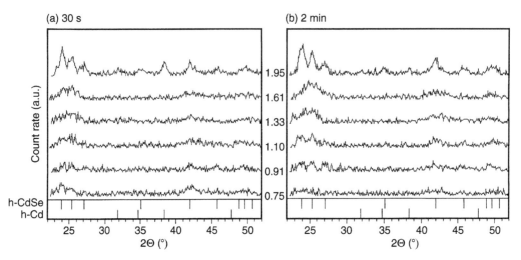

FIGURE 18.9 XRD diffraction pattern of samples annealed at 700°C for 30 s and 2 min, between the graphs the Cd:Se dose ratio is indicated.[41] The Se dose was 2.7×10^{16} at/cm².

TABLE 18.2
Mean CdSe and Cd Precipitate Diameters for a Fixed Se Dose of 2.7×10^{16} at/cm² after 30 s and 2 min at 700°C[41]

	CdSe Crystallite Size (nm)		Cd Crystallite Size (nm)	
Implanted Cd:Se Ratio	30 s	2 min	30 s	2 min
1.95	11	9	20	—
1.61	4	4	—	—
1.33	4	4	—	—
1.10	4	7	—	—
0.91	—	—	—	—
0.75	5	—	—	—

the width of (110)-peak by Scherer's formula a mean CdSe crystallite diameter between 4 and 5 nm results. For a Cd:Se ratio of 1.95 the mean CdSe crystallite size reaches 11 nm. In addition diffraction peaks of hexagonal Cd precipitates appear. Here the width of the (101)-diffraction peak at 38.35° gives a mean crystallite diameter of 20 nm.

When the annealing time is increased from 30 s to 2 min a slight increase of the intensity of the diffraction peaks is found. For a Cd:Se dose ratio of 1.95 a decrease of the Cd precipitate size is indicated by a reduction of the (110)-diffraction peak of Cd. The determined crystallite sizes are summarized in Table 18.2. Annealing at 1000°C results after 30 s in the formation of CdSe crystallites with a mean diameter larger than approximately 20 nm. Annealing at 1000°C results after 30 s in the formation of large CdSe crystallites (Table 18.3). XTEM images of Figure 18.10 show that the large precipitates are surrounded by smaller ones. The diffraction patterns depicted in Figure 18.11 reveal that there is an increase in the mean precipitate diameter with increasing Cd:Se dose ratio up to 1.61. Moreover this is accompanied by a steadily increasing peak intensity indicating an increase of the crystalline volume of the precipitates. In the case of Cd:Se dose ratios of more than 1.54 the diffraction peak intensities reduce and with it their diameter. This is most probably the

TABLE 18.3
Mean CdSe and Cd Precipitate Diameters for a Fixed Se Dose
of 2.7×10^{16} at/cm² after 30 s and 32 min at 1000°C[41]

Implanted Cd:Se Ratio	CdSe Crystallite Size (nm)		Cd Crystallite Size (nm)	
	30 s	32 min	30 s	32 min
2.14	27	23	27	21
1.61	44	13	16	14
1.33	27	21	—	—
1.10	21	17	—	—
0.91	—	12	—	—
0.68	25	21	—	—

result of material loss by out-diffusion of CdSe. Many surfaces near large precipitates disappear, leaving behind voids in the SiO_2 matrix. In addition the formation of Cd nanocrystalline precipitates takes place at Cd:Se dose ratios of 1.61 and above for both annealing times 30 s and 30 min. A prolongation of the annealing time to 32 min results in a decrease in the mean crystallite diameter due to dissolution of the larger precipitates for all dose ratios investigated. This manifests in a reduction of the diffraction peak intensities and is particularly pronounced for the Cd:Se dose ratio of 1.61.

Analogous to the series of samples with a Se dose of 2.7×10^{16} at/cm² a series of samples was synthesized with a Se dose of 4.8×10^{16} at/cm². From the XTEM image depicted in Figure 18.12 of a sample with a Cd:Se ratio of 1.33 annealed at 1000°C for 30 s can be seen that large precipitates are surrounded by smaller ones. The electron diffraction pattern in Figure 18.12(c) indicates that the CdSe precipitates formed have a hexagonal crystalline structure.

With increasing annealing duration the diffraction peak intensities are decreasing. After 30 s annealing time the CdSe diffraction pattern vanishes for a dose ratio of 1.95 and for 2 min and 16 min this occurs already slightly above 1.46. At the same time the diffraction peaks of the crystalline Cd precipitates develop at a dose ratio 1.77 and 1.61 for an annealing time of 16 and 32 min, respectively.

At an annealing temperature of 1000°C out-diffusion of material sets in already at a Cd:Se dose ratio of 1.33 which results in a severe reduction of the diffraction peak intensity (Figure 18.13(a)). The loss of material could also be confirmed by quantitative RBS measurements. This loss intensifies for longer annealing durations and the reduction of relevant diffraction peaks shifts to lower Cd:Se dose ratios (Figure 18.13(b)–(d)). Distinct diffraction peaks which appear after annealing for 32 min (Fig 18.13(e)) at Cd:Se dose ratios of 1.61 and above indicate the existence of Cd precipitates ((101)-reflex at 38.35°).

18.4 PHOTOLUMINESCENCE

Compared to the corresponding bulk materials QDs can have superior properties in the field of optical and optoelectronic applications, such as gain media for light-amplifying devices. These prospects for potential applications have initiated several attempts to grow QDs of various semiconductor materials.

The unique optoelectronic properties of QDs are based on the restricted free-electron and/or hole motion, which gives rise to a substantial change in the density of states. An ensemble of QDs, embedded in a host, possesses a huge interfacial area, which has a decisive influence on the finally attained optical and electrical properties. In practically all cases a passivation of the QD surface is essential in order to achieve maximal photoluminescence yield. For example, compound semiconductor QDs of CdSe synthesized by a chemical route have to be capped with organic compounds in

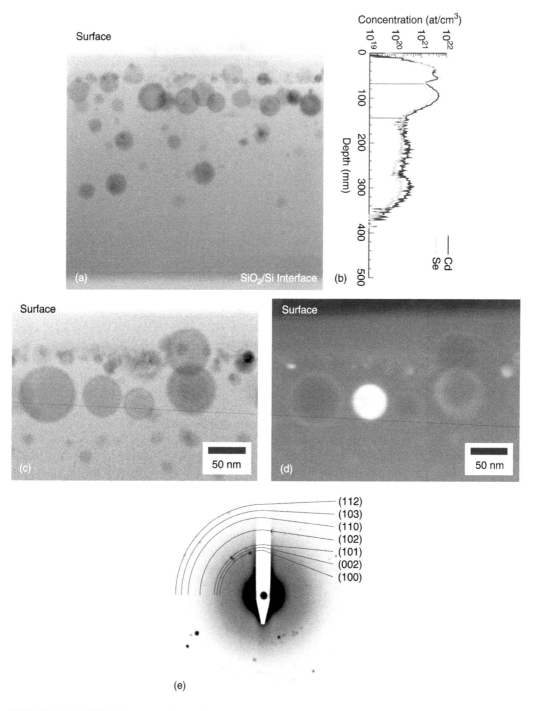

FIGURE 18.10 XTEM images of a sample annealed at 1000°C for 30 s. The CdSe dose ratio was 1.28 with a Se dose of 2.7×10^{16} at/cm². (a) Bright field, (b) SIMS depth profile of Cd and Se, (c) bright field enlargement, (d) dark field enlargement image and (e) electron diffraction pattern with the positions of the diffraction rings of the hexagonal CdSe phase indicated.[41]

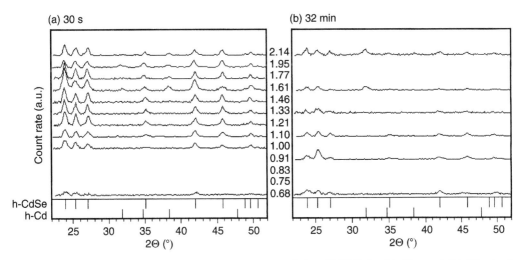

FIGURE 18.11 XRD diffraction pattern of samples annealed at 1000°C for (a) 30 s and (b) 32 min. In between the graphs the Cd:Se dose ratio is indicated.[41] The Se dose was 2.7×10^{16} at/cm^2.

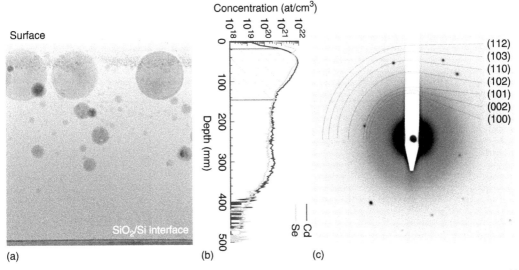

FIGURE 18.12 (a) XTEM images of a sample annealed at 1000°C for 30 s. The Se dose was 4.78×10^{16} at/cm^2 and the CdSe dose ratio was 1.33, (b) SIMS depth profile of Cd and Se, (c) electron diffraction pattern with the positions of the diffraction rings of the hexagonal CdSe phase indicated.[41]

order to passivate the QD's surface.[28] Even core shell QD, i.e., QDs of CdSe capped with a thin layer of larger bandgap CdS, can be fabricated by the chemical route and have been successfully used as laser active material in the form of a close-packed assembly.[29] It was found by Rutherford backscattering spectroscopy (RBS)[30] that the CdSe nanoclusters formed were Cd-rich, independent of the ratio of reactants used for their chemical synthesis. Taylor et al.[30] concluded from PL measurements, and the absence of vacancy defect luminescence, that the excess Cd must be located at the surface of the CdSe nanoclusters passivating Se dangling bonds, which otherwise would act as efficient non-radiative charge traps. These surface chemistry effects are leading to a deviation of the stoichiometry compared to the corresponding bulk material and have to be taken into account.

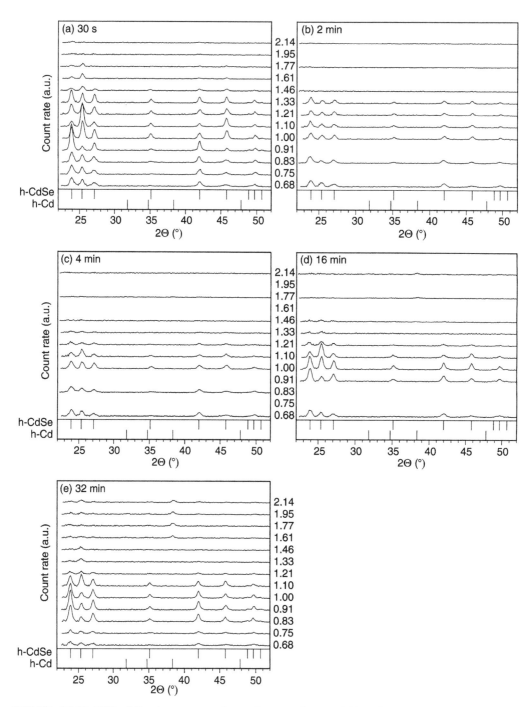

FIGURE 18.13 XRD diffraction pattern of samples annealed at 1000°C for different annealing times: (a) 30 s, (b) 2 min, (c) 4 min, (d) 16 min, (e) 32 min. In between the graphs the Cd:Se dose ratio is indicated.[41] The Se dose was 4.8×10^{16} at/cm².

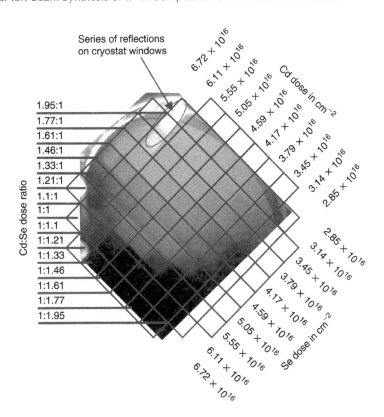

FIGURE 18.14 (See color insert following page 172) A luminescent wafer at 15 K (annealing condition 1000°C, 30 s in pure Ar atmosphere) with implantation pattern. Each of the 88 areas of the dose raster was implanted with a constant and homogeneous Cd:Se dose ratio.

Hence the preparation of compositionally varying samples by a combinatorial method on a single wafer or chip, followed by a fast screening method, can significantly accelerate the identification of optimal conditions and provide correlated and easily comparable data for interpretation and with it further development.

Fast screening of a photoluminescent materials library was performed by illumination with a Hg-vapor lamp. A picture together with the corresponding dose pattern is shown in Figure 18.14. The temperature of the wafer was 15 K. The implanted dose ratio pattern of Cd and Se in 500-nm-thick SiO_2 thermally grown on a 4-inch silicon wafer annealed at 1000°C for 30 s in pure flowing Ar atmosphere.

PL spectra from selected wafer areas according to the implanted dose pattern were measured with an experimental arrangement depicted in Figure 18.15. The wafer was mounted into a movable optical cryostat. The location of measurement on the wafer was positioned in a step and scan mode into an Ar-ion laser beam. The generated PL light was collected by an off-axis parabolic mirror and guided into a grating spectrometer with a camera cooled with CCD liquid nitrogen.

The matrix of PL spectra measured on the wafer depicted in Figure 18.14 is shown in Figure 18.16. The temperature of the wafer was 15 K. The implanted dose pattern is reorganized so that constant Se doses and the Cd:Se dose ratios form columns and lines respectively. The maximum intensities of the spectra are normalized to the maximum intesity of the sample with the parameters 4.59×10^{16} at/cm^2 and the Cd:Se ratio of 1.33. Optical excitation was achieved by a continuous wave Ar-ion laser at a wavelength of 488 nm. The spot diameter was defined by an aperture to be 0.5 mm. In order to avoid laser beam heating of the sample the laser power was limited to 1 mW. the decrease in the width of the PL signal with increasing Cd content is Striking. The decrease of the width of the

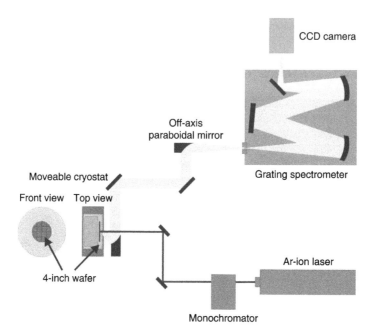

FIGURE 18.15 Schematic diagram for a screening PL measurement. The optical cryostat with a 4-inch window can be moved in front of the laser beam in order to address different locations of the cooled wafer surface.

PL spectra is accompanied by an increase of the PL peak height. This trend is present for the complete range of Se doses. In this dose ratio regime a sharp increase in the PL signal height can be observed when the Cd:Se dose ratio reaches 1.33. If a greater Cd surplus is implanted the peak intensity decreases slightly and shifts towards lower energies.

For comparison a PL spectrum of a CdSe single crystal is depicted in the lower left corner of Figure 18.16. The band edge luminescence of bulk CdSe at the energy of 1.83 eV is indicated by a vertical line.

In accordance with findings made for CdSe particles synthesized by a chemical route, passivation of non-radiative surface states by Cd might explain this strong increase in light emission at an energy of 1.815 eV, but it cannot be excluded that the additional Cd creates either a radiative state at the CdSe nanocrystal surface, in the surrounding matrix, or acts as a donor state built into the CdSe nanocrystal lattice.

Inhomogeneous broadening towards lower energies observed on samples at the position 5.05×10^{16} at/cm^2 and 5.56×10^{16} at/cm^2 for a dose ratio of 0.9 of Figure 18.16 can be attributed to emission of phonons, which manifests in a "phonon wing" in the PL spectrum. Equidistant phonon satellites can be resolved for the sample in position 4.59, 5.05, and 5.56×10^{16} at/cm^2 and Cd:Se dose ratio of 0.9. This observation correlates with the formation of larger CdSe crystallites by growth and coalescence of smaller ones confirmed by X-ray diffraction and XTEM measurements.[16,19,31]

In order to prove that the optical transition has its origin at least in the co-existence of CdSe, Cd, and Se, separate samples each individually implanted with Cd or Se followed by the same annealing procedure were produced. For comparison the PL-spectra together with the spectrum of a CdSe single crystal are shown in Figure 18.17. The maximum peak height of the spectra are normalized with respect to the ion beam synthesized CdSe sample (curve c) (Cd:Se dose ratio 1.33 annealed at 1000°C for 30 s). The scaling factors are given in brackets. Despite the small amount of CdSe material present in the ion beam synthesized CdSe sample (corresponding to an approximately 30-nm-thick CdSe layer) an intensive luminescence peak close to the band edge luminescence of bulk CdSe was found. The single implanted Cd and Se samples show only a very weak PL response.

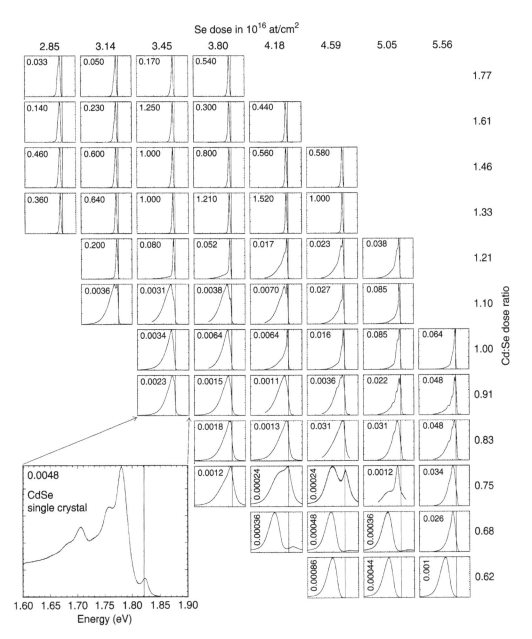

FIGURE 18.16 Normalized PL spectra measured at 15 K after annealing at 1000°C for 30 s in pure Ar atmosphere. The reference spectrum in the lower left corner stems from a CdSe single crystal. The spectra are normalized to spectrum at the position of column 4.59×10^{16} at/cm^2 and row 1.33. The scaling factors relative to the normalized spectrum are indicated in the upper left corner of each spectrum.

18.5 SUMMARY AND OUTLOOK

The application of combinatorial and high-throughput methodology to materials synthesis by multiple element ion implantation has been demonstrated. Since high-dose ion implantation is a time-consuming process the interaction of multiple implanted elements is barely investigated. This can change with the introduction of the combinatorial approach in this field. There are a multitude

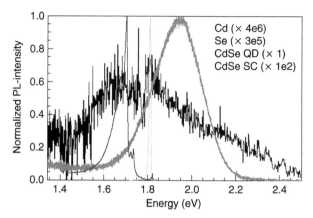

FIGURE 18.17 PL spectra of single implanted (a) Cd, (b) Se, (d) single crystalline (SC) CdSe and (c) ion beam synthesized CdSe QDs in SiO_2 on silicon. Post-implantation thermal treatment of the samples was identical (30 s at 1000°C in flowing Ar-atmosphere). The individual scaling factors are written in brackets.

of systems to which combinatorial ion implantation can be applied successfully. Among them is the classical field of doping by ion implantation where the interaction of different dopants can now be effectively assessed. In the field of materials synthesis of buried compounds and nanoparticles, where complex interdependencies determine the finally attained material properties, combinatorial implantation schemes can help to accelerate the identification of key process parameters and with it the understanding of the underlying physical and chemical processes. The test of models becomes much more reliable due to complete data sets as function of the parameters under investigation.

With a fast synthesis the need of fast screening techniques is inherently connected. Even in the case of high-dose ion implantation the amount of material synthesized is very small, which puts high demands on the fast screening techniques. It was demonstrated in this work that very fast parallel screening can be done by illuminating the whole wafer at low temperatures. In this way the most efficient luminescent materials library elements can be easily identified. However, details of the PL spectra were obtained by measuring the samples in succession. It was shown that optically active CdSe nanoclusters can be synthesized by ion implanation in technologically important thin films of SiO_2. Moreover it was also shown that highly efficient luminescent CdSe material is obtained in the Cd overstoichiometric implanted regime.

REFERENCES

1. Efros, Al. L. and Efros, A. L., Interband absorption of light in a semiconductor sphere, *Soviet Phys. Semicond.*, 16(7), 772, 1982.
2. Norris, D. J., Efros, Al. L., Rosen, M. and Bawendi, M. G., Size dependence of exciton fine structure in CdSe quantum dots, *Phys. Rev. B*, 53(24) 16347, 1996.
3. Shimizu-Iwayama, T., Hole, D. E. and Townsend, P. D., Optical properties of interacting Si nanoclusters in SiO_2 fabricated by ion implantation and annealing, *Nucl. Inst. Methods Phys. Res. B*, 147, 350, 1999.
4. Mikhailovsky, A., Malko, A. V., Hollingsworth, J. A., Bawendi, M. G. and Klimov, V. I., Multiparticle interactions and stimulated emission in chemically synthesized quantum dots, *Appl. Phys. Lett.* 80(30), 2380, 2002.
5. Kagan, C. R., Murray, C. B. and Bawendi, M. G., Long-range resonance transfer of electronic excitations in close-packed CdSe quantum-dot solids, *Phys. Rev. B*, 54(12) 8633, 1996.
6. Gindele, F., Westphaling, R., Woggon, U., Spanhel, L. and Platschek, V., Optical gain and high quantum efficiency of matrix-free, closely packed CdSe quantum dots, *Appl. Phys. Lett*, 71, 2181, 1997.

7. Artemyev, M. V., Woggon, U., Jaschinski, H., Gurinovich L. J. and Gaponenko, S. V., Spectroscopic Study of Electronic States in an Ensemble of Close-Packed CdSe Nanocrystals, *J. Phys. Chem. B*, 104, 11617, 2000.

8. Markwitz, A., Rebohle, L., Hofmeister, H. and Skorupa, W., Homogeneously size distributed Ge nanoclusters embedded in SiO_2 layers produced by ion beam synthesis, *Nucl. Inst. Methods Phys. Res. B*, 147, 361, 1999.

9. Arnold, G. W. and Borders, J. A., Aggregation and migration of ion-implanted silver in lithia-alumina-silica glass, *J. Appl. Phys.*, 48, 1488, 1977.

10. Lindner, J. K. N. and Stritzker, B., Controlling the density distribution of SiC nanocrystals for the ion beam synthesis of buried SiC layers in silicon, *Nucl. Inst. Methods Phys. Res. B*, 147, 249, 1999.

11. Mantl, S., Ion-beam synthesis of buried metallic and semiconducting silicides, *Nucl. Inst. Methods Phys. Res. B*, 80(1), 895, 1993.

12. Meldrum, A., Haglund, R. F., Boatner, L. A. and White, C. W., Nanocomposite materials formed by ion implantation, *Adv. Mater.*, 13, 19, 1413, 2001.

13. Shiryaev S. Yu, and Nylandsted Larsen, A., High-dose mixed Ga/As and Ga/P ion implantations in silicon single-crystals, *Nucl. Inst. Methods Phys. Res. B*, 80(81), 846, 1993.

14. Ziegler, J. F., Biersack, J. P. and Littmark, U, *The Stopping and Range of Ions in Solids*, Pergamon, New York, 1985.

15. Möller, W. and Posselt, M., *Wissenschaftlich-Technische Berichte FZR-317*, Forschungszentrum Rossendorf, 2001.

16. Karl, H., Großhans, I. and Stritzker, B., Combinatorial ion beam synthesis of semiconductor nanocluster, *Meas. Sci. Technol.*, 16, 32–40, 2004.

17. Chen, Ch., Pan, Ch., Zhu, D., Hu, J. and Li, M., Combinatorial ion synthesis and ion beam analyses of materials libraries on thermally grown SiO_2, *Mat. Sci. Eng. B*, 72, 113, 2000.

18. Sato, Y., Sakaguchi, I., Suzuki, M. and Haneda, H., Development of combinatorial ion implantation system, *Jpn. J. Appl. Phys.*, 42, 5867, 2003.

19. Großhans, I., Karl, H. and Stritzker, B., Advanced apparatus for combinatorial synthesis of buried II-VI nanocrystals by ion implantation, *Mat. Sci. Eng. B*, 101, 212, 2003.

20. Sakaguchi, I., Hishita, S. and Haneda, H., Combinatorial ion implantation techniques application to optical characteristics of ZnO, *Jpn. J. Appl. Phys.*, 43, 8A, 5562, 2004.

21. Großhans, I., Karl, H. and Stritzker, B., Combinatorial synthesis of ZnTe nanocrystals in SiO_2 on silicon by ion implantation, *Nucl. Instr. Meth. B*, 190, 865, 2002.

22. Karl, H., Hipp, W., Großhans, I. and Stritzker, B., Ion beam synthesis of buried CdSe nanocrystallites in SiO_2 on (100)-silicon *Mat. Sci. Eng. C*, 19, 55, 2002.

23. Karl, H., Großhans, I., Attenberger, W., Schmid, M. and Stritzker, B., Buried ZnTe nanocrystallites in thermal SiO_2 on silicon synthesized by high dose ion implanation, *Nucl. Instr. Meth. B*, 178, 126, 2002.

24. Lifshitz, I. M. and Syyozov, V. V., The kinetics of precipitation from supersaturated solid solutions, *J. Phys. Chem. Solids*, 19(1–2), 35, 1961.

25. Wagner, C., Theorie der Alterung von Niederschlägen durch umlösen (Ostwald-Reifung), *Zeitschrift für Elektrochemie*, 65(7–8), 581, 1961.

26. Karl, H., Großhans, I. and Stritzker, B., Self Organized Compound Semiconductor Nanocrystallite Distributions in SiO_2 on Silicon Synthesized by Ion Implanation, *Mat. Res. Soc. Symp. Proc.*, 794, 83, 2004.

27. Desnica-Frankovic, I. D., Desnica, U. V., Dubcek, P., Buljan, M., Bernstorff, S., Karl, H., Großhans, I. and Stritzker, B., Ion beam synthesis of buried Zn-VI quantum dots in SiO_2-grazing incidence small-angle X-ray scattering studies, *J. Appl. Cryst.*, 36, 439, 2003.

28. Murray, C. B., Norris, G. J. and Bawendi, M. G., Synthesis and characteriszation of nearly monodisperse CdE (E= S, Se, Te) semiconductor nanocrystallites, *J. Am. Chem. Soc.*, 115, 8706, 1993.

29. Eisler, H.-J., Sundar, V. C., Bawendi, M. G., Walsh, M., Smith, H. I. and Klimov, V., Color-selective semiconductor nanocrystal laser, *Appl. Phys. Lett.*, 80(24), 4614, 2002.

30. Taylor, J., Kippeny, T., Bennett, J. C., Huang, M., Feldman, L. C. and Rosenthal, S. J., *Mat. Res. Soc. Symp. Proc.*, 536, 413, 1999.

31. Großhans, I., Karl, H. and Stritzker, B., Combinatorial ion implantation – a smart technique applied to synthesize CdSe-nanocrystals, *Nucl. Instr. Meth.*, 216, 396, 2004.

32. Chopard, B., Luthi, P. and Droz, M., Micorscopic approach to the formation of Liesegang patterns, *J. Stat. Phys.*, 76, 661, 1994.

33. Liesegang, R., Über einige Eigenschaften von Gallerten, *Naturwissenschaften Wochenschr.*, 11, 353, 1896.
34. Polezhaev, A. and Müller, S., Complexity of precipitation patterns: Comparison of simulation with experiment, *Chaos*, 4, 631, 1994.
35. Venzl, G. and Ross, J., Nucleation and colloidal growth in concentration gradients (Liesegang rings), *J. Chem. Phys.*, 77, 1302, 1982.
36. Brechet, Y. and Kirkaldy, J. S., Contribution to the theory of diffusion-reaction controlled Liesegang patterns, *J. Chem. Phys.*, 90, 1499, 1989.
37. Gálfi, L. and Rácz, Z., Properties of the reaction front in an A=B -> C type reaction-diffusion process, *Phys. Rev. A*, 38, 3151, 1988.
38. Landolt-Börnstein, *Numerical Data and Functional Relationships in Science and Technology*: New Series IV/19A, Springer Verlag, Berlin, Heidelberg, New York, 1999.
39. Großhans, I., Karl, H. and Stritzker, B., Temporal evolution of ion implanted CdSe distribution in SiO_2 on silicon during annealing, *Nucl. Instr. Meth.*, 219–220, 820, 2004.
40. Karl, H., Großhans, I., Wenzel, A., Claessen, R., Cirlin, G. E., Egorov, V. A., Polyakov, N. K., Petrov, V. N., Ustinov, V. M., Ledentsov, N. N., Alferov, Zh. I., Strocov V. N. and Stritzker, B., Stoichiometry and absolute atomic concentration profiles obtained by combined Rutherford backscattering spectroscopy and secondary-ion mass spectroscopy: InAs nanocrystals in Si, *Nanotechnology*, 13, 631, 2002.
41. Grobhans, I., Strukturelle und optische Untersuchungen an ionenstrahlsynthetisierten CdSe-Nanokristallen, Dissertation, Universität Augsburg, 2004.

19 Preparation of Dielectric Thin-Film Libraries by Sol–Gel Techniques

Virginie Jéhanno, Berit Wessler, Wolfgang Rossner, Gerald Frenzer, and Wilhelm F. Maier

CONTENTS

19.1 INTRODUCTION

The short innovation cycles in communication technology require the development and optimization of high-performance dielectrics for passive integration and miniaturization, utilized, for example, in bandpass filters, antennas, or amplifiers. Conventional one-at-a-time techniques for developing materials are slow and expensive. Thus acceleration of the search by high-throughput technologies is attractive. High-throughput experimentation (HTE) enables a rapid exploration of multidimensional composition and processing spaces and is on the way to becoming an efficient tool for materials discovery and optimization. It has been successfully applied to homogeneous and heterogeneous catalysis[1] as well as to a broad range of materials. The HTE process is composed of design of experiment,[2] library synthesis, and property analysis.[3] In the field of catalysis, synthesis techniques for powder or thick films have mainly adopted variations of wet chemical processes, transferred to pipetting robots or ink-jet dispensers,[1] while in the search for electronic materials gas-phase routes for thin films such as physical or chemical vapor deposition in combination with masking or gradient techniques are commonly applied.[4] Due to its complexity the conventional way of preparation of dielectric materials as ceramics does not lend itself to high-throughput experimentation. Wet synthesis techniques in analogy to the catalysis work, however, seemed more attractive, because it is a low-cost technique and principally allows to broadly vary composition at extremely mild synthesis conditions. However, HTE sol–gel techniques for producing arrays of defined thin films of identical thickness (< 2 µm) on libraries are not known, especially not for the elemental combinations of interest for dielectric applications.

The present study focuses on the synthesis of libraries, consisting of sol–gel deposition of chemically diverse thin films followed by calcination and sintering. The films are prepared out of a set of precursor solutions of 20 elements commonly used in dielectric materials. The influence of different parameters on the film quality, such as the type of precursors, their miscibility with each other, thermal processing, and techniques to prestructure the substrate is investigated. Films are characterized by high-throughput X-ray diffraction, scanning electron microscopy, and high-throughput X-ray fluorescence.

19.2   LIBRARY SYNTHESIS

Typically, the synthesis of thin-film libraries is accomplished by gas-phase methods such as sputtering,[5] physical vapor deposition (PVD), molecular beam epitaxy (MBE), or pulsed-laser deposition (PLD).[5,6] The application of sol–gel techniques is much less common.[7–9] The main disadvantage of gas-phase methods is the lack of a bulk synthesis for discovered new materials and its high costs, of advantage is the facile access of most elements of the periodic table. With gas-phase methods there is no principal limit to chemical composition, although the number of elements is often limited to four or less by the deposition equipment used. The advantages of the sol–gel approach are the facilitated scale-up and the use of low-cost equipment for dosing of liquids. The disadvantage are the larger library and sample size and the dependence on suitable recipes.

In order to use sol–gel techniques for the combinatorial synthesis of libraries for materials with interesting dielectric properties, appropriate elements had to be selected and associated precursors for each of those chosen elements based on sol–gel recipes had to be developed, which not only allows to widely vary the range of compositions, but also tolerates the use of dopants.

Sufficiently stable precursor sols for the chosen element combinations need to allow the handling with a pipetting robot, the formation of multinary mixtures in desired quantities and the formation of a thin film after deposition onto a substrate surface. For the deposition of these liquid precursors onto a substrate, such as a 6″ silicon wafer, an appropriate structuring of the substrate is needed. Accurate measurements of the dielectric properties of thin films require low porosity and roughness, crack-free formation and a minimum film thickness of ~1 µm. Thin film libraries are commonly screened for dielectric properties, i.e., dielectric constant and loss factor, with a scanning evanescent microwave microscope.[10–12]

19.2.1   Precursor Set

19.2.1.1   Synthesis Techniques

Sol–gel precursors are based on solutions of metal alkoxides, acetates, or acetyl acetonates in solvent mixtures, which allow the preparation of stable multinary combinations without precipitation or rapid gelation. For each element an individual single-element precursor sol was synthesized. To enable miscibility of all precursors with each other in any composition at room temperature, the same solvent basis and complexing agent had to be achieved, i.e., propionic acid and iso-propanol. The new precursor solutions mixed by the pipetting robot have to be stable for at least 5 h, which is the time required for the preparation of the library. Gelation, e.g., the condensation, is supposed to take place in the as-deposited film rather than in the mixture of different precursors. Some individual precursor recipes are summarized in Table 19.1 and described below.

TABLE 19.1
Preparation of Single-Element Precursors for the Library Synthesis

Metal Alkoxide	Propionic Acid (vol%)	Alcohol (vol%)	Ethylene Glycol (vol%)	Water (mol eq.)
Ba ethoxide	60	30 (ethanol)	10	
Ca i-propoxide	80	20 (i-propanol)	0	2
Sr i-propoxide	30	50 (i-propanol)	20	
Ta ethoxide	40	50 (i-propanol)	10	
Ti n-propoxide	40	60 (i-propanol)	0	
Zr i-propoxide	80	20 (i-propanol)	0	

The sols (0.25–0.5 mol L^{-1}) were prepared adapting the methods described by Shimooka et al.[13,14] and Schnöller et al.[15] by dissolving the selected precursors in alcohol. Propionic acid or ethylene glycol was added to stabilize the sol against condensation. In some cases water was added to initiate the hydrolysis. Alkoxides used were $Ba(OEt)_2$ (Chemat Technology, 99.5%), $Ca(O^iPr)_2$ (ABCR, 98%), $Sr(O^iPr)_2$ (Chemat Technology, 99% and ABCR, 97%), $Ta(OEt)_5$ (Aldrich, 99.98%), $Ti(O^nPr)_4$ (Chemat Technology, low-purity grade), $Zr(O^iPr)_4$ (Gelest, medium grade), respectively, as precursors for barium, calcium, strontium, tantalum, titanium, and zirconium sols. The solvents used were all high-purity grade (supplier: Merck).

Later a set of compatible sol recipes for 20 different elements, commonly used in dielectric materials, e.g., Na, Co-, or Wions, has been developed successfully in addition to the above-mentioned starting set of six elements. These are also based on different starting materials with different ratios of iso-propanol as solvent and propionic acid as complexing agent prepared under individual heating conditions.

19.2.1.2   Binary and Ternary Combinations of the Single-Element Precursors

Binary and ternary combinations of the described single-element sols were obtained by deposition of a defined volume of each sol into the wells of the library plate by a pipetting robot. A description of the systematic elemental combination is presented in Table 19.2. The mixtures were stable in air for at least 6 h. Precipitation in some mixtures could be overcome by changing the sequence of the sol addition. It was found that stable sol mixtures in a wide range of compositions can only be obtained if the mixed sol preparation starts with the addition of the sols containing the precursors of Zr, Ti, and Ta, in some cases Ca, while the Ba sol should always be added last.

For the additional set of the 20 element precursors binary and ternary mixtures in a 1:1 ratio and 1:1:1 ratio, respectively, were carried out as well showing rapid precipitation or gelation only in few mixtures.

19.2.2   PRESTRUCTURING OF THE SUBSTRATES

The common techniques used to form thin films from precursor solutions are spin- or dip-coating, which does not lend itself readily to combinatorial library preparation with highly diverse materials. Here the simple deposition of defined volumes of premixed sols onto the substrate surface was chosen. To avoid mixing of adjacent sol-depositions, reliable spatial separation of deposited sols is required. Preoxidized 6″ Si wafers were selected as substrate materials. Several approaches have been tested. One approach was to use a synthesis reactor, in which a mask with drill holes as reactor volume was pressed against the Si wafer. The sealing was provided by Viton fittings (O-rings) as described in reference 15. The sols were positioned in the drill holes of the masks and dried. After

TABLE 19.2
Binary and Ternary Mixtures of Sol–Gel Precursors

	1	2	3	4	5	6	7	8	9	10	11	12
1	Ba Ca –	Ba Sr Ti	Ba Ti Ca	Ba Zr Ti	Ca Sr Ta	Ca Ti –	Ca Zr Ta	Sr Ca Ba	Sr Ta Zr	Sr Zr Ca	Ta Ca –	Ta Sr Ti
2	Ba Ca Sr	Ba Sr Zr	Ba Ti Sr	Ca Ba –	Ca Sr Ti	Ca Ti Ba	Ca Zr Ti	Sr Ca Ta	Sr Ti –	Sr Zr Ta	Ta Ca Ba	Ta Sr Zr
3	Ba Ca Ta	Ba Ta –	Ba Ti Ta	Ca Ba Sr	Ca Sr Zr	Ca Ti Sr	Sr Ba –	Sr Ca Ti	Sr Ti Ba	Sr Zr Ti	Ta Ca Sr	Ta Ti –
4	Ba Ca Ti	Ba Ta Ca	Ba Ti Zr	Ca Ba Ta	Ca Ta –	Ca Ti Ta	Sr Ba Ca	Sr Ca Zr	Sr Ti Ca	Ta Ba –	Ta Ca Ti	Ta Ti Ba
5	Ba Ca Zr	Ba Ta Sr	Ba Zr –	Ca Ba Ti	Ca Ta Ba	Ca Ti Zr	Sr Ba Ta	Sr Ta –	Sr Ti Ta	Ta Ba Ca	Ta Ca Zr	Ta Ti Ca
6	Ba Sr –	Ba Ta Ti	Ba Zr Ca	Ca Ba Zr	Ca Ta Sr	Ca Zr –	Sr Ba Ti	Sr Ta Ba	Sr Ti Zr	Ta Ba Sr	Ta Sr –	Ta Ti Sr
7	Ba Sr Ca	Ba Ta Zr	Ba Zr Sr	Ca Sr –	Ca Ta Ti	Ca Zr Ba	Sr Ba Zr	Sr Ta Ca	Sr Zr –	Ta Ba Ti	Ta Sr Ba	Ta Ti Zr
8	Ba Sr Ta	Ba Ti –	Ba Zr Ta	Ca Sr Ba	Ca Ta Zr	Ca Zr Sr	Sr Ca –	Sr Ta Ti	Sr Zr Ba	Ta Ba Zr	Ta Sr Ca	Ta Zr –

	1	2	3	4	5	6	7	8	9	10	11	12
1	Ta Zr Ba	Ti Ba Zr	Ti Sr Ca	Ti Zr –	Zr Ba Ta	Zr Sr Ba	Zr Ta Ti					
2	Ta Zr Ca	Ti Ca –	Ti Sr Ta	Ti Zr Ba	Zr Ba Ti	Zr Sr Ca	Zr Ti –					
3	Ta Zr Sr	Ti Ca Ba	Ti Sr Zr	Ti Zr Ca	Zr Ca –	Zr Sr Ta	Zr Ti Ba					
4	Ta Zr Ti	Ti Ca Sr	Ti Ta –	Ti Zr Sr	Zr Ca Ba	Zr Sr Ti	Zr Ti Ca					
5	Ti Ba –	Ti Ca Ta	Ti Ta Ba	Ti Zr Ta	Zr Ca Sr	Zr Ta –	Zr Ti Sr					
6	Ti Ba Ca	Ti Ca Zr	Ti Ta Ca	Zr Ba –	Zr Ca Ta	Zr Ta Ba	Zr Ti Ta					
7	Ti Ba Sr	Ti Sr –	Ti Ta Sr	Zr Ba Ca	Zr Ca Ti	Zr Ta Ca						
8	Ti Ba Ta	Ti Sr Ba	Ti Ta Zr	Zr Ba Sr	Zr Sr –	Zr Ta Sr						

Legend: ☐ Precipitation   ▨ Gel   ☐ Stable sol

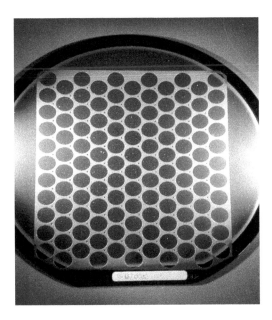

FIGURE 19.1 Prestructured substrate with ceramic mask for libraries.

drying the masks, the O-rings could be removed. All films showed inhomogeneities of the thickness as a result of adhesion of the sol to the reactor walls. Due to the high thickness of the as-deposited film cracks occur easily. This technique is thus not suitable for the desired application.

In another approach, screen-printing techniques were used to deposit a glass ceramic mask with circular openings onto the Si wafer in order to create 121 separated circular synthesis wells as shown in Figure 19.1. Up to ten printing steps were necessary to obtain a mask thickness of 70 μm after sequential calcinations of the printed paste at 1000°C. This method yielded high-quality films, but leakage of the sols through the interface between the wafer and the ceramic mask could not be eliminated and thus the well separation was not satisfying.

The most promising prestructuring of the substrate turned out to be a Si wafer with 9 × 10 circular grooves (Figure 19.2). The grooves were produced by sandblasting Si wafers after coverage of their surface with a photolithographically structured mask.[17] This particular groove structure, whose cross-section is shown schematically in Figure 19.3, serves as a sink for excess sol deposited by a pipetting robot. This arrangement allows to produce high-quality films with a large variety of sols. It successfully prevents interference with other sols. With respect to the large shrinkage of the film due to evaporation of the solvent and pyrolysis of the organics the as-deposited films should not exceed a certain critical thickness to avoid cracks in the thin films. The resulting films show a much higher film quality than the films reported above. Cracks are avoided to a large extent. To prepare thicker films multiple coatings are necessary.

19.2.3   DEPOSITION OF THE SAMPLES

A pipetting robot was used to dose and mix the single precursor sols to create chemically diverse precursor solutions for binary and ternary mixed oxides. These mixtures were prepared in microplates and transferred onto the prestructured substrate. Depending on the sol concentration, several successive depositions (~2 μL), followed by rapid calcinations on a hotplate, were necessary to obtain the desired film thickness of about 1 μm. Single-precursor sol concentrations of 0.25 mol L^{-1} were found optimal for this process. It allowed the formation of low-porosity films of the desired

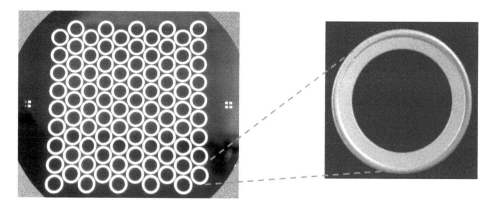

FIGURE 19.2 Prestructured Si wafer with 9 × 10 grooves, prepared by sand blasting.

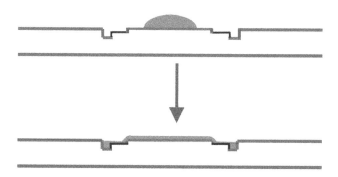

FIGURE 19.3 Schematic cross-section of sol spreading in the grooves of the prestructured Si wafer.

film thickness after only five deposition–calcination cycles. In Figure 19.4 the cross-section of a film prepared from five coatings of a 0.25 mol L^{-1} sol mixture of neodymium, magnesium, and tantalum precursors[16] is shown. A film thickness of 1 μm is obtained.

Finally, the calcined films are sintered at temperatures above 700°C to initiate phase formation, crystallization, and grain growth.

The composition of the samples on a typical library, the library design, was based on the distribution of two ternary composition spreads on 36 cavities of the substrate. The composition was systematically varied in 20% steps covering the whole composition ranges as shown in Figure 19.5.

19.2.3.1   Characterization of the Thin Films

The samples on the libraries were analyzed for elemental composition and phase composition as well as microstructure and film thickness obtained by the deposition methods used in this study.

19.2.4   ELEMENTAL COMPOSITION

The elemental composition of the samples was tested on a 36-element library by using XRF. The library composition was based on the distribution of Sr-Ca-Ti and Ba-Ca-Ti ternary diagrams. Several measurements were performed on each sample. The expected composition of the samples as well as their distribution on the substrate are presented in Table 19.3.

FIGURE 19.4 SEM image of the cross-section of a sintered film made of five depositions of 0.25 mol L^{-1} sol.

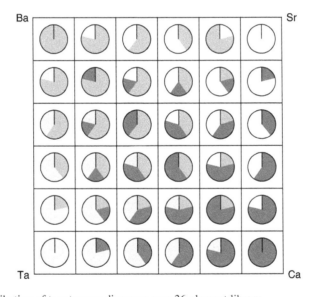

FIGURE 19.5 Distribution of two ternary diagrams on a 36-element library.

TABLE 19.3
Composition of the Test Library Measured by XRF

1	70 Sr	7	2 Ca, 8 Sr	13	4 Ca, 6 Sr	19	6 Ca, 4 Sr	25	8 Ca, 2 Sr	31	70 Ca
2	2 Ba, 8 Sr	8	8 Sr, 2 Ti	14	2 Ca, 6 Sr, 2 Ti	20	4 Ca, 4 Sr, 2 Ti	26	6 Ca, 2 Sr, 2 Ti	32	8 Ca, 2 Ti
3	4 Ba, 6 Sr	9	2 Ba, 6 Sr, 2 Ti	15	6 Sr, 4 Ti	21	2 Ca, 4 Sr, 4 Ti	27	4 Ca, 2 Sr, 4 Ti	33	6 Ca, 4 Ti
4	6 Ba, 4 Sr	10	4 Ba, 4 Sr, 2 Ti	16	2 Ba, 4 Sr, 4 Ti	22	4 Sr, 6 Ti	28	2 Ca, 2 Sr, 6 Ti	34	4 Ca, 6 Ti
5	8 Ba, 2 Sr	11	6 Ba, 2 Sr, 2 Ti	17	4 Ba, 2 Sr, 4 Ti	23	2 Ba, 2 Sr, 6 Ti	29	2 Sr, 8 Ti	35	2 Ca, 8 Ti
6	70 Bu	12	8 Ba, 2 Ti	18	6 Ba, 4 Ti	24	4 Ba, 6 Ti	30	2 Ba, 8 Ti	36	10 Ti

Compositional homogeneity of the samples was also characterized with this method by measuring several points on each sample. A relatively good homogeneity was then observed, considering the errors on measurement and calibration of the device. Table 19.4 summarizes the experimental results. Large deviations between expected and obtained compositions were observed for most samples containing Sr.

To test for potential segregation phenomena already in the sol, a sol with the composition Ca_2Sr_8 was prepared. The upper half of the sol was decanted and both fractions were calcined and analyzed by XRF. The upper half showed a content of 53% Sr and the lower half 76% Sr. Furthermore, the well-stirred sol Ca_2Sr_8 was filled in a well with 4 mm in diameter of a polycarbonate library. The expected composition was confirmed by XRF (82% Sr, 18% Ca). After evaporation of the solvent overnight, the film covering the bottom of the well was analyzed by a line scan analysis. The center concentration (40% Sr, 60% Ca) deviated significantly from the concentrations on rim (83% Sr, 17% Ca) of the well. Those results, independent of different recipes used for the synthesis of the sols, indicate unexpected phase separations already in the sol of the system Ca/Sr.

This phenomenon was studied further for other elements M in the mixtures Ca/M = 2/8, 5/5, and 8/2. The films and the associated sols were calcined at 600°C and studied by XRF. Figure 19.6 shows selected measured elemental composition of a variety of oxides doped with 20mol% calcium oxide. The results indicate significant deviations between expected (80%) and measured elemental content in the films as well as in its associated sols.

TABLE 19.4
Elementary Compositions Measured with XRF

1	10Sr	7	4.5Ca-5.5Sr	13	8Ca-2Sr	19	9Ca-1Sr	25	9Ca-1Sr	31	10Ca
2	3Ba-7Sr	8	4Sr-6Ti	14	4.9Ca-1.1Sr-4Ti	20	5.4Ca-0.7Sr-3.9Ti	26	6Ca-1Sr-3Ti	32	4.5Ca-5.5Ti
3	7.4Ba-2.6Sr	9	2Ba-2.4Sr-5.6Ti	15	2Sr-8Ti	21	2.3Ca-0.9Sr-6.7Ti	27	3.8Ca-0.3Sr-5.9Ti	33	4.2Ca-5.8Ti
4	8.8Ba-1.2Sr	10	7Ba-1Sr-2Ti	16	4.5Ba-4.5Sr-1Ti	22	0.9Sr-9.1Ti	28	3Ca-0.2Sr-6.8Ti	34	3Ca-7Ti
5	9.4Ba-0.6Sr	11	8.8Ba-0.6Sr-0.6Ti	17	6.5Ba-0.5Sr-3Ti	23	4.4Ba-0.4Sr-5.2Ti	29	0.4Sr-9.6Ti	35	2.5Ca-7.5Ti
6	10Ba	12	8Ba-2Ti	18	6Ba-4Ti	24	5Ba-5Ti	30	3Ba-7Ti	36	10Ti

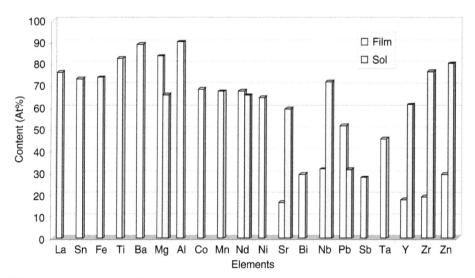

FIGURE 19.6 XRF results of Ca_2/element$_8$ compositions on Si-wafer library (measurement in the middle position of film) compared with the results for selected calcined sols.

While some of this deviation may be due to the above-mentioned problem of phase separation, some may also be due to the general problem of changes in elemental composition, film thickness, and film density affecting the XRF signal intensity of different elements. This change in signal intensity (or emission depth) of XRF can be calibrated and corrected in well-defined systems,[18] but it results in additional problems in combinatorial approaches, where doping with different elements results in a large variety of materials rendering quantitative corrections nearly impossible. For reliable thickness measurements conventional reflectometry has been used successfully, details of which will be reported later.

Another Al-specific problem was encountered with the interaction of the support material, the Si wafer (Si Kα line). Overlapping emission lines cause an increase in the Al content, which is dependent on the layer thickness (see Figure 19.7), here represented by the change in sol volume used for film preparation. That this is a special effect of the interaction of Al with Si could be confirmed by a variety of other studies (Ca_2Sr_8, Ca_2Ti_8, Ca_2Zn_8, Ca_2Ta_8), where film thickness does not correlate with signal intensity.

Since in high-throughput or combinatorial studies the film composition, density, and final film thickness may vary greatly, XRF should not be used for reliable quantification. Nevertheless, XRF is very useful to confirm the presence or absence of elements in samples and to determine the homogeneity of elemental composition in individual samples on libraries through line scans.

19.2.5 DETERMINATION OF FILM PROPERTIES

Thickness, density, and the surface roughness are important parameters for the measurement of the dielectric properties of the films. A minimal film thickness of 1 μm is required by scanning evanescent microwave microscopy (not shown here) to allow reliable differentiation from the substrate data. The samples should moreover present comparable film quality to allow a comparison of the measured dielectric constants and quality factors. Figure 19.8 shows selected sintered thin films with different element compositions. In Figure 19.9 a roughness profile of a barium magnesium tantalate mixed oxide film obtained by alpha step measurements is depicted. An average roughness R_a of 5 to 30 nm is observed in high-quality films, an average roughness more than 100 nm might cause inaccuracy in measuring the dielectric properties. Film thickness was determined reliably by

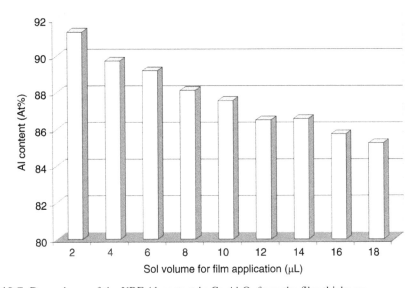

FIGURE 19.7 Dependence of the XRF Al content in $Ca_2Al_8O_x$ from the film thickness.

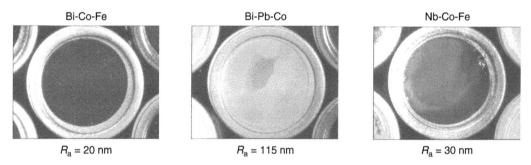

Bi-Co-Fe   Bi-Pb-Co   Nb-Co-Fe

R_a = 20 nm R_a = 115 nm R_a = 30 nm

FIGURE 19.8 Light microscopic image of some selected thin films of ternary composition and values for the average roughness R_a.

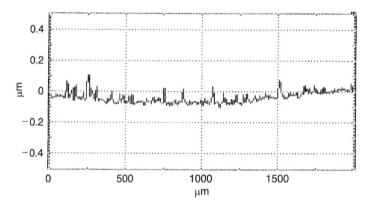

FIGURE 19.9 Roughness profile of a $BaMg_{1/3}Ta_{2/3}O_3$ thin films (2 mm scan) showing an average roughness of 25 nm.

SEM images from cross-sections of library samples. Since this implies the destruction of the libraries, it was only performed on test libraries. For characterization of dielectric properties though an automated non-destructive measurement is needed for measuring the thickness. Samples were prepared by breaking the library and subsequent sputtering with gold–palladium films (Blazers Union SCD 040). Typical images (SEM 4100S, Hitachi) of cross-sections of films generated on the wafer structured by sand blasting are shown in Figure 19.10, Figure 19.11 and Figure 19.12. Ta-Nd-Mg in the ratio 20-40-30 shows a very smooth and homogeneous film. Ti-Nd-Mg shows a rougher surface, Ti-Ta-Nd shows peeling off at the edges of the films. A certain roughness at the edges cannot be avoided completely due to the deposition technique. Thus, property measurement will have to be carried out in the middle of the sample. In all films studied, the thickness varied between 0.5 to 1.5 μm. The porosity seems to be very low (also compare Figure 19.4).

The film quality strongly depends on the properties of the precursors such as viscosity or gelation behavior. If the precursor tends to gel rapidly, the viscosity increase will lead to a higher film thickness in the as-deposited film thus leading to a higher surface roughness and even cracks. It is, however, extremely difficult to control each individual precursor property, since all different mixtures will react differently. The rate of pyrolysis, phase formation, grain growth, and size will also influence the film thickness significantly. Empirical testing of basic recipes is therefore essential.

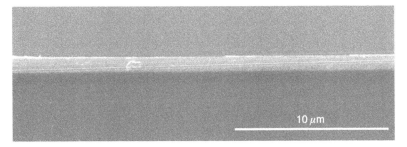

FIGURE 19.10 Cross-section SEM micrograph of a Ta-Nd-Mg (20-40-30) thin film (five coatings, sintered at 800°C/3 h).

FIGURE 19.11 Cross-section SEM micrograph of a Ti-Nd-Mg (20-40-30) thin film (five coatings, sintered at 800°C/3 h).

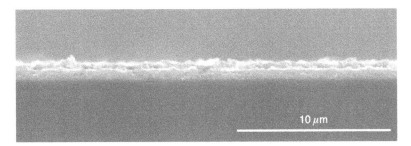

FIGURE 19.12 Cross-section SEM micrograph of a Ti-Ta-Nd (20-40-30) thin film at film edge (five coatings, sintered at 800°C/3 h).

19.2.6 PHASE COMPOSITION

High-resolution X-ray diffraction was used to characterize the phase compositions of the samples on the library structured by ceramic printing. To study the reproducibility of the sample synthesis and to compare the microstructures of the thin films with expected phases obtained from bulk preparation, samples from the ternary diagram Ba-Sr-Ti distributed on a library according to Figure 19.6 were investigated. Measurements were performed with the spatially resolved high-resolution X-ray diffraction system GADDS (General Area Detection Diffraction System; Bruker-AXS, D8, equipped with

TABLE 19.5
Dominating Phase Composition of the Different Samples of the L18 Library

Sample	Synthesis Composition	Major Phases Identified
1	Ba	$BaCO_3$
2	0.8 Ba, 0.2 Ti	$BaTiO_3$
3	0.6 Ba, 0.4 Ti	$BaTiO_3$
4	0.4 Ba, 0.6 Ti	$BaTiO_3$
5	0.2 Ba, 0.8 Ti	none
6	Ti	TiO_2 (Anatase + Rutile)
7	0.8 Ba, 0.2 Ti	$BaTiO_3$
8	0.8 Ba, 0.2 Sr	none
9	0.6 Ba, 0.2 Sr, 0.2 Ti	BaSrTi oxide
10	0.4 Ba, 0.2 Sr, 0.4 Ti	BaSrTi oxide
11	0.2 Ba 0.2 Sr, 0.8 Ti	BaSrTi oxide
12	0.2 Ba, 0.8 Ti	BaSrTi oxide
13	0.6 Ba, 0.4 Ti	$BaTiO_3$
14	0.6 Ba, 0.3 Sr, 0.2 Ti	BaSrTi oxide
15	0.6 Ba, 0.4 Sr	none
16	0.4 Ba, 0.4 Sr, 0.2 Ti	none
17	0.2 Ba, 0.4 Sr, 0.4 Ti	BaSrTi oxide
18	0.4 Sr, 0.6 Ti	$SrTiO_{2.6}$
19	0.4 Ba, 0.6 Ti	$BaTiO_3$
20	0.4 Ba, 0.2 Sr, 0.4 Ti	BaSrTi oxide
21	0.4 Ba, 0.4 Sr, 0,2 Ti	none
22	0.4 Ba, 0.6 Sr	none
23	0.2 Ba, 0.6 Sr, 0.2 Ti	none
24	0.6 Sr, 0.4 Ti	none
25	0.2 Ba, 0.8 Ti	TiO_2 (Anatase)
26	0.2 Ba, 0.4 Sr, 0.4 Ti	BaSrTi oxide
27	0.2 Ba, 0.4 Sr, 0.4 Ti	BaSrTi oxides, $SrTiO_3$
28	0.2 Ba, 0.6 Sr, 0.2 Ti	none
29	0.2 Ba, 0.8 Sr	$SrCO_3$ SrO
30	0.8 Sr, 0.2 Ti	none
31	Ti	TiO_2 (Anatase)
32	0.2 Sr, 0.8 Ti	SrTi oxide
33	0.4 Sr, 0.6 Ti	SrTi $O_{2.6}$
34	0.6 Sr, 0.4 Ti	SrTi oxide
35	0.8 Sr, 0.2 Ti	none
36	Sr	none

area detector). A program was written to monitor the movements of the library on the xy-table. The X-rays (Cu $K\alpha$) were focused by a 500-μm collimator. Selected diffractograms measured from 15–50 2θ are summarized in Figure 19.13. The samples are numbered vertically in this example. Comparing the phase composition identified from XRD and listed in Table 19.5, it is concluded that most film compositions are well reproduced (see sample numbers 2 and 7, 3 and 13, 9 and 14, 10 and 20, 12 and 32, 16 and 21, 17 and 27, 18 and 33, 30 and 35). Some exceptions were found in which the peak positions did not match the reproduction (sample numbers 4 and 19, 5 and 25, 6 and 31, 24 and 34, 11 and 26, 23 and 28). These differences could be traced to accuracy of the pipetting robot.

In this library, the precursor combinations were dispensed in a microplate by disposable tips in multipipetting mode. This might be a source of error. As the thermal treatment was identical for all samples on the plate, temperature differences during drying, calcination, or annealing can be

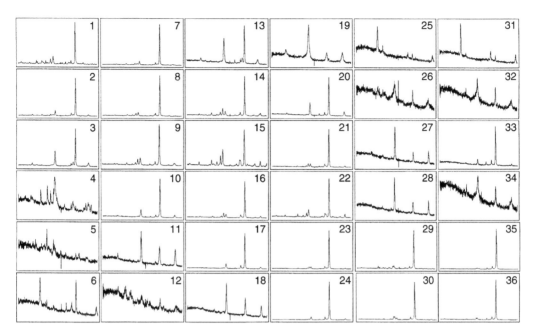

FIGURE 19.13 X-ray diffractograms of the samples on the library prepared according to Figure 19.5.

excluded. The changes in crystallization will be the result of compositional variations, but could also be affected by impurities in the wells.

The principal phases present in the samples are listed in Table 19.5. Some secondary phases were present in most of the samples, which were partially difficult to identify. Some samples were also strongly amorphous, as can be seen from the diffractograms in Figure 19.13. Considering the phase composition of the samples, the ternary diagram of oxides in the system Ba-Sr-Ti seems to be reproducible with HTE. For further applications in the domain of dielectric ceramics, only the well-crystallized mixed oxides present in the middle of the ternary diagram (samples 9, 10, 11, 16, 17, and 23) seem to be of interest as expected from studies about this system.[19,20]

For the measurement technique itself, this experiment proved the difficulty to handle a large quantity of data delivered by a measurement device. Analysis of the diffractograms is complicated and time consuming and still has to be performed manually. Here, automated analysis of diffractograms for main components, as offered now by various manufacturers, would greatly improve the HTE. Therefore adapted software should be able to compare X-ray diffraction data and references. Without an appropriate tool, the phase analysis measurements methods cannot be introduced in the combinatorial workflow for the development of ceramics, as it would be necessary for such a work.

Finally, a remark should be given about the device itself and the measurement of the libraries prepared in this work. As the samples were thin films, the material quantity considered during the measurements was low. As the material quantity is low, the answer signal obtained in diffraction is weak. Moreover, thin films often present preferred crystalline orientations, which could introduce errors in the identification of the phase. Longer measuring times, thus decreasing the throughput, might improve the accuracy of the XRD measurement.

19.3   DISCUSSION ON HIGH-THROUGHPUT CHARACTERIZATIONS

Chang et al.[20] present libraries prepared by sputtering, which were scanned for dielectric properties by evanescent microwave microscope. In this case the surface smoothness is guaranteed by the deposition technique. In contrast to common PVD and CVD preparation of thin films for the study of electronic properties, sol–gel techniques may have the advantage of facile up-scaling and facile reproduction of the bulk materials in laboratory quantities. Considering the electron microscope images presented in this work, roughness and thickness variation are serious problems during thin film preparation with sol–gel processing. This is mainly a result of the synthesis technique applied.

Although it has been shown that in most cases film composition matches that of the composition planned by the synthesis, some cases of significant deviations have been detected. Since little is known about the origin of these deviations, thin film composition should be controlled for every new sol–gel procedure used for thin film preparation. Danielson et al.[21] presented a synthesis of phosphors by sol–gel with a pipetting robot, in which the phase composition was controlled a posteriori with X-ray diffraction. In reference 22, the samples were measured with EDX.

Here, it should be mentioned that validation experiments with measurement of dielectric constants of identical barium magnesium tantalate films on such a sol–gel library with a microwave microscope provided correct dielectric constants with a standard deviation <10%. Details will be reported later.

19.4   CONCLUSION

A new technique for the preparation of dielectric thin film libraries by a parallel liquid phase synthesis technique has been presented. After the development of appropriate sol–gel recipes for 20 common elements in dielectric ceramics, films of binary and ternary combinations were deposited on prestructured substrates using a pipetting robot. A new structuring technique using photolithographic masking and sand blasting was established successfully to separate the thin films on a silicon wafer. A suitable thermal treatment was then developed to obtain about 1-μm-thick homogeneous films of high density. The quality of the samples on library was controlled using either high-throughput or conventional technologies. Careful analyses of the materials on the libraries revealed that the preparation of homogeneous films with a smooth surface for a wide variety of compositions is not a simple task. The individual precursors formed by mixing the element precursors in various compositions all show individual properties such as viscosity, gelation, and pyrolysis behavior, which influence the film quality significantly. XRF has shown unexpected separation phenomena for certain elements. The different dependence of the backscattered X-ray intensity on precise film thickness and element, especially with thin films, render the use of XRF obsolete for quantitative film analysis. Due to the many problems identified with, XRF, this method should not be used for quantification of diverse materials on combinatorial libraries. However, as a rapid high-throughput analysis method with high spatial resolution, XRF is still valuable to test for sample quality and qualitative confirmation of expected composition. The characterization results indicate that thin film preparation for high-throughput studies of electronic properties requires high standards and well-optimized synthesis procedures. On the other hand, the sol–gel approach presented here for library preparation allows highly automated library preparation and thus opens the door to combinatorial variation of composition and doping.

REFERENCES

1. Maier, W. F., (Guest Editor), Special issue: Combinatorial Catalysis, *Appl. Catal. A*, 254/1, 1–167, 2003.
2. Cawse, J. N., (Ed.), *Experimental Design for Combinatorial and High Throughput Materials Development*, Wiley-Interscience, Hoboken, 2003.
3. Potyrailo, R. A. and Amis, E. J., (Eds.), *High-Throughput Analysis*, Kluwer Academic, New York, 2003.
4. Xiang, X.-D. and Takeuchi, I., (Eds.), *Combinatorial Materials Synthesis*, Marcel Dekker Inc., New York, 2003.
5. Koinuma, H., Aiyer, H. N. and Matsumoto, Y., Combinatorial solid state materials science and technology, *Sci. Technol. Adv. Mater.* 1, 1, 2000.

6. Sun, X., Gao, C., Wang, J. and Xiang, X. D., Identification and optimization of advanced phosphors using combinatorial libraries, *Appl. Phys. Lett.*, 70, 3353, 1997.
7. Rantala, J. T., Kololuoma, T. and Kivimäki, L., *Combinatorial and Composition Spread Techniques in Materials and Device Development*, in Proceedings of SPIE, Jabbour, G. E., (Ed.), 2000, 3941.
8. Funakubo, H., He, G. and Iijima, T., *Combinatorial and Composition Spread Techniques in Materials and Device Development II*, in Proceedings of SPIE, Jabbour, G. E., and Koinuma, H., (Eds.), 2001, 4281.
9. Wessler, B., Jéhanno, V., Rossner, W. and Maier, W.F., Combinatorial synthesis of thin film libraries for microwave dielectrics, *Appl. Surf. Sci.*, 223, 30, 2004.
10. Wei, T., Xiang, X.-D., Wallace-Freedman, W. G. and Schultz, P. G., Scanning tip microwave near-field microscope, *Appl. Phys. Lett.*, 68, 3506, 1996.
11. Gao, C., Wei, T., Duewer, F., Lu, Y. and Xiang, X.-D., High spatial resolution quantitative impedance microscopy by scanning tip microwave near-field microscope, *Appl. Phys. Lett.*, 71, 1872, 1997.
12. Steinhauer, D.E., Vlahacos, C.P., Wellstood, F.C., Anlage, S.M., Canedy, C., Ramesh, R., Stanishevsky, A. and Melngailis, J., Quantitative imaging of dielectric permittivity and tunability with a near-field scanning microwave microscope, *Rev. Sci. Instrum.*, 71, 2751, 2000.
13. Shimooka, H. and Kuwabara, M., Crystallinity and stoichiometry of nano-structured sol-gel-derived $BaTiO_3$ monolithic gels, *J. Am. Ceram. Soc.*, 79, 2983, 1996.
14. Shimooka, H. and Kuwabara, M., Preparation of dense $BaTiO_3$ ceramics from sol-gel-derived monolithic gels, *J. Am. Ceram. Soc.*, 78, 2849, 1995.
15. Frantzen, A., Scheidtmann, J., Frenzer, G., Maier, W.F., Jockel, J., Brinz, T., Sanders, S. and Simon, U., High-throughput method for the impedance spectroscopic characterization of resistive gas sensors, *Angew. Chem. Int. Ed.*, 43, 752, 2004.
16. Schnöller, M. and Wersing, W., *Sol-Gel Proceedings of Advanced Dielectric Ceramics for Microwave Applications*, in Material Research Soc., Symposium Proceedings, 1989, 45.
17. Rossner, W. and Wessler, B., *Substrat mit einem planen Oberflächenabschnitt, Verfahren zum Herstellen eines Materialfilms auf dem Oberflächenabschnitt des Substrats und Verwendung des Substrats*, Patent No. DE 10245675.5.
18. Van der Haar, L. M., Sommer, C. and Stoop, M. G. M., New developments in X-ray fluorescence metrology, *Thin Solid Films*, 450, 90, 2004.
19. Baumert, B. A., Chang, L.-H., Matsuda, A. T., Tracy, C. J., Cave, N. G., Gregory, R. B. and Fejes, P. L., A study of barium strontium titanate films for use in bypass capacitors, *J. Mat. Res.*, 13(1), 197, 1998.
20. Chang, H., Takeuchi, I. and Xiang, X.-D., A low-loss composition region identified from a thin-film composition spread of $(Ba_{1-x-y}Sr_xCa_y)TiO_3$, *Appl. Phys. Lett.*, 74, 1165, 1999.
21. Danielson, E., Devenney, M., Giaquinta, D. M., Golden, J. H., Haushalter, R. C., McFarland, E. W., Poojary, D. M., Reaves, C. M., Weinberg, W. H. and Wu, J., A rare earth phosphor containing one-dimensional chains identified through combinatorial methods, *Science*, 279, 837, 1998.
22. Klein, J., *Konventionelle und kombinatorische Methoden bei der Suche nach hochtemperaturstabilen Mischoxiden*, PhD-thesis, Wiss. Monog., MPI für Kohlenforschung, Verlag Mainz, Band D 032, 1999.

Section 6

Optic Materials

20 Combinatorial Fabrication and Screening of Organic Light-Emitting Device Arrays

Joseph Shinar, Gang Li, Kwang Ohk Cheon, Zhaoqun Zhou, and Ruth Shinar

CONTENTS

20.1   INTRODUCTION

Combinatorial approaches for discovery and optimization of new materials have emerged as a powerful tool for a wide range of applications in fields such as catalysis, optoelectronics, and luminescence.[1] The multiplicity of parameters that affect device performance makes combinatorial approaches an important tool for device development as well. Indeed, in recent years, such approaches have proven useful in fabrication, optimization, and basic studies of organic light-emitting devices (OLEDs).[2–10] This chapter reviews these studies, with particular emphasis on the studies conducted by the authors, which include two-dimensional (2-D) combinatorial arrays of UV–violet OLEDs,[5,11] 1-D arrays of red-to-blue OLEDs,[7] 1-D combinatorial screening of white OLEDs,[8] and 1-D combinatorial arrays fabricated to study Förster energy transfer in doped OLEDs.[9]

The performance of small molecular and polymer red-to-blue and white OLEDs has improved dramatically over the past decade.[12–19] However, OLED performance depends on many variables such as layer thickness, composition, device configuration, electrode material, and interfaces. Therefore, determining the optimal parameters in complex; multilayer structures using the conventional approach of fabricating one device at a time, varying only one parameter, is tedious and introduces both systematic errors and random variations from batch to batch. Combinatorial fabrication of OLEDs using, for example, a sliding shutter and substrate rotation technique, which allows variations in parameters such as layer thickness or doping, removes these errors and random variations,[2,3,5] as it enables fabrication of 2-D arrays of OLEDs in which two parameters are varied systematically across the array. Apart from these two parameters, other fabrication conditions remain essentially identical for all the pixels, enabling a more reliable and far more efficient OLED optimization procedure.

20.2   COMBINATORIAL SCREENING OF LUMINESCENT MATERIALS

Combinatorial processes were used to accelerate the experimental search for luminescent materials such as refractive oxides, doped with rare earth or transition metal ions, needed for photonic technologies.[1,6,20,21] A combinatorial process is advantageous for fabrication and screening of such luminescent materials, as the properties of these materials are very sensitive to dopant and host characteristics and processing. Thus, utilizing combinatorial techniques, it becomes possible to identify materials with superior attributes through systematic screening of large material libraries. Phosphor libraries were generated using shadow masks,[1,6,22,23] lithographic masks,[1,6,20] movable-shutter mask strategies,[20,24] and chemical ink-jet liquid dispensing.[6,25] These approaches were used for identifying phosphor systems, optimizing material composition of the phosphors, and generating phosphor libraries. Recently libraries with 100 to > 1000 different chemical compounds were generated on a $1'' \times 1''$ substrate using such techniques.[6] Combinatorial synthesis has also proven valuable in discovery of luminescent transition-metal complexes.[26] Design of such novel complexes is a key synthetic goal for OLED application.[27]

20.3   COMBINATORIAL SCREENING OF ELECTRON AND HOLE
TRANSPORT LAYERS

The movable shutter approach for generating the phosphor libraries was also used to investigate and optimize the electron transport layer (ETL) in OLEDs, specifically the effect of tris(8-hydroxy quinolinato) Al (Alq$_3$) layer thickness on device performance.[6] Using a movable shutter, controlled by a robotic arm, various Alq$_3$ layer thicknesses were deposited in a single experiment. The OLED was based on a constant ~55 nm hole-transport layer (HTL) composed of 5,6,11,12-tetraphenyl-naphthacene (rubrene)-doped N,N'-diphenyl-N,N'bis(3-methylphenyl)-[1-1$'$-biphenyl]-4-4$'$-diamine (TPD) and polycarbonate, and an ETL of Alq$_3$. A peak external quantum efficiency $\eta_{ext} = 1.2\%$ was observed for an ~60 nm thick Alq_3 layer at an applied voltage of 10.8 V. The technique also enabled monitoring of the evolution of the OLED recombination zone as a function of the Alq_3 layer thickness. It was observed, from the normalized electroluminescence (EL) spectra at the peak η_{ext}, that the recombination zone extends significantly into the Alq_3 layer for thicknesses in the range 50–80 nm. For thicknesses of 30 and 40 nm, the emission zone shifts toward the rubrene-doped HTL region, though Alq_3 emission was still observed.[6]

Other reports have also described such combinatorial fabrication.[2,3,10] Rapid screening of the ETL in multilayer OLEDs was achieved by combinatorial fabrication of 49 devices using a movable sledge, carrying a shutter, and a substrate holder, positioned on top of the movable sledge, which could be rotated stepwise.[3] This experimental setup enabled fabrication of sectors with different material composition and device configuration on a single substrate by shutter movement and substrate rotation. Linear gradients of Alq_3 film thickness in ITO/TPD/Alq_3/Al OLEDs and orthogonal gradients of Alq_3 and an additional layer, hole-blocking spiro-quinoxaline ether (spiro-Qux), which was deposited on top of the Alq_3 layer, were evaluated, while keeping the TPD layer thickness and other fabrication parameters constant.[3] It was observed in this single experiment that a 67 nm thick Alq_3 layer resulted in the maximal photometric efficiency $\eta_{photo} \approx 2.8$ Cd/A. The addition of the spiro-Qux layer did not improve the power efficiency η_{power}, though certain combinations of Alq_3/spiro-Qux thicknesses improved η_{photo}.

The influence of the HTL thickness on the efficiency was also studied.[2] A 0–100 nm linear gradient of TPD was first deposited, followed by a 90° rotation of the substrate and subsequent deposition of Alq_3 as an orthogonal 0–150 nm linear gradient. It was observed that the dependence of η_{photo} on the Alq_3 thickness was much larger than that of the TPD, while η_{power} depended almost equally on the thickness of both layers.

The power of combinatorial device fabrication was demonstrated for the widely studied [copper phthalocyanine (CuPc)]/[N,N'-diphenyl-N,N'-bis(1-naphthylphenyl)-1,1$'$-biphenyl-4,4$'$-diamine ((α-)NPB or (α-) NPD)]/Alq_3 OLEDs by Riess et al.[28] Using a rotating substrate holder with a shutter in front of the substrate, the NPB and Alq_3 layer thicknesses were varied from 10 to 100 nm in 10 nm increments, while the CuPc layer thickness was kept constant at 15 nm. The highest η_{photo} was observed for a device with 60 nm Alq_3 and 40 nm NPB. A drop in the efficiency observed for Alq_3 thinner than 30 nm was explained by exciton-quenching effects due to coupling to the metal cathode. The effect of a thin HTL layer was less pronounced due to the presence of the CuPc layer separating the emission zone from the conducting anode.

The results also showed, as expected, that the voltage needed to obtain a current density $J = 20$ mA/cm^2 increases for increasing total device thickness.[28] At 20 mA/cm^2 the dependence of the voltage on both the ETL and HTL layer thicknesses was comparable, suggesting that the voltage drop across the Alq_3 and NPB layers is similar at that current density. At other current densities, however, a redistribution of the electric field between the two layers occurs.

By depositing a second layer following 180° shutter rotation after deposition of the first layer, and using a step gradient for deposition of both layers, a two-layer device array with constant total thickness was generated (e.g., [15 nm CuPc]/[x nm NPB]/[(110 − x) nm Alq_3]).[28] This differs from the case where the shutter was rotated at 90°, which enables all thickness combinations. With the total thickness constant, the devices differed only in the position of the NPB/Alq_3 interface. Analysis of this type of array provides insight into issues such as the relative thicknesses of the HTL and ETL that would result in balanced carrier injection, i.e., the minimal number of carriers that would reach the opposite electrode without recombination, which may be a significant loss mechanism.

20.4   COMBINATORIAL SCREENING OF DOPING IN OLEDS

The movable shutter approach was also used for fast screening of the effect of doping Alq_3 host with a red-light-emitting 4-(dicyanomethylene)-2-t-butyl-6(1,1,7,7-tetramethyljulolidyl-9-enyl)-4H-pyran (DCJTB) dopant in CuPc/NPD/[Alq_3:DCJTB]/Alq_3 OLEDs.[6] It was observed that the charge transport properties, light emission, and η_{ext} are affected by the dopant level. The drive voltage increased, the emission spectrum red-shifted, and η_{ext} decreased with increasing dopant concentration.[6]

A combinatorial approach was also used for rapid screening of the doping (oxidation) level of polymeric anodes for optimal hole injection in polyfluorene OLEDs.[4] A 2-D array in which the thickness of the poly(4,4′-dimethoxy-bithiophene) (PDBT) anode layer increased in one direction, and the electrochemical equilibrium potential E_{eq} (and hence the workfunction ϕ_w as well) increased in the perpendicular direction, was fabricated. Screening of this array demonstrated, as expected, that at a given bias, J and EL (and hence η_{power}) decreased with increasing PDBT thickness (due to the increasing resistance of the device) and increased with increasing E_{eq} (due to the increasing ϕ_w).

20.5   CASE STUDY 1. TWO-DIMENSIONAL COMBINATORIAL FABRICATION AND SCREENING OF UV–VIOLET OLEDS[5,11]

20.5.1   INTRODUCTION

UV/violet OLEDs are highly desirable as, for example, excitation sources for other red-to-blue fluorescent films and luminescent sensors. This section describes the combinatorial fabrication and screening of 2-D arrays of UV/violet ITO/CuPc/[4,4′-bis(9-carbazolyl)biphenyl (CBP)]/[2-(4-biphenylyl)-5-(4-*tert*-butylphenyl)-1,3,4-oxadiazole (Bu-PBD)]/CsF/Al OLEDs, as described by Zou et al.,[5] and ITO/CuPc/CBP/[2,9-dimethyl-4,7-diphenyl 1,10-phenanthrolin (BCP)]/CsF/ Al OLEDs, which are currently being studied by Zhou et al.[11] The OLEDs were fabricated combinatorially using a sliding shutter technique similar to that described by Schmitz et al,[2,3] to screen them for the optimal thicknesses of the CuPc hole injecting layer and the electron transport or hole blocking layer. The emission from these CBP-based OLEDs peaked at 380–390 nm and originated from the bulk of the CBP layer.

20.5.2   EXPERIMENTAL PROCEDURE

The structures of CuPc, CBP, Bu-PBD, BCP, and the OLEDs are shown in Figure 20.1. Details on the treatment of the 150–200 nm thick Applied Films Corp. ITO-coated glass prior to deposition of the organic layers are given elsewhere.[5,29,30] The organic layers were deposited at ~1 Å/s by thermal evaporation in a vacuum chamber (background pressure ~10^{-6} Torr) installed in a glove box. To vary the thickness of the CuPc and Bu-PBD layers (t_{CuPc} and t_{Bu-PBD}, respectively), a sliding shutter, placed ~2 mm in front of the substrate, was used. By opening the shutter stepwise, seven different CuPc thicknesses ranging from 0 to 36 nm were deposited; this was followed by depositing the CBP layer over the whole substrate. The sample was then rotated by 90°, the shutter was closed, and the stepwise deposition was repeated for the Bu-PBD layer. Next, an ~ 10 Å thick CsF layer,[31] and finally, the ~200 nm Al cathode were deposited; the latter through a mask containing a 21 × 21 matrix of ~1.5 mm diameter holes. Hence, 3 × 3 = 9 nominally identical OLED pixels were obtained for each pair of t_{CuPc}, t_{Bu-PBD} values (see Figure 20.1). Details on the procedure used to characterize the array are given in reference 5.

20.5.3   ITO/CuPc/CBP/Bu-PBD/[CsF or AlOₓ]/Al OLEDs

Figure 20.2 shows the photoluminescence (PL) spectrum of the CBP film and the EL spectrum of a typical device. As clearly seen, both spectra peak at ~ 385 nm, and their short-wavelength bands are nearly identical. However, the EL wing at $\lambda > 390$ nm is broader and stronger than that of the PL, with shoulders at ~450, ~480, and ~520 nm. These shoulders are probably due to perylene contamination.[32,33] Since Förster energy transfer from CBP to perylene should contribute identically to the EL and PL, the perylene-related EL bands are probably due to direct electron–hole recombination on the perylene. This is consistent with the highest occupied molecular orbital (HOMO) energy

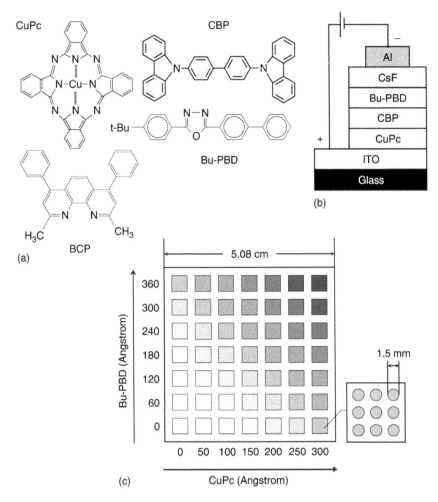

FIGURE 20.1 (a) Molecular structure of the materials used to fabricate the combinatorial UV–violet OLEDs described by Zou et al.[5] and by Zhou et al.[11] (b) The structure of the OLEDs. (c) The structure of the combinatorial matrix array of OLEDs.

level (E_{HOMO}) of CBP ($E_{HOMO} = 6.3$ eV)[34] and perylene ($E_{HOMO} = 5.3$ eV),[35] which suggests that perylene should be a strong hole-trapping center in CBP.

Measurements of the current density J and radiance R vs bias voltage showed clearly that adding CuPc enhances the current injection up to a thickness of 20 nm. This contrasts with the behavior reported by Aziz et al.,[36] but is in agreement with a study by Yu et al,[37] which yielded an optimal $t_{CuPc} \sim 50$ nm for their polyfluorene-based OLEDs. To account for this behavior, we note that the equivalent resistance of the ITO/CuPc layers must be lower for the optimized t_{CuPc} than for thinner t_{CuPc}. This lower resistance might be achieved if (1) the annealing that the CuPc layer undergoes during its deposition reduces its resistivity significantly, or (2) the equivalent resistance of the ITO/CuPc is affected by the surface roughness of the ITO. In particular, if the roughness is 15–50 nm, a t_{CuPc} in this range may be required to optimize hole injection. A study by Forsythe et al. does indeed suggest the latter scenario.[38]

While the optimal value for t_{CuPc} was ~ 20 nm, at any given voltage, J was highest for $t_{Bu\text{-}PBD} = 0$. Yet while Bu-PBD decreases J, it increases R, and hence η_{ext}. This is actually expected, since the addition of Bu-PBD moves the electron–hole recombination zone away from the cathode. This reduces the non-radiative quenching of the singlet excitons (SEs) by defects at the organic–cathode

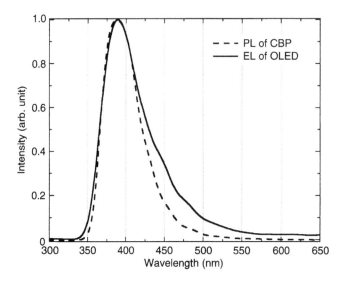

FIGURE 20.2 The PL spectrum of CBP and the EL spectrum of ITO/CuPc/Bu-PBD/CsF/Al OLEDs.

interface and by the metal mirror cathode itself.[39,40] The observed optimal $t_{Bu\text{-}PBD}$ was ~18 nm, in good agreement with the ~20 nm range of this non-radiative energy transfer from the SE to the metal cathode.[39,40]

Figure 20.3(a) shows R at $J = 10$ mA/cm^2 for the whole combinatorial array. This value of J was chosen because it is below the saturation level of the $R(V)$ curves of all the OLEDs. The peak $R = 0.38$ mW/cm^2 was obtained with $t_{CuPc} = 15$ nm and $t_{Bu\text{-}PBD}$ ~18 nm. These values of R are very high and are currently being reexamined.[11] However, regardless of the actual values of R, Figure 20.3 demonstrates the effectiveness of the combinatorial screening procedure to determine the optimal values of the thickness of the hole injecting layer and ETL in these UV/violet OLED arrays, and its ability to provide insight into their behavior.

Figure 20.3(b) shows the behavior of R at $J = 10$ mA/cm^2 for a similar combinatorial array of OLEDs fabricated with a nominal 12 Å Al$_2$O$_3$, rather than 10 Å CsF, buffer layer.[29] It was fabricated by thermally evaporating a nominal 12 Å-thick layer of Al followed by exposure to dry oxygen. We note, however, that ellipsometry measurements suggested that such a 12 Å-thick Al film results in a ~25 Å-thick oxide layer.[41] As clearly seen, the overall dependence of the devices' EL on t_{CuPc} and $t_{Bu\text{-}PBD}$ is similar to Fig. 20.3(a), but the peak R is much lower than that obtained with a CsF buffer layer. It is believed that charged defects, probably dianions,[42] whose concentration in and near the AlO$_x$ buffer layer is much higher than around the CsF layer, induce an additional non-radiative quenching mechanism for the SEs.[43]

In summary, through combinatorial fabrication, rapid evaluation and optimization of UV–violet ITO/CuPc/CBP/Bu-PBD/CsF/Al OLEDs matrix arrays was achieved. The EL spectrum, which peaked at ~385 nm, was apparently due to bulk emission from CBP. At a current density 10 mA/cm^2, a peak radiance of 0.38 mW/cm^2, corresponding to an external quantum efficiency of 1.25%, was obtained in devices with 15 nm CuPc and 18 nm Bu-PBD.

20.5.4   ITO/CuPc/CBP/BCP/CsF/Al OLEDs

Similar 2-D combinatorial arrays of CBP OLEDs, but with the Bu-PBD replaced by the hole-blocking material BCP, are currently being studied by Zhou et al.[11] Figure 20.4 compares the $J(V)$ curves of an optimal CBP/Bu-PBD OLED with a similar CBP/BCP OLED; the devices are identical with the notable exception that the 18 nm Bu-PBD layer is replaced by a 25 nm BCP layer. As clearly

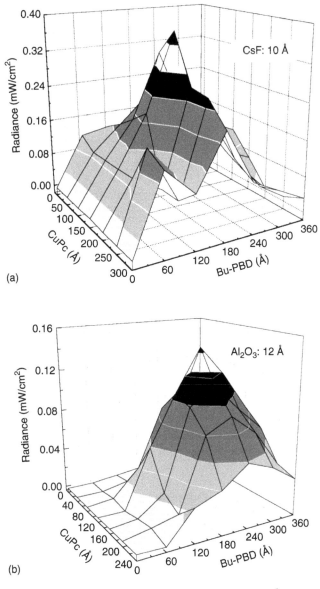

FIGURE 20.3 (a) The 2-D mapping of the radiance of the OLEDs with a 10 Å CsF/Al cathode as a function of the CuPc and Bu-PBD thicknesses. (b) The similar 2-D mapping of the radiance of the OLEDs with a 12 Å Al_2O_3/Al cathode.

seen, at any given bias higher than the turn-on voltage, the current through the CBP/BCP device is much higher than that through the CBP/Bu-PBD device. It is therefore suspected that the efficiency of CBP OLEDs with a BCP ETL would be significantly higher than that of CBP devices with a Bu-PBD ETL. Yet the maximal R of 2-D combinatorial arrays of CBP/BCP OLEDs at $J = 10$ mA/cm² is only 0.07 mW/cm², i.e., a factor of 5 less than the maximal R of the similar CBP/Bu-PBD array described above.[11]

Figure 20.5 shows that the behavior of the ITO/CuPc/CBP/BCP/CsF/Al OLEDs differs from that of the ITO/CuPc/CBP/BuPBD/CsF/Al OLEDs in another important aspect. While the EL spectrum of the latter devices red-shifts with increasing bias (see Figure 20.5 inset) that of the former

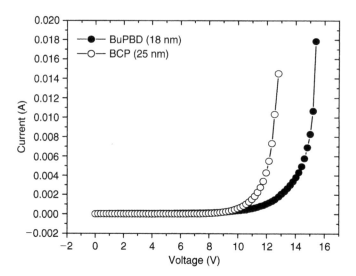

FIGURE 20.4 Current density–voltage $J(V)$ curves for ITO/[15 nm CuPc]/[50 nm CBP]/[18 nm Bu-PBD]/
[1 nm CsF]/Al (solid circles) and ITO/[15 nm CuPc]/[50 nm CBP]/[25 nm BCP]/[1 nm CsF]/Al (open circles)
OLEDs. Note the superior performance of the CBP/BCP devices over the CBP/Bu-PBD devices.

remains stable. The red shift of the latter devices may contribute to the behavior of their radiance,
which we now consider in detail.

The values of R of the recent CBP/BCP OLEDs were determined as follows: (i) The total luminance was measured using a Minolta L110 Luminance Meter; (ii) the luminance of a 10-nm-wide
band of visible emission (e.g., the 420–430 nm band) was determined by its contribution to the total
integral of the emission, weighted by the phototopic curve (i.e., the spectral response of the normal
human eye); (iii) The luminance of that band was converted to R; (iv) the total R was determined
from the contribution of that band to the total emission.

20.6 CASE STUDY 2. ONE-DIMENSIONAL COMBINATORIAL FABRICATION OF BLUE-TO-RED OLEDs[7]

20.6.1 INTRODUCTION

Color modification in thermal vacuum evaporated small molecular OLEDs has attracted strong
attention due to the applications of such OLEDs for full-color flat panel displays[44] and general lighting applications.[45] One well-known method to modify the color is the guest–host (G–H) or molecular doping approach, where both the doping concentration[44,46–48] and applied bias[48,49] affect the
emission spectrum. The guest emission spectra are shifted by the doping concentration,[46–48] and the
relative intensity of guest and host emission vary with applied bias.[48,49] Yet control of the doping
concentration is problematic due to instabilities in the fabrication procedure. In contrast, color
tuning via changes in the thickness of the doped layer may provide a facile fabrication process for
vacuum-evaporated devices. This section demonstrates this approach by describing 1-D combinatorial blue-to-red G–H OLEDs;[7] The emission shifted from blue to red as the thickness of the doped
layer increased from 0 to 35 Å.

In the typical G–H film, the guest molecule, with a relatively low gap between the HOMO and
lowest unoccupied molecular orbital (LUMO), is doped into the higher HOMO–LUMO gap host
molecule. If the guest absorption spectrum overlaps the host emission spectrum, then the excitons
of the host are transferred non-radiatively to the guest fluorophores,[50,51] which emits with its own
spectrum. However, trap emission is another possible dopant emission mechanism. Since either the

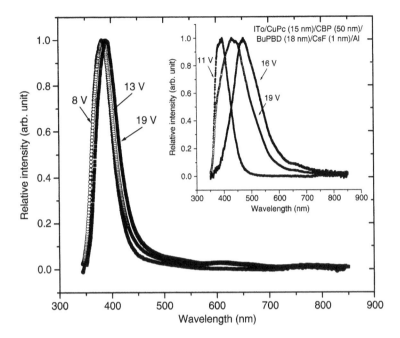

FIGURE 20.5 Evolution of the EL spectra of an ITO/[15 nm CuPc]/[50 nm CBP]/[25 nm BCP]/[1 nm CsF]/ Al OLED with applied bias. Inset: Evolution of the EL spectra of an ITO/[15 nm CuPc]/[50 nm CBP]/ [18 nm Bu-PBD]/[1 nm CsF]/Al OLED. Note the stability of the EL spectrum of the former vs the red-shift in the emission spectra of the latter.

HOMO or the LUMO level of the guest is in the host HOMO–LUMO gap, the guest molecule traps charge carriers,[52] becoming a trap emission center.[53]

The guest emission energy may change considerably with doping concentration c_d. In 2-methyl-6-[2-(2,3,6,7-tetrahydro-1H,5H-benzo[i,j]quinolizin-9-yl) ethenyl]-4H-pyran-4-ylidene] propane-dinitrile (DCM2)-doped NPD the emission peak red-shifts from 2.18 eV (570 nm) at 0.5 wt% to 1.96 eV (632 nm) at 5 wt%.[48] This solid-state solvation effect (SSSE)[44,46] red shift is due to dipole–dipole interaction between the excited guest molecules and surrounding dipoles: with increasing polarization of the host by increased doping with the highly polar DCM2 molecule, the DCM2 emission itself red-shifts.[46] In this section, we show that the doping thickness t_d, rather than c_d, may affect the host polarization, if t_d is less than the dipole–dipole interaction range.

20.6.2 EXPERIMENTAL PROCEDURE

The fabrication procedure of the combinatorial OLED array is described in Section 20.5.2. The OLEDs' structure was ITO/[5 nm CuPc]/[38 nm NPD]/[t_d nm 5 ± 0.6 wt% DCM2-doped NPD]/ [4,4′-bis(2,2′-diphenyl-vinyl)-1,1′-biphenyl (DPVBi)]/[10 nm Alq$_3$]/[1 nm CsF]/[150 nm Al], where t_d = 1, 2, 3, 4, 6, 10, 20, and 35 Å (see Figure 20.6). For color tuning, DCM2 and NPD were co-deposited after the neat NPD layer; the ratio of their depositions rates corresponded to c_d = 5 ± 0.6 wt% DCM2 in NPD. It should be noted that the ~2 mm gap between the substrate and shutter may have affected the real t_d due to a shadow effect. However, the variation in t_d was still systematic. To generate the host (i.e., donor) fluorescence, a 40 nm thick DPVBi blue-emitting layer was deposited, followed by a 10 nm thick Alq$_3$ ETL, a 1 nm thick CsF layer, and a 150 nm thick Al cathode. The Al was deposited through the 2″ × 2″ mask that contains 21 × 21 hole pixels used for the study described in Section 20.5. The maximum EL of some of the pixels in this array exceeded 50,000 cd/m^2,[48] as measured by the Minolta LS110 luminance meter and/or the Hamamatsu 3456 PMT.

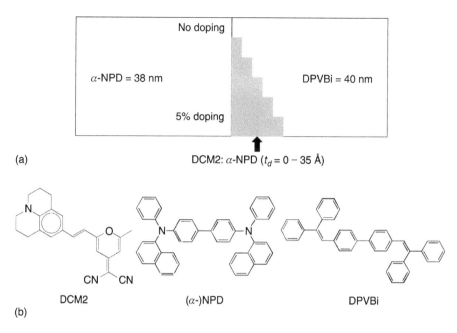

(a)

(b)

FIGURE 20.6 (a) Schematic device diagram of the DCM2 doped α-NPD layer. Nominal doping thickness t_d varies from 0 to 35 Å. (b) Molecular structures of DCM2, (α-)NPD, and DPVBi.

Due to the spring-loaded In-ball contact to the Al disk cathodes, the foregoing 2-D arrays could not be encapsulated. To enable encapsulation, 1-D arrays were fabricated by etching stripes of ITO, evaporating the organic layers, and then evaporating Al stripes perpendicular to the ITO stripes. The arrays were encapsulated by applying a narrow strip of epoxy around the perimeter of the ITO-coated glass substrate, and covering the devices with a glass slide.

20.6.3 PROPERTIES OF THE RED-TO-BLUE ARRAYS

Figure 20.7 shows the peak emission energy E_{max} of the DCM2 emission versus t_d at $V_{app} = 10$ V. E_{max} of a device with $t_d = 100$ Å was adopted from previous work.[48] As clearly seen, E_{max} red-shifts from 2.15 eV (575 nm) for $t_d = 1$ Å to 2.0 eV (620 nm) for $t_d = 35$ Å. The circles are the experimental values, and the linear line is the logarithmic fit to the data. The error bars represent the asymmetry and the ambiguity of the peaks of the spectra due to the vibronic modes.

The brightness and efficiency of the devices initially increased rapidly with t_d, peaking at $t_d = 2$ Å; $L_{max} = 120$ Cd/m² for $J = 1$ mA/cm², 1050 Cd/m² for $J = 10$ mA/cm²; $\eta_{ext,\,max} \approx 4.4\%$ and $\eta_{power,max} \approx 5$ lumens/W. Beyond the maximum, these values decreased gradually with increasing t_d, to $\eta_{ext} \approx 1.4\%$ and $\eta_{power} < 1$ lumens/W at $t_d = 35$ Å.

Figure 20.8 shows the evolution of the color coordinates of the devices with t_d at $V_{app} = 8$ and 10 V. The coordinates start from blue, cross the white region, and quickly approach orange and red. At $V_{app} = 10$ V the color of the undoped device is blue [$(x, y) = (0.15, 0.15)$], for the $t_d = 1$ Å device it is white [$(x, y) = (0.35, 0.33)$], and for the $t_d = 35$ Å it is red [$(x, y) = (0.61, 0.37)$].

Figure 20.9 shows an encapsulated array of pixels with different t_d on the $2'' \times 2''$ substrate, with $V_{app} = 8$ V. There was no intentional DCM2 doping in the first two columns, t_d was 1 Å in the third column, and beyond that column t_d increased nonlinearly to 35 Å in the last columns. Importantly, all the pixels are comparably bright at 8 V.

The t_d-dependence of both E_{max} (Figure 20.7) and the efficiencies are very similar to their doping concentration (c_d)-dependence described previously.[44,46–48] First, the logarithmic dependence

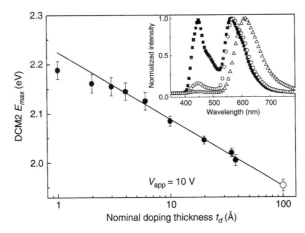

FIGURE 20.7 Spectral peak energy E_{max} of the DCM2 EL vs. $\log(t_d)$ for the fixed doping level of $5 \pm 0.6\%$. Inset: EL spectra for $t_d = 1$ Å (solid squares), 2 Å (open circles), and 35 Å (open triangles). The emission peaks at 460 and around 600 nm are due to DPVBi and DCM2, respectively.

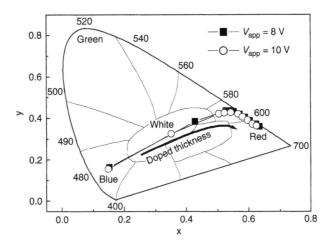

FIGURE 20.8 Doping-thickness-dependence of the color coordinates for $V_{app} = 8$ V, and 10 V.

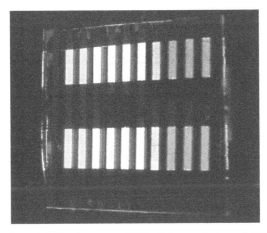

FIGURE 20.9 (See color insert following page 172) Color evolution of OLEDs with varying DCM2-doped-layer thickness at $V_{app} = 8$ V.

of E_{max} on t_d is similar to its dependence on c_d in DCM2-doped Alq_3,[46] where this behavior was explained by local ordering of polar molecules, and yielded a maximum domain radius $R_c \sim 110$ Å. The typical intermolecular spacing is known to be ~ 8 Å in amorphous molecular thin films,[46,47] so c_d is directly proportional to t_d for $t_d \leq 10$ Å. Thus the dependence of E_{max} on t_d and c_d should be the same for $t_d \leq 10$ Å. For $t_d > 10$ Å, E_{max} should become independent of t_d if a fluorophore interacts only with its nearest neighbors, because the number of nearest-neighbor DCM2 molecules is determined by c_d. Yet the behavior of E_{max} in Figure 20.7 shows that its logarithmic dependence on t_d is still effective up to at least 100 Å. This implies that a DCM2 fluorophore can interact via the dipole–dipole interaction with remote DCM2 molecules which are 100 Å away. Thus the dipole–dipole interaction range is ≥ 100 Å, similar to the domain size $R_c \sim 110$ Å resulting from the local ordering model of DCM2 in Alq_3.[46] Within the dipole–dipole interaction range or in an ordered domain, all of the DCM2 molecules interact collectively. For $t_d \leq R_c$, the total number of DCM2 molecules in a domain can be controlled either by c_d or by t_d. Hence, for $t_d \leq R_c$, the logarithmic dependence of E_{max} on t_d should be, and is, the same as its dependence on c_d.

Second, for 4-(dicyanomethylene)-2-methyl-6-(p-dimethyl aminostyryl)-4H-pyran (DCM1) in Alq_3, the PL[47,54] and EL[53,54] efficiencies were found to be maximal at low doping levels $0.2 \leq c_d \leq 3$ mol%. The t_d-dependence of the efficiencies of DCM2 in this work is similar: it rapidly rises to the maximum, and then monotonically decreases with t_d up to 100 Å; in the 100 Å device, $\eta_{ext} \approx 0.5$ % at $J = 1$ mA/cm[2].[48] Thus the efficiency is affected by doping up to at least $t_d \sim 100$ Å, and the explanation for the t_d-dependence of the energy shift can apply to the efficiency, and both are similar to the c_d-dependence for $t_d \leq R_c$.

In conclusion, the emission color of [(α-)NPD]/[5 wt% DCM2 in α-NPD]/[DPVBi] OLEDs is found to vary from blue to red as t_d increases from 0 to 35 Å, due largely to the effect of the thickness on the guest emission spectrum. This behavior is due to the dipole–dipole interaction among the DCM2 molecules, whose range exceeds 100 Å. Hence the t_d-dependence of the emission color and efficiency is similar to the doping concentration c_d dependence. At $J = 1$ mA/cm[2], the $t_d = 2$ Å nominal doping thickness device exhibited the highest performances in brightness ($L \sim 120$ cd/m[2]) and in efficiency ($\eta_{ext} \sim 4.4$ %). At 8 V, however, all the blue-to-red pixels are comparably bright, suggesting a particularly facile application of this fabrication procedure for multicolor display applications.

The facile fabrication of bright and efficient combinatorial blue-to-red OLED arrays could also be promising for the emerging new platform of luminescent chemical and biological sensors, in which the sensing elements are excited by OLED pixels.[55–59] This platform provides a uniquely simple scheme for structural integration of the excitation source and the sensing elements, and should lead to compact, low-cost, multianalyte sensor (micro)arrays. The different, individually addressable OLEDs pixels, emitting at different wavelengths, would then selectively excite the corresponding sensing elements.

20.7 CASE STUDY 3. ONE-DIMENSIONAL COMBINATORIAL SCREENING OF INTENSE WHITE OLEDs (WOLEDs)[8]

20.7.1 INTRODUCTION

WOLEDs are of growing interest as a low-cost alternative for backlights in liquid-crystal displays, and more recently as general white solid-state light sources.[45,48,60,61] For displays, $L \leq 300$ Cd/m[2] is sufficient, but lighting requires $L \sim 2000$ Cd/m[2]. In addition, stability exceeding 10,000 h and η_{power} superior to those of fluorescent tubes ($\eta_{power} \sim 50$ lm/W), both at $L \sim 2000$ Cd/m[2], must be achieved.

By lightly doping a host material with a dye, incomplete energy transfer from the host to the guest results in emission from both.[60,61] In addition, since the emission spectra of organic molecules are usually broad, if the host is a blue emitter and the guest is a red or orange emitter, this single doped layer can yield white emission. This strategy has yielded bright ($L = 42\,000$ Cd/m^2) and efficient ($\eta_{power} \sim 2.9$ lm/W) WOLEDs.[61] Since the white emission is from one layer, the device shows relatively good color stability.[60,61] This section describes very bright and efficient WOLEDs based on orange-emitting layers of 2–10 nm thickness and 0.25 and 0.5 wt% rubrene-doped DPVBi, with surprisingly stable color coordinates.

To fabricate efficient WOLEDs, both the host and dopant need to be efficient emitters. DPVBi is one of the most efficient blue emitters, yielding blue OLEDs with $\eta_{ext} = 3.5\%$.[48] The PL quantum yield of red-emitting rubrene is $\eta_{PL} = 100\%$ when doped properly into hosts.[62] The Förster energy transfer from an excited DPVBi molecule to rubrene is proportional to the overlap between the host emission spectrum and dopant absorption spectrum.[50,63] The DPVBi emission peaks at $\sim$460 nm; the rubrene absorption spectrum peaks at 460, 490, and 529 nm.[62] Thus, efficient energy transfer should occur.

20.7.2 EXPERIMENTAL PROCEDURE

Figure 20.10 shows the HOMO and LUMO levels of CuPc, α-NPD, DPVBi, and Alq$_3$, and the devices' structure. For figure clarity, we only show results for 2-, 4-, and 10-nm-thick doped layers, doped with 0.25 and 0.5 wt% rubrene (devices 1–3 and 4–6, respectively). Device 7 is the control device, i.e., without a doped layer. All of the devices with the same doping concentration were deposited as a single 1-D combinatorial array. The organic layers were deposited at $\sim$1 Å/s; the rubrene-doped DPVBi layer was deposited at $\sim$3 Å/s. The CsF layer was deposited at $\sim$0.1 Å/s, and the Al was deposited at $\sim$5 Å/s through the same mask used to fabricate the UV/violet arrays described above.[5] The $J(V)$ and $L(V)$ curves, and the EL spectra, were measured as described above. η_{power} was calculated assuming a Lambertian intensity distribution.

20.7.3 PROPERTIES OF THE WOLEDs

The EL spectra of devices 1–7 are shown in Figure 20.11. As expected, the emission is due to both rubrene and DPVBi. As the thickness of the doped layer or the doping concentration decreases, the DPVBi emission increases relative to that of rubrene. However, there is a difference between devices with thin (2 or 4 nm) vs thick (10 nm) doped layers. Since the emission zone is only a few nm thick,[60] in the devices with thin doped layers the emission zone includes both the doped layer and part of the adjacent pure DPVBi layer; in the devices with a 10-nm-thick doped layer, the emission is due only to the doped layer.

Color stability is an important issue in WOLEDs. Figure 20.12(a) shows the evolution of the EL with bias for device 1 (0.25 wt%, 2 nm) from 5 to 12 V, in which L changed from 12 to nearly 45 000 Cd/m^2 and J from 0.1 to $> 10^3$ mA/cm^2. Note that the blue emission band increases by $\sim$40% relative to the orange-red emission band over this range. However, from 1 to 1027 mA/cm^2 (6 to 12 V), it increased by only 20%, and from 22.6 to 1027 mA/cm^2 (8–12 V) there was no observable change. Hence, while L increased $\sim$1000-fold, the spectra barely changed.

Figure 20.12(b) shows the evolution of the color coordinates of devices 1–6 from 6 to 12 V, in 2 V steps. Since there is no color variation due to dopant concentration, in contrast to the case of DCM2-doped (α-)NPD,[60] all the coordinates are along the line connecting the coordinates of the DPVBi emission (0.15, 0.16) with those of rubrene emission (0.46, 0.53).

Although the emission from the 10-nm-thick doped layer devices is due mostly to the doped layer, the devices still exhibit a small color variation. This variation is attributed to saturation of the dopants in these lightly doped devices (see Section 20.8).

The role of the emission zone in determining color stability can be assessed by comparing the change in the color coordinates of devices 4 and 6 from 6 to 12 V: The change in the coordinates of Device 6 [from (0.422, 0.466) to (0.398, 0.444)] is much smaller than that of device 4 [from (0.345, 0.406) to (0.298, 0.347)]. Thus, it appears that an emission zone contained in the doped

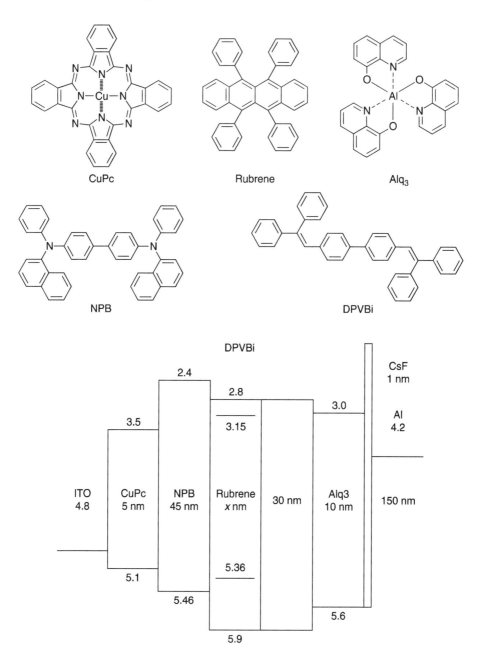

FIGURE 20.10 Molecular structures, HOMO and LUMO energies of the molecular films, and the devices' structure. $x = 0, 2, 4, 6$ or 10 nm for 0.25 wt% or 2, 4, 6, 8 or 10 nm for 0.5 wt% rubrene in DPVBi.

layer results in improved color stability. This effect is smaller in devices 1–3, which are more lightly doped.

Figure 20.12(b) also shows that all devices are relatively color stable. From 5 to 12 V, corresponding to $L \sim 10$ Cd/m^2 to $L > 40\ 000$ Cd/m^2 for devices 1–3 and $L < 30$ Cd/m^2 to $L \sim 25\ 000$ Cd/m^2 for devices 4–6, the coordinate changes are all well within the white region, varying by only $\sim 10\%$ over a ~ 30 db change in L. It can be seen that although the emission zone probably extends from the doped layer into the pure DPVBi layer in devices 1, 2, 4, and 5, those devices

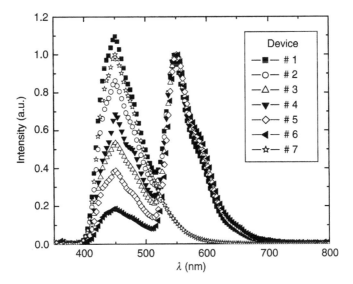

FIGURE 20.11 EL spectra of the WOLEDs at 10 V, normalized to the rubrene emission band. As the doping concentration or doping layer thickness increases, the DPVBi emission decreases with respect to rubrene emission.

still demonstrate relatively good color stability. By considering the Förster radius (~3 nm; see Section 20.8) and the width of the emission zone (~5 nm),[60] we conclude that this is probably due to comparably efficient energy transfer from DPVBi to rubrene in the doped layer and in the region of the pure DPVBi layer that is adjacent to the doped layer.

The $J(V)$ and $L(V)$ characteristics of the devices are shown in Figure 20.13. The $J(V)$ curves shift systematically to higher voltage from device 1 to device 6, due to the thicker doped layer and increasing carrier trapping in the more heavily doped devices.

Figure 20.14(a) and Figure 20.14(b) show η_{ext} and η_{power} vs L. The maximal η_{ext} reported here is less than that of the phosphorescent WOLED, but at 100 mA/cm² these WOLEDs yield $2.6 \leq \eta_{ext} \leq 3.1\%$, which is comparable to the phosphorescent WOLEDs. Moreover, the roll-off in the efficiency of these fluorescent WOLEDs at high L is slower than that of the phosphorescent OLEDs.[64] This is probably due to triplet–triplet annihilation in the phosphorescent WOLEDs. Note, however, that the device with the highest η_{ext} (device 1 at 4.6%) is not the device with the highest brightness (device 4 at 74 200 Cd/m²). This is due to the response of the human eye, since in device 4 the orange component (to which the eye is more sensitive) is larger than that in device 1.

Doping DPVBi with rubrene is also advantageous for the stability of the OLEDs. The glass transition temperature of DPVBi is a relatively low $T_g = 64°C$, which renders it susceptible to crystallization and consequent failure of the device. Doping with rubrene prevents crystallization and thus improves the device stability and lifetime significantly.[65]

In summary, highly bright and efficient fluorescent WOLEDs based on rubrene doped DPVBi were efficiently screened through combinatorial fabrication of the devices. Their color coordinates were tuned by varying the thickness and doping concentration of the doped layer. The highest brightness achieved was $L > 74\,000$ Cd/m², and the maximal η_{ext} was 4.6%. All of the devices exhibited a maximal brightness $L > 50,000$ Cd/m², and $\eta_{ext} > 3.1\%$ at $L = 1000$ Cd/m². The blue DPVBi emission increased relative to the orange rubrene emission as the bias increased, but the devices exhibited relatively good color stability with color coordinate variations of $< 10\%$ over a brightness range of 30 db.

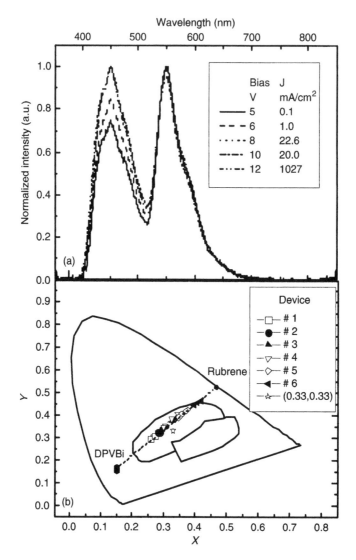

FIGURE 20.12 (a) Variation of the EL Spectra with driving voltage for device 1 (0.25 wt%, 2 nm). At 5 V, the current density was 0.1 mA/cm² and the brightness was 12 Cd/m². At 12 V, they were 1027 mA/cm² and 44,900 Cd/m², respectively. (b) The color coordinates of the WOLEDs from 6 V to 12 V, in 2 V steps. The line connects the coordinates of DPVBi and rubrene emissions. Note that except for device 6, all coordinates are inside the white region. All devices exhibited relatively good color stability with weak dependence on the bias.

20.8   CASE STUDY 4. ONE-DIMENSIONAL COMBINATORIAL STUDY OF FÖRSTER ENERGY TRANSFER IN GUEST–HOST OLEDs[9]

20.8.1   INTRODUCTION

This section analyzes energy transfer in highly efficient doped DPVBi OLEDs. A region of the HTL adjacent to the host (H) DPVBi layer is doped with an efficient guest (G) red dye. The H-to-G energy transfer probability is determined by comparing the emission from the two fluorophores and its dependence on the applied field. It decreases with increasing field, probably due to an increasing fraction of positively charged dye molecules (i.e., which trap a hole). It is also estimated that at

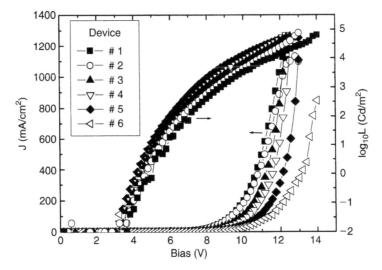

FIGURE 20.13 $J(V)$ and $L(V)$ curves of the white OLEDs. Note that the maximal $L > 74,000$ Cd/m^2 in device 4. Note that the turn-on voltage V_{on}, defined as the voltage at $L = 1$ Cd/m^2, increases from 4.0 V in devices 1 and 2 to 5.0 V in device 6. The maximal L of devices 1 to 6 are 50,100 (at 12.8 V), 54,500 (at 12.8 V), 64,400 (at 13.2 V), 57,200 (at 12.6 V), 74,200 (at 13.0 V), and 63,800 (at 14.0 V), respectively.

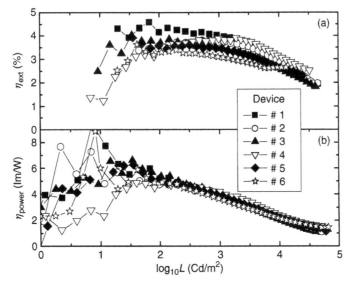

FIGURE 20.14 (a) η_{ext} and (b) η_{power} vs. L of the WOLEDs. Device 1 has maximal $\eta_{power} = 6.0$ lm/W, $\eta_{ext} = 4.6\%$, and $\eta_{photo} = 11.0$ Cd/A at $J = 0.62$ mA/cm^2 ($L = 68$ Cd/m^2). At 1000 Cd/m^2, devices 1 to 6 yield $3.6 \leq \eta_{power} \leq 4.1$ lm/W, and $3.1 \leq \eta_{ext} \leq 4.0\%$.

fields as low as 0.4 MV/cm, ~50% of the dye emission is due to trap emission rather than Förster energy transfer. The analysis yields a Förster energy transfer radius $R_0 = 33.5 \pm 3.5$ Å.

As mentioned above, molecularly doped G–H blends have been exploited to improve the efficiency[66,67] or modify the emission color[44,66,68] of OLEDs. The doped guest emission is usually due to energy transfer from the host molecules,[44,67,68] and the energy transfer rate depends on the overlap between the host emission and guest absorption spectra.[50,63] In most of the efficient G–H OLEDs the

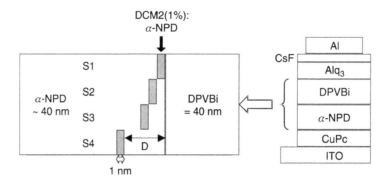

FIGURE 20.15 Structure of the devices with combinatorial variations in DCM2-doped α-NPD/DPVBi interfaces. The nominal thickness D of the undoped α-NPD "gap" layer is 0, 1, 2, and 5 nm for devices S1, S2, S3, and S4, respectively.

HOMO and LUMO levels of the guest are inside the host HOMO–LUMO gap. This situation satisfies the spectral overlap condition if the Stokes shifts between the absorption and emission spectra of the host and guest are not too large. However, since the guest HOMO and LUMO levels are inside the host HOMO–LUMO gap, they trap carrier, reducing the current flowing through the devices.[48,52]

In most of the efficient G–H OLEDs the overall emission color varies with bias because the ratio of guest-to-host emission generally decreases with increasing bias.[48,49] This decrease is generally believed to be related to electric field-induced SE quenching, although the details of this process are not clear. In general, due to field-induced quenching, efficient OLED operation is limited to average electric fields below ~2 MV/cm. Indeed, PL measurements on the blend of poly(phenyl-p-phenylene vinylene) (PPPV) and bisphenol-A-polycarbonate (PC) demonstrated that the PL intensity was quenched by ~30% at 2 MV/cm.[69] However, in many of the efficient devices the strong evolution of the color with bias cannot be explained by electric field quenching.[48] Indeed, one would expect the higher-energy host SEs, normally with a lower binding energy, to be more susceptible to quenching than the lower-energy guest SEs. This is in contrast to the observed behavior of these efficient devices. The following results and their analysis show that the observed behavior can be explained by increased trapping of holes in the guest molecules with increased field, as the host energy cannot be transferred to a guest molecule containing a hole.

To study the color evolution and Förster energy transfer mechanism in such OLEDs, the selective doping method[70] was used to dope the α-NPD HTL in efficient blue DPVBi OLEDs.[71] The combinatorial fabrication of the devices was critical, due to the small variations in the structure of the OLED pixels of the array.

20.8.2 Experimental Procedure

The 21 × 21 combinatorial 1.5-mm-diameter OLED pixel arrays were fabricated as described in Sections 20.5.2, 20.6.2, and 20.7.2, and elsewhere.[48] The OLED structure was TO/(5 nm CuPc)/ (40 nm α-NPD)/(1 nm DCM2:α-NPD)/(D nm α-NPD)/(40 nm DPBVi)/(10 nm Alq$_3$)/(1 nm CsF)/ (150 nm Al), with $D = 0$, 1, 2, and 5 nm (devices S1, S2, S3, and S4, respectively; see Figure 20.15). The nominal doping level of DCM2 in α-NPD was 1 wt.%, which corresponds to 1.6 mol%. The brightness and EL spectra were measured as described in Sections 20.6.2 and 20.7.2.

20.8.3 Förster Energy Transfer Model

In these OLEDs, electron–hole recombination occurs in the DPVBi layer. Hence emission from the guest is due to Förster resonant energy transfer (FRET) from DPVBi to DCM2,[48] and information on this energy transfer can be deduced from systematic variations in the distance D.

In the Förster process, a SE on the donor induces a dipole in the acceptor molecule, and the inducing donor field can interact with the induced acceptor dipole.[50,63] The interaction is proportional to D^{-6} and to the overlap integral R_0^6 of the host emission and guest absorption spectra. The predicted transfer probability from host to guest P_{HG}^Q can then be expressed as[44]

$$P_{HG}^Q = \frac{\alpha \cdot Q}{1 + \alpha \cdot Q} \qquad (20.1)$$

where Q is the doping concentration and $\alpha \equiv (R_0/D)^6$. On the other hand, the actual transfer probability P_{HG}^η can be estimated from the number of photons emitted by the host N_H and by the guest N_G as follows.[44] N_G is proportional to P_{HG}^η, the guest PL quantum yield η_G^{PL}, and the total number of SEs N_{tot}^{ex} generated in the host molecules, i.e.,

$$N_G \propto \eta_G^{PL} \cdot P_{HG}^\eta \cdot N_{tot}^{ex} \qquad (20.2)$$

Similarly, for the host emission,

$$N_H \propto \eta_H^{PL} \cdot (1 - P_{HG}^\eta) \cdot N_{tot}^{ex} \qquad (20.3)$$

If the outcoupling efficiency ξ is the same for guest and host emission, then the proportionality constants in Equations 20.2 and 20.3 are the same and the equations yield

$$P_{HG}^\eta = \frac{\eta_H^{PL}}{(N_H/N_G) \cdot \eta_G^{PL} + \eta_H^{PL}} \qquad (20.4)$$

The measurements described below yield P_{HG}^η, and comparison with Equation 20.1 then yields the Förster energy transfer radius R_0.

20.8.4 Properties of the Combinatorial Array of DCM2-Doped α-NPD/DPVBi OLEDs

Figure 20.16 shows the EL spectra of device S2 at $V = 7$, 10, and 14 V. The 460-nm band is due to DPVBi;[48] the 570-nm band, due to the DCM2 dopant, depends on the doping concentration.[60] At low voltage the DPVBi emission is barely noticeable, but its intensity increases rapidly with increasing V and it approaches the DCM2 intensity at 14 V; the other devices behave similarly.

The host/guest emission intensity ratio N_H/N_G of device S2 is analyzed in the inset of Figure 20.16. To calculate N_H/N_G, the spectra were converted to energy scale (eV) and deconvoluted to two Gaussians; at low bias ($V < 5$ V), N_H/N_G was estimated by linear and exponential extrapolation based on higher voltage data. In determining the average field E in the device, the built-in potential $V_{bi} = 2.1$ V[72] was subtracted from V/d, where d is the total thickness of the organic layers.

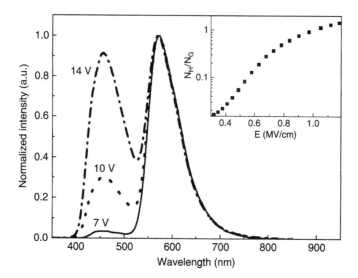

FIGURE 20.16 Spectral response of device S2 for different applied voltages. Inset: The ratio of photons emitted by the host to photons emitted by the guest.

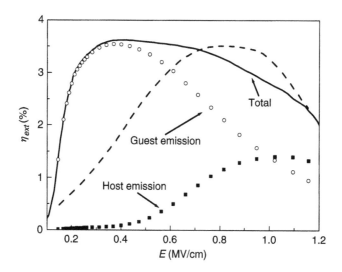

FIGURE 20.17 η_{ext} of device S2, and separate efficiencies of host and guest emission. The dashed line is the efficiency of the pure undoped α-NPD/DPVBi device.

As clearly seen, N_H/N_G increases dramatically with increasing E, from ~ 0.015 at $0.3\,\text{MV/cm}$ to ~ 1.5 at $1.2\,\text{MV/cm}$.

Figure 20.17 shows η_{ext} of device S2 (solid line), separate host and guest efficiencies ($\eta_{ext,H}$ and $\eta_{ext,G}$, respectively), and η_{ext} of the undoped device (dashed line) vs E. The maximum η_{ext} are $\sim 4\%$ for device S1 (not shown) and $\sim 3.5\%$ for S2, which are similar to the maximal $\eta_{ext,} \sim 3.5\%$ of the undoped device. At low field, DCM2 has very high $\eta_{ext,G}$ compared to $\eta_{ext,H}$. With increasing field $\eta_{ext,H}$ increases but $\eta_{ext,G}$ decreases gradually when $E > 0.3\,\text{MV/cm}$.

In small molecular OLEDs, it is generally believed that 25% of the excitons are SEs, and a fraction given by the PL quantum yield (η^{PL} or η_{PL}) of these decay radiatively. In devices in which the electron and hole injection is balanced, η^{PL} can be estimated from η_{ext} and the outcoupling efficiency $\xi \approx 1/n^2$ (where n is the index of refraction of the organic layers).[73] In the undoped α-NPD/DPVBi device, $\eta_{ext} \approx 3.5\%$ yields $\eta^{PL}_{DVPBi} \sim 45\%$. For DCM2, assuming 100% energy transfer from DPVBi in device S1 at the maximal η_{ext}, we obtain $\eta^{PL}_{DCM2} \sim 52\%$. This latter value is in striking agreement with $\eta_{PL} = 53\%$ for DCM2 in CBP.[67]

Given $\eta^{PL}_{DCM2} \sim 52\%$, $\eta^{PL}_{DVPBi} \sim 45\%$, and N_H/N_G values in the inset of Figure 20.16, and assuming that the DCM2 is excited only by Förster energy transfer from the host, Equation 20.4 now yields P_{HG} vs. E, which is plotted in Figure 20.18(a). As clearly seen, P_{HG} is not constant. In devices S1 and S2 it is close to 1 below 0.6 MV/cm, but decreases at higher fields. In devices S3 and S4, it decreases at low fields and levels off at high fields.

DCM2 is known to be a hole trap in α-NPD.[48] With increasing hole injection, the DCM2 trap-sites inside the α-NPD layer become increasingly filled by trapped holes. In general, the detrapping process for these trapped holes takes a longer time than SE decay on the host. Hence, the trapped holes affect the energy transfer to the guest molecules, since Förster energy transfer to a positively ionized guest molecule (i.e., a trap-filled guest) is not possible. Thus we attribute the decrease in P_{HG} vs E to the increased fraction of DCM2 molecules occupied by holes. In Equation 20.1, the guest concentration Q should be replaced by the reduced concentration of neutral guest fluorophores, but the evaluation of the modified Q is outside the scope of this treatment due to the complexity of the trapping dynamics.

At low field, the concentration of available fluorophores is expected to be relatively close to the original doping concentration Q. Hence, in order to minimize the trapped charge effect, low fields are selected to estimate R_0. Figure 20.18(b) shows the experimentally determined behavior of P_{HG} vs D. The solid squares and open circles are the values at 0.4 and 0.6 MV/cm, respectively; the error bars are due to the uncertainty in D. Fitting Equation 20.1 to the results then yields R_0. Note that even though there is no gap between the doped layer and host emitting DPVBi layer in device S1, the minimum G–H distance in that sample should be more than the molecular distance between G and H at contact ($\sim$ 5 Å).[74] For the other devices, instead of the distance to the middle of the doped layer, the nominal D are used due to the D^{-6} dependence of the energy transfer. The lines in Figure 20.18(b) are calculated from Equation 20.1 using $R_0 = 37$ Å for 0.4 MV/cm (solid line), and $R_0 = 30$ Å for 0.6 MV/cm (dashed line). Hence we conclude that $R_0 = 33.5 \pm 3.5$ Å. This value is in good agreement with previously reported values of 30–40 Å.[60]

η_{ext} of the undoped device peaks at 0.8 MV/cm (Figure 20.17), but that of device S2 peaks at $\sim$0.4 MV/cm. This behavior of η_{ext} is believed to be partially due to direct trap guest emission, i.e., emission from SEs formed by recombination of a hole trapped on a guest molecule and an electron, or vice versa. Indeed, since DCM2 molecules in α-NPD are trap sites for both holes and electrons, trap emission can occur. To estimate the level of trap emission, we note that the substantial efficiency of the pure α-NPD/DPVBi device is about half of the DCM2 efficiency of device S2 at 0.4 MV/cm (see Figure 20.15). Thus it appears that at 0.4 MV/cm, 50% of the guest emission does not occur by energy transfer from the host. Indeed, the open triangles in Figure 20.18(b) were obtained by assuming that 50% of DCM2 emission is due to carriers trapped by DCM2; their values yield $R_0 = 33$ Å, which is very close to the value obtained above. At higher fields, P_{HG} deviates considerably from the low-field value due to other effects, so the analysis of its behavior is more complex.

In summary, the Förster energy transfer probability P_{HG}, and radius R_0, were evaluated through combinatorial variations in the DCM2-doped α-NPD/DPVBi interfaces. P_{HG} was found to decrease with increasing field, apparently due to an increasing fraction of positively ionized guest molecules. At moderate fields of $\sim$0.4 MV/cm it was also concluded that approximately 50% of the guest emission was due to trap emission. R_0 was concluded to be 33.5 ± 3.5 Å at low fields.

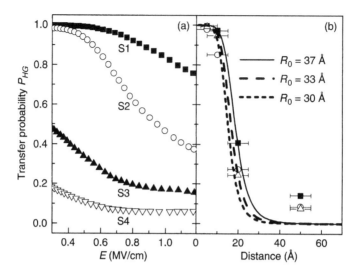

FIGURE 20.18 Energy transfer probability P_{HG}: (a) electric field dependence, and (b) distance dependence.

20.9   SUMMARY AND CONCLUDING REMARKS

Combinatorial fabrication results in multilayer OLED libraries in a single experiment. Linear and step gradients can be prepared by shutter movement, and different sectors can be generated using masks together with substrate rotation. The resulting OLEDs are of different material combinations, layer thicknesses, dopant level, and device configurations. They can therefore be screened and optimized in a fast and reliable manner, eliminating errors stemming from data obtained from multiple experiments changing one parameter at a time. This chapter reviewed the various reports on the fabrication and screening of OLED libraries, with emphasis on the studies conducted by the authors, showing that combinatorial screening of OLEDs has become a powerful tool for screening various OLEDs, and for studying their basic optoelectronic properties. The studies reviewed in detail included screening of two-dimensional UV/violet arrays of 4,4′-bis(9-carbazolyl)biphenyl (CBP), with peak emission at ~385 nm, and three different one-dimensional arrays. (1) Arrays of efficient blue-to-red OLEDs based on a blue-emitting 4,4′-bis(2,2′-diphenyl-vinyl)-1,1′-biphenyl (DPVBi) layer and a red-emitting DCM2-doped DPVBi; the OLED color evolved from blue to red as the thickness of the doped layer increased from 0 to 3.5 nm. Importantly, the brightness of all of the OLEDs was similar at a given bias. (2) Intense white OLEDs based on rubrene-doped DPVBi. The maximal brightness of these arrays exceeded 74,000 Cd/m². (3) Arrays of DCM2-doped NPD, followed by a layer of pure NPD of varying thickness, interfaced to a DPVBi layer. This sequence of layers enabled a detailed study of Förster energy transfer from DPVBi to DCM2, its evolution with bias, charge trapping by DCM2 and its effect on the Förster energy transfer, and an accurate determination of the transfer radius $R_0 = 33.5 \pm 3.5$ Å. Additional combinatorial OLED screening should therefore further enhance the performance and understanding of this class of devices, thus impacting the commercialization of OLED technology.

ACKNOWLEDGMENTS

Ames Laboratory is operated by Iowa State University for the US Department of Energy under Contract W-7405-Eng-82. This work was supported by the Director for Energy Research, Office of Basic Energy Sciences.

REFERENCES

1. Sun, T. X., Combinatorial search for advanced luminescence materials, *Biotechnol. Bioengineer.* (*Combin. Chem.*), 61(4), 193–201, 1999.
2. Schmitz, C., Thelakkat, M. and Schmidt, H.-W., A combinatorial study of the dependence of organic LED characteristics on layer thickness, *Adv. Mat.*, 11, 821, 1999.
3. Schmitz, C., Pösch, P., Thelakkat, M. and Schmidt, H. W., Efficient screening of electron transport material in multi-layer organic light emitting diodes by combinatorial methods, *Phys. Chem. Chem. Phys.*, 1, 1777–1781, 1999.
4. Gross, M., Muller, D. C., Nothofer, H.-G., Scherf, U., Neher, D., Brauchle, C. and Meerholz, K., Improving the performance of doped p-conjugated polymers for use in organic light-emitting diodes, *Nature*, 405, 661–665, 2000.
5. Zou, L., Savvate'ev, V., Booher, J., Kim, C.-H. and Shinar, J., Combinatorial fabrication and studies of intense efficient ultraviolet-violet organic light emitting device arrays, *Appl. Phys. Lett.*, 79, 2282, 2001.
6. Sun, X. and Jabbour, G. E., Combinatorial screening and optimization of luminescent materials and organic light-emitting devices, *Mat. Res. Soc. Bull.*, 27(4), 309–315, 2002.
7. Cheon, K. O. and Shinar, J., Combinatorial fabrication and study of doped-layer-thickness dependent color evolution in bright small molecular organic light-emitting devices, *Appl. Phys. Lett.*, 83, 2073, 2003.
8. Li, G. and Shinar, J., Combinatorial fabrication and studies of bright white organic light-emitting devices based on emission from rubrene-doped 4,4′-bis(2,2′-diphenylvinyl)-1,1′-biphenyl, *Appl. Phys. Lett.*, 83, 5359, 2003.
9. Cheon, K. O. and Shinar, J., Förster energy transfer in combinatorial arrays of selective doped organic light-emitting devices, *Appl. Phys. Lett.*, 84, 1201, 2004.
10. Thelakkat, M., Schmitz, C., Neuber, C. and Schmidt, H.-W., Materials screening and combinatorial development of thin film multilayer electro-optical devices, *Macromol. Rapid. Comm.*, 25, 204–223, 2004.
11. Zhou, Z., Shinar, R., Wu, H. S. and Shinar, J., unpublished results.
12. Friend, R. H., Gymer, R. W., Holmes, A. B., Burroughes, J. H., Marks, R. N., Taliani, C., Bradley, D. D. C., Dos Santos, D. A., Brédas, J. L., Lögdlund, M. and Salaneck, W. R., Electroluminescence in conjugated polymers, *Nature*, 397, 121–128, 1999.
13. Shinar, J., (Ed.), *Organic Light Emitting Devices: A Survey*, Springer Verlag, NY, 2003.
14. Kafafi, Z. H. and Lane, P. A. (Eds), *Organic Light Emitting Materials and Devices VII,* SPIE Conf. Proc. 5214, SPIE, Bellingham, WA, 2004.
15. Baldo, M. A., O'Brien, D. F., You, Y., Shoustikov, A., Sibley, S., Thompson, M. E. and Forrest, S. R., Highly efficient phosphorescent emission from organic electroluminescent devices, *Nature*, 395, 151–154, 1998.
16. Baldo, M. A., Lamansky, S., Burrows, P. E., Thompson, M. E. and Forrest, S. R., Very high efficiency green organic light emitting devices based on electrophosphorescence, *Appl. Phys. Lett.*, 75, 4, 1999.
17. O'Brien, D. F., Baldo, M. A., Thompson, M. E. and Forrest, S. R., Improved energy transfer in electrophosphorescent devices, *Appl. Phys. Lett.*, 74, 442, 1999.
18. Adachi, C., Baldo, M. A., Forrest, S. R. and Thompson, M. E., High-efficiency organic electrophosphorescent devices with tris(2-phenylpyridine)iridium doped into electron-transporting materials, *Appl. Phys. Lett.*, 77, 904, 2000.
19. Burrows, P. E., Forrest, S. R., Zhou, T. X. and Michalski, L., Operating lifetime of phosphorescent organic light emitting devices, *Appl. Phys. Lett.*, 76, 2493, 2000.
20. Wang, J., Yoo, Y., Gao, C., Takeuchi, I., Sun, X., Chang, H., Xiang, X.-D. and Schultz, P. G., *Science*, 279, 1712–1714, 1998.
21. Danielson, E., Devenney, M., Giaquinta, D. M., Golden, J. H., Haushalter, R. C., McFarland, E. W., Poojary, D. M., Reaves, C. M., Weinberg, W. H. and Wu, X. D., *Science*, 279, 837–839, 1998.
22. Xiang, X.-D., Sun, X., Briceno, G., Lou, Y., Wang, K.-A., Chang, H., Wallace-Freedman, W. G., Chen, S.-W. and Schultz, P. G., *Science*, 268, 1738–1740, 1995.
23. Briceno, G., Chang, H., Sun, X., Schultz, P. G. and Xiang, X.-D., *Science*, 270, 273–275, 1995.
24. Danielson, E., Golden, J. H., McFarland, E. W., Reaves, C. M., Weinberg, W. H. and Wu, X. D., *Nature*, 389, 944–948, 1997.

25. Sun, X.-D., Wang, K.-A., Yoo, Y., Wallace-Freedman, W. G., Cao, C., Xiang, X.-D. and Schultz, P., *Adv. Mater.*, 9, 1046, 1998.

26. Lowry, M. S., Hudson, W. R., Pascal, Jr., R. A. and Bernhard, S. J., *Am. Chem. Soc.*, 126, 14129–14135, 2004.

27. Pohl, R., Montes, V. A., Shinar, J. and Azenbacher, P., Red-green-blue emission from tris (5-aryl-8-quinolinolate) Al(III) complexes, *J. Org. Chem.*, 69, 1723–1725, 2004; Montes, V. A., Li, G., Pohl, R., Shinar, J. and Anzenbacher, Jr., P., Effective color tuning in OLEDs based on aluminum tris(5-aryl-8-hydroxyquinoline) complexes, *Adv. Mat.*, 16, 2001, 2004.

28. Riess, W., Beierlein, T. A. and Riel, H., Optimizing OLED structures for a-Si display applications via combinatorial methods and enhanced outcoupling, *Phys. Stat. Sol. (a)*, 201, 1360–1371, 2004.

29. http://www.appliedfilms.com/.

30. Li, F., Tang, H., Shinar, J., Resto, O. and Weisz, S. Z., Effects of aquaregia treatment of indium tin oxide substrates on the behavior of double-layered organic light emitting diodes, *Appl. Phys. Lett.*, 70, 2741, 1997.

31. Hung, L. S., Tang, C. W. and Mason, M. G., Enhanced electron injection in organic electroluminescent devices using an Al/LiF electrode, *Appl. Phys. Lett.*, 70, 152, 1997.

32. Kozlov, V. G., Parthasarathy, G., Burrows, P. E., Forrest, S. R., You, Y. and Thompson, M. E., Optically pumped blue organic semiconductor lasers, *Appl. Phys. Lett.*, 72, 144, 1998. In this work the emission bands of perylene are redshifted by 10–15 nm relative to those found in the work by Zou et al. (reference 5).

33. Choudhury, B., Kim, C.-H., Zou, L. and Shinar, J., unpublished results.

34. Adachi, C., Baldo, M. A. and Forrest, S. R., Electroluminescence mechanisms in organic light-emitting devices employing a Europium chelate doped in a wide energy gap bipolar conducting host, *J. Appl. Phys.*, 87, 8049, 2000.

35. Zhang, Z.-L., Jiang, X.-Y., Xu, S.-H. and Nagamoto, T., In: *Organic Electroluminescent Materials and Devices*, S. Miyata and H. S. Nalwa (Eds), Gordon and Breach, Amsterdam, 1997.

36. Aziz, H., Popovic, Z. D., Hu, N.-X., Hor, A.-M. and Xu, G., Degradation mechanism of small molecule-based organic light-emitting devices, *Science*, 238, 1900, 1999.

37. Yu, W.-L., Pei, J., Cao, Y. and Huang, W., Hole-injection enhancement by copper phthalocyanine (CuPc) in blue polymer light-emitting diodes, *J. Appl. Phys.*, 89, 2343, 2001.

38. Forsythe, E. W., Abkowitz, M. A., Gao, Y. and Tang, C. W., Influence of copper phthalocynanine on the charge injection and growth modes for organic light emitting diodes, *J. Vac. Sci. Tech. A*, 18, 1869, 2000.

39. Drexhage, K. H., In: *Progress in Optics*, E. Wolf (Ed.), 12, 165, North Holland, Amsterdam, 1974.

40. Becker, H., Burns, S. E. and Friend, R. H., Effect of metal films on the photoluminescence and electroluminescence of conjugated polymers, *Phys. Rev. B*, 56, 1893, 1997.

41. Junge K. E. and Shinar, J., unpublished results.

42. Liao, L. S., Fung, M. K., Lee, C. S., Lee, S. T., Inbasekaran, M., Woo, E. P. and Wu, W. W., Electronic structure and energy band gap of poly (9,9-dioctylfluorene) investigated by photoelectron spectroscopy, *Appl. Phys. Lett.*, 76, 3582, 2000; Liao, L. S., Cheng, L. F., Fung, M. K., Lee, C. S., Lee, S. T., Inbasekaran, M., Woo, E. P. and Wu, W. W., Interface formation between poly(9,9-dioctylfluorene) and Ca electrode investigated using photoelectron spectroscopy, *Chem. Phys. Lett.*, 325, 405–410, 2000; Greczynski, G., Fahlman, M. and Salaneck, W. R., An experimental study of poly(9,9-dioctyl-fluorene) and its interfaces with Li, Al, and LiF, *J. Chem. Phys.*, 113, 2407, 2000.

43. Li, G., Kim, C.-H., Lane, P. A. and Shinar, J., Magnetic resonance studies of tris-(8-hydroxyquinoline) aluminum-based organic light-emitting devices, *Phys. Rev. B*, 69, 165311, 2004.

44. Bulovic, V., Shoustikov, A., Baldo, M. A., Bose, E., Kozlov, V. G., Thompson, M. E. and Forrest, S. R., Bright, saturated, red-to-yellow organic light-emitting devices based on polarization-induced spectral shifts, *Chem. Phys. Lett.*, 287, 455–460, 1998.

45. Duggal, A. R., Shiang, J. J., Heller, C. M. and Foust, D. F., *Appl. Phys. Lett.*, 80, 3470, 2002.

46. Baldo, M. A., Soos, Z. G. and Forrest, S. R., *Chem. Phys. Lett.*, 347, 297, 2001.

47. Zhong, G. Y., He, J., Zhang, S. T., Xu, Z., Xiong, Z. H., Shi, H. Z., Ding, X. M., Huang, W. and Hou, X. Y., *Appl. Phys. Lett.*, 80, 4846, 2002.

48. Cheon, K. O. and Shinar, J., Bright white small molecular organic light-emitting devices based on a red-emitting guest–host layer and blue-emitting 4,4'-bis(2,2'-diphenylvinyl)-1,1'-biphenyl, *Appl. Phys. Lett.*, 81, 1738, 2002.

49. Kalinowski, J., Di Marco, P., Fattori, V., Giulietti, L. and Cocchi, M., *J. Appl. Phys.*, 83, 4242, 1998.

50. Pope, M. and Swenberg, C. E., *Electronic Processes in Organic Crystals,* 2nd edition, Oxford University Press, Oxford, 1999.
51. Cheon, K. O. and Shinar, J., unpublished results.
52. von Malm, N., Steiger, J., Schmechel, R. and von Seggern, H., *J. Appl. Phys.*, 89, 5559, 2001.
53. Littman, J. and Martic, P., *J. Appl. Phys.*, 72, 1957, 1992.
54. Kalinowski, J., Picciolo, L. C., Murata, H. and Kafafi, Z. H., *J. Appl. Phys.*, 89, 1866, 2001.
55. Aylott, J. W., Chen-Esterlit, Z., Friedl, J. H., Kopelman, R., Savvateev, V. and Shinar, J., *Optical Sensors and Multisensor Arrays Containing Thin Film Electroluminescent Devices*, US Patent No. 6,331,438, December 2001.
56. Savvate'ev, V., Chen-Esterlit, Z., Aylott, J. W., Choudhury, B., Kim, C.-H., Zou, L., Friedl, J. H., Shinar, R., Shinar, J. and Kopelman, R. Integrated organic light emitting device/fluorescence-based chemical sensors, *Appl. Phys. Lett.*, 81, 4652, 2002.
57. Choudhury, B., Shinar, R. and Shinar, J., Luminescent chemical and biological sensors based on the structural integration of an OLED excitation source with a sensing component, in: *Organic Light Emitting Materials and Devices VII*, Z. H. Kafafi and P. A. Lane (Eds), SPIE Conf. Proc. 5214, 64, 2004.
58. Choudhury, B., Shinar, R. and Shinar, J., Glucose biosensors based on organic light emitting devices structurally integrated with a luminescent sensing element, *J. Appl. Phys.*, 96, 2949, 2004.
59. Shinar, R., Choudhury, B., Zhou, Z., Wu, H.-S., Tabatabai, L. and Shinar, J., Structurally integrated organic light-emitting device-based sensors for oxygen, glucose, hydrazine, and anthrax, in: *Smart Medical and Biomedical Sensor Technology II*, Brian M. Cullum (Ed.), SPIE Conf. Proc. 5588, 59, 2004.
60. Deshpande, R. S., Bulovic, V. and Forrest, S. R., White-light-emitting organic electroluminescent devices based on interlayer sequential energy transfer, *Appl. Phys. Lett.*, 75, 888, 1999.
61. Chuen, C. H. and Tao, Y. T., *Appl. Phys. Lett.*, 81, 4499, 2002.
62. Mattoussi, H., Murata, J., Merritt, C. D., Lizumi, Y., Kido, J. and Kafafi, Z. K., Photoluminescence quantum yield of pure and molecularly doped organic solid films, *J. Appl. Phys.*, 86, 2642, 1999.
63. Förster, T., *Discuss. Faraday Soc.*, 27, 7, 1959.
64. D'Andrade, B., Thompson, M. E., and Forrest, S. R., Controlling exciton diffusion in multilayer white phosphorescent organic light emitting devices, *Adv. Mat.*, 14, 147, 2002.
65. Hamada, Y., Sano, T., Shibata, K. and Kuroki, K., Influence of the emission site on the running durability of organic electroluminescent devices, *Jpn. J. Appl. Phys.*, 34, L824, 1995.
66. Tang, C. W., Van Slyke, S. A. and Chen, C. H., Electroluminescence of doped organic thin films, *J. Appl. Phys.*, 65, 3610, 1989.
67. D'Andrade, B. W., Baldo, M. A., Adachi, C., Brooks, J., Thompson, M. E. and Forrest, S. R., High-efficiency yellow double-doped organic light-emitting devices based on phosphor-sensitized fluorescence, *Appl. Phys. Lett.*, 79, 1045, 2001.
68. Hamada, Y., Kanno, H., Tsujioka, T., Takahashi, H. and Usuki, T., *Appl. Phys. Lett.*, 75, 1682, 1999.
69. Kersting, R., Lemmer, U., Deussen, M., Bakker, H. J., Mahrt, R. F., Kurz, H., Arkhipov, V. I., Bässler, H. and Göbel, E. O., *Phys. Rev. Lett.*, 73, 1440, 1994.
70. Lam, J., Gorjanc, T. C., Tao, Y. and D'Iorio, M., *J. Vac. Sci. Technol. A,* 18(2), 593, 2000.
71. Spreitzer, H., Schenk, H., Salbeck, J., Weissoertel, F., Riel, H. and Riess, W., *Proc. SPIE*, 3797, 316, 1999.
72. Brütting, W., Riel, H., Beierlein, T. and Riess, W., *J. Appl. Phys.*, 89, 1704, 2001.
73. Kim, J. S., Ho, P. K. H., Greenham, N. C. and Friend, R. H., Electroluminescence emission pattern of organic light-emitting diodes: Implications for device efficiency calculations, *J. Appl. Phys.*, 88, 1073, 2000.
74. Karlsson, H. S., Read, K. and Haight, R., *J. Vac. Sci. Technol. A*, 20(3), 762, 2002.

21 Combinatorial Approach to Advanced Luminescent Materials

Jun Bao and Chen Gao

CONTENTS

21.1 INTRODUCTION

A luminescent material, or phosphor, is a solid that converts certain types of energy into electromagnetic radiation over thermal radiation. The electromagnetic radiation emitted by a luminescent material is usually in visible, but it could also be in other spectral regions, such as the ultraviolet or infrared. Luminescence can be classified into several types according to the method of excitation: photoluminescence (excited by electromagnetic radiations, e.g., in fluorescent lamps), cathodoluminescence (by electron beams, e.g., in cathode ray tubes), electroluminescence (by electric field), triboluminescence (by mechanical energies, e.g., grinding), X-ray luminescence (by X-rays), and chemoluminescence (by chemical energies).[1]

A phosphor usually consists of host and activator materials. The host material could be oxide, garnet, sulfide, oxysulfide, vanadate, germanate, etc. The activator material, also known as dopant or impurity, is usually rare-earth or transition metal elements. The activator material may act either as a luminescent center when excited, or as a sensitizer that absorbs and transfers the energy to the luminescent center.[2] Phosphors have been widely used for decades in cathode ray tubes (CRTs), plasma display panels (PDPs), field emission, flat-panel displays, as well as in fluorescent lamps. In recent years, new phosphors with advanced luminescent properties are in great demand due to the rapid development of modern photonic technologies such as mercury-free lamps,[3] flat-panel displays,[4] and computed tomography.[5] These advanced luminescent properties include high quantum efficiency, good absorption of the excitation energy, adequate colors, long lifetime, and low cost.

However, the discovery of new phosphors with part or all of these advanced luminescent properties remains a difficult problem, because these properties are highly sensitive to the change in dopant composition, host stoichiometry, and processing conditions.[1,2] Therefore, although the physical mechanism of luminescence is relatively well understood, there is no reliable theory to guide the search of advanced phosphors. Consequently, the discovery of phosphors that meet the requirements of a given application is highly empirical. Despite more than 100 years of intensive studies,[1] only fewer than 100 useful commercial phosphor materials have been discovered through the conventional one-at-a-time synthesis and characterization. To accelerate the search for advanced phosphors for various photonic applications, a combinatorial strategy has been developed to synthesize and screen new phosphors with superior properties. The aim of this chapter is to provide a brief overview of the combinatorial synthesis and high-throughput screening techniques, followed by a comprehensive review of recent progresses on the discovery and optimization of new phosphors based on this method.

21.2   COMBINATORIAL SYNTHESIS OF PHOSPHORS: THIN FILM DEPOSITION COMBINED WITH MASKS AND SOLUTION-BASED SYNTHETIC METHODS

Luminescent materials can be synthesized in either thin films or powders. Combinatorial methods have been developed to allow a parallel or fast-sequential synthesis of phosphor libraries in both forms.[6]

Thin-film deposition conjunction with a combinatorial masking strategy can be used to synthesize thin-film "spatially addressable" libraries, where each sample in the library is formed from multiple-layer precursors.[7,8] After the deposition of precursor layers, a low-temperature treatment is necessary to enable the interdiffusion of the precursor layers. Several high-vacuum thin-film deposition methods, such as sputtering, laser ablation, thermal, and e-beam evaporation have been used for the synthesis of thin-film libraries. Masking strategy, including the choice of mask form and masking schemes, is the key to the art of making a thin-film library.

Physical shadow masks[7,8] and photolithographic masks[9] are the most commonly used mask forms for generating a phosphor library. The physical shadow masking method allows a parallel synthesis of low-density (100–1000 compositions/in^2) library.[6] The primary mask first separates different samples on the substrate, and then a sequence of secondary masks is overlaid on top of it. Through this method, the thickness of the thin-film precursors can be controlled. The sequence and pattern of the secondary masks determine the final stoichiometry of materials in the library. Increasing the spatial density of distinct compositions in a given library can improve the effectiveness of the library. Compared with the physical shadow mask technique, photolithographic mask technique is well suited for generating high-density thin-film libraries ($> 1000/in^2$) because of its high spatial resolution, alignment accuracy, and neglectable shadowing effect.[6] In this procedure, before each thin-film deposition, a photoresist is coated and patterned, leaving behind open windows. After thin-film deposition, the remaining photoresist and overlying film are lifted off with acetone, leaving behind films only in the open window regions.[9]

The choice of masking schemes, to a large extent, determines the efficiency of libraries. Binary masking, quaternary masking and their hybrids are widely applied in the design of a thin-film library. In binary masking, one half of the total primary masking area is covered for each deposition, and every binary mask has a different pattern of coverage (Figure 21.1(a)). The resulting array contains 2^n different compositions with n deposition masking steps.[7] The disadvantage of the binary mask is that there is no grouping or discrimination of precursors according to their properties in sample compositions of the library, which may generate mixtures of precursors that do not form uniform phases. To increase the selectivity and effectiveness in synthesis and screening of phosphor libraries, a quaternary combinatorial masking scheme was designed,[9] as shown in Figure 21.1(b).

It involves n different masks, which successively subdivide the substrate into a series of self-similar patterns of quadrants. The rth ($1 \leq r \leq n$) mask contains 4^{r-1} windows and each window exposes one-quarter of the area deposited with the previous mask. Within each window is an array of 4^{n-r} gridded sample sites. Each mask is used in up to four sequential depositions and rotated by 90° each time. This process produces 4^n different compositions with $4n$ deposition steps and can be used to efficiently survey materials consisting of up to n precursor components. The advantage of the quaternary mask is that precursors in the same mask group will not spatially overlap, which provides selectivity in the design of library. In comparison with the binary or gradient masking techniques, the quaternary combinatorial masking strategy represents a substantial improvement to screen large landscapes of diverse composition and could be applicable to many classes of materials.

A high-density thin-film phosphor library containing 1024 different compositions on a 2.5-cm-wide square substrate was synthesized using thin-film deposition combined with quaternary photolithographic masks.[9] As illustrated in Figure 21.1(b), five masks were used with a sequence of masking and precursor deposition as following: A_1: Ga_2O_3 (355 nm); A_2: Ga_2O_3 (426 nm); A_3: SiO_2 (200 nm); A_4: SiO_2 (400 nm); D_1: CeO_2 (3.5 nm); D_2: EuF_3 (11.3 nm); D_3: Tb_4O_7 (9.2 nm); E_1: Ag (3.8 nm); E_2: TiO_2 (6.9 nm); E_3: Mn_3O_4 (5.8 nm); B_1: Gd_2O_3 (577 nm); B_2: ZnO (105 nm); B_3: ZnO (210 nm); C_1: Gd_2O_3 (359 nm); C_2: Y_2O_3 (330 nm); and C_3: Y_2O_3 (82.5 nm). The numbers in parentheses are film thicknesses, and the notation X_i represents a deposition step with mask X rotated

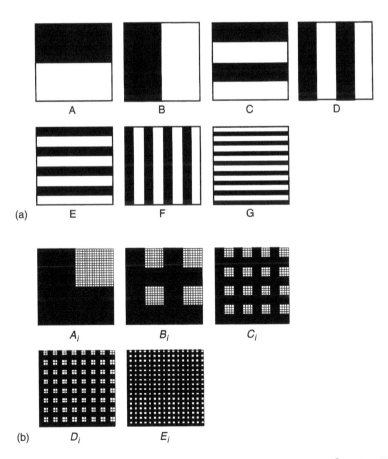

FIGURE 21.1 (a) A binary mask group. After seven masking and deposition steps, $2^7 = 128$ different material can be generated. (b) A "self similar" quaternary mask group. For each mask, up to four different patterns can be generated by rotating the mask counterclockwise by $(i-1) \times 90°$. (Reprinted from Xiang, X. D. et al., *Science*, 268, 1738, 1995. With permission.)

$(i-1)$ 90° counter-clockwise relative to the position depicted in Figure 21.1(b). The steps B_4, C_4, D_4, and E_4 were omitted to include compounds consisting of less than five precursors in the libraries to study the effects of dopants. Figure 21.2 shows the photo-luminescent photograph of the library under ultraviolet irradiation. A scanning spectrophotometer was used to measure the excitation and emission spectra of individual luminescent samples in the library. Optimal compositions were identified with the use of gradient libraries. This process led to the identification of an efficient blue photoluminescent composite material: $Gd_3Ga_5O_{12}/SiO_2$. Experimental evidences suggest that the strong blue luminescence in this material may arise from interfacial effects between SiO_2 and $Gd_3Ga_5O_{12}$.

The thin-film combinatorial approach is effective in discovering new phosphors. However, most phosphors used in industry are in the powder form, which can be prepared by solution-phase methods. By considering that the properties of thin film may be different from those of the corresponding power sample, it is necessary to develop the parallel powder-synthesis techniques to complement the thin-film combinatorial method. In addition, solution-based methods allow mixing of precursors at the molecular level, thereby reducing the need for high-temperature interdiffusion compared with that in the solid-state synthesis. In solution-based parallel synthesis, droplets of precursor solutions are laid into predefined microreactors (usually microwells drilled into a ceramic substrate) using microdispensers (usually piezoelectric-driven ejectors). Then, the precursors are mixed and reacted in the microreactors to form a library of powder materials. The ability to control droplet number and volume accurately is critical for determining the stoichiometry of final materials. Ink-jet delivery shows many advantages over other microdispense apparatus, e.g., nanoliter dispensing capability, high accuracy, etc.[10] The technique is well suited for generating powder libraries with a density of about 100/in², which is much lower than that of thin-film deposition method because its ability of compositional control is not as good as the latter. Sun et al.[11] have first verified

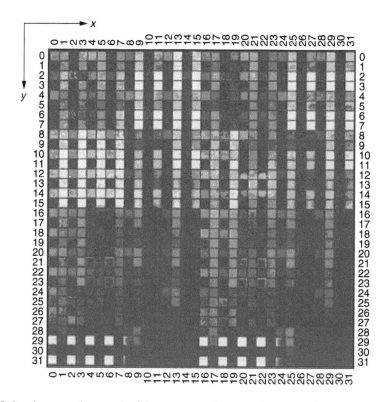

FIGURE 21.2 Luminescent photograph of the quaternary library under UV excitation.

the applicability of the ink-jet delivery system to the microscale chemical synthesis of libraries by generating a library of rare-earth-doped metal-oxide phosphors. Using the ink-jet delivery systems, phosphor and catalyst materials libraries have been fabricated from solution precursors.[11–13]

Figure 21.3(a) shows a schematic diagram of the drop-on-demand ink-jet delivery system for combinatorial synthesis of the powder-material libraries.[14] The eight independent piezoelectric ink-jet heads and X–Y stage are controlled by a computer via the driving circuit and motion controller. Each ink-jet head is connected to a solution reservoir through a tube, and the substrate with a microreactor array is fixed on the stage. The ejection of ink-jet heads and the position of stage are automatically coordinated by software according to the concentrations of solutions, interval between reactors, substrate size, and the composition map, etc. Driven by software, the ink-jet heads that connected to different solutions precursors in turn focus on a same microreactor in the substrate by the movement of X–Y stage and eject appropriate amount of droplets of solutions. Then these heads in turn focus on the next microreactor and implement the ejection. The process is repeated until the whole ejection is completed and generates a library with different solution-mixing precursors.

The ink-jet head consists of a sapphire nozzle, a stainless steel diaphragm, and a piezoelectric disk as shown in Figure 21.3(b). High-voltage electric pulses coming from the driving circuit are applied to the piezoelectric disk to produce a mechanical vibration. The vibration crosses the diaphragm and propagates toward the nozzle in the form of acoustic waves. The positive pressure of the vibration accelerates the liquid around the nozzle to overcome the surface tension to form an ejected drop. Details of the formation of the drops can be found in reference 15. Due to the intrinsic instability of the ejection system, the volume of the droplets has a statistic distribution. By measuring the weights of 25,000 de-ionized water drops ejected at different moments within 1 h, the average droplet volume was estimated to be about 10 nL with a standard deviation of ~10%. The ejection repeatability is 0.5–2 kHz.

However, a significant limitation of the above-mentioned ink-jet delivery technique is the applicability: only soluble compounds could be used as the precursors. Based on the fact that a lot of suspensions have been used with ink-jet technique in ceramic freeforming,[16] fabrication of electronic devices[17] and functionally graded material,[18] Chen et al. developed a universal preparation method of stable insoluble oxide suspensions using a wet ball-milling technique so that insoluble oxides can be used as precursors in the form of nano-suspensions.[14] These suspensions are suitable for ink-jet combinatorial synthesis due to the ultrafine/nano particle size, high surface tension, low viscosity, and high concentration. To verify the applicability, a photo-luminescent library containing $Y_2O_3{:}Eu_x^{3+}$ and $Y_2O_3{:}Tb_y^{3+}$ was synthesized by ejecting Y_2O_3, Eu_2O_3 and Tb_4O_7 suspensions using the above-mentioned "drop-on-demand" ink-jet delivery system. Figure 21.4 shows the photo-luminescent photograph of the library under 254 nm UV excitation. The photo-luminescent intensity of red emission varies with content of Eu^{3+}, and maximum luminescent intensity appears at about 5% Eu^{3+} concentration, which fairly agrees with reports in the literature.[1] This result indicates that the delivered Eu_2O_3 is properly doped into Y_2O_3 matrix and homogeneous materials were formed.

21.3　HIGH-THROUGHPUT SCREENING OF PHOSPHORS LIBRARIES: PHOTOGRAPHY AND SCANNING SPECTROMETER SYSTEM

After phosphor libraries are synthesized, they are ready to undergo a high-throughput screening for the properties of scientific and technical application interests. Photography using either film cameras or digital charge-coupled devices (CCD)[19] has the advantages of parallel analysis and excellent spatial resolution (~10 μm). This simple technique can be used to characterize the luminescent intensities and color coordinate of a phosphors library by taking the photoluminescent photograph of the library under UV or other high-energy excitations. In most cases, human eyes are sensitive

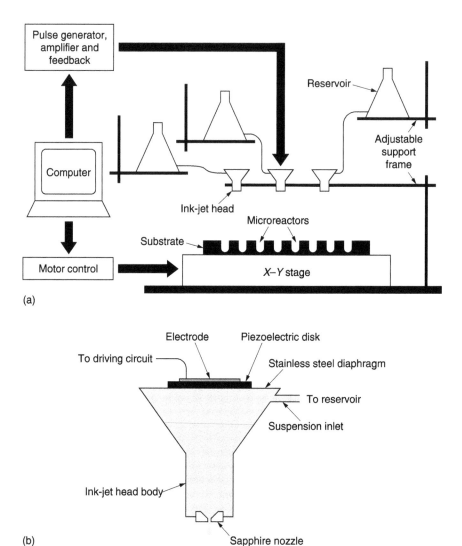

FIGURE 21.3 (a) Schematic diagram of the drop-on-demand ink-jet delivery system. (b) Schematic of the ink-jet head.

enough to pick up the phosphor lead of interest directly under UV-lamp excitation. An example of such a photoluminescent photograph is shown in Figure 21.2. For more quantitative analysis, an automatic scanning spectrometer system is usually used to measure the emission spectra of luminescent samples in the library (Figure 21.5).[14] This system consists of a mercury lamp, a portable optical fiber spectrometer (Ocean Optics, Inc., Model SD2000), and an X–Y stage. The spectrometer is equipped with a 25/200-μm slit, a 600-grooves/mm grating blazed at 400 nm and covering a spectral range from 200 to 850nm with the efficiency higher than 30%, and a 2048-element linear silicon CCD-array detector, so that a whole visible spectrum can be collected in one shot. The materials library was fixed on the moving table of the X–Y stage. Emission spectrum from each sample in the library was measured when the fiber-optic probe was focused on it. A light shelter was used to shield the interference from other samples. Using the scanning spectrometer, the emission spectra of Y_2O_3:Eu_x^{3+}, varying, with the content of Eu^{3+} in the library (Figure 21.4), were measured and are shown in Figure 21.6.

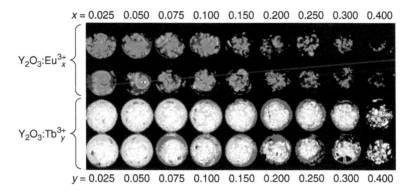

FIGURE 21.4 (See color insert following page 172) Composition map and photoluminescent photograph of the library under UV excitation.

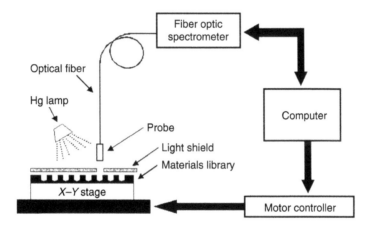

FIGURE 21.5 Schematic diagram of the setup for measuring the emission spectra of the materials in library.

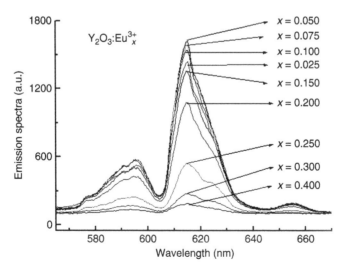

FIGURE 21.6 Emission spectra of the samples with different Eu^{3+} content in the library.

21.4   COMBINATORIAL SEARCH FOR ADVANCED LUMINESCENT MATERIALS

21.4.1   Ultraviolet Luminescent Materials

Using the combinatorial method, Symyx Technologies, Inc. identified a new red phosphor from a thin-film library containing 25,000 different samples, $Y_{0.845}Al_{0.070}La_{0.060}Eu_{0.025}VO_4$, which has a superior quantum efficiency compared with the existing commercial red phosphors. This library was deposited on an unheated $3''$ silicon wafer through electron beam evaporation.[20] A stainless steel primary mask consisting of 230-μm square elements spaced 420 μm apart was attached to the substrate to separate individual library elements. The spatial variation of materials deposited on the library was created using stationary and movable physical masks to control the thickness of specific evaporants in selected regions of the substrate (Figure 21.7(a)). Two mask sets were used, one with a single 19.1-mm-wide rectangular slit and the other with four 4.8-mm-wide rectangular slits. Four constant-thickness columns consisting of SnO_2, V, Al_2O_3 + V (15:8 molar ratio), and Al_2O_3 were first deposited. On top of these layers four rows with linearly varying thicknesses of La_2O_3, Y_2O_3, MgO, and $SrCO_3$ were then deposited to divide the substrate into 16 host lattice subregions. Finally, within each subregion, columns of rare earths Eu_2O_3, Tb_4O_7, Tm_2O_3, and CeO_2 were deposited in linearly varying thicknesses, which resulted in approximately 600 different chemical compositions per square centimeter. As the constituents of each library element were deposited in layers, subsequent oxidative thermal processing at various temperatures was necessary for interplanar mixing and formation of the homogeneous materials.

High-throughput screening for ultraviolet-excited photoluminescence was performed by photographing the visible emission of the library under UV excitation using a CCD camera (Figure 21.7(b)). Quantitative measurements of chromaticity were carried out from three images obtained using red, green, and blue tristimulus filters. The highest efficiency materials with desirable chromaticity identified in the initial high-density library were red phosphors (Eu^{3+} doped) with $Y_{1-m}Al_mVO_4$ as the host. To investigate and optimize the best host compositions, a second library was designed and synthesized. This library included La as a third Group III host. First, both Eu_2O_3 and V were deposited uniformly (26.3 nm and 189.6 nm, respectively) over the entire substrate, followed by Y_2O_3, Al_2O_3, and La_2O_3 deposited as linear gradients along three axes rotated by $120°$, thereby exploring all possible $Y_{0.95-m-n}Al_nLa_mEu_{0.05}VO_4$ compounds. The measured results of photoluminescent intensity show that the composition with maximum intensity and red chromaticity suitable for commercial application is $Y_{0.82}Al_{0.07}La_{0.06}Eu_{0.05}VO_4$. The activator concentration was subsequently optimized with the third library in which Eu^{3+} concentration was varied from 0 to 20% using a single movable mask. Screening of this optimization library identified 2.5% Eu as the most efficient dopant composition, which gave a new phosphor composition $Y_{0.845}Al_{0.07}La_{0.06}Eu_{0.025}VO_4$ with improved red chromaticity ($x = 0.67$, $y = 0.32$) compared with the more orange standard commercial red phosphor, $Y_{1.95}O_3Eu_{0.05}$ ($x = 0.64$, $y = 0.35$). Synthesis of powder bulk samples by conventional methods confirmed the performance of the identified compositions.

Using the same combinatorial synthesis and screening techniques, researchers from Symyx Technologies, Inc. identified a novel luminescent oxide Sr_2CeO_4, which has an unusual one-dimensional chain structure.[21] The discovery of this fundamentally new blue–white phosphor in an unexpected region of composition space validates the combinatorial methodology of materials science. The new luminescent phase Sr_2CeO_4 was identified from a library of over 25,000 compositionally independent elements prepared using electron beam evaporation with multiple targets and moving masks.[20] Screening for candidate materials was performed using a CCD camera, as previously described.[20] The comparison of the visible emission of a subsection of the library under 254-nm excitation with that of known luminescent materials identified a potentially new phosphor material with blue–white emission in the region of the combinatorial library containing Sr, Sn, and Ce. A subsequent ternary library containing combinations of these three elements revealed that Sn is not needed for the observed emissive properties, and that maximum luminosity was observed at Sr:Ce

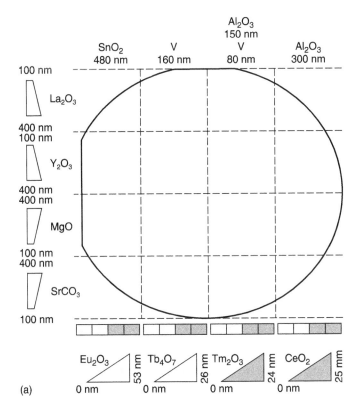

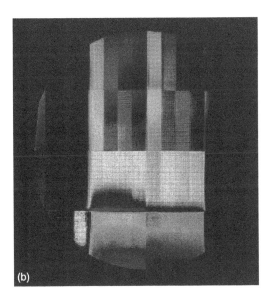

FIGURE 21.7 (a) Deposition map of diverse phosphor discovery library. Thicknesses of target materials deposited on the 3-inch-diameter substrate are shown. (b) Photograph of the library under 254 nm ultraviolet excitation. (Reprinted from Danielson, E. et al., *Nature*, 389, 944, 1997. With permission.)

ratios greater than unity. To study the optical and other physical properties, a bulk powder sample with Sr:Ce ratio of 2:1 was prepared from the reaction of CeO_2 with $SrCO_3$ at 1000°C in air for 48 h with four intermediate regrindings. The excitation and emission spectra of Sr_2CeO_4 displayed broad maxima at 310 and 485 nm, with the emission appearing blue–white to the eye (CIE 1931 chromaticity coordinates $x = 0.198$, $y = 0.292$) and with a quantum yield of 0.48. This phosphor can be effectively excited by X-rays and exhibits efficient cathodoluminescence (5.1 lm/W at 20 kV and 1 μA/cm^2). Unlike most other rare earth-based oxide phosphors, the emission from Sr_2CeO_4 is quite broad and has an uncharacteristically long excited-state lifetime (51.3 ± 2.4 μs) compared with d–f transitions within the Ce^{3+} excited states.

The structure of Sr_2CeO_4 determined from powder X-ray diffraction is highly anisotropic and is an unusual phosphor containing one-dimensional chains, which was not previously found in rare earth-based oxide phosphors. The structure consists of linear chains of edge-sharing CeO_6 octahedra with two terminal oxygen atoms per cerium center, which are isolated from each other by Sr^{2+} cations. It is believed that this low-dimensional structure, with its terminal Ce–O bonds, is critical for the luminescence in Sr_2CeO_4, and that the mechanism of this luminescence is based on the ligand-to-metal charge transfer from O^{2-} to Ce^{4+}, not the transitions arising from isolated valence transitions from Ce^{3+} defect centers as in all known cerium-based phosphors.

To ascertain if there were other luminescent phases within the M_2CeO_4 (M = Ba, Ca, and Mg) phase space, a triangular library was prepared with pure Ba_2CeO_4, Ca_2CeO_4, and Mg_2CeO_4 at each corner. This library was prepared by robotically dispensed sol–gel precursors and followed by heating to 900°C. The Sr-containing region has the brightest emission under 254-nm excitation (Figure 21.8), and there is no appreciable luminosity from the Ba and Mg regions.

The above works illustrate that the combinatorial approach can identify fundamentally new and unexpected structures with properties arising from new mechanisms.

21.4.2 VACUUM ULTRAVIOLET LUMINESCENT MATERIALS

Recently, driven by the advances in the PDP and mercury-free fluorescent lamp techniques, the demand on high-efficiency vacuum ultraviolet (VUV) phosphors has increased significantly. Liu et al.[22] adopted the combinatorial method to search for a set of promising phosphors for PDP or

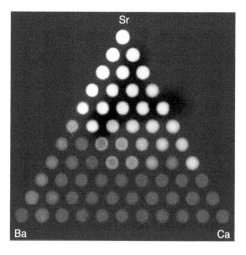

FIGURE 21.8 Photograph of a ternary M_2CeO_4 (M = Ca, Sr, and Ba) library under 254-nm excitation. The Ce concentration in each element is constant, and the M gradient decreases linearly from the corner of the pure M_2CeO_4 composition in a direction along the bisector of the opposite edge. (Reprinted from Danielson, E. et al., *Science*, 279, 837, 1998. With permission.)

mercury-free fluorescent lamp applications. In their research, (Y,Gd)BO$_3$:Eu was chosen as the starting material, and the influence of PO$_4$, Mg, Sr substitutions on the emission of Eu^{3+} was studied. The purpose of this work was to find a set of leads for high-efficiency VUV phosphors study through systematic search.

The library was synthesized using the solution-based multi-ink-jet delivery system as shown in Figure 21.3(a). The precursors were Y$_2$O$_3$, Gd$_2$O$_3$, Eu$_2$O$_3$, MgO nitric acid solutions, and Sr(NO$_3$)$_2$, Al(NO$_3$)$_3$, H$_3$BO$_3$, and NH$_4$H$_2$PO$_4$ aqueous solutions. A 50-mm × 50-mm corundum plate predrilled with 11 × 11 hemispherical holes was used as the substrate. After all precursor solutions are delivered, the library was dried and then fired. To implement the high-throughput screening of the synthesized phosphor library, a vacuum cathode discharge lamp simulating the environments of the PDP discharging cell and the mercury-free fluorescent lamp was designed (Figure 21.9). They are filled with a Xe–He or Xe–Ne gas mixture at similar condition. The discharge of the gas emits Xe radiation in the VUV range, including a line at 147 nm (Xe monomer) and a band around 172 nm (Xe dimer band). Coupling the discharge lamp with a vacuum characterization chamber through a differential pumping system, the sample library is excited by the VUV emission passing through the center hole of the differential pumping system, and the photoluminescence is photographed outside the chamber through an optical window. In measurement, the library was mounted in the characterization chamber, and the screening system was pumped down to a background of < 10^{-2} Pa. Then, 1:5 Xe:Ne(He) gas mixture was fed into the system, and the differential pumping system was used to keep the vacuum of the sample chamber at about 2 Pa, while the vacuum in the lamp reached approximately 10 Pa. After high-voltage trigger, the lamp maintained a stable discharge current and the VUV radiation emitted from the discharging gases. This condition is almost the same as that of the PDP discharge cell and the mercury-free lamp. The photoluminescence from samples in library was imaged in a parallel way with a commercial CCD camera. When detailed spectrum information is desired, the commercial CCD camera can be replaced with a scientific grade CCD and a set of band-pass filters (ranging from 400 to 650 nm with FWHM of 10 nm) driven by a motorized filter wheel.

The emission picture of the library under the excitation of the Xe's VUV radiation is obtained through red filter (Figure 21.10). From the picture, two sets of good red phosphors are identified: GdSr(B$_x$P$_{2-x}$)O$_{5.5-7.5}$:Eu$_{0.1}$, and Sr$_2$(B$_x$P$_{2-x}$)O$_{5-7}$:Eu$_{0.1}$. The chromaticity of the former is better than that of the latter. It is believed that the dominant transition in GdSr(B$_x$P$_{2-x}$)O$_{5-7}$:Eu$_{0.1}$ is $^5D_0 \rightarrow {}^7F_0$, so that the wavelength of the main emission of GdSr(B$_x$P$_{2-x}$)O$_{5.5-7.5}$:Eu$_{0.1}$ is longer than that of Sr$_2$(B$_x$P$_{2-x}$)O$_{5.5-7.5}$:Eu$_{0.1}$ where the dominant transition is supposed to be $^5D_0 \rightarrow {}^7F_j$. The scale-up experimental results show that the bulk GdSr(B$_x$P$_{2-x}$)O$_{5.5-7.5}$:Eu$_{0.1}$ sample prepared by conventional methods is brighter than that of the famous red commercial phosphor Y$_2$O$_3$:Eu under the same excitation.

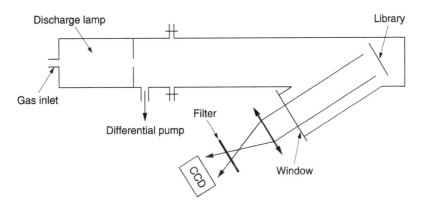

FIGURE 21.9 VUV phosphors library screening system.

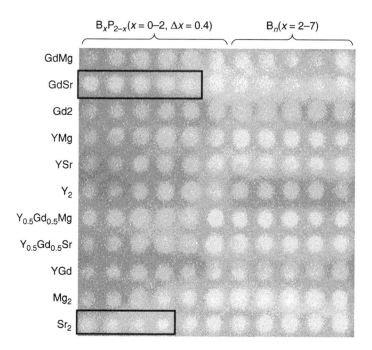

FIGURE 21.10 Emission from the library under VUV excitation.

Sohn et al.[23] identified a new red phosphor $Y_{0.9}(P_{0.92}V_{0.03}Nb_{0.05})O_4$:$Eu^{3+}$ at VUV excitation through a three-step combinatorial screening. The starting point of the investigation lies in the possibility that yttrium-based YRO_4 compounds such as $YAsO_4$, $YNbO_4$, YPO_4, and YVO_4 could be potential candidates for PDP.

The phosphor libraries were synthesized using the solution-based combinatorial chemistry method with a computer-programmed injection system. The photoluminescence efficiency of phosphors at VUV excitation was measured using a system consisting of a Kr_2 excimer lamp producing a 147-nm emission band as the excitation source, a vacuum chamber, a CCD detector, an automatically maneuverable X–Y stage on which the sample plates can be mounted and a computer-controlling unit. The excitation spectrum was examined using another system that consists of a deuterium lamp, a vacuum chamber, a vacuum monochromator to monochromatize excitation light, an emission monochromator, a photomultiplier tubes, and a computer-controlling unit. The luminance was calculated by integrating the product of the emission spectrum and the standard visual spectral efficiency curve by obeying the CIE regulation.

The quaternary library in terms of luminance at 147 nm excitation for a $Y(As,Nb,P,V)O_4$ system is shown in Figure 21.11 with all the inner shells. The Eu^{3+} doping content was fixed at 0.1 mol. The grayness of each sphere in the library represents the luminance of Eu^{3+} fluorescence. The maximum luminance is obtained at the composition $Y_{0.9}(P_{0.83}As_{0.06}V_{0.06}Nb_{0.06})O_4$:$Eu_{0.1}^{3+}$, which is the apex compound of the second shell. The CIE chromaticity is $x = 0.6603$ and $y = 0.3394$ for this material. These values are acceptable due to the fact that the required chromaticity for PDP application is $x = 0.67$ and $y = 0.33$. To search for more promising materials, an additional screening was carried out in the vicinity of this composition, which involves $Y(P_{1-a-b-c}As_aV_bNb_c)O_4$ compositions with a, b, and c increment of 0.03 from 0 to 0.15 mol. The results show that the incorporation of As is not favorable for enhancing the luminance at all. This could be ascribed to the volatilization of the As ion during the firing. In addition, the luminance deteriorates if the total doping content of V and Nb exceeds 0.1 mol. Consequently, a ternary combinatorial library was

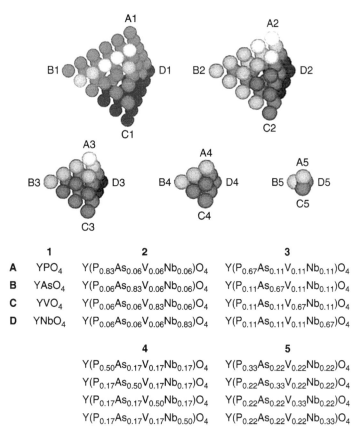

FIGURE 21.11 The quaternary combinatorial library in terms of luminance at 147-nm excitation for a $Y(As,Nb,P,V)O_4$ system fired at 1200°C, showing all the inner shells. (Reprinted from Sohn, K. S. et al., *Chem. Mater.*, 714, 2140, 2002. With permission.)

used for the final screening in which the apex compounds were $YPO_4:Eu^{3+}$, $Y(P_{0.9}V_{0.1})O_4:Eu^{3+}$, and $Y(P_{0.9}Nb_{0.1})O_4:Eu^{3+}$. The screening result in Figure 21.12 shows that the maximum luminescence is at $Y(P_{0.92}V_{0.03}Nb_{0.05})O_4$ and the luminance is comparable to the commercially available $(Y,Gd)BO_3:Eu^{3+}$ phosphor. The luminescent mechanism study in association with the density of states calculation revealed that the absorption near the band edges in YRO_4 (R = P, V, and Nb) involves excitations from the oxygen 2p-like states near the top of the valence band (VB) to the cations nd-like states near the bottom of the conduction band (CB). The band structure of YPO_4 consists of oxygen 2p-like states for VB and yttrium 4d-like states for CB, whereas the CB of YVO_4 and $YNbO_4$ consist of the vanadium and niobium 4d-like states, respectively. The incorporation of a small amount of V and Nb into the YPO_4 host should act as a bridge for energy transfer and thereby enhance the luminescent efficiency. The transfer route of the VUV excitation energy is the Y–O charge transfer→R–O charge transfer in the RO_4 (R = V, Nb) group → Eu^{3+} center.

21.4.3 Cathodoluminescent (CL) Phosphors

The rapid development of advanced display and lighting technologies, such as field-emission displays (FED) and PDP, requires the phosphors with advanced properties. In particular, FED phosphors should show efficient and durable cathodoluminescence at low voltages and high current densities. Conventional phosphors for CRTs are no longer suitable for these applications, especially the most

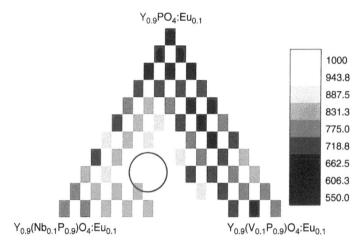

FIGURE 21.12 Ternary combinatorial library around the YPO_4-rich region. (Reprinted from Sohn, K. S. et al., *Chem. Mater.*, 14, 2140, 2002. With permission.)

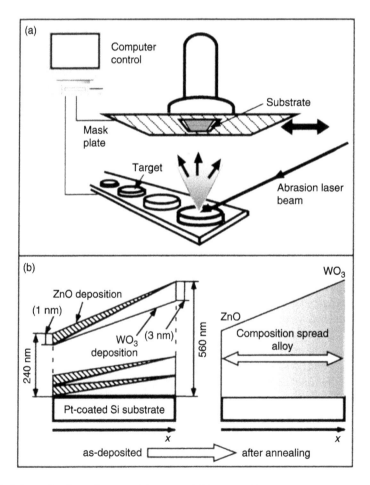

FIGURE 21.13 Schematic view of composition-spread film deposition using the combinatorial PLD with moving mask: (a) experimental setup and (b) deposition sequence for the composition-spread film consisting of the ZnO and WO_3 layers with atomic-layer-thickness gradation. (Reprinted from Hayashi, H. et al., *Appl. Phys. Lett.*, 82(9), 1365, 2003. With permission.)

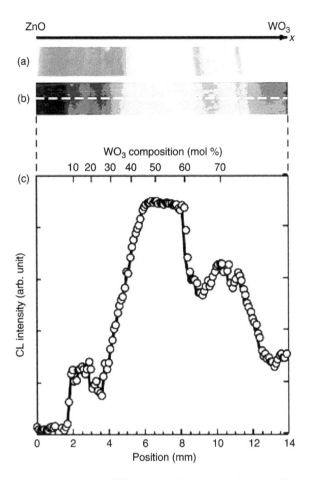

FIGURE 21.14 (a) Photograph of the ZnO–WO$_3$ composition-spread film. (b) Distribution of cathodolumi-nescence image, and (c) change in the cathodoluminescence intensity as a function of WO$_3$ molar fraction. (Reprinted from Hayashi, H. et al., *Appl. Phys. Lett.*, 82(9), 1365, 2003. With permission.)

popular sulfide-based phosphors, which undergo fast degradation at high current densities.[24] The development of new oxide-based cathodoluminescent phosphors has become an immediate task for the materials science community.[25]

Hayashi et al.[26] screened for blue phosphor materials in a ZnO–WO$_3$ binary system through combinatorial methodology that involved fabricating composition spread samples and imaging their cathodoluminescence. The deposition of the ZnO–WO$_3$ composition-spread film on a Pt-coated Si (Pt/Si) substrate (14 mm × 310 mm × 30.5 mm) at room temperature, using a combinatorial pulsed laser deposition (PLD) system, is shown in Figure 21.13.

The cathodoluminescence image of the composition spread film was measured with an acceleration voltage of 600 V and a current density of 1 mA/cm^2 at room temperature. Figure 21.14a shows a photograph of the sample. The distribution of the blue emission obtained by monitoring the cathodoluminescence along the x axis of the sample is also shown in Figure 21.14(b). The brightness of the blue emission changes discontinuously with the W composition. Although there is a black area near the pure ZnO side, a blue emission can be seen throughout most of the spread. In particular, there is a bright blue emission around the middle. In addition to this, there is a green emission between the pure ZnO side and the middle. Figure 21.14(c) presents the blue emission intensity that varies with the WO$_3$ molar fraction. The blue emission intensity is analyzed along the

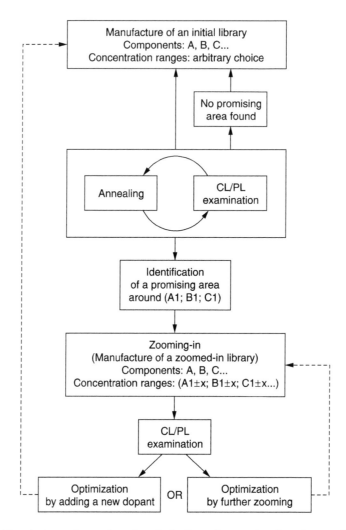

FIGURE 21.15 The strategy of search and optimization of new phosphors by combinatorial method. (Reprinted from Mordkovich, V. Z. et al., *Adv. Funct. Mater.*, 13(7), 519, 2003. With permission.)

white dashed line in Figure 21.14(b). From the distribution of blue emission intensity obtained by monitoring the cathodoluminescence of the composition-spread, the brightest emissive phase near stoichiometric $ZnWO_4$ is in the region of 45–60 mol% WO_3, including stoichiometric $ZnWO_4$ (50 mol%). Other emissive phases were also found at WO_3 compositions of 10–20 mol% and 65–75 mol%. These phases have not yet been reported.

A flexible combinatorial search-and-optimization strategy was employed for new cathodoluminescent phosphor identification in four ZnO-based quasi-binary and quasi-ternary systems: ZnO:W, ZnO:V, ZnO:(W,Mg), and ZnO:(Y,Eu).[25] A strategy chart for combinatorial search and optimization of new phosphor materials is presented in Figure 21.15. The PLD system was employed to fabricate thin-film combinatorial libraries of ZnO-based phosphors on the substrates of sapphire or Pt-coated silicon wafers. Screening of the libraries included examination of both PL and low-voltage CL properties. The measurements were carried out on the pixels of different compositions or on the spread with continuous concentration change. CL was excited by electron beam with acceleration

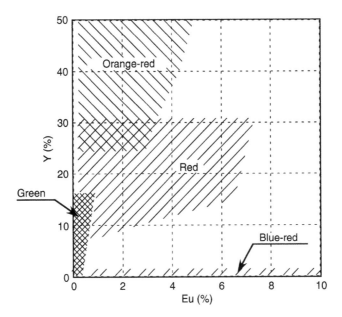

FIGURE 21.16 The phase diagram of CL in ZnO:(Y,Eu) libraries. The red area corresponds to Eu^{3+} luminescence in the mixed zinc-yttrium-europium oxide with the main emission line at 617 nm. The orange–red area corresponds to Eu^{3+} luminescence in the europium-activated yttrium oxide Y_2O_3 with the main emission line at 611 nm. The blue–red area corresponds to a very weak Eu^{2+} blue (440 nm) and Eu^{3+} red (694 nm) luminescence in ZnO. The green area corresponds to a weak intrinsic green luminescence of ZnO at 505 nm. (Reprinted from Mordkovich, V. Z. et al., *Adv. Funct. Mater.*, 13(7), 519, 2003. With permission.)

voltage from 10 to 1200 V and the current densities of up to 20 mA/cm². A fiber-optic probe was used to collect the emission for the spectroscopic measurements.

Four new and efficient phosphor compositions have been discovered and optimized. The ZnO:W system shows strong blue luminescence for a broad range of W concentrations. The highest efficiency is observed at 45–47% W, where the concentration of $ZnWO_4$ phase is close to 100%, while W deficiency is still assured. At higher W concentrations, the blue luminescence is suppressed due to the formation of W-rich tungstates, including tungsten bronzes. The optimization of the blue ZnO:W phosphor by co-doping with Mg was also investigated. The strongest luminescence is observed at 44% W and 16% Mg. The newly found ZnO:(W,Mg) exhibits narrower emission band that is situated at 470 nm, providing a better chromaticity than that of ZnO:W system. Strong yellow luminescence was found in the ZnO:V system over a broad range of V concentrations. At V content below 2%, the libraries still show weak ZnO-like green luminescence. The rise in V content results in the appearance and steady rise of yellow 560-nm luminescence. As soon as the system reaches the $Zn_3V_2O_8$ stoichiometric point at 40% V, the luminescence efficiency suddenly drops. The measurements of the luminescence efficiency vs V concentration clearly show a broad maximum at 33% V, or in the range of 32–40% V in more general terms. The CL screening results of ZnO:(Y,Eu) libraries are summarized in Figure 21.16. Four areas with different luminescent properties can be seen in the diagram. The areas are assigned to "red," "orange–red," "blue–red," and "green." "Zooming in" on the red area locates the optimized composition of the ZnO:(Y,Eu) phosphor as 12–15% Y and 2–3% Eu. The optimized ZnO:(Y,Eu) exhibits better chromaticity but lower luminescence intensity compared with that of commercial Y_2O_3:Eu.

21.5   SUMMARY AND OUTLOOK

Combinatorial methods have been applied to search and optimize the advanced phosphors with superior properties, such as high luminescent efficiency, pure chromaticity, etc. The phosphor libraries can be synthesized in either thin-film or powder forms through thin-film deposition combined with masks or solution-based synthetic methods. Various fluorescent properties, including luminescent intensity, color coordinate, and excitation or emission spectra can be screened by high-throughput techniques. Using the combinatorial methods, a few new phosphors with advanced properties were discovered rapidly and efficiently, such as a red phosphor $Y_{0.845}Al_{0.070}La_{0.060}Eu_{0.025}VO_4$ with excellent chromaticity, a blue phosphor Sr_2CeO_4 containing a novel one-dimensional chain structure, a high-efficiency phosphor $Y_{0.9}(P_{0.83}As_{0.06}V_{0.06}Nb_{0.06})O_4{:}Eu_{0.1}^{3+}$, etc.

The progress of combinatorial methodologies, to a large extent, determines the efficiency of phosphors discovery and optimization processes. The latest technique advances, such as visible Fourier transform and time-resolved fluorescence, etc., may integrate with the combinatorial methodology in the near future and contribute significantly to the combinatorial study of phosphors. The developed combinatorial synthesis and high-throughput screening techniques that have been successfully applied in the discovery of new phosphors may also be useful in the search and optimization of other materials systems, providing that the properties of interest could be converted into fluorescence signals.

REFERENCES

1. Blasse, G. and Grabmaier, B. C., *Luminescent Materials*, Springer-Verlag: New York, 1994, Ch. 1.
2. Jandeleit, B., Schaefer, D. J., Powers, T. S., Turner, H. W. and Weinberg, W. H., Combinatorial material and catalysis, *Angew. Chem. Int. Ed.*, 38, 2494, 1999.
3. Justel, T., Nikol, H. and Ronda, C., New developments in the field of luminescent materials for lighting and displays, *Angew. Chem. Int. Ed.*, 37, 3085, 1998.
4. Ronda, C. R., Phosphors for lamps and displays – an applicational view, *J. Alloys. Comp.*, 225, 534, 1995.
5. Grabmaier, B. C., Rossner, W. and Leppert, J., Ceramic scintillators for x-ray computed tomography, *Phys. Status Solidi A-Appl. Res.*, 130, K183, 1992.
6. Sun, T. X., Combinatorial search for advanced luminescence materials, *Biotechnol. Bioeng.*, 61 (4), 193, 1999.
7. Xiang, X. D., Sun, X. D., Briceño, G., Lou, Y. L., Wang, K. A., Chang, H., Wallace-Freedman, W. G., Chen, S. W. and Schultz, P. G., A combinatorial approach to materials discovery, *Science*, 268, 1738, 1995.
8. Briceño, G., Chang, H., Sun, X. D., Schultz, P. G. and Xiang, X. D., A class of cobalt oxide magneto-resistance materials discovered with combinatorial synthesis, *Science*, 270, 273, 1995.
9. Wang, J. S, Yoo, Y., Gao, C., Takeuchi, I., Sun, X. D., Chang, H., Xiang, X. D. and Schultz, P. G., Identification of a blue photoluminescent composite material from a combinatorial library, *Science*, 279, 1712, 1998.
10. Lemmo, A. V., Rose, D. J. and Tisone, T. C., Ink-jet dispensing technology: applications in drug discovery, *Current Opinion in Biotech.*, 9, 615, 1998.
11. Sun, X. D., Wang, K. A., Yoo, Y., Wallace-Freedman, W. G., Gao, C., Xiang, X. D. and Schultz, P. G., Solution-phase synthesis of luminescent materials libraries, *Adv. Mater.*, 9, 1046, 1997.
12. Reddington, E., Sapienza, A., Gurau, B., Viswanathan, R., Sarangapani, S., Smotkin, E. S. and Mallouk, T. E., Combinatorial electrochemistry: a highly parallel, optical screening method for discovery of better electrocatalysts, *Science*, 280, 1735, 1998.
13. Reichenbach, H. M. and McGinn, P. J., Combinatorial solution synthesis and characterization of complex oxide catalyst powders based on the $LaMO_3$ system, *Appl. Catal. A-Gen.*, 244, 101, 2003.
14. Chen, L., Bao, J., Gao, C., Huang, S. X., Liu, C. H. and Liu, W. H., Combinatorial synthesis of insoluble oxide library from ultrafine/nano particle suspension using a drop-on-demand ink-jet delivery system, *J. Comb. Chem.*, 6, 699, 2004.
15. Le, H. P., Progress and trends in ink-jet printing technology, *J. Imaging Sci. Techn.*, 42 (1), 49, 1998.
16. Slade, C. E. and Evans, J. R. G., Freeforming ceramics using a thermal jet printer, *J. Mater. Lett.*, 17, 1669, 1998.

17. Ridley, B. A., Nivi, B. and Jacobson, J. M., All-inorganic field effect transistors fabricated by printing, *Science*, 286, 746, 1999.
18. Mott, M. and Evans, J. R. G., Zirconia/alumina functionally graded material made by ceramic ink jet printing, *Mat. Sci. Eng. A-Struct.*, 271, 344, 1999.
19. Sun, T. X. and Jabbour G. E., Combinatorial screening and optimization of luminescent materials and organic light-emitting devices, *MRS Bulletin,* 27 (4), 309, 2002.
20. Danielson, E., Golden, J. H., McFarland, E. W., Reaves, C. M., Weinberg, W. H. and Wu, X. D., A combinatorial approach to the discovery and optimization of luminescent materials, *Nature*, 389, 944, 1997.
21. Danielson, E., Devenney, M., Giaquinta, D. M., Golden, J. H., Haushalter, R. C., McFarland, E. W., Poojary, D.M., Reaves, C. M., Weinberg, W. H. and Wu, X. D., A rare-earth phosphor containing one-dimensional chains identified through combinatorial methods, *Science,* 279, 837, 1998.
22. Liu, X. N., Cui, H. B., Tang, Y., Huang, S. X., Liu, W. H. and Gao, C., Combinatorial screening for new borophosphate VUV phosphors, *Appl. Surf. Sci.,* 223, 144, 2004.
23. Sohn, K. S., Zeon, Il. W., Chang, H., Kwon Lee, S. and Park, H. D., Combinatorial search for new red phosphors of high efficiency at VUV excitation based on the YRO_4 (R = As, Nb, P, V) system, *Chem. Mater.*, 14, 2140, 2002.
24. Shiononoya, S. and Yen, W. M., *Phosphor Handbook*, CRC Press, Boca Raton, FL, 1998.
25. Mordkovich, V. Z., Hayashi, H., Haemori, M., Fukumura, T. and Kawasaki, M., Discovery and optimization of new ZnO-Based phosphors using a combinatorial method, *Adv. Funct. Mater.*, 13 (7), 519, 2003.
26. Hayashi, H., Ishizaka, A., Haemori, M. and Koinuma, H., Bright blue phosphors in $ZnO-WO_3$ binary system discovered through combinatorial methodology, *Appl. Phys. Lett.,* 82 (9), 1365, 2003.

22 Combinatorial Screening and Optimization of Phosphors for Flat Panel Displays and Lightings

Kee-Sun Sohn and Namsoo Shin

CONTENTS

22.1 INTRODUCTION

A large number of unexplored chemical compositions that could yield materials with potentially useful properties that could be used in a myriad of applications exist. In fact, fewer than 1% of all possible ternary compounds and fewer than 0.01% of all possible quaternary compounds have been synthesized so far. Of the conventional inorganic methods available, combinatorial chemistry has the most potential for significantly increasing the rate at which new compositions are synthesized and characterized. Furthermore, trends in physical properties and performance over large regions of compositional space may yield important guidance in developing the fundamental understanding and theories that are, at present, lacking.

Combinatorial chemistry is an established procedure that offers efficient experimental methodology for the systematic screening of a variety of functional materials. In particular, the applicability of combinatorial chemistry has recently been extended to inorganic luminescent materials. In cases where the combinatorial chemistry technique has been applied to phosphor materials, most of the approaches were concerned with thin-film technologies, the fundamental goal of which is the development of a larger library of small substrates.[1–6] Conventional solid-state combinatorial chemistry based on thin-film technology has been carried out on a very large scale in terms of the number of available compositions in a batch, namely, a huge number of compounds are given on the small substrate and the products are synthesized one at a time. In fact, it is now possible to prepare more than a 2000-composition library on a 1-inch square substrate using masking technology. It should, however, be noted that such an approach has limitations. First, the amount of each compound is too small to permit its conventional characterization, and the luminescent properties in the thin-film state may be different from the powder properties. Because phosphors in powdered form are typically used in most application systems, a powder-based approach could be more promising for phosphor research.

In this regard, we are concerned as to whether solution-based combinatorial chemistry can be applied to the synthesis of phosphor powders. A systematic experimental procedure has been developed to generate solution-based libraries of inorganic luminescent materials. In solution-based combinatorial chemistry, the reagents are typically delivered individually to each sample site. In order to create these libraries in an efficient manner, with respect to process speed and accuracy, we developed a scanning multi-injection delivery system for the rapid and accurate delivery of several hundred microliter volumes of precursor solutions to the sample sites. As a result, a material in powdered form can be obtained in sufficient quantities that will permit a conventional characterization process to be carried out, without the need for any special additional instrumentation. Thus we were able to construct a several hundred-composition library in a batch and only two or three days were required to complete the synthesis and characterization of a batch. However, it is true that our solution combinatorial chemistry system of a small library (only about 100 compositions are available per batch) is not comparable to a thin-film combinatorial chemistry system, which is typically comprised of more than 2000 compositions in view of screening efficiency. However, the creation of these libraries is relatively more practical from the standpoint of accuracy.

Even though the most advanced combinatorial chemistry methodology developed so far was adopted for actual screening processes, it would be practically impossible to examine the complete composition range of a multicomponent system without a new optimization process that surpasses any other conventional high-throughput combinatorial chemistry method in terms of efficiency. In order to improve these shortcomings, computational optimization processes such as genetic algorithm-assisted combinatorial chemistry (GACC) has recently attracted interest due to its ability to compensate for the weak points in the traditional high-throughput combinatorial chemistry, thereby enhancing the efficiency of exploring new materials for specific purposes.[7–11] In fact, this strategy and methodology originated in the pharmaceutical research area.[7–9] When the area of inorganic material synthesis is concerned in association with GACC, these strategies were recently used for the development of heterogeneous catalytic materials.[12,13] Such approaches were found to be very efficient and promising in the context of a search for new inorganic catalytic materials. It is our opinion that inorganic phosphors are more suitable for the GACC method, based on the fact that the screening process can be greatly facilitated compared to that for catalytic materials. Even though our solution combinatorial chemistry system of a small library is not comparable to a thin-film combinatorial chemistry system, from the standpoint of screening efficiency, the introduction of the GACC to our small library solution combinatorial chemistry system made it possible to promote the efficiency of ours to a level comparable to that of the high-throughput thin-film combinatorial chemistry system.

In an attempt to search for new promising phosphors for plasma display panel (PDP) and light-emitting diodes (LED), a solution-based combinatorial chemistry method involving high-throughput synthesis and characterization has been employed.[14–21] In addition to this conventional combinatorial chemistry approach, a more advanced screening strategy, in which combinatorial chemistry and computational heuristics are hybridized, such as genetic algorithm, was also employed.[22,23] Several examples of a phosphor search based on the above-described conventional high-throughput combinatorial chemistry as well as GACC are addressed in this chapter. Sections 22.2 and 22.3 deal with the conventional high-throughput combinatorial search for new phosphors for PDP applications without employing any computational optimization strategy, and thereby providing the results from the combinatorial screening based on a Eu^{3+}-doped $Y(As,Nb,P,V)O_4$ quaternary system and a Tb^{3+}-doped $CaO\text{-}Gd_2O_3\text{-}Al_2O_3$ ternary system. In particular, phase identifications for optimum compounds obtained from combinatorial screening are addressed in Section 22.3. Section 22.4 is focused more on computational optimization rather than conventional high-throughput combinatorial chemistry. The GACC process was applied to a search for new red phosphors for tricolor light-emitting diodes (LED) based on the alkali earth boro-silicate system described in Section 22.4.

22.2 Eu^{3+}-DOPED $Y(As,NB,P,V)O_4$ QUATERNARY SYSTEM

22.2.1   Introduction

Based on the possibility that there might exist a mixed composition of $Y(As,Nb,P,V)O_4$ system of better luminescent performance than the single-compound phosphors at the VUV excitation, a solution combinatorial chemistry synthesis and characterization was employed. Quaternary and ternary combinatorial libraries were designed to implement an efficient screening process.

Several Eu^{3+} doped borates such as $YBO_3{:}Eu^{3+}$, $LuBO_3{:}Eu^{3+}$, $ScBO_3{:}Eu^{3+}$, and $(YGd)BO_3{:}Eu^{3+}$ have been considered as red phosphors for the PDP application. The starting point of the present investigation lies in the possibility that some other oxide compounds could be also applied to the PDP. In this regard, yttrium-based YRO_4 compounds such as $YAsO_4$, $YNbO_4$, YPO_4, and YVO_4 could be potential candidates and worth investigating. These compounds were developed long ago and the luminescent properties have been studied extensively.[24,25] But the former investigations dealing with these materials were focused either on the photoluminescence under 254 nm excitation or on the cathodoluminescence. One has no doubt that the development of new phosphors to be used for the PDP application is of primary concern in the present investigation, so that the photoluminescence at the 147 nm excitation, which is adopted as an excitation source in the PDP application, should be taken into consideration. In fact, each single compound has been well developed in terms of luminescence efficiency at the 254 nm excitation (PL) and at the excitation by the electron bombardment (CL). But there has been little consideration with regard to the luminescent property of mixed compound. By considering the assumption that the highest luminescence is not always elicited from a single-phase line compound, a mixed compound could be more suited to the host of Eu^{3+} emission.

A quaternary library was developed in terms of the luminance and color chromaticity at the 147 nm excitation, so that it could be possible to screen all the composition that the four above-mentioned compounds constitute. A finer screening using a ternary library was also carried out around the optimum composition obtained from the quaternary library. As a result of the screening process, a final composition was obtained, which shows almost tantamount luminescent efficiency to the commercially available red phosphor.

22.2.2   Spectral and Structural Analyses of Base Compounds

Prior to the combinatorial approach, spectral analysis was performed for the base compounds (the apex compounds of the quaternary library). Figure 22.1 shows the emission and excitation spectra of $YAsO_4{:}Eu^{3+}$, $YNbO_4{:}Eu^{3+}$, $YPO_4{:}Eu^{3+}$, and $YVO_4{:}Eu^{3+}$. The excitation spectra of ${}^5D_0\text{-}{}^7F_2$ emission were measured in the range from 140 to 350 nm. The results shows that only the $YPO_4{:}Eu^{3+}$

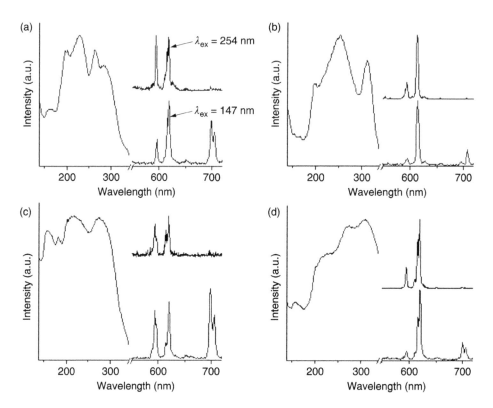

FIGURE 22.1 Excitation and emission spectra of each constituent compound (a) $YAsO_4:Eu^{3+}$, (b) $YNbO_4$:Eu^{3+}, (c) $YPO_4:Eu^{3+}$, and (d)$YVO_4:Eu^{3+}$. The excitation spectra were detected with the emission probe fixed at 613 nm and emission spectra was measured both under the 254 (upper) and 147 (lower) nm excitations.

compound exhibits a considerable emission at the VUV excitation. On the other hand, the excitation spectra of all the other compounds show a dramatic drop in the VUV range below 200 nm.

If the 147 nm excitation were concerned, it would be quite reasonable to assume that the excitation energy is absorbed firstly by the host lattice, which involves the transition between 4d-like states of Y and 2p-like states of O. The absorbed energy may then be transferred to RO_4 groups and lastly transferred to Eu^{3+} center. Otherwise, the excitation energy is absorbed by the host lattice and transferred directly from the host lattice to the Eu^{3+} center. Among the four YRO_4s, only the YPO_4 host pertains to the latter case, while all the others belong to the former case, namely, the YPO_4 host never undergoes the energy transfer from the host to PO_4 group en route, since the 2p- and 3d-like states of P are located far above the 4d-like states of Y. The electronic structure of the YPO_4 host is discriminated from the other compounds, which is associated with the information taken from the excitation spectra, wherein only the YPO_4 host shows strong absorption in the VUV range. More detailed discussions were presented in reference 17 where the first principal calculation of density of state (PDOS) was dealt with.

As for the emission spectra, we found that the relative intensity of $^5D_0$-^7F_j emission peaks, so-called branch ratio, varies with the host lattice. Such a variation is well known to be due to the different crystal field effect. The relative intensity of $^5D_0$-^7F_j emission peaks is an important factor that plays a decisive role in calculating the color chromaticity from the practical point of view. First of all, the smaller the $^7F_1/^7F_2$ ratio, the closer to the optimum value the color chromaticity. On the other hand, the branch ratio varies with not only the host lattice but also with the excitation light wavelength. The $^5D_0$-$^7F_4$ emission peaks at around 700 nm become conspicuous for all the host lattices when the 147 nm excitation is adopted.

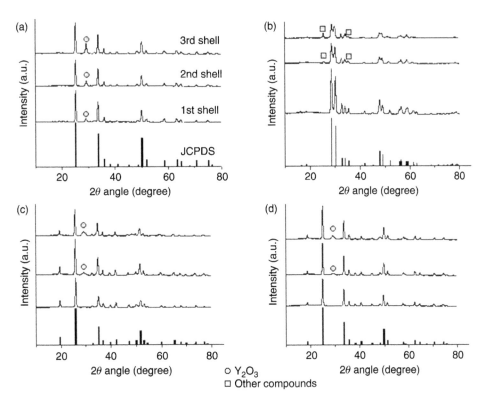

FIGURE 22.2 The XRD patterns of the apex compounds from the first to third shells by the side of (a) $YAsO_4:Eu^{3+}$, (b) $YNbO_4:Eu^{3+}$, (c) $YPO_4:Eu^{3+}$, and (d) $YVO_4:Eu^{3+}$.

The crystalline structures of base compounds were monitored. Even though the $YNbO_4$ among the four base compounds has a different crystalline structure (Fergusonite structure[26]), the other three base compounds such as $YAsO_4$, YPO_4 and YVO_4, are all isomorphous (Xenotime structure[27,28]). Figure 22.2(a)–(d) shows the XRD patterns of the apex compounds from the first to third shells on each constituent's side. The samples used for the XRD are all taken from the batch fired at 1200°C. The XRD patterns of three isomorphous compounds are quite similar, whereas $YNbO_4$ is discriminated from the others. The inclusion of $YNbO_4$ may constitute an obstacle to make single-phase solid solutions. In fact, the trace of $YNbO_4$ can be observed in most apex compounds of inner shells even if the peak intensity is not high. The XRD pattern of Nb-rich compounds (Figure 22.2(b)) has also the trace of the other apex compounds. In addition, a weak Y_2O_3 peak is also detectable for all the apex compounds of inner shells even though it does not appear in the apex compounds of the outermost shell except for the case of $YAsO_4$. What interests us most is that the XRD pattern of As-rich compounds includes a relatively high Y_2O_3 peak. It is most likely that the volatilization of As is responsible for the Y_2O_3 peak. Arsenic is more volatile than the other species, so that we predict that the incorporation of $YAsO_4$ will not be favorable for enhancing the luminescent efficiency. $YAsO_4$ will be omitted from the final screening step for such a reason.

22.2.3 Screening Results

Figure 22.3 shows the quaternary combinatorial library in terms of luminance at the 147 nm excitation for a $Y(As,Nb,P,V)O_4$ system fired at 1200°C. The Eu^{3+} doping content was fixed as 0.1 mol, which was determined from the independent concentration quenching experiment for each base

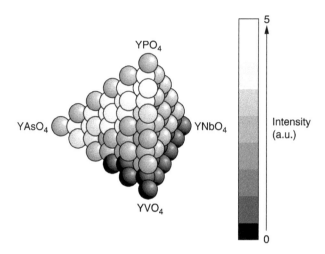

FIGURE 22.3 The quaternary combinatorial library in terms of luminance at the 147 nm excitation for a $Y(As,Nb,P,V)O_4$ system fired at 1200°C.

compound. Fortunately the optimum Eu^{3+} contents are not considerably different from one another as far as the 147 nm is adopted as an excitation light wavelength, i.e., they are determined within the range from 0.08 to 0.12 and the concentration quenching curves do not show a sharp peak in the vicinity of the critical point.

The grayness of each sphere in the library represents the measured luminance of Eu^{3+} fluorescence either under 254 or 147 nm excitation. The brighter the sphere, the higher the luminance. Figure 22.4 shows the inner shells of Figure 22.3 and the apex compositions summarized in the table below the schematics. In fact, although more libraries were obtained from some other batches fired at 1100 and 1300°C, they are omitted because the maximum luminance was found in the 1200°C batch and the overall luminescence level of them is lower than the 1200°C batch. The maximum luminance is obtained at the composition $Y_{0.9}(P_{0.83}As_{0.06}V_{0.06}Nb_{0.06})O_4{:}Eu_{0.1}{}^{3+}$, which is the apex compound of the second shell.

Not to mention the luminance, the color chromaticity also plays a significant role in the assessment of phosphors. Two quaternary combinatorial libraries in terms of the CIE color chromaticity x and y, which are calculated using the spectrum data under the 147 nm excitation, are represented in Figure 22.5(a) and (b), respectively. In order to inquire into the reliability of our calculation, the CIE color chromaticity x and y values of the commercial $(Y,Gd)BO_3{:}Eu^{3+}$ phosphor are also estimated to be 0.6507 and 0.349, which are in good agreement with the reported values. The x and y value in the library varies in the range 0.6561–0.6721 and 0.3277–0.3436, respectively. The $Y(P_{0.83}As_{0.06}V_{0.06}Nb_{0.06})O_4$ composition of the maximum luminance exhibits $x = 0.6603$ and $y = 0.3394$. These values are acceptable by the fact that the required values for the PDP application are $x = 0.67$ and $y = 0.33$.

Strictly speaking, the quaternary combinatorial library adopted in the present investigation is too sparse to search for an optimum composition at a time. There should be more promising compositions than the composition $Y(P_{0.83}As_{0.06}V_{0.06}Nb_{0.06})O_4$, if a finer screening was carried out at the vicinity of this composition. As a preliminary step prior to the final combinatorial screening, several compositions were examined around the $Y(P_{0.83}As_{0.06}V_{0.06}Nb_{0.06})O_4$, all of which are not provided by the quaternary combinatorial library in Figure 22.3. The solution combinatorial chemistry method was also used in this additional experiment, which involves $Y(P_{1-a-b-c}As_aV_bNb_c)O_4$ compositions with a, b, and c increased by 0.03 from 0 to 0.15 mol. Even though the resultant

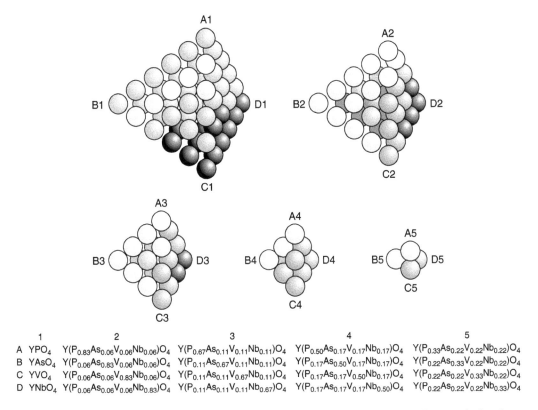

	1	2	3	4	5
A	YPO_4	$Y(P_{0.83}As_{0.06}V_{0.06}Nb_{0.06})O_4$	$Y(P_{0.67}As_{0.11}V_{0.11}Nb_{0.11})O_4$	$Y(P_{0.50}As_{0.17}V_{0.17}Nb_{0.17})O_4$	$Y(P_{0.33}As_{0.22}V_{0.22}Nb_{0.22})O_4$
B	$YAsO_4$	$Y(P_{0.06}As_{0.83}V_{0.06}Nb_{0.06})O_4$	$Y(P_{0.11}As_{0.67}V_{0.11}Nb_{0.11})O_4$	$Y(P_{0.17}As_{0.50}V_{0.17}Nb_{0.17})O_4$	$Y(P_{0.22}As_{0.33}V_{0.22}Nb_{0.22})O_4$
C	YVO_4	$Y(P_{0.06}As_{0.06}V_{0.83}Nb_{0.06})O_4$	$Y(P_{0.11}As_{0.11}V_{0.67}Nb_{0.11})O_4$	$Y(P_{0.17}As_{0.17}V_{0.50}Nb_{0.17})O_4$	$Y(P_{0.22}As_{0.22}V_{0.33}Nb_{0.22})O_4$
D	$YNbO_4$	$Y(P_{0.06}As_{0.06}V_{0.06}Nb_{0.83})O_4$	$Y(P_{0.11}As_{0.11}V_{0.11}Nb_{0.67})O_4$	$Y(P_{0.17}As_{0.17}V_{0.17}Nb_{0.50})O_4$	$Y(P_{0.22}As_{0.22}V_{0.22}Nb_{0.33})O_4$

FIGURE 22.4 The quaternary combinatorial library in terms of luminance at the 147 nm excitation for a $Y(As,Nb,P,V)O_4$ system fired at 1200°C, showing all the inner shells.

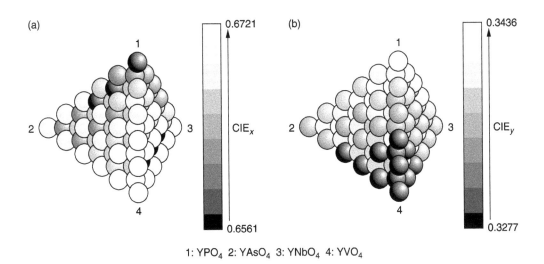

1: YPO_4 2: $YAsO_4$ 3: $YNbO_4$ 4: YVO_4

FIGURE 22.5 The quaternary combinatorial library in terms of color chromaticity (a) x and (b) y at the 147 nm excitation.

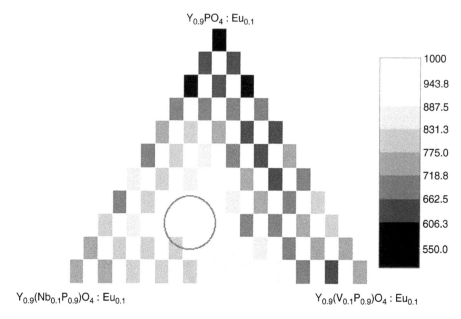

FIGURE 22.6 Ternary combinatorial library around the YPO_4-rich region.

combinatorial library is not present here for the compactness of the chapter, two very important points were confirmed by this additional experiment. The most important point is that the incorporation of As is not favorable for enhancing the luminance at all, so that the As was removed from the library. This could be ascribed to the volatilization of the As ion during the firing. The other point is that if the total doping content of V and Nb exceeds 0.1 mol, then the luminance is deteriorated no matter what combination is taken. Consequently, a ternary combinatorial library was used for the final screening, in which the apex compounds are $YPO_4:Eu^{3+}$, $Y(P_{0.9}V_{0.1})O_4:Eu^{3+}$, and $Y(P_{0.9}Nb_{0.1})O_4:Eu^{3+}$. The screening result in Figure 22.6 shows that the maximum luminescence was detected at $Y(P_{0.92}V_{0.03}Nb_{0.05})O_4$ and the luminance is comparable to the commercially available $(Y,Gd)BO_3:Eu^{3+}$ phosphor.

22.2.4 The Optimum Composition

Now that the optimum composition was obtained to be $Y(P_{0.92}V_{0.03}Nb_{0.05})O_4$, the next step is to inquire into the reason why this composition shows better luminescent characteristics. As discussed already, it is postulated that the YPO_4-based compounds show a more efficient absorption in the VUV range. The incorporation of a small amount of V and Nb constitutes VO_4 and NbO_4 groups, which could act as a bridge between the host absorption by the Y-O charge transfer and the Eu^{3+} center. Such a bridging effect could facilitate the energy transfer process. Thus the incorporation of a small amount of V and Nb enhances the luminescence of Eu^{3+} center by the assist of the VO_4 and NbO_4 bridges. In fact, there has already been an attempt to make a mixed compound $(Y(P,V)O_4)$ with the intension of enhancing the blue (or red) emissions of VO_4 group (or Eu^{3+} center) under the UV excitation.[24,25] The $Y(P_{1-x}V_x)O_4$ phosphors, with $x = 0.25$–0.35 mol, was found to promote the absorption (and emission) efficiency at the 254 nm excitation by the VO_4 formation. In the presence of Eu^{3+} center, the VO_4 group in the YPO_4 host could be considered by absorbing the 254 nm UV light and then down-converting it to an energy of blue light and eventually transferring it to the Eu^{3+} center non-radiatively. On the other hand, the present investigation deals with the 147 nm VUV light excitation, so that neither VO_4 nor NbO_4 group directly absorbs the excitation energy but act as a bridge connecting the YPO_4 host and the Eu^{3+} center. There has been no consideration as regards the role of the YPO_4 host in the case of the 254 nm excitation. The YPO_4 host cannot absorb the 254 nm UV light efficiently, since both the charge transfer bands, i.e., Y-O and P-O charge transfer, are located in much higher energy region than the 254 nm. It is certain that the direct energy transfer from the Y-O group to the Eu^{3+} center is inefficient. That is why the non-mixed YPO_4 host does not allow the Eu^{3+} center to make efficient luminescence even if the YPO_4 host has an energy absorption band appropriate for the VUV excitation.

It should be noted that the phosphors of interest exhibit significantly different responses at the UV and VUV excitations. For instance, the optimum V doping content has been known to be relatively high ($x = 0.25$–0.35 mol) at the 254 nm excitation,[24,25] whereas the optimum doping content of V is determined to be 0.08 mol at the 147 nm excitation, which can be chosen on the YPO_4-$Y(P_{0.9}V_{0.1})O_4$ line in Figure 22.6. The total doping content of V and Nb for the best composition does not also exceed 0.1 mol at the 147 nm excitation. Such a difference could be attributed to a different route through which the excitation energy reaches the fluorescent level Eu^{3+} of activator. The excitation light wavelength-dependent concentration quenching behavior has been also discussed when the optimum Eu^{3+} doping content was dealt with in the preceding subsection, along with the previously observed similar behavior in Mn^{2+} doped Zn_2SiO_4 phosphors.[25,26]

The conventional concentration quenching has been interpreted only by considering the energy transfer (or migration) between VO_4 groups. If the direct excitation on the VO_4 group was adopted, i.e., 254 nm excitation, the conventional analysis would take effect. But the present investigation is dealing with the 147 nm excitation so that another effect such as the host–VO_4 transfer would contribute more to the concentration quenching of VO_4 groups, which leads to the lower optimum VO_4 content at the 147 nm excitation. This means that the host–VO_4 transfer should be a rate-controlling step for the whole process.

22.2.5 Conclusion

An optimum red phosphor for the PDP application was found to be $Y_{0.9}(P_{0.92}V_{0.03}Nb_{0.05})O_4:Eu^{3+}$ by using the combinatorial chemistry method. The luminescent efficiency and CIE color chromaticity at the 147 nm VUV excitation is comparable to the conventional $(Y,Gd)BO_3:Eu^{3+}$ phosphor. It was suggested that the incorporation of a small amount of V and Nb into the YPO_4 host should act as a bridge for the energy transfer and thereby enhance the luminescent efficiency. The transfer route of the VUV excitation energy was Y-O charge transfer $\rightarrow$ R-O charge transfer in the RO_4 (R = V, Nb) group Eu^{3+} center.

22.3 Tb^{3+}-DOPED CaO-Gd$_2$O$_3$-Al$_2$O$_3$ TERNARY SYSTEM

22.3.1 Introduction

In an attempt to apply newly found phosphors to the plasma display panel (PDP), Tb^{3+} activated CaO-Gd_2O_3-Al_2O_3 ternary system was screened in terms of the photoluminescence (PL) at the vacuum ultraviolet (VUV) excitation (147 nm). In particular, a polymerized-complex precursor route was adopted, which has been widely applied to the synthesis of a variety of materials such as various inorganic functional materials.[30] The polymeric precursor method has been known to be a simple, cost-effective and versatile low-temperature route. The general idea of the method is the automatic distribution of cations throughout the polymer structure, so that we can inhibit their segregation and precipitation from the solution. Upon heating these organic materials, multicomponent oxides are obtained at a relatively low temperature. As a result, we synthesized the phosphors of fine particles and low agglomerated particles. We examined the luminance, crystallinity, and morphology of the powder samples produced by the combinatorial polymerized-complex method.

A ternary library was developed based on the relative PL efficiency (or relative luminance) at the 147 nm excitation, so that it could be possible to screen all the compositions in the CaO-Gd_2O_3-Al_2O_3 system. Even though the Gd_2O_3 has been well known as a good host for Eu^{3+} emission,[31,32] it is rather inappropriate for the Tb^{3+} emission. This is because the crystal field around Tb^{3+} ions in the Gd_2O_3 host is not favorable for the major radiative transition, $^5D_4 \rightarrow {}^7F_j$. In an attempt to improve the Tb^{3+} emission, we introduced co-dopants such as Ca and Al into the Gd_2O_3 host since we had expected that the Ca and Al co-doping might have altered the crystal field effect. But the conspicuous improvement could not be found at the co-doping level. The co-doping level represents the range of $0 < x < 0.2$ in the case of $Gd_{2-x}(Ca$ or $Al)_xO_3$. On the contrary, we achieved a significant enhancement in luminance beyond the co-doping level. New stoichiometric compounds, which are composed of Ca, Al, and Gd, were created beyond the co-doping level, and a certain mixture of theses compounds exhibited the highest PL efficiency.

22.3.2 SCREENING RESULTS AND PHASE IDENTIFICATION

Figure 22.7 shows a ternary library at 1100°C in terms of the relative luminance obtained from the corresponding emission spectrum measured at the 147 nm excitation. The luminance level is represented as shaded circle, wherein the darker the circle, the higher the luminance. The promising composition range was determined as $0.2 < Ca < 0.25$, $0.25 < Gd < 0.3$, and $0.45 < Al < 0.5$, regardless of firing temperature. Several promising compositions in the library are numbered for indexing.

The highest luminance was obtained at the composition of $Ca_{0.25}(GdTb)_{0.25}Al_{0.5}O_\delta$ in the library, which was numbered as 3. An additional experiment revealed that the firing temperature of solid-state reactions should be raised to at least 1500°C to produce a similar level of luminance. There is no doubt that the low firing temperature is the most essential advantage of the polymeric complex method. The precursor solutions are mixed at the molecular level, thereby leading to

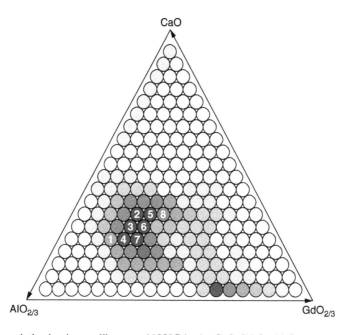

FIGURE 22.7 The relative luminance library at 1100°C in the CaO-Gd_2O_3-Al_2O_3 ternary system.

the homogeneous distribution of reactants and the drop in reaction temperature. The maximum luminance that we could attain in the 1100°C library is as good as 70% of the commercially available $Zn_2SiO_4:Mn^{2+}$ green phosphor currently used for PDP application.

The numbered phosphor samples were examined by XRD so as to understand the stoichiometry of the constituent. Figure 22.8 shows the XRD pattern of sample '3', which exhibits the highest luminance. We found that sample '3' consists mostly of $CaGdAlO_4$ and $CaGdAl_3O_7$ phases. We also examined the other numbered samples by XRD and confirmed that the numbered compounds consist of two major constituent compounds, $CaGdAlO_4$ and $CaGdAl_3O_7$, and also a minor compound identified as $GdAlO_3$.

In order to take into account the spectral behavior of the major constituent compounds, we synthesized the constituent compounds by the conventional polymeric precursor method. Figure 22.9 shows the emission and excitation spectra of $CaGdAlO_4$ and $CaGdAl_3O_7$. Unlike these two compounds, the emission and excitation spectra of $GdAlO_3$ compound could not be measured properly because of the very low level of emission intensity. Thus, the presence of $GdAlO_3$ phase would not be favorable for enhancing the luminance. As can be seen in Figure 22.9, the excitation spectra of $CaGdAlO_4$ and $CaGdAl_3O_7$ show that certain degrees of absorption exist in the VUV range below 150 nm. The excitation spectra of $GdAlO_3$ reveal no absorption band in the range below 150 nm. Irrespective of whether the $GdAlO_3$ phase is included or not, a more important key point that we have to address is the ratio of $CaGdAlO_4$ and $GdAl_3O_7$. In order to assess the exact value of the volume fraction of each compound, a quantitative estimation could be obtained using the Rietveld method.[33] We used the Fullprof software[34] for the correct assessment. Figure 22.8 shows the observed and calculated XRD patterns for sample 3. As can be seen, both the observed and calculated results are very well matched. The exact volume fraction is summarized in Table 22.1. The result in Table 22.1 shows that we have to pursue a certain ratio of $CaGdAlO_4$ and $CaGdAl_3O_7$ containing as small an amount of $GdAlO_3$ as possible in order to enhance the luminance of the mixed compounds. In particular, due to the fact that sample 3 shows the highest luminance, the ratio of $CaGdAl_3O_7$ to $CaGdAlO_4$ phase should be around 3 to achieve high luminance.

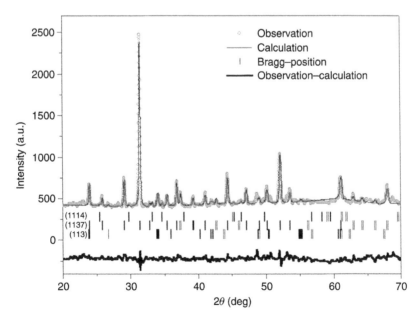

FIGURE 22.8 X-ray diffraction patterns of $Ca_{0.25}(GdTb)_{0.25}Al_{0.50}O_\delta$ (sample 3). The solid line represents the calculated result based on the Rietveld method and the Bragg position of $CaGdAlO_4$ (1114), $CaGdAl_3O_7$ (1137), and $GdAlO_3$ (113) are also indicated at the bottom.

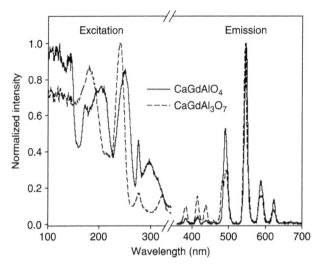

FIGURE 22.9 Excitation and emission spectra of Tb-doped $CaGdAlO_4$ and $CaGdAl_3O_7$.

TABLE 22.1
Constituent Compounds and Relative Luminance for Several Compositions

	Composition			Constituent Compound			
Sample No.	Ca	Gd + Tb	Al	$CaGdAlO_4$	$CaGdAl_3O_7$	$GdAlO_3$	Relative Luminance
1	0.2	0.20	0.60	2.11	85.4	12.5	78
2	0.3	0.25	0.45	51.0	48.0	1.00	72
3	0.25	0.25	0.50	23.2	73.3	3.51	100
4	0.20	0.25	0.55	5.76	76.4	17.8	87
5	0.30	0.30	0.40	61.0	35.5	3.50	75
6	0.25	0.30	0.45	32.5	44.0	23.5	82
7	0.20	0.30	0.50	12.6	51.7	35.7	85
8	0.30	0.35	0.35	73.8	19.1	7.10	50

22.3.3   THE OPTIMUM COMPOSITION

The luminance of the samples adjacent to sample '3' is much higher than the rest of the composition in the library. More importantly, their luminance is also higher than any of the single-phase constituent compounds such as $CaGdAlO_4$ and $CaGdAl_3O_7$. It is necessary to know the reason why the mixed compounds show much higher luminance than each constituent compound. Firstly, we suspected that a misleading choice of Tb^{3+} concentration was responsible for the higher luminance of the mixed compound. If Tb^{3+} ions were distributed unevenly to each constituent compound during the synthesis, this might lead to the optimum Tb^{3+} concentration in each constituent compound. However, the concentration quenching experiments both in the $CaGdAlO_4$ and $CaGdAl_3O_7$ compounds, varying the Tb^{3+} concentration from 0.001 to 0.1 mol, revealed that this suspicion was ruled out. Both the concentration quenching plots are nearly coincident and the plots are so broad around the optimum Tb^{3+} concentration that the above speculation of uneven partitioning is groundless.[16] In addition, we synthesized separately the $CaGdAlO_4$ and $CaGdAl_3O_7$ compounds with an optimum

Tb^{3+} concentration for each, and the luminance of them was compared to the mixture of them. This confirmative, auxiliary experiment also revealed that the luminance of constituent compounds prepared separately was not comparable to their mixture.

There are several extrinsic factors that could have a great influence on the luminance, for example, particle size and shape, the surface state, and so on. We scrutinized the SEM micrographs of mixed and constituent compounds to find any clue for the higher luminance of mixed compound. Even though we did not present the micrographs, there was no conspicuous difference between mixed and constituent compounds (spheroidal particle of size ~0.5 μm). Accordingly we have to find another reason for the higher luminance of the mixed compound. One of plausible reasons for the higher luminance of mixed compound could be radiation trapping. Radiation trapping has been known to arise in the powder form and leads to the luminescence quenching. But this hypothesis is unlikely, because no direct evidence was obtained from the decay time measurement for several Tb^{3+} concentrations. If radiation trapping was active in the constituent compounds, the decay time should have increased with the Tb^{3+} concentration. However, such behavior was not detected in the experiment. In summary, none of the above-mentioned hypotheses were validated at the current stage. Thus the higher luminance of mixed compound is still in question.

22.3.4 CONCLUSIONS

A combinatorial polymerized-complex method was used in order to search for new Tb^{3+}-activated green phosphors based on CaO-Gd_2O_3-Al_2O_3 ternary system. The highest luminance was obtained at the composition of $Ca_{0.25}(GdTb)_{0.25}Al_{0.5}O\delta$ in the 1100°C library. We confirmed that the newly found phosphors consist of two major constituent compounds such as $CaGdAlO_4$ and $CaGdAl_3O_7$ and also containing a minor compound identified as $GdAlO_3$. The mixed compounds show a lot higher luminance than any of the constituent compounds. There exists an optimum fraction of $CaGdAlO_4$ and $CaGdAl_3O_7$ phases, that is, a $CaGdAl_3O_7/CaGdAlO_4$ ratio of 3 was found to be favorable for high luminance. We confirmed that a non-stoichiometric compound could show much higher luminance than the stoichiometric line compounds in the $CaGdAlO_4/CaGdAl_3O_7$ mixture system.

22.4 GENETIC ALGORITHM-ASSISTED COMBINATORIAL CHEMISTRY (GACC)

22.4.1 INTRODUCTION

An evolutionary optimization process hybridizing a genetic algorithm and combinatorial chemistry (combi-chem) was originated in the pharmaceutical research field.[7–10] Baerns et al.[12,13] recently used GACC for the development of heterogeneous catalytic materials. The Baerns et al. approach was found to be very efficient and promising in a search for new inorganic catalytic materials. We also employed GACC, which was tailored exclusively for the development of LED phosphors with a high luminescent efficiency, when excited by soft ultraviolet irradiation. The ultimate goal of our study was to develop oxide red phosphors, which are suitable for tricolor white light-emitting diodes (LED). To accomplish this, a computational evolutionary optimization process was adopted to screen a Eu^{3+}-doped alkali earth borosilicate system. The genetic algorithm is a well-known, very efficient heuristic optimization method and combi-chem is also a powerful tool for use in an actual experimental optimization process. Therefore the combination of a genetic algorithm and combi-chem would enhance the searching efficiency when applied to phosphor screening.

Light-emitting diodes (LEDs) have attracted interest based on the possibility of being used as general lighting devices. The development of white color is significant to expand LED applications towards general lighting. One of the very promising ways of producing white light is a tricolor white LED consisting of an LED device that emits either soft UV or blue light and RGB phosphors that down-convert the soft UV or blue light into visible light.[35–37] Our objective was to develop new oxide RGB phosphors. As a first step red oxide phosphors were developed in this chapter. Blue and green phosphors will be also developed in the near future based on the same methodology and a similar chemical composition system. We initially adopted a Eu^{3+} activated alkali earth borosilicate system as a red phosphor for tricolor white LED. This system includes seven cations, including Eu, Mg, Ca, Sr, Ba, B, and Si. There are a large number of stoichiometric compounds (single-phase line compounds) and an infinite number of their solid solutions in this system as well.

It is impossible to track down all the compositions in this seven-dimensional system, so that we need an optimization strategy, which can relieve experimental burdens. In this regard, there have been some efforts to develop a quantitative structure property relationship (QSPR) based on trained artificial neural networks (TANN).[38–40] It should be, however, noted that the TANN was versatile for QSPR but useless for global optimization. The prediction by the TANN can never yield much higher activity than the initial data used for training by considering the nature of artificial neural networks. This means that the TANN can never predict an optimum compound, which is hugely better than the best member in the initial training data. On the other hand, the GACC that we adopted here is capable of finding a real, global optimum irrespective of initial data set, even though much more experimental implementation is required. Therefore, the GACC could be a promising optimization strategy in terms of screening efficiency and accuracy. In this section, vertical GACC simulations and an actual synthesis were carried out and promising red phosphors for tricolor white LED applications were optimized.

22.4.2 GACC AND SIMULATIONS

The difference between conventional high-throughput combi-chem and GACC is obvious. The regular combinatorial synthesis method is based on a systematic combinatorial library (= large number of orderly arrayed compositions), in which the composition of adjacent members is in a certain sequence. Thus, screening should be carried out over a huge number of compositions in order, and thereby covering a certain range of composition. As the number of constituent elements increases, however, the number of compositions will increase drastically. Thus, it would be practically impossible to track down the whole range of compositions that are of concern. On the other hand, the GACC begins with non-orderly arrayed compositions. Namely, the combinatorial library was constituted randomly in advance. It could be acceptable even if the population of this library is much smaller than the regular high-throughput thin-film combi-chem system. A library of 108 random compositions was adopted in the present GACC process, whereas the typical population of regular combi-chem is several thousand compositions or more. This randomly chosen library is called "the first generation". Evolutionary operations such as elitism, selection, crossover, and mutation were then applied to this first generation using the actually measured luminance values of all the members in the first generation. This computational evolutionary process yielded another new library of the same number of compositions as the first one, which is called the second generation. The second generation will show a somewhat improved luminance. The same processing was done on the second generation and yielded the third generation and so on. This process will improve the luminance of all the members in the generation as the generation number increases, finally leading us to the optimum. Namely, the GACC includes repetitions of the synthesis/measurement based on the combi-chem technique and the computational evolutionary operation based on the measurement results. It can be thus summarized that the GACC consists of two parts; one is the experiment part and the other is the intermediate computation part.

The root of GACC method is "the randomness", so that there was no specific reason to use the Eu-doped alkali borosilicate system as the base composition in the present investigation. Our understanding is that it could be acceptable even if several more starting elements were added to this system. This addition, however, would lead to deterioration in the efficiency of the method. If the experimental situation allowed, it would be possible to employ all the elements in the periodic table as a base composition. If so, the GACC would weed out a lot of useless elements automatically by the evolutionary principle. In fact, such a weeding out took place in the present investigation. Namely, Sr was removed at an early stage in this respect, which will be confirmed below in the results section. Prior to the experiment involving synthesis, characterization, and computational evolutionary process, simulations were carried out using three different hypothetical objective functions. It is not efficient to examine the effect of the calibration of model parameters, such as selection, crossover, and mutation method, etc., on optimization speed and correctness because the optimum choice of model parameters is strongly dependent on the objective function, but we are totally unaware of the objective function of concern in the actual experiment. Baerns et al.[12,13] have already investigated the effect of the above-described model parameters based on a certain objective function and found that the variations between them had little effect on the outcome.[13] An investigation of how the whole optimization process is altered by the different objective functions is significant. In this regard, three different hypothetical objective functions were introduced and optimization speed and correctness was then monitored with respect to the objective function while keeping all the model parameters fixed. In parallel, the effect of population size was also taken into account for all the objective functions. The selection, crossover, and mutation rates were all set at 100%. The roulette wheel selection was adopted and the highest two compositions in the former generation were elicited and copied to the next generation. The simple principle of roulette wheel selection is that the higher the luminance the higher the opportunity of being selected. Nonetheless it does not mean that only the high luminance members have an opportunity of being selected. It should be noted that some members of low luminance level could give a refreshing effect occasionally. In this respect, the roulette wheel selection has been considered to be the best selection method. The single-point crossover was adopted and the crossover point was determined randomly. The composition was normalized after the crossover. The mutation was achieved by adding and subtracting 0.01 mol for two arbitrary chosen components, respectively. The operation of crossover and mutation were described schematically in Figure 22.10. Two parent members chosen by the roulette wheel selection method were represented as composition bands as can be seen in Figure 22.10. One of them was shaded in order to make discrimination between them and hence to trace them out after the crossover. They were treated as chromosomes and the element sectors consisting of the composition band were regarded as genes that have some information affecting the luminance of the member. The crossover created two offspring by exchanging the genes of the parents and the subsequent mutational operation were executed on these offspring.

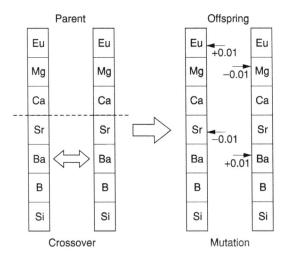

FIGURE 22.10 Schematic description of crossover and mutation in the genetic algorithm used for both simulation and experiment. The crossover and mutation position was determined randomly.

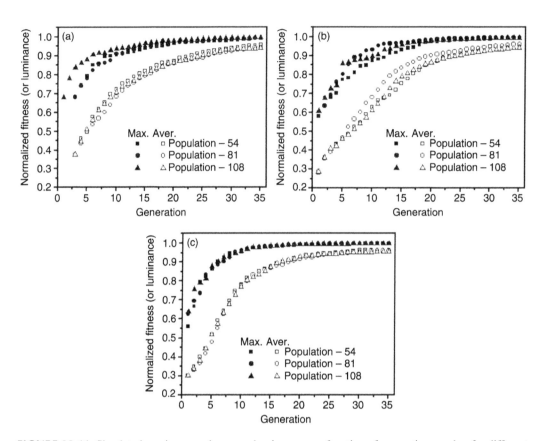

FIGURE 22.11 Simulated maximum and average luminance as a function of generation number for different population sizes. Simulations (a), (b), and (c) were based on three hypothetical objective functions: Equations 22.1, 22.2, and 22.3, respectively. The choice of objective function and population size has little influence on the optimization.

Figure 22.11(a)–(c) shows the simulation results based on three hypothetical objective functions (= hypothetical luminance function), in which the maximum and average fitness (luminance, in our case) are plotted as a function of the generation number for three different population sizes (54, 81, and 108). All the data shown in Figure 22.11(a)–(c) are the averaged results from ten runs. The hypothetical objective function means that the luminance was assumed to be an arbitrary defined function of seven variables (the mol fractions of seven constitutive elements could be these variables). The objective functions used for Figure 22.11(a)–(c) were all analytically soluble, so that all the possible global optimum and local optimum values along with their precise locations could be analytically obtainable. Equation 22.1 shows the uniquely convex objective function with no local maximum point, which was used for the simulation in Figure 22.11(a).

$$\text{Objective Function: } \sum_{i=1}^{7} x_i^2$$

$$\text{Global Optimum Point: } \left(\frac{1}{\sqrt{7}}, \frac{1}{\sqrt{7}}, \frac{1}{\sqrt{7}}, \frac{1}{\sqrt{7}}, \frac{1}{\sqrt{7}}, \frac{1}{\sqrt{7}}, \frac{1}{\sqrt{7}}\right) \quad (22.1)$$

Number of Local Optimum Points: 0

On the other hand, Equations 22.2 and 22.3 show somewhat more complex-shaped functions with several or more local maximum points, which were used for the simulation in Figure 22.11(b) and (c), respectively.

$$\text{Objective Function: } \sum_{i=1}^{7} a_i \sin\left(\frac{\pi x_i}{2a_i}\right) a_{1\sim3} = \frac{1}{9} a_{4\sim7} = \frac{1}{6}$$

$$\text{Global Optimum Point: } \left(\frac{1}{9}, \frac{1}{9}, \frac{1}{9}, \frac{1}{6}, \frac{1}{6}, \frac{1}{6}, \frac{1}{6}\right)$$

(22.2)

Number of Local Optimum Points: 32

$$\text{Objective Function: } \sum_{i=1}^{7} \sin\left(\frac{7\pi}{2} x_i\right)$$

$$\text{Global Optimum Point: } \left(\frac{1}{7}, \frac{1}{7}, \frac{1}{7}, \frac{1}{7}, \frac{1}{7}, \frac{1}{7}, \frac{1}{7}\right)$$

(22.3)

Number of Local Optimum Points: 6

Equations 22.1–22.3 also include the exact location of a global optimum point (= a hypothetical optimum composition) and the number of local optimum points, the existence of which might hamper getting close to global optimum point. The function values, which are below zero, were set as zero to permit the simulation of a more plausible objective function with discontinuities. The simulation results in Figure 22.11(a)–(c) were useful in judging how the objective function affects the efficiency of the optimization process and in determining how many generations will be needed to reach the global optimum point. Consequently, this simulation played a significant role for an appropriate planning of experiments before the implementation.

The most significant finding from the simulation is that the optimization speed and correctness did not vary significantly with the objective function. It was also found that the population size had little influence on the optimization speed and correctness at least in the population size range that we adopted in the simulation. In fact, there was no difference in the simulation result between population sizes for two objective functions (Figure 22.11(a) and (c)). Unlike the others, however, Figure 22.11(b) showed a slight effect of population size on the optimization speed. As can be seen in Figure 22.11(b), the steepest increment in the maximum and average luminance was not observed in the 108-population but in the 81-population, even though the difference was not so significant. This was a thoroughly opposite result to a prediction based on our intuition that the large population size would have been favorable for a fast approach to the global optimum point. The simulation results revealed that the optimization speed and correctness had no consistent relationship with the population size when the population size was bigger than 54. Additional simulations, the results of which were not presented here though, also revealed that the threshold population size was about 50 for all these three hypothetical objective functions. The threshold population size was defined as a population size for which we reached 99% of the global optimum value in less than ten generations. Accordingly, it can be summarized that the change in objective function and population size had no conspicuous influence on the optimization speed and correctness in our simulation condition. Even though a trivial effect of the population size was noticeable in Figure 22.11(b), there would be no systematic way of unveiling it at the present stage. It should be noted that the simulation result could be applicable to our actual experimental optimization but might be useless because an actual objective function that we have to cope with in the experiment is definitely different from the simple hypothetical objective functions that we used for the simulation. This simulation only provided information for reaching the optimum point in ten generations irrespective of the choice of objective function and population size under the present model parameter setting.

22.4.3 EXPERIMENTAL RESULTS AND DISCUSSION

The population size adopted for the experimental optimization process was 108. Considering the fact that the simulation result showed that the optimization efficiency was not affected significantly by the population size, it would be favorable for the population size to be as small as possible from the economical point of view, i.e., the smallest population size (54) should have been adopted in the experimental optimization process. It was, however, the largest population size (108) that we actually adopted in the experimental optimization process for the following reasons; firstly the simulations were carried out under a certain limited circumstance, which might be somewhat different from the actual experimental situation that includes some kind of mistakes. In this regard the larger population may be more favorable for the reliability. More importantly, our combi-chem synthesis system is so well organized that the enhanced population size never causes extra effort and cost. The composition in the first generation was not completely random but some knowledge based on the material science was involved into the library design by taking into account the nature and characteristics of phosphors. The confinements employed in the first generation are as below; Eu^{3+} doping did not exceed 50 mol% of the sum of alkali earth elements and compositions that do not contain any alkali earth elements were non-existent.

Figure 22.12 shows the actual experimental results, in which the maximum and average luminance values are plotted as a function of generation number. As can be seen in Figure 22.12, both values increase with increasing generation number and the maximum luminance is reached only after the sixth generation. The inset in Figure 22.12 shows the first and tenth generation photographed under an excitation of 365 nm. Even though the lamp light was not illuminated evenly over the library in this case and the excitation light wavelength of the lamp (365 nm) differs from the 400 nm excitation that was adopted for subsequent quantitative measurements, it can be seen that the tenth generation contains much more promising compositions than the first. The maximum luminance began to saturate from the sixth generation. Consequently, the composition of the highest luminance was fixed at $Eu_{0.14}Mg_{0.18}Ca_{0.07}Ba_{0.12}B_{0.17}Si_{0.32}O_\delta$ after that. The first generation contained a considerable number of glassy members (= 38 of 108). These were all precluded prior

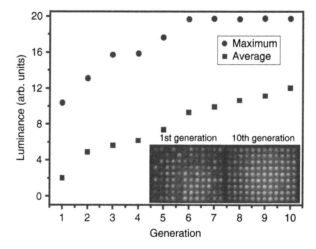

FIGURE 22.12 (See color insert following page 172) Experimental maximum and average luminance as a function of generation number. The inset shows libraries of both the first and tenth generation at the 365 nm excitation.

to the computation by setting their luminance values at zero, irrespective of the intensity of their luminescence, thereby reducing the number of glassy members in the next generation. As a result, there were very few glassy members in the tenth generation ($= 4$ of 108). It should be noted that the discontinuities that we introduced into the hypothetical objective functions reflected the occurrence of these glassy members. Figure 22.13 shows the average mol fraction of each component for the top 11 compositions for each generation, which represents the top 10%. Even though we started with a seven-element system, strontium was rapidly weeded out (the composition of the highest luminance is approximated to $Eu_{0.14}Mg_{0.18}Ca_{0.07}Ba_{0.12}B_{0.17}Si_{0.32}O_{\delta}$). The top 11 compositions do not deviate greatly from the composition of the highest luminance. This proved that the evolutionary optimization certainly took place during our experimental process.

Figure 22.14 shows the XRD data of the highest luminance sample, the composition of which is $Eu_{0.14}Mg_{0.18}Ca_{0.07}Ba_{0.12}B_{0.17}Si_{0.32}O_{\delta}$. It was found that this sample had an oxyapatite structure ($M_{10}(SiO_4)_6O_{\delta}$, M = Ca, Eu, Mg, Ba, P63/m). A more stringent examination revealed that the XRD pattern of $Eu_{0.14}Mg_{0.18}Ca_{0.07}Ba_{0.12}B_{0.17}Si_{0.32}O_{\delta}$ composition was analogous to the calcium europium oxide silicate structure ($Ca_2Eu_8Si_6O_{26}$ or $Ca_2Eu_8(SiO_4)_6O_2$) in terms of the intensity distribution. The standard data of $Ca_2Eu_8(SiO_4)_6O_2$, which almost reproduced the observed diffraction peaks, were also presented in Figure 22.14. The lattice parameters were estimated to be $a = 9.428$, $c = 6.912$ Å from the XRD data of $Eu_{0.14}Mg_{0.18}Ca_{0.07}Ba_{0.12}B_{0.17}Si_{0.32}O_{\delta}$ composition, which were slightly smaller than the $Ca_2Eu_8(SiO_4)_6O_2$ compound ($a = 9.440$, $c = 6.918$ Å). The overall feature of the XRD pattern of $Eu_{0.14}Mg_{0.18}Ca_{0.07}Ba_{0.12}B_{0.17}Si_{0.32}O_{\delta}$ composition, however, does not appear to indicate a single-phase compound such as $M_{10}(SiO_4)_6O_{\delta}$ because there are several unmatched peaks, the intensity of which is negligible though. It can be thus deduced that $Eu_{0.14}Mg_{0.18}Ca_{0.07}Ba_{0.12}B_{0.17}Si_{0.32}O_{\delta}$ consists of a major phase, the structure of which is the oxyapatite and a mixture of several minor unknown phases. Because of the large number of candidates that could constitute the $Eu_{0.14}Mg_{0.18}Ca_{0.07}Ba_{0.12}B_{0.17}Si_{0.32}O_{\delta}$ composition, it was out of the scope of the present investigation to identify all the minor constituent phases in detail and obtain their precise proportion.

Neglecting the presence of such minor phases, the results from the phase identification were interpreted based on the oxyapatite structure. It was manifest that the $Ca_2Eu_8(SiO_4)_6O_2$ compound was not a phosphor at any rate, which was also revealed by an auxiliary experiment. Namely, the $Ca_2Eu_8(SiO_4)_6O_2$ compound was synthesized using a solution pyrolysis method, which is identical to the synthesis method adopted in the GACC process. It was found that the $Ca_2Eu_8(SiO_4)_6O_2$ compound did not exhibit a strong PL response. Another auxiliary experiment was also performed in order to make the $Ca_2Eu_8(SiO_4)_6O_2$ compound have active PL response. Such an experiment was associated with the doping of some other elements into the $Ca_2Eu_8(SiO_4)_6O_2$ compound. Magnesium,

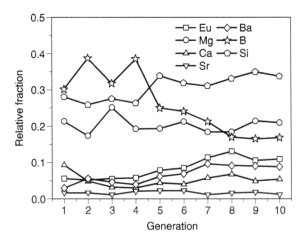

FIGURE 22.13 Average mol fraction of each component for the top 11 compositions of each generation.

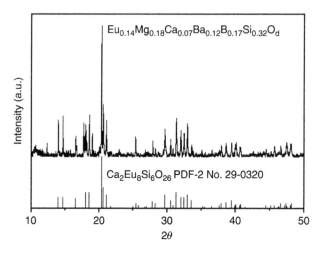

FIGURE 22.14 XRD pattern of $Eu_{0.14}Mg_{0.18}Ca_{0.07}Ba_{0.12}B_{0.17}Si_{0.32}O_\delta$ composition along with the standard data of $Ca_2Eu_8(SiO_4)_6O_2$.

barium and boron were such elements that constitute the $Eu_{0.14}Mg_{0.18}Ca_{0.07}Ba_{0.12}B_{0.17}Si_{0.32}O_\delta$ composition but were not included in the $Ca_2Eu_8(SiO_4)_6O_2$ compound. Boron was thought of as a fluxing agent, so that it was excluded from the auxiliary experiment. The other elements such as Mg and Ba were introduced into the $Ca_2Eu_8(SiO_4)_6O_2$ either one by one or in groups. As a result, it was found that Mg-doping in place of Eu up to Mg/Eu = 1 enhanced PL efficiency. Similarly, the effect of Ba and Ba + Mg doping was also examined. It was also found that the Ba and Ba + Mg doping increased PL efficiency to a certain extent. It should be noted that the ratio of the sum of molar fraction of alkali earths and Eu^{3+} activator (0.51) to the Si molar fraction (0.32) is 1.59 in the $Eu_{0.14}Mg_{0.18}Ca_{0.07}Ba_{0.12}B_{0.17}Si_{0.32}O_\delta$ composition. This figure is close to the ratio (Ca+Eu)/Si (= 1.67) in the $Ca_2Eu_8(SiO_4)_6O_2$ compound. Consequently, this confirms that the $Eu_{0.14}Mg_{0.18}Ca_{0.07}Ba_{0.12}B_{0.17}Si_{0.32}O_\delta$ composition made an oxyapatite structure ($M_{10}(SiO_4)_6O_\delta$, M = Ca, Eu, Mg, Ba).

From the practical point of view, it is more interesting to investigate how promising the luminance level of the $Eu_{0.14}Mg_{0.18}Ca_{0.07}Ba_{0.12}B_{0.17}Si_{0.32}O_\delta$ could be rather than a phase identification. Figure 22.15 shows the emission and excitation spectra of $Eu_{0.14}Mg_{0.18}Ca_{0.07}Ba_{0.12}B_{0.17}Si_{0.32}O_\delta$, along with some other commercially available phosphors for comparison. The luminance of $Eu_{0.14}Mg_{0.18}Ca_{0.07}Ba_{0.12}B_{0.17}Si_{0.32}O_\delta$ was much higher than the commercially available $Y_2O_3:Eu^{3+}$ and $Y_2O_2S:Eu^{3+}$ red phosphors at 400 nm excitation. However, these data were not compared to the ZnCdS:Ag red phosphor, which has been considered to be a good candidate for the red component of a tricolor white LED.[3] Instead, a commercially available ZnS:Cu,Al green phosphor was used as a reference by considering the fact that the quantum efficiency of the ZnS:Cu,Al green phosphor is approximately twice as high as the ZnCdS:Ag red phosphor. Consequently, it was found that the luminance of $Eu_{0.14}Mg_{0.18}Ca_{0.07}Ba_{0.12}B_{0.17}Si_{0.32}O_\delta$ at 400 nm excitation was lower than the ZnCdS:Ag red phosphor by a factor of 10%.

22.4.4 Conclusions

In summary, the validity and applicability of the GACC method was confirmed with regard to the development of inorganic phosphor materials. Through the simulation, using three different hypothetical objective functions, it was found that the global optimum point could be reached in at least ten generations, even though slight differences might exist with respect to the objective function and population size. In the actual experiment, the maximum luminance was enhanced by about 100% only six generations after the onset of the GACC process. The maximum luminance was obtained

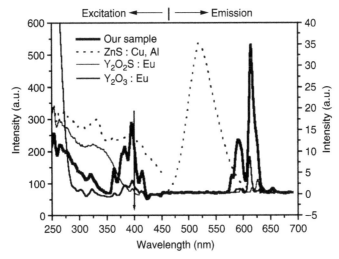

FIGURE 22.15 Emission and excitation spectra of $Eu_{0.14}Mg_{0.18}Ca_{0.07}Ba_{0.12}B_{0.17}Si_{0.32}O_{\delta}$ and commercially available Y_2O_3:Eu^{3+}, Y_2O_2S:Eu^{3+}, and ZnS:Cu,Al phosphors for comparison.

at a composition of $Eu_{0.14}Mg_{0.18}Ca_{0.07}Ba_{0.12}B_{0.17}Si_{0.32}O_{\delta}$. The result from the structural analysis showed that the structure of optimum composition $Eu_{0.14}Mg_{0.18}Ca_{0.07}Ba_{0.12}B_{0.17}Si_{0.32}O_{\delta}$ was an oxy-apatite $(M_{10}(SiO_4)_6O_{\delta}, M = Ca, Eu, Mg, Ba)$ structure and some minor unknown phases also exist. The luminance of $Eu_{0.14}Mg_{0.18}Ca_{0.07}Ba_{0.12}B_{0.17}Si_{0.32}O_{\delta}$ was higher than any other well-known red phosphors at 400 nm excitation. Consequently, this new phosphor could be used as a red phosphor for tricolor white LED applications.

REFERENCES

1. Xiang, X.-D., A combinatorial approach to materials discovery, *Science,* 268, 1738, 1995.
2. Briceno, G., Chang, H., Sun, X., Schultz, P. G. and Xiang, X.-D., A class of cobalt oxide magnetoresistance materials discovered with combinatorial synthesis, *Science,* 270, 273, 1995.
3. Senkan, S. M., High-throughput screening of solid-state catalyst libraries, *Nature,* 394, 350, 1998.
4. Danielson, E., Golden, J. H., McFarland, E. W., Reaves, C. M., Weinberg, W. H. and Wu X. D., A combinatorial approach to the discovery and optimization of luminescent materials, *Nature,* 389, 944, 1997.
5. Danielson, E., Devenney, M., Giaquinta, D. M., Golden, J. H., Haushalter, R. C., McFarland, E. W., Poojary, D. M., Reaves, C. M., Weinberg, W. H. and Wu, X. D., A rare-earth phosphor containing one-dimensional chains identified through combinatorial methods, *Science,* 279, 837, 1998.
6. Sun, X.-D., Wang, K.-A., Yoo, Y., Wallace-Freedman, W. G., Gao, C., Xiang, X.-D. and Schultz, P. G., Solution-phase synthesis of luminescent materials libraries, *Adv. Mater.,* 9, 1046, 1997.
7. Weber, L., Wallbaum, S., Broger, C. and Gubernator, K., Optimization of the biological activity of combinatorial compound libraries by a genetic algorithm, *Angew. Chem. Int. Ed. Engl.,* 34, 2280, 1995.
8. Singh, J., Ator, M. A., Jaeger, E. P., Allen, M. P., Whipple, D. A., Soloweij, J. E., Chodhary, S. and Treasurywala, A. M., Application of genetic algorithms to combinatorial synthesis: a computational approach to lead identification and lead optimization, *J. Am. Chem. Soc.,* 118, 1669, 1996.
9. Weber, L., Evolutionary combinatorial chemistry: application of genetic algorithms, *Drug Discov. Today,* 3, 379, 1998.
10. Bures, M. G. and Martin, Y. C., Computational methods in molecular diversity and combinatorial chemistry, *Curr. Opin. Chem. Biol.,* 2, 376, 1998.
11. Gillet, V. J., Khatib, W., Willett, P., Fleming, P. J., and Green, D. V. S., Combinatorial library design using a multiobjective genetic algorithm, *J. Chem. Ihf. Comput. Sci.,* 42, 375, 2002.
12. Buyevskaya, O. V., Bruckner, A., Kondratenko, E. V., Wolf, D. and Baerns, M., Fundamental and combinatorial approaches in the search for and optimization of catalytic materials for the oxidative dehydrogenation of propane to propene, *Catal. Today,* 67, 369, 2001.

13. Wolf, D., Buyevskaya, O. V. and Baerns, M., An evolutionary approach in the combinatorial selection and optimization of catalytic materials, *Appl. Catal. A-Gen.*, 200, 63, 2000.
14. Sohn, K.-S., Lee, J. M., Jeon, I. W. and Park, H. D., Combinatorial searching for Tb^{3+}-activated phosphors of high efficiency at vacuum UV excitation, *J. Electrochem. Soc.*, 150, H182, 2003.
15. Sohn, K.-S., Lee, J. M., Jeon, I. W. and Park, H. D., Optimization of red phosphor for plasma display panel by the combinatorial chemistry method, *J. Mater. Res.*, 17, 3201, 2002.
16. Kim, C. H., Park, S. M., Park, J. G., Park, H. D., Sohn, K.-S. and Park, J.T., Combinatorial synthesis of Tb-activated phosphors in the CaO-Gd$_2$O$_3$-Al$_2$O$_3$ system, *J. Electrochem. Soc.*, 149, H21, 2002.
17. Sohn, K.-S., Jeon, I. W., Chang, H., Lee, S. K. and Park, H. D., Combinatorial search for new red phosphors of high efficiency at VUV excitation based on the YRO$_4$ (R=As, Nb, P, V) system, *Chem. Mater.*, 14, 2140, 2002.
18. Seo, S.Y., Sohn, K.-S., Park, H. D. and Lee, S., Optimization of Gd$_2$O$_3$-based red phosphors using combinatorial chemistry method, *J. Electrochem. Soc.*, 149, H12, 2002.
19. Sohn, K.-S., Seo, S. Y. and Park, H. D., Search for long phosphorescence materials by combinatorial chemistry method, *Electrochem. Sol. Stat. Lett.*, 4, H26, 2001.
20. Sohn, K.-S., Seo, S. Y., Kwon, Y. N. and Park, H. D., Direct observation of crack tip stress field using the mechanoluminescence of SrAl$_2$O$_4$:(Eu,Dy,Nd), *J. Am. Ceram. Soc.*, 85, 712, 2002.
21. Sohn, K.-S., Park, E. S., Kim, C. H. and Park, H. D., Photoluminescence behavior of BaAl$_{12}$O$_{19}$:Mn phosphor prepared by pseudocombinatorial chemistry method, *J. Electrochem. Soc.*, 147, 4368, 2000.
22. Sohn, K.-S., Kim, B. I. and Shin, N., Genetic algorithm-assisted combinatorial search for new red phosphors of high efficiency at soft ultraviolet excitation, *J. Electrochem. Soc.*, 151, H243, 2004.
23. Sohn, K.-S., Lee, J. M. and Shin, N., A search for new red phosphors using a computational evolutionary optimization process, *Adv. Mater.*, 15, 2081, 2003.
24. Agrawal, D. K. and White, W. B., The luminescence studies of the intermediate compounds in the system Y$_2$O$_3$-P$_2$O$_5$ and Gd$_2$O$_3$-P$_2$O$_5$, *J. Electrochem. Soc.*, 133, 1261, 1986.
25. Blasse, G. and Bril, A., Luminescence of phosphors based on host lattices ABO$_4$ (A is Sc, In; B is P, V, Nb), *J. Chem. Phys.*, 50, 2974, 1969.
26. Nishiyama, K., Abe, T., Sakaguchi, T. and Momozawa, N., Damping properties of YNbO$_4$–Nb$_2$O$_5$–Y$_2$O$_3$ ceramics, *J. Alloys Comp.*, 355, 103, 2003.
27. Milligan, W. O., Mullica, D. F., Beall, G. W. and Boatner, L. A., Structural investigations of YPO$_4$, ScPO$_4$, and LuPO$_4$, *Inorg. Chim. Acta*, 60, 39, 1982.
28. Jayaraman, A., A high-pressure Raman study of yttrium vanadate (YVO$_4$) and the pressure-induced transition from the zircon-type to the scheelite-type structure, *J. Phys. Chem. Solid*, 48, 755, 1987.
29. Shionoya, S. and Yen, W. M., *Phosphor Handbook*, CRC Press, Boca Raton, 2000, p. 17.
30. Pechini, M. P., Method of preparing lead and alkaline earth titanates and niobates and coating method using the same to form a capacitor, US Patent No. 3,330,697, 1967.
31. Park, J.-C., Moon, H.-K., Kim, D.-K., Byeon, S.-H., Kim, B.-C. and Suh, K.-S., Morphology and cathodoluminescence of Li-doped Gd$_2$O$_3$:Eu^{3+}, a red phosphor operating at low voltages, *Appl. Phys. Lett.*, 77, 2162, 2000.
32. Seo, S. Y., Sohn, K.-S., Park, H. D. and Lee, S., Optimization of Gd$_2$O$_3$-based red phosphors using combinatorial chemistry method, *J. Electrochem. Soc.*, 149, H12, 2002.
33. Young, R. A., *The Rietveld Method*, Oxford University Press, Oxford, 1993, pp. 95–101.
34. Rodriguez-Carvajal, J., FULLPROF: A program for Rietveld refinement and pattern matching analysis, abstract of the satellite meeting on powder diffraction of the XV congress of the IUCr, Troulouse, France, 1990, pp. 127–128.
35. Schlotter, P., Schmidt, R. and Schneider, J., Luminescence conversion of blue light emitting diodes, *Appl. Phys. A: Mater. Sci.. Process.*, 64, 417, 1997.
36. Sato, Y., Takahashi, N. and Sato, S., Full-color fluorescent display devices using a near-UV light-emitting diode, *Jpn. J. Appl. Phys. Part 2.*, 35, L838, 1996.
37. Huh, Y.-D., Shim, J.-H., Kim, Y. and Do, Y. R., Optical properties of three-band white light emitting diodes, *J. Electrochem. Soc.*, 150, H57, 2003.
38. Sasaki, M. H., Hamada, H., Kintaichi, Y. and Ito, T., Application of neural network to the analysis of catalytic reactions analysis of no decomposition over Cu/ZSM-5 zeolite, *Appl. Catal. A-Gen.*, 132, 261, 1995.
39. Hattori, T. and Kito, S., Neural network as a tool for catalyst development, *Catal. Today*, 23, 347, 1995.
40. Hou, Z.-Y., Dai, Q., Wu, X.-Q. and Chen, G.-T., Artificial neural network aided design of catalyst for propane ammoxidation, *Appl. Catal. A-Gen.*, 161, 183, 1997.

23 Technology of PLD for Photodetector Materials

Arik G. Alexanian, Nikolay S. Aramyan, Karapet E. Avjyan, Ashot M. Khachatryan, Romen P. Grigoryan, and Arsham S. Yeremyan

CONTENTS

23.1 INTRODUCTION

In the field of synthesis and discovery of multicomponent systems (chemical, biological, etc.) the range of materials and properties of interest is constrained mainly by the availability of measurement and detection methods and devices. In this respect, photodetectors and sensors are of special importance. Application of systems of high-performance (high-sensitivity, fast-operation) radiation detectors with widest possible range of detectable wavelengths would provide high-throughput synthesis and development of various chemical compositions, drugs, catalysts, allowing the information to be retrieved and processed already in the process of synthesis.

In this chapter, we review the results of our research towards the development of pulsed-laser deposition (PLD) technique for synthesis of various materials and structures for photodetector applications.

The technology of PLD has recommended itself as a reliable and flexible method for fabrication of high-quality thin films and structures of various materials. Besides the simplicity of control of technological parameters in the synthesis process, the attractiveness of this method is also that it allows a large variety of experimental designs to be applied for combinatorial and high-throughput synthesis and description, while maintaining the main advantages of the technique such as quality and reproducibility of properties of synthesized materials.

Several research groups have developed a number of refined experimental designs applied to PLD technique (see references 1–3 and references therein) depending on the class of synthesized materials and their expected properties, and, of course, depending on which characteristics of the PLD process are put on the basis of technological approach for synthesis. As examples of such approaches, one can mention the application of precursor method for production of large compositional libraries of materials with various properties,[2] the creation of continuous spread of compositions using the (profile) non-uniformity of the plume of laser-ablated material,[3] etc.

Since steady qualitative and quantitative progress is foreseen in such designs with application of state-of-the-art control/description techniques and equipment, we will not describe in much detail the constructional–schematic solutions for high throughput. Instead, we will mainly concentrate on physical factors related to the peculiarities of PLD, as well as with specifics of materials and structures considered, which affect the characteristics and performance of radiation detectors. On one hand, the exploration of these factors is at the starting point of any experimental design development; on the other hand, the same factors determine the class of possible and important parameters for control and optimization of performance.

Concerning the radiation detectors considered here, our particular approach to the experimental setup is determined by such parameters as: the microstructure of heterojunction interface, the composition of "participating" materials, the free carrier concentration in thin-film materials, the energy band diagram of quantum-well structures.

23.2 PECULIARITIES OF PLD TECHNOLOGY

The typical PLD process for film growth represents pulsed laser ablation of target material and deposition of laser-induced plasma on the substrate. This is shown schematically in Figure 23.1 (see also Figure 23.2a).

The regimes of laser influence, geometry of location of targets, substrates, masks, etc., as well as the post-deposition processes of growth initiation and film processing are determined by the given task.[1] Here, we mention several characteristic features of the process, which will determine our following approaches to synthesis of certain materials and structures.

1. The first peculiarity is associated with the thickness of layer, d, of material incoming to the substrate per pulse of evaporating laser. This quantity depends on the energy and mechanisms of interaction of laser pulse with target material, as well as on the kinetic and dynamic characteristics of the plume of ablated material during its expansion in vacuum. There are a number of extensive reviews on this issue published in literature.[4–6] One particular manifestation of laser–matter interaction, which presents practical interest for us is that it is possible to achieve a regime when the amount of deposited material does not depend on the radiation intensity, I. During the evaporation by nanosecond pulses, the resulting plasma screens the surface of the target and reduces the dependence of d on I. The latter approximately can be written as $d \sim I^{1/2}$ or $d \sim \ln I$ (instead of $d \sim \exp[-\text{const}/I]$ or $d \sim I$, which corresponds to the initial stage of the thermal regime of ablation).[5,6] On the other hand, the thermal energy accumulated in the plasma bunch

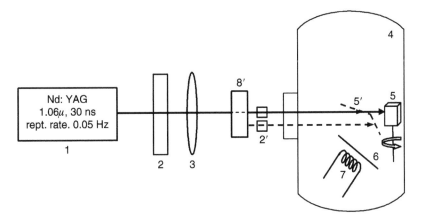

FIGURE 23.1 Typical PLD setup. 1 – Q-switched laser, 2 – attenuation filter, 3 – focusing lens, 4 – deposition chamber (vacuum 2×10^{-6} mmHg), 5 – targets, 6 – substrate, 7 – resistance heater, 8 – laser beam splitting system. The components denoted by primed numbers replace those without prime in case of simultaneous deposition from different targets.

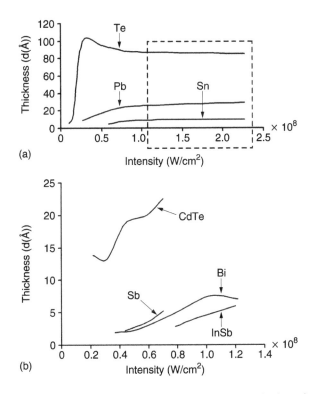

FIGURE 23.2 Dependence of the thickness of deposited layer on the laser pulse intensity for various materials.

increases with I. As a result, the thickness d remains constant in a wide range of radiation intensities. In Figure 23.2, we presented the characteristic dependences of d on I, which we then use for deposition of various films and structures. The existence of a stable regime (the isolated region on Figure 23.2) becomes of special importance when dealing with requirements of programmability and controllability of parameters – such as stoichiometry and thickness – for the films obtained from single targets of materials.

2. Another peculiarity of PLD is related to the growth rate of the film. In order to provide formation of two-dimensional nuclei the flux velocity of atoms incoming to substrate must be larger than the rate of their escape from the interaction region due to the diffusion. Pulsed methods of evaporation are the convenient approach to provide high deposition rates. In the case of PLD the deposition rate is limited only by overheating of the substrate, and the limiting values of about 10^6 Å/s exceed on several orders of magnitude those achievable in other deposition techniques. The high growth rate provides the continuity of layers even at monomolecular thicknesses.

3. The third peculiarity is the energy spectrum of particles. Depending on their impact on the surface and subsurface regions of the substrate, particles can be distinguished into two groups: those with energies less than 20–25 eV, which do not result in defects in substrate, and those with higher energies. The latter dislodge the atoms from the surface layer resulting in vacancy-type defects. The proportion of slow and "defect-inducing" particles determines the processes accompanying the irradiation of surface by plasma. Reduction of the amount of rapid particles corresponds to the possibility of production of defect-free crystals. In contrast, using only rapid ions causes effective generation of vacancies in near-substrate region. A very interesting case is the intermediate regime, when the condensation occurs on the surface already irradiated by a dose of rapid ions. The irradiation creates an array of additional centers of crystallization on single crystalline substrates. This provides the possibility of epitaxial (repeating the structure of underlying substrate) growth of films at comparatively low mobility of adsorbed atoms, i.e., at temperature lower than that required in other techniques.

4. Due to the above-mentioned "energy dependence" of growth mechanisms, PLD provides a number of parameters for compositional and structural control of multicomponent films during sequential, as well as simultaneous, deposition of constituents from various targets. The latter may consist of both elementary and composite materials. In the case of composite targets, due to the extremely rapid processes of ablation and deposition (with duration less than the characteristic time describing the escape of particles of the most volatile component from the reaction region) in PLD, it becomes possible to transfer (without violation) to the substrate the composition stoichiometry of the target. This is an advantage of PLD over other vacuum deposition techniques, in which the strong difference in vapor pressures of the initial components makes difficult stoichiometric control, especially in the case of ultrathin films.

23.3   SYNTHESIS OF QUATERNARY SOLID SOLUTIONS $Ga_xIn_{1-x}As_ySb_{1-y}$ (FABRICATION OF SEMICONDUCTOR STRUCTURES WITH SPATIAL VARIATION OF ENERGY BAND GAP)

23.3.1   TECHNOLOGICAL APPROACHES: COMPOSITIONAL AND STRUCTURAL CONTROL AND DESCRIPTION

We start from the synthesis of single-phase solid solutions $Ga_xIn_{1-x}As_ySb_{1-y}$ with variable x (y) (i.e., with varying lattice parameter) using the method of laser mixing of source materials GaAs and InSb. While pointing out the possibility of such mixing by simultaneous PLD of materials from different targets, we consider here another method, which is analogous to high-energy ion implantation: an initial film of GaAs or InSb is deposited on the substrate, which is then implanted by high-energy particles of the other laser-ablated material. In this, we use two approaches which are both based on the possibility of congruent stoichiometry transfer mentioned above.

23.3.1.1 Approach A

In the first case, using comparatively low intensities of evaporating pulses (0.7 J per pulse, 30 ns duration) a continuous layer of InSb with certain thickness is initially deposited on a KBr or KCl substrate, and the library of various compositions is obtained by implantation of laser-ablated GaAs while varying the substrate temperature and laser fluence. The scheme providing the creation of such discrete library is illustrated in Figure 23.3a.

After deposition of an array of ~100 Å thick InSb layer cells, temperature gradient is created on the substrate in the vertical direction (this corresponds to direction (y) in Figure 23.3). Then the plasma flux of evaporated GaAs is sequentially directed to these cells by simultaneously changing the position of the substrate and light attenuation filters. In this way the two-dimensional coordinate system of technology parameters (laser intensity, I, and substrate temperature, T) is mapped onto the coordinate plane (xy) of the substrate plate on which the array of rectangles is formed with area 1×1 mm^2 each.

The structural investigations of synthesized solutions were carried out using the high-energy electron diffraction method, though it should be noted that various developments of the X-ray diffraction method such as concurrent X-ray diffractometer,[7] the two-dimensional phase-mapping X-ray microdiffractometer[8] were successfully used by other groups for more rapid structural characterization of composites.

Electron-diffraction analysis shows that single-phase solid solutions with lattice parameters different from those of InSb and GaAs are obtained in a certain range of temperature and laser fluences. We demonstrate this in Figure 23.3b presenting sample diffraction patterns from various cells including one that corresponds to the immiscible phase.

The reverse mapping of (xy) plane onto the coordinate system (I,T) shows almost linear dependence of lattice parameter on technology parameters. This is shown in Figure 23.3c, where the range of lattice parameters corresponding to single-phase solutions is presented.

The described scheme allows one to obtain discrete library of compositions $Ga_xIn_{1-x}As_ySb_{1-y}$ instead of continuous spread, which would be obtained if we fully exploit the gradient of substrate temperature. However, such approach facilitates and simplifies the information retrieval during the following description of structural and optical characteristics of composites.

23.3.1.2 Approach B

In the second used approach, InSb films with thicknesses in the range 10–40 Å are deposited at substrate temperature corresponding to monocrystalline growth of these films ($T = 260°C$). Then ablated GaAs is deposited in the same thickness range (taking into account the amount of material deposited per pulse). As distinct from the approach A, the technology parameters here are the thicknesses (more precisely, amounts) of two mixing materials.

Structural measurements have shown that, compared to the case A, single-phase films with a wider range of lattice parameters are obtained when using approach B (see Table 23.1).

In order to verify that the indicated ranges of values of lattice parameter correspond to quaternary solid solutions, i.e., to determine the compositions of the obtained materials, the structural investigations should be complemented by other measurements. In other words, the same value of lattice parameter may correspond to two ternary phases with either $x = 0$ or $y = 0$, and a quaternary phase $Ga_xIn_{1-x}As_ySb_{1-y}$ with certain $x(y)$. Therefore, for adequate description of synthesized materials we have performed optical measurements of energy band gap of materials. Comparison of results of these measurements with existing nomograms[9] confirmed the origin of quaternary solutions.

Table 23.1 shows the dependence of lattice parameter on the composition $x = y$, as well as the corresponding measured values of energy band gap. These dependences were subsequently used for fabrication of band-gap engineered structures and heterojunctions (see below).

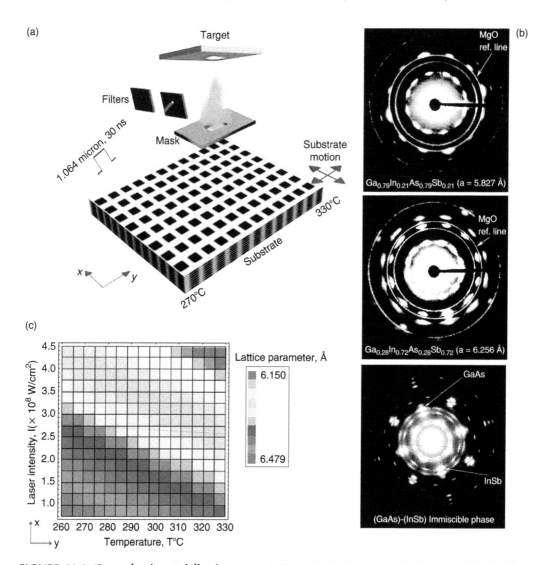

FIGURE 23.3 (See color insert following page 172) Synthesis of composition library of single-phase $Ga_{1-x}In_xAs_{1-y}Sb_y$. (a) PLD configuration; (b) electron-diffraction patterns taken from three cells of the library (corresponding to three compositions); (c) dependence of the lattice parameter of synthesized composites on the technology parameters.

It has been established[10] that in this particular technological approach the GaAs and InSb are interchangeable as long as we mix small amounts of these materials for synthesis of quaternary solutions, i.e., one can first grow a monocrystalline GaAs as underlying film, and then use ablated InSb as bombarding plasma. In strongly non-equilibrium and non-stationary processes of ablation and deposition, the crucial technology parameters for these two materials, namely, monocrystalline growth temperature and energy distribution of ablation products overlap in a wide range.

It should be mentioned that in both described approaches (A and B) there is some limitation imposed on the thickness of initially deposited film, after which one is not able to achieve the complete mixing of materials while maintaining proportion $x = y$ over the volume of the film. The maximum thickness of a single-phase film $Ga_xIn_{1-x}As_ySb_{1-y}$ (with $x = y$) which can be obtained

TABLE 23.1
Lattice Parameters of Various Compositions and Corresponding Values of Measured Energy Band-Gaps

Lattice Parameter (Å)	Composition, $x = y$	Energy Band-Gap (eV)
5.742	0.894	1.143
5.827	0.792	0.916
6.016	0.558	0.568
6.1	0.468	0.476
6.143	0.416	0.404
6.256	0.276	0.247
6.344	0.168	0.181

after one period of "laser implantation" was ~100 Å depending on the initial film material, substrate temperature, and laser fluence. Therefore, larger thicknesses required for various applications are achieved by repetition of the synthesis process with alternation of InSb and GaAs targets. At this, in order to control the composition over the film volume, the initial continuous layer in each following stage is deposited in the regime of immiscibility determined beforehand (this regime corresponds to the 3rd diffraction pattern in Figure 23.3b).

23.3.2 FABRICATION OF GRADED GAP SEMICONDUCTOR STRUCTURES

The developed method of laser mixing was used to fabricate ($Ga_xIn_{1-x}As_ySb_{1-y}$)-based graded gap semiconductor structures in which the energy band gap is varied along the growth direction of the film. Such systems have found widespread optoelectronic application, and can be used, particularly, for creation of tunable and switchable radiation detectors. Varying the energy band gap of the structure we assign in advance the spectral range of interband transitions of carriers, thus predetermining the characteristics of spectral response of the optoelectronic device based on it.

Figure 23.4a illustrates the realized graded gap structure based on five compositional phases of $Ga_xIn_{1-x}As_ySb_{1-y}$, which correspond to five regimes of deposition (here we implemented approach B described above). As a result of deposition of first five layers the band-gap energy of the structure varies continuously from the value 0.247 eV to 0.916 eV. After that, the process of deposition is continued in the reverse order of alternation of layers until sufficient thickness of the structure (~2500 Å) is attained which is necessary for optical measurements.

In Figure 23.4b we presented the spectral dependence of absorption coefficient for a more complicated structure including a wider range of compositions (which is demonstrated by additional energy ranges of response: (0.11–0.247) eV and (0.916–1.1) eV. In the range of photon energies 0.47–0.92 eV (wavelengths 2.1–1.35 microns) the value of absorption coefficient is changed only on 2%. Therefore, from the point of view of detector applications, namely as a simple photoresist, uniform response characteristics of this system are expected in this wavelength range.

23.4 PLD-PRODUCED HETEROJUNCTIONS (THE ROLE OF HETEROJUNCTION INTERFACE)

In this section, we extend the results of technology development described in the previous section to the synthesis of semiconductor heterojunctions for photodetector applications. Along with technology peculiarities we explore in more detail the physical relation of various photodetector characteristics to the structural properties of these systems, as this relationship allows to establish and determine the technology parameters of optimization and improvement.

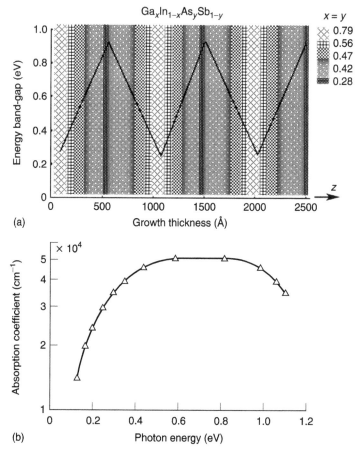

FIGURE 23.4 Graded gap semiconductor structures based on $Ga_xIn_{1-x}As_ySb_{1-y}$. (a) Spatial variation of bandgap energy, (b) absorption coefficient of a structure with a wider range of compositions.

23.4.1 ADVANTAGES OF DETECTORS BASED ON HETEROSTRUCTURES (AN OVERVIEW)

Traditional semiconductor p–n junctions and photoresists on InSb, HgCdTe-based systems have found the most widespread industrial use.[11] The spectral range covered by these two materials is 0.2–5.4 μm for InSb and 2–15 μm for HgCdTe. At the same time, a steady interest exists to detectors based on semiconductor heterojunctions (HJ), which form at the interface of two materials with different energy band-gaps. By alternation of thin layers of narrow and wide band-gap materials quantum well (QW) heterostructures are obtained, which are characterized with (quasi-) discrete spectrum of carriers. Although both HJ- and QW-based detectors operate in a relatively narrow range of wavelengths, they surpass the detectors on usual p–n junctions by many other characteristics.[12–15] Among physical reasons underlying the potential advantages of HJ and QW detectors we will mention here only those common for these two systems. Firstly, as distinct from detectors on usual the p–n junction, in heterojunctions the absorption of incident radiation in the narrow-gap material takes place immediately in the region of space charge (in the case of a QW detector, immediately in the quantum well) due to the existence of a "window" of the wide band-gap material. This reduces the losses related with the recombination in the "non-active" volume and increases the speed of operation. Secondly, choosing appropriate materials (constructing the profile of potential barriers) and/or varying the thickness of the active layer one is able to rebuild the electron energy structure and tune the spectrum of registered wavelengths to the given task. Finally, quantum well heterostructures, namely, arrays of quantum dots with discrete spectrum of carriers, are considered as best candidates for devices with temperature-independent characteristics,[16] which is connected with the possibility of suppression of electron-phonon relaxation mechanisms in quantum dots by appropriate reconstruction of their electron energy spectrum.[17,18] Thus, efficient operation of these systems is expected at room temperatures, while cooling to ~77 K is usually required for best operation of detectors available now.[11]

The spectral characteristics and performance of detectors based on semiconductor heterojunctions are determined, in general, by the choice of the constituent materials: energetic and microscopic states of the heterojunction which lay on the basis of detector operation are predetermined by intrinsic properties of materials such as electron energy affinity, crystal lattice parameter, etc. The relationship between these same parameters prescribes the technological conditions for fabrication of heterojunction and often can be the main limiting factor for application of a given technology approach.

As was described earlier (see Section 23.2), the plasma flux in the process of PLD is characterized with a quite wide energy distribution of particles. Using only the low-energy part of the plasma provides the means for deposition of defect-free films with parameters close to those obtained by molecular beam epitaxy. On the other hand, the plasma containing rapid ions in an amount sufficient to decrease the epitaxial growth temperature (see Section 23.2) allows one to produce ultra-thin single-crystalline films and their combinations, which would be impossible to obtain by other techniques.[1,19]

Thus, once we have a certain level of such technology control (in the sense mentioned above), we can consider the heterojunction interface as the quality factor. In other words, further approaches for optimization and discovery of detector properties are brought to the technological capability of construction and improvement of the heterojunction interface.

From here onwards in this section, we present the results of investigations of physical bases and photodetector characteristics of three types of heterojunctions obtained by PLD: abrupt and smooth junctions nInSb–nGaAs, and junctions pInSb–nCdTe. On the example of these systems we deduce the qualitative dependence of detection characteristics (spectral response, operation speed) on the origin and state of heterojunction interface.

23.4.2 TECHNOLOGY AND STRUCTURAL FEATURES

The first system studied as photodetector material is the isotype heterojunction nInSb–nGaAs, which is referred to the class of so-called "non-ideal" heterojunctions and is characterized by a large mismatch of lattices of constituent materials. Such lattice mismatch (~14% for the system in question) induces a large number of states in the junction interface (Figure 23.5(a)).

Fabrication of single-crystalline "non-ideal" heterojunctions with abrupt interfaces is difficult with traditional film growth technologies, in which deposition of materials occurs at low deposition rates and in equilibrium conditions. Particularly, this is so for the nInSb–nGaAs system considered here.[13,20] The reason for this is that due to the large lattice mismatch the temperature required for oriented growth of films is high, which results in interdiffusion of materials in a rather large interface region. Due to the rapid introduction of the energy into the system during the PLD, the film growth occurs under strongly non-equilibrium conditions, and the temperature of oriented growth decreases significantly. In these circumstances, it is possible to fabricate abrupt heterojunctions in regimes, when interdiffusion of constituent materials is practically excluded.[19]

On the other hand, if we are interested in smooth junctions based on these materials, then it is desirable to achieve some degree of control on the structure of interface region, which is practically impossible when the latter forms as a result of interdiffusion of materials in high-temperature conditions.

The developed method of synthesis of quaternary solutions described in Section 23.3 can serve us an effective tool in achieving this goal. Since the method allows us to control the composition

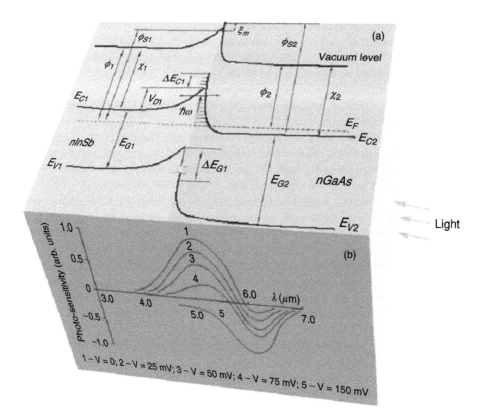

FIGURE 23.5 Abrupt isotype heterojunction nInSb–nGaAs. (a) Energy band diagram (traditional notations from literature are used), (b) photoresponse spectra at various bias voltages.

(and hence, the lattice parameter and electron affinity) of solution at very small thickness levels we can achieve the smoothing of potential barrier between InSb and GaAs layers by means of deposition of intermediate layers of GaInAsSb, with lattice parameters changing monotonically from larger (InSb) to smaller (GaAs) value (Figure 23.6(a) cf. Figure 23.5(a)). In this way we are capable of tracking the dependence of heterojunction (photodetection) characteristics on the interface microstructure.

Heterojunctions pInSb–nCdTe (Figure 23.7) represent another class which is interesting to us from the point of view of peculiarities of interface microstructure. As distinct from the previous case, the materials InSb and CdTe have similar crystalline structure with close lattice parameters: $a_{Insb} = 6.4793$ Å, $a_{CdTe} = 6.4829$ Å. Therefore, it is expected that only a small number of defects will form at the heterojunction interface.

In heterojunctions based on A_3B_5 and A_2B_6 type compounds, namely, in the system in question, a semi-insulating layer is usually formed due to the diffusion of the V group element into the A_2B_6 material. This affects significantly the (photoelectric) characteristics of the heterojunction. Once again, the orientation growth of CdTe on InSb substrates occurs at sufficiently low temperatures (200–250°C) in PLD method, allowing one to exclude unwanted diffusion during the growth process. Moreover, using PLD it is possible to obtain uniform and single-crystalline films as thin as 20 Å, which allows one to study the variation of electrical and photoelectric characteristics depending on the thickness of deposited layers.

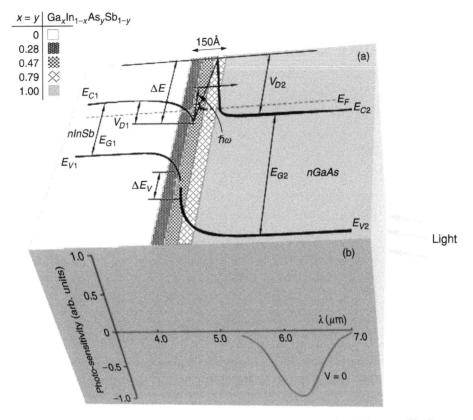

FIGURE 23.6 Smooth isotype heterojunction nInSb–nGaAs. (a) Energy band diagram. Single-arrow path shows the mechanism for charge separation. (b) Photoresponse spectrum at 0 bias voltage.

23.4.3 PHOTODETECTOR CHARACTERISTICS AND THEIR RELATION WITH HETEROJUNCTION STRUCTURE

23.4.3.1 Abrupt Heterojunction nInSb–nGaAs

Theoretical calculations show[21] the possibility of the sign-reversal of photoresponse on the spectral curve of a "non-ideal" heterojunction with large number of interface states, and we will discuss this on our samples obtained by PLD-technique.

Figure 23.5 shows the photoresponse of the heterojunction in the spectral range 3.5–6.5 μm at various applied bias voltages, when an area of 4×4 mm^2 of the sample surface is illuminated from the side of wide bandgap n-type GaAs. As seen from the figure, the photoresponse sign-reversal is observed in the range of fundamental absorption of InSb at bias voltages 0–75 mV, which is related to the existence of lattice-mismatch-induced interface states, as described above.

With application of reverse voltages, the maximum value of the negative photosignal increases, while the maximum value of positive signal decreases. The wavelength λ_0 at which the signal vanishes depends linearly on the magnitude of the applied voltage and shifts towards the shorter wavelengths as the latter increases. Maximum values of positive and negative signals (while having a trend to shift to longer wavelengths) remain within the experimental accuracy at the same photon energies. Only at large values $V \geq V_{D1}$ (see Figure 23.5a) when the barrier V_{D1} is entirely removed by the external field and the positive signal vanishes, the maximum of negative signal shifts towards shorter wavelengths.

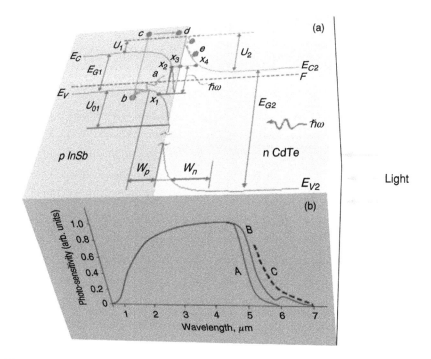

FIGURE 23.7 Heterojunction pInSb–nCdTe. (a) Energy band diagram. Possible mechanisms of space charge separation are indicated: path (abcd) – over-barrier transition of photo-created carriers ($\nabla\omega > E_{G1} + U_1$); path $x_1 x_2 x_3 x_4$ – sub-barrier transition ($\nabla\omega < E_{G1}$); (b) Spectral distribution of photosensitivity at various doping levels of pInSb: $A - N_A = 2 \times 10^{13}$ cm^{-3}, $B - N_A = 4.17 \times 10^{14}$ cm^{-3}, $C - N_A < 10^{13}$ cm^{-3} with applied external voltage ($U_B = 0.1$ V). (From Alexanian, A.G. et al. *Meas. Sci. and Technol.*, 16, 167, 2005, with permission.)

Here we present the qualitative explanation of the observed reversal of the photoresponse sign. In order to do this we proceed from the energy band diagram of the nInSb–nGaAs heterojunction (Figure 23.5a), in which traditional notations for corresponding quantities are used (see Table 23.2). The heterojunction interface is considered consisting of two spatially separated semiconductor surfaces, each with quasi-continuously spaced energy states which act as trapping centers with large cross-section. Taking into account the large number of the partially occupied states, which make the interface metallic-like, we represent the heterojunction as consisting of three different junctions: (i) a Schottky barrier consisting of the narrow-gap InSb and its metallic-like surface; (ii) a Schottky barrier formed of the metallic-like surface of GaAs and the GaAs layer; and (iii) a metallic contact between two planes of surface states.

In the considered spectral range 3.5–6.5 μm the contribution to the photocurrent comes from the photoemission of electrons from the surface states of nInSb and nGaAs, from the valence band of narrow-gap nInSb, and from the unbound states in the conduction band of nInSb. The relations between the current and the voltage can be then written as:

$$I_1 = I_{S1}\left[\exp(A_1 V_1) - 1\right] - I_{R1}$$

$$I_2 = I_{S2}\left[\exp(-A_2 V_2) - 1\right] - I_{R2}, \tag{23.1}$$

where I_{R1}, I_{R2} are the densities of photocurrents through the Schottky diodes 1 and 2, $I_{S1}[\exp(A_1 V_1) - 1]$ and $I_{S2}[\exp(-A_2 V_2) - 1]$ are the densities of dark currents through diodes 1 and 2

TABLE 23.2
Notations Used in Figure 23.5, Figure 23.6 and Figure 23.7

Notation	Quantity
Φ_i	Work function
$E_{c(v)i}$	Conduction (valence) band minimum (maximum)
$\Delta E_{c(v)i}$	Conduction (valence) band discontinuity
E_{Gi}	Energy band gap
E_F	Fermi energy
$\hbar\omega$	Photon energy
χ_i	Electron affinity
$U; V_{Di}$	Band bending
W_i	Space charge region

when a reverse bias voltage ($\ll+\gg$ to GaAs), $V = V_1 + V_2$, is applied to the junction. At small values of the signal, when $I_{R1}/I_{S1} \ll 1$ and $I_{R2}/I_{S2} \ll 1$, one obtains from (23.1) for the photovoltage

$$U_p = \frac{I_{R1}}{A_1 I_{S1}} - \frac{I_{R2}}{A_2 I_{S2}}. \tag{23.2}$$

Since I_{R1} and I_{R2} depend strongly on the incident photon energy, the applied bias, and the parameters characterizing the barriers, it follows from (23.2) that U_p can change the sign depending on the incident photon energy, and the energy corresponding to the sign reversal depends on the applied voltage.

The observed dependence of photovoltaic properties of nInSb–nGaAs heterojunction (with two-side depletion) on the applied bias was used to calculate a pyrometer with null indication. The principle of operation of the pyrometer lies in the following: when radiation from a black body falls on the junction from the side of wide band-gap GaAs, the photovoltage vanishes at certain value of the bias voltage. Since the spectral distribution of incident radiation depends on the temperature of black body, the bias voltage can be calibrated directly in temperature values. Calculations show that device can operate as a null indicator (i.e., a device recording the wavelength at which the signal vanishes) in the range 5.2–5.7 μm with a zero-point sensitivity, $dV/d\lambda = 0.387$ V/μm, in the linear range, and as an IR pyrometer with sensitivity 0.5 mV/K at 110–200 K; 0.11 mV/K at 300 K; 40 μV/K at $T = 500$ K.

23.4.3.2   Smooth Heterojunction InSb–GaAs

The change in the photoresponse sign is not observed in structures with a transition region (Figure 23.6).

Such transition region was produced by deposition of three layers with various compositions: $(GaAs)_{0.28}(InSb)_{0.72}$; $(GaAs)_{0.47}(InSb)_{0.53}$; $(GaAs)_{0.79}(InSb)_{0.21}$ – with a total thickness of 150 Å. The lattice parameters of these compositions were 6.26 Å, 6.1 Å, and 5.83 Å, respectively. Thus, the difference of lattice parameters decreases from layer to layer and the number of interface states decreases.[19b] In the case of existence of transition region, the energy diagram of heterojunction changes significantly (Figure 23.6(a), cf. Figure 23.5(a)) and a quantum well is formed at the junction interface. Hence, the mechanisms of spatial separation of charged carriers and of the onset of photovoltage also change; the electrons absorb photons and make transition from the quantum state to unoccupied states in InSb with subsequent tunneling to GaAs.

23.4.3.3   Heteronjunctions pInSb–nCdTe

Heterojunctions pInSb–nCdTe (Figure 23.7) were fabricated by PLD of nCdTe layers with various thicknesses on the polished pInSb substrates, which were partially compensated (N_A–$N_D \approx 4.17 \times 10^{14}$ cm^{-3} at $T = 77$ K). Deposition regimes are determined according to the corresponding dependence presented in Figure 23.1b. The thickness of deposited cadmium telluride films were in the range 0.1–1.2 μm and the surface area of obtained heterojunctions varied in the range 0.25–0.36 cm^2.

The cut-off voltage, V_C – corresponding to the total band bending in heterojunction materials – depends on the CdTe thickness and was determined from the family of current–voltage characteristics at $T = 77$ K.[4] The value of V_C saturates at CdTe thicknesses $d \geq 0.5$ μm. The maximum width of space charge region in CdTe corresponds to $W_n = 0.45$ μm (see Figure 23.7(a)). At smaller thicknesses, the CdTe layer is completely depleted. Using the measured values e$V_C = U_D = U_{01} + U_2 = 0.45$ eV and $W_n = 0.45$ μm, the concentration of donor impurities, the space charge region in InSb, as well as the band bending for each heterojunction material were determined within the Anderson model. The corresponding values are: $N_D = 8.55 \times 10^{14}$ cm^{-3}, $W_p = 1.025$ μm, $V_{Dp} = 0.2$ eV, and $V_{Dn} = 0.25$ eV.

The photoelectric properties of the junction were studied at liquid nitrogen temperature and at normal incidence of radiation from the side of wide band-gap CdTe. The typical spectral distributions of measured photosensitivity are shown in Figure 23.7b. The spectral band registered in the photovoltaic regime (without application of external voltage) was 1–5.5 μm at the photosignal level 0.1 (in arbitrary units). The maximum signal corresponds to the wavelength 4.8 μm.

Observation of such spectra is due to the changes of the absorption and the charge separation mechanism in the formed energy structure.[14a] In the range λ < 4.8 μm, the photosignal results from the interband absorption in narrow-gap InSb with subsequent transition (diffusion) of created electrons over the barrier formed by the wide-gap CdTe (the path abcde in Figure 23.7a). Towards the short wavelengths, the photosignal decreases due to the reduction of the extinction region in InSb. At λ > 4.8 μm, the following mechanisms of signal origin are possible: (i) sub-barrier tunneling of photocreated electrons from the conduction band of InSb to conduction band of CdTe; (ii) resonance phototunneling (in certain conditions) of carriers from the valence band of InSb to the conduction band of CdTe through a virtual state in the band gap of InSb (path $x_1x_2x_3x_4$ in Figure 23.7a).

As can be seen from Figure 23.7b, the long wavelength edge of photoresponse band depends on the doping level of pInSb. At concentrations $N_A < 10^{13}$ cm^{-3} the signal vanishes at a wavelength 5.5 μm corresponding to the fundamental absorption edge of pInSb. However, application of a voltage (0.1 V) to such sample results in the observation of the signal beyond this edge (~7 μm, Figure 23.7b). The maximum of photosignal on the tail of the curve B corresponds to resonance phototunneling of carriers through the quantum well with barriers formed at the interface.[14]

Thus, this structure can operate in both photovoltaic and diode (with application of external voltage) regimes and the photosignal spectrum can be tuned over a wide range.

An important feature of this detector is its speed of operation. The measurement of the switch-off time of the heterojunction was carried out using a λ = 1.06 μm wavelength laser pulse with pulse duration of 30 ns. The pulse was simultaneously directed to the sample and to the photoelectric multiplier with time constant 3 ns, and the responses were registered in the real time mode by an x–y recorder. The switch-off time was also determined in the detection regime, and it was found from these two independent measurements that the switch-off time of the junction is $\tau_s < 15$ ns and does not change with the scanning of the laser beam over the heterojunction surface. Nevertheless, the measured limit for τ_s depended on our experimental capabilities, and the actual value τ_s may be less; estimations show an expected ultra-speed response of this system: $\tau_s \leq 10^{-11}$ s.

Now, we turn to another important property of obtained heterojunctions, pInSb–nCdTe, which is of practical importance. When illuminated with a monochromatic radiation at certain wavelength from the side of wide band-gap CdTe, optical memory effect is observed in samples in the wavelength range 0.37–1.37 μm.[14b] The excitation of HJ results in a photo-electromotive force (photo-emf), which

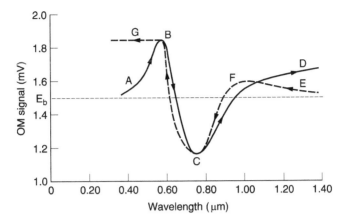

FIGURE 23.8 Wavelength dependence of optical memory signal (arrows indicate the scanning direction of external radiation). (From Alexanian, A.G. et al., *Meas. Sci. and Technol.*, 16, 167, 2005. With permission.)

persists when the excitation is turned off. The effect was observed in a temperature range 77–165 K, under the steady integral background radiation, the photo-emf for which was 1.5 meV (idling regime).

Figure 23.8 shows the spectral dependence of optical memory (OM) signal at direct and reverse scanning of the incident light spectrum, showing a hysteresis in this dependence. The studied wavelength range can be divided into three subranges:

a) $\lambda = 0.37$–0.575 μm, where the OM signal enhances with increasing incident wavelength, i.e., a "recording effect" is observed (curve AB in Figure 23.8)

b) $\lambda = 0.575$–0.768 μm, in which the OM reduces with increasing wavelength, i.e., "clearing effect" (BC curve)

c) $\lambda = 0.768$–1.37 μm, where the "recording effect" is again observed (CD curve)

The maximum of OM signal is observed at $\lambda = 0.575$ μm (B), and the minimum is at $\lambda = 0.768$ μm (C). At large thicknesses of the CdTe layer the OM signal increases reaching a maximal value at $d = 0.45$ μm. The OM signal decreases with further increase of the thickness. The photomemory effect was observed also, when an external bias was applied on the structure. In biased structure, the photovoltage was by two orders of magnitude higher than that observed in idling regime. The OM signal persists in both biased and unbiased junctions for no less than 10^5 s.

It is also interesting that if, after the first act of illumination the heterojunction, instead of being returned to its initial state is additionally irradiated, the memory voltage increases from the initial to a new value. This continues until this voltage reaches a certain maximum value, which corresponds to the saturation of traps at the junction interface. Thus, the obtained heterostructure possesses a property of integration of incident radiation.

For "recording" samples the achieved sensitivity to the incident radiation was ~0.66 μJ/mm^2, and the "recording time" was $\tau_0 \sim 8.3 \times 10^{-4}$ s, though the latter is not limited to this value and depended on our measurement capabilities.

23.5 THIN FILMS PbTe, PbSnTe, AND PERIODIC STRUCTURES PbTe-PB^A-PbTe

Up to this point, we considered films and structures which were obtained from compound targets. The composition and structure of films was controlled based on the possibility of congruent transfer of target material to the substrate. Nevertheless, the advantages of the PLD method are truly revealed when we use elementary sources (single targets) of materials. Using this approach (also in combination with the previous one) gives additional technology parameters such as particle fluxes from different targets, energy state, and chemical activity of particles in different fluxes, which result in practically unlimited spectrum of synthesized materials and compositions with a higher level of technology control.

In this section, we use the method of sequential PLD from elementary sources for the synthesis of semiconductor thin films and periodic structures based on chalcogenide materials.

The detector characteristics of these systems (e.g., spectral range of sensitivity) can be governed with such physical parameters as concentration of free carriers and thickness of the quantum well in periodic structure. These parameters are directly related with the crystallinity, composition, and stoichiometry of thin-film systems. Therefore, in order to establish this relationship it is suitable to consider first the physical conditions which are responsible for the film growth process and determine the appropriate technology approach.

23.5.1 TECHNOLOGY

Compounds of type A_4B_6 usually crystallize with significant violation of stoichiometry.[22,23] The relatively low enthalpy of formation of vacancies (0.3–0.8 eV) and high substrate temperature ($T \geq 500°C$) result in unavoidable generation of intrinsic defects (mainly, vacancies) which are electrically active. The energy levels of vacancies do not lie in the energy band-gap; therefore, the free carriers can exist in allowed energy bands at arbitrarily low temperatures. The equilibrium concentrations of vacancies in the sublattice of lead (acceptors) and in sublattice of chalcogen (donors) generally have the order 10^{18}–10^{19} cm^{-3}. Only special technological methods allow one to reduce the free carrier concentrations to the level 10^{16}–10^{17} cm^{-3}.[23]

During the deposition of material, chemical endothermic reaction occurs on substrate with formation of PbTe and SnTe molecules. The binding energy of these so-called chemically adsorbed atoms reaches several electron volts. Besides the chemisorbed atoms, physically adsorbed atoms may accumulate on the substrate surface, which have energy of several hundredths electron-volts. Since the growth of single-crystalline films of these chalcogenides requires a high substrate temperature ($> 500°C$) in conventional growth methods the heat energy corresponding to this temperature turns out to be high enough to generate point defects. The prolonged thermal annealing[23] is also ineffective in decreasing the concentration of point defects; therefore, it is practically impossible by these traditional methods to obtain concentrations less than 10^{16} cm^{-3}.

Below, the technological principles are described, based on which we have developed the PLD technology for synthesis of films PbTe, $Pb_{1-x}Sn_xTe$ with controllable stoichiometry and composition.

Conditions in which a defect-free film is grown can be expressed as follows: $\varepsilon_{pha} < \varepsilon < \varepsilon_{cha}$, $\varepsilon < \varepsilon_{thd}$, where ε_{pha} and ε_{cha} are desorption energies of physically and chemically adsorbed atoms, ε_{thd} is the energy of formation of thermal defects, $\varepsilon = \varepsilon_T + \varepsilon_{chr} + \varepsilon_{epl}$ is the reduced substrate energy, ε_{chr} is the liberation energy of the chemical reaction, ε_{epl} is the energy given to substrate by erosion plasma. Thus, the reduced substrate energy, ε, consists of three components which can be (except ε_{chr}) varied depending on the requirements to the grown film (purity, diffusion, interface, etc.). Due to the high epitaxy rate and, consequently, lower epitaxial temperature in PLD (200°C), there is possibility to exclude defect formation resulting from the high temperature of the substrate. Moreover, excess amount of physically adsorbed atoms can be removed from the surface by changing the energy state of erosion plasma.

We based upon the principles described above to develop the method of sequential PLD of chalcogenide materials.

The experimental setup was schematically shown in Figure 23.1. In this case, source components of deposited material are located on the holder, which is rotated to sequentially bring the targets under the evaporating laser beam. The resulting fluxes of particles (electrons, ions, atoms) possess an energy spectrum, which is governed by the laser intensity for each material.

Rapid particles influence every previously deposited layer inducing and promoting the surface migration of chemisorbed atoms and hastening desorption of physically adsorbed atoms. As was already discussed in Section 23.2 and shown in Figure 23.2(a), the thickness of deposited mono-layer is practically independent of the laser intensity in a wide range of intensities of technology laser. On the other hand, the energy state of erosion plasma does depend on the intensity. Based on this, deposition is carried out in this laser fluence range. In this way, the thickness of the layer (the deposited amount of each material) is controlled by the number of laser pulses, while the role of technology variable is played by energy of erosion plasma, which governs the reduced substrate temperature, ε (see discussion above).

Complex structural, electrophysical, and photoelectric analyses were carried out for films and structures obtained by this method.[24] Not presenting detailed description of these investigations, we give here only some important results.

Using this method it was possible to grow films of PbTe, $Pb_{1-x}Sn_xTe$ with concentrations of free carriers as low as about 10^{12} cm^{-3} at 77 K, i.e., close to the value inherent to the intrinsic semi conductor. Thus, due to the simultaneous structural and stoichiometry control we are able to vary this concentration in a wide range (10^{12}–10^{19} cm^{-3}).

An especially important result is that single-crystalline high-purity films were obtained by this method on a variety of substrates – KBr, KCl, mica, polycrystalline corundum, silicon, etc., which significantly broadens the range of applications of these materials.

Finally, due to the high level of control and programmability of the thickness and composition new opportunities open for using this technique for "quantum engineering" of structures based on these materials. This would result in new device applications, including photodetectors.

23.5.2 PHOTOELECTRIC CHARACTERISTICS

In Figure 23.9, we present the spectral dependence of photoconductance of various samples meas-ured near the absorption edge at 100 K, referring to other publications on electrophysical and other photoelectric investigations of these materials.[24-26]

As can be seen from Figure 23.9 the photoconductance spectra depend on the composition and the free carrier concentration.

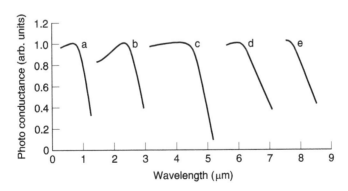

FIGURE 23.9 Spectral dependence of photoconductance of deposited layers at 100 K. (a) A periodic structure (70 Å) PbTe -Pb[a] (50 Å)-PbTe(70 Å) – (20 periods); (b) n-type PbTe (n = 10^{19} cm^{-3} at 77 κ); (c) High-purity, high-stoichiometry PbTe; (d) $Pb_{0.85}Sn_{0.15}Te$; (e) $Pb_{0.82}Sn_{0.18}Te$. The thickness of samples b, c, d, and e is 0.2 μm. (From Alexanian, A.G. et al., *Meas. Sci. and Technol.*, 16, 167, 2005. With permission.)

The photoconductance edge in a periodic structure PbTe-Pb^a-PbTe (curve a) is observed at the wavelength 1.5 μm, which corresponds to the blue shift of the absorption edge of PbTe film due to the quantum confinement:

$$\Delta\varepsilon = \varepsilon_1^c + \varepsilon_1^v = 0.6 \text{ eV},$$

where $\varepsilon_1^{c,v} = \pi^2\hbar^2/2m_{c,v}^\perp d^2$ ($m_{c,v}^\perp = 0.02m_e$, $d = 7$ nm). Thus,

$$\varepsilon = \varepsilon_G + \Delta\varepsilon = 0.83 \text{ eV or } 1.5 \text{ μm}.$$

The blue shift of the photoconductance edge in doped PbTe film (curve b), compared with the pure sample (curve c), agrees well with the shift of the absorption edge in heavily doped semiconductors (the Burstein–Moss effect).

Thus, PLD-produced thin films PbTe, Pb$_{1-x}$Sn$_x$Te with controllable parameters, as well as the periodic structures on their basis can be used as materials for detectors covering the wavelength range from 1 to 10 μm.

23.6 SYNTHESIS OF SOLID SOLUTIONS Bi$_{1-x}$Sb$_x$

Another interesting approach for the synthesis of composite materials is the PLD method of simultaneous deposition of components from different targets. Here, we will discuss this approach applied for the deposition of thin-film Bi$_{1-x}$Sb$_x$ on KBr substrates.[27] In the range of compositions $0.085 < x < 0.22$ the solutions Bi$_{1-x}$Sb$_x$ represent narrow-gap semiconductors and are of special interest from the point of view of optoelectronics applications in the far infrared. The prospective photonics application of these materials is connected with the possibility of their thin-film implementation and conjugation with other systems on the required substrates. The configuration of the PLD setup in this case is slightly different compared to the previous case of sequential deposition. As shown in Figure 23.1, the main modification is the change of the geometry of target locations and an additional optical system for splitting of the initial laser beam. The split beams are directed simultaneously to different targets of Bi and Sb. The laser intensity in the irradiation zone of Bi and Sb targets is adjusted in the ranges $q_{Bi} = (1.65 - 3.3) \times 10^8$ W/cm^2 and $q_{Sb} = (0.38 - 1.4) \times 10^8$ W/cm^2, respectively.

The dependence of the thicknesses on the pulse intensity is different for two materials and is presented in Figure 23.2(b).

An important technology parameter is the dimensionless parameter $Q = q_{Sb}/q_{Bi}$, characterizing the deposition process. In the regulated range of intensities Q varies from the value 0.11 to 0.86. The specific compositions of Bi$_{1-x}$Sb$_x$ layers are obtained by means of regulation of Bi and Sb fluxes (i.e., varying the parameter Q) in the process of simultaneous ablation of materials.

Deposition of layers is carried out on KBr (001) substrate at room to 170°C temperatures. Electron-diffraction studies have shown that in the whole range of variation of Q, single-phase layers corresponding to solid solution Bi$_{1-x}$Sb$_x$ with various x were obtained independently on the structural performance. At small values of Q (near to 0.12), the first signs of textured growth are observed at temperature 85°C. Started from temperature 100°C single-crystalline growth is observed with hexagonal structure. Bi$_{1-x}$Sb$_x$ crystallites are ordered and distributed with orientation {00.1} parallel to the orientation (001) of KBr surface. The crystalline structure does not change up to growth temperature 170°C, while with further increase of temperature deposited layers are deteriorated, which is due to the non-controllable re-evaporation of bismuth from the KBr surface. As parameter Q increases the onset of temperature of single-crystalline growth shifts towards higher values, and at $Q > 0.36$ single-crystalline growth is observed at 170°C. At $Q > 0.4$ and 170°C textured growth of layers occurs.

The lattice parameters of obtained layers are determined by electron-diffraction method. The composition x (in percent) is calculated assuming that Vegard's law holds for $Bi_{1-x}Sb_x$, i.e.,

$$a_x = a_0 - k \cdot x,$$

where $a_0(Bi) = 4.56$ Å, $k = 0.26$ Å. The dependence of x on the parameter Q at temperature 170°C is shown in Figure 23.10.

23.7 CONCLUSIONS

We have presented a review of the results of investigations into the applications of pulsed-laser deposition for the synthesis of various materials and structures that have important optoelectronics applications, namely, as infrared detectors. Pulsed-laser deposition allows diverse approaches to be used for the development of experimental design, depending on which physical and chemical characteristics of the growth process are used as the basis for synthesis. Using the example of quaternary semiconductor solutions $Ga_{1-x}In_xAs_{1-y}Sb_y$, we have demonstrated an experimental design which allows a compositional library to be created by controllable laser mixing of parent compound materials, based on the possibility of congruent composition transfer from target to substrate. In our approach, the substrate temperature and laser intensity play the role of technology variables for composition control. Results of this development were extended for the fabrication of semiconductor graded-gap structures and heterojunctions with abrupt and smooth interfaces. The physical bases of three types of heterojunctions were investigated and the qualitative dependence of their photoresponse characteristics on the interface structure was deduced. Based on these dependences we conclude that since the developed method of laser mixing provides a sufficient level of control over the interface microstructure, the latter can be considered as the quality and optimization parameter for detectors.

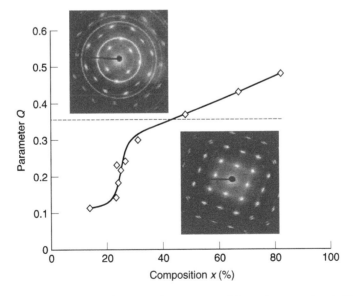

FIGURE 23.10 Dependence of composition x on the parameter Q (growth temperature is 170°C). The dashed line separates the single-crystalline (bottom) and textured (top) growth regions. Insets show sample electron-diffraction patterns in corresponding regions.

The PLD method of sequential deposition from elementary sources of materials was employed for fabrication of thin-film PbTe, PbSnTe, as well as periodic quantum-well structures based on them. The method provides the means for control of stoichiometry and concentration of free carriers in chalcogenide films obtained on various types of substrates. Concentration of free electrons is the quality parameter of obtained films and determines the range of their photosensitivity. This concentration can be varied in wide range of values – from intrinsic to high-doping levels of semiconductor. Photoelectric properties of quantum-well structures based on chalcogenide materials were discussed.

We also presented results of PLD synthesis of solid solutions $Bi_{1-x}Sb_x$ by simultaneous deposition of components from elementary sources. The composition of films can be controlled directly by a technology parameter determined by the relative ratio of laser fluences on targets.

Finally, we have not given detailed descriptions of experimental designs for each of the considered methods but concentrated mainly on physical fundamentals. However, inherent to all employed techniques is that there are no fundamental limitations on combination of possible designs for optimization and discovery, nor on integration and broadening of the class of synthesized materials and structures. This shows that there is a wide spectrum of technological tasks that are the subject for ongoing and future research.

REFERENCES

1. Chrisey, D. B. and Hubler, G. K., *Pulsed Laser Deposition of Thin Films*, John Wiley and Sons, New York, 1994.
2. Takeuchi, I., van Dover, R. B. and Koinuma, H., Combinatorial synthesis and evaluation of functional inorganic materials using thin-film techniques, *MRS Bullet.*, 27, 301, 2002.
3. Christen, H. M., Silliman, S. D. and Harshavardhan, K. S., Continuous compositional-spread technique based on pulsed-laser deposition and applied to the growth of epitaxial films, *Rev. Sci. Instr.*, 72, 2673, 2001.
4. Miller, J. C., *Laser Ablation: Principles and Applications*, Springer Series in *Mater. Sci.*, Vol. 28, Springer-Verlag, Berlin, 1994.
5. Amoruso, S., Bruzzese, R., Spinelli, N. and Velotta, R., Characterization of laser-ablation plasmas, *J. Phys. B*, 32, R131–R172, 1999.
6. Anisimov, S. I. and Luk'yanchuk, B. S., Selected problems of laser ablation theory, *Physics-Uspekhi*, 45, 301, 2002.
7. Omote, K., Kikuchi, T., Harada, J., Kawasaki, M., Ohtomo, A., Ohtani, M., Ohnishi, T., Komiyama, D. and Koinuma, H., In: *Proc. SPIE, Vol. 3941*, SPIE—The International Society for Optical Engineering, Bellingham, WA, 2000, p. 84.
8. Isaacs, E. D., Marcus, M., Aeppli, G., Xiang, X.-D., Sun, X.-D., Schultz, P., Kao, H.-K., Cargill III, G. S. and Haushalter, R., Synchrotron x-ray microbeam diagnostics of combinatorial synthesis. *Appl. Phys. Lett.*, 73, 1820, 1998.
9. Casey, H. C. and Panish, M. B., *Heterostructure Lasers. Part B.*, Academic Press, New York–San Francisco–London, 1978.
10. Alexanian, A.G., Kazarian, R. K. and Matevossian, L. A., Elektronnaya Promyshlennost, *Soviet Electron Industry*, 1, 55, 1982.
11. Miseo E. V. and Wright, N. A., Developing a chemical-imaging camera, *The Industrial Physicist, Phys.*, 9, 29, 2003.
12. Levine, B. F., Quantum-well infrared photodetectors, *J. Appl. Phys.*, 74, R1, 1993.
13. Venkataraghavan, R., Rao, K. S. R. K., Hegde, M. S. and Bhat, H. L., Influence of growth parameters on the surface and interface quality of laser deposited InSb/CdTe heterostructures, *Phys. Stat. Sol.*, (a) 163, 93, 1997.
14. (a) Alexanian, A. G., Alexanian, Al. G., Kazarian, R. K., Matevossian, L. A. and Nickogossian, H. S., On resonance phototunneling and photoelectric properties of pInSb-nCdTe heterojunction, *Inter. J. Infrared and Millimeter Waves*, 14, 2203, 1993; (b) Alexanian, A. G., Aramyan, N. S., Grigoryan, R. P., Khachatrian, A. M., Matevossian, L. A. and Yeremyan, A. S., On the optical memory of a thin-film

pInSb-nCdTe heterojunction obtained by laser pulsed deposition, In: *Progress in Semiconductor Materials for Optoelectronic Applications*, Jones, E. D., Manasreh, M. O., Choquette, K. D. and Friedman, D., Eds., *Mat. Res. Soc. Symp. Proc.*, 692, H9.38, 2002.

15. Liu, A. W. K. and Santos, M. B., *Thin Films: Heteroepitaxial Systems*, World Scientific Publishing Co Pte Ltd, Singapore, New Jersey, London, 1998.

16. Jiang, H. and Singh, J., Self-assembled semiconductor structures: electronic and optoelectronic properties, *IEEE J. Quant. Electron.*, 34, 1188, 1998.

17. Aleksanyan, A.G., On the influence of electron-phonon interactions on inverse population in semiconductor superlattices and size-quantized film in magnetic field. *Kvantovaya Elektronika,* [Soviet Quantum Electronics (in Russian)] 12, 837, 1985.

18. Bockelman, U. and Bastard, G., Phonon scattering and energy relaxation in two-, one-, zero-dimensional electron gases, *Phys. Rev. B*, 42, 8947, 1990.

19. (a) Avdzhyan, K. E., Aleksanyan, A. G., Kazaryan, R. K., Matevosyan, L. A. and Mirzabekyan, G. E., The laser deposition and optical investigation of the $Ga_xIn_{1-x}As_ySb_{1-y}$ thin films of various compositions, *Kvantovaya Elektronika,* [Soviet Quantum Electronics (in Russian)] 15, 181, 1988; (b) Avdzhyan, K. E., Quantum size effects in the InSb-GaAs periodic structure and in $Ga_xIn_{1-x}As_ySb_{1-y}$ films produced by laser spraying, ibid. 11, 1264, 1984.

20. Hinkley, E. D. and Rediker, R. H., GaAs-InSb graded-gap heterojunction, *Solid State Electron.*, 10, 671, 1967.

21. Milnes, A. G. and Feucht, D. L., *Heterojunctions and Metal-Semiconductor Junctions,* Academic Press, New York, 1972.

22. Jakobus, T. and Hornung, J., Influence of the VPE parameters on the epitaxial film properties for the (Pb1-xSnx)1+yTe1-y system. *Journ. Cryst. Growth,* 45, 224, 1978.

23. Sizov, F. F. and Plyatsko, S.V., Homogeneity range and nonstoichiometric defects in IV-VI narrow-gap semiconductors, *Journ. Cryst. Growth*, 92, 571, 1988.

24. Alexanian, A. G. and Khachatryan, A. M., On the possibility of optaining highly stoichiometric PbTe and PbSnTe films and periodic structures by laser-pulse epitaxy method, *Proceed. of NAS Armenia, "Fizika"*, 32, 44, 1997.

25. Parker, S. G. and Johnson, R. W., Preparation and properties of PbSnTe. In: *Preparation and Properties of Solid State Materials*, Wilcox, W. R., Ed., Marcel Dekker Inc., New York, 1981, p. 1.

26. Dolzhenko, D., Ivanchik, I., Khokhlov, D. and Kristovskiy, K., Lead telluride-based far-infrared photodetectors – a promising alternative to doped Si and Ge, In: *Progress in Semiconductor Materials for Optoelectronic Applications*, Jones, E. D., Manasreh, M. O., Choquette, K. D. and Friedman, D., Eds., *Mat. Res. Soc. Symp. Proc.*, 692, H4.6, 2002.

27. Alexanian, A. G., Avetisyan, H. N., Avjyan, K. E., Aramyan, N. S., Aleksanyan, G. A., Grigoryan, R. P., Khachatryan, A. M. and Yeremyan, A. S., "Pulsed laser deposition of Bi- and Sb-based solid solutions and multilayer structures", In: *Progress in Semiconductor Materials III- Electronic and Optoelectronic Applications*, Friedman, D., Manasreh, M. O., Buyanova, I., Auret, F. D. and Munkholm, A, Eds., *Proceedings of MRS*, 799, Z 2.9.1, 2003.

Index

A

Additive, 4, 95, 116, 195, 196, 199, 202–217, 277, 286–288, 296, 297, 299, 314, 320
Adsorption, 30, 37, 38, 101, 174
Algorithm, 40, 50, 116
  genetic, 425, 427, 437, 439
Alloy, 5, 29, 33, 95, 150, 262, 263, 267, 271, 279, 285, 293, 305, 306, 308, 314, 316, 318, 319, 330, 333
Automation, 6, 8, 26, 29, 62, 100, 152, 168, 349

B

Band gap, 447, 450, 451–454, 459–462
Battery material, 279–302
Bead, 18, 22–28, 33–40, 154
  alumina, 38
  polymer, 18
  resin, 22
  single-bead reactor, 40–42
Binary, 23, 24, 29, 70, 120, 138–142, 150, 153, 156, 175, 176, 180, 184, 260, 264, 265, 324, 325, 341, 361, 363–365, 374, 406, 407, 419, 420

C

Calcination, 29, 30, 32, 37, 38, 104, 108, 109, 111, 154, 155, 158, 297, 362, 365, 366, 372
Calibration, 10, 40, 119, 204, 368, 439
  multivariate, 203
Catalyst
  activity, 8, 12, 100, 104, 108, 115–127, 153–155, 160, 165, 166, 173–175, 179–190, 259–260, 264–268, 272–273
  electro-, 5, 259–273
  enantioselective, 5
  fuel cell, 5
  heterogeneous, 5, 11, 116, 149–171, 173–192, 426, 437
  homogeneous, 5, 361
  polymerization, 5
  reference, 11
  selectivity, 9, 116–125, 132, 133, 144–145, 187
Cathodoluminescence, 405, 414, 417, 419, 420, 427
Chemical vapor deposition (CVD), 149, 160, 374
Chemoluminescence, 178, 405
Chemometrics, 13

Chip, 355
  combinatorial, 324
  DNA-on-the-chip, 19
  materials, 40
  memory, 323
  micro-, 26
  synthesis, 23
Chromatography
  gas, 133–135, 153, 156, 165, 168, 222
  gel-permeation (GPC), 222
Classification, 47–49, 52, 151
Combinatorial reactor, 132
Combinatorial
  screening, 150, 239–256
  synthesis, 17–45, 150, 239–256, 284, 288–289, 299, 338, 362, 380, 406, 409, 412, 422, 438
Composite, 293, 450–452
  electrode, 288, 293, 295, 299
  film, 284, 288, 293
  material, 27, 296, 408, 450, 464
  nano-, 338
Composition spread, *see* Spread
Conducting polymer, 239–256
Contamination, 133, 134
Cross-talk, 291
Cross-validation, 50, 203, 204
Crystallization, 27, 72, 79–80, 151, 330, 332, 366, 373, 393, 450

D

Data management, 10, 100
Data mining, 6, 10, 12, 13, 50, 57, 58
Defect, 27, 28, 30, 101, 298, 353, 383, 384, 414, 450, 455, 456, 462
Deposition
  catalyst, 276
  chemical vapor deposition (CVD), 149, 175, 361
  composition-spread, 418–419, 422
  electrochemical, 271, 285–287
  electron beam evaporation, 150
  gradient, 381
  ink jet, 29, 152, 222, 260, 261, 262, 361, 380, 408–410, 415
  library, 28, 29, 271
  liquid, 150, 363
  material, 28
  metal, 154